The Origins and Spread of Domestic Plants in Southwest Asia and Europe

edited by

Sue Colledge and James Conolly

LONDON AND NEW YORK

University College London Institute of Archaeology Publications

Publications of the Institute of Archaeology, University College London

Director of the Institute: Stephen Shennan
Publications Series Editor: Peter J. Ucko

The Institute of Archaeology of University College London is one of the oldest, largest and most prestigious archaeology research facilities in the world. Its extensive publications programme includes the best theory, research, pedagogy and reference materials in archaeology and cognate disciplines, through publishing exemplary work of scholars worldwide. Through its publications, the Institute brings together key areas of theoretical and substantive knowledge, improves archaeological practice and brings archaeological findings to the general public, researchers and practitioners. It also publishes staff research projects, site and survey reports, and conference proceedings. The publications programme, formerly developed in-house or in conjunction with UCL Press, is now produced in partnership with Left Coast Press, Inc. The Institute can be accessed online at http://www.ucl.ac.uk/archaeology.

ENCOUNTERS WITH ANCIENT EGYPT Subseries, Peter J. Ucko, (ed.)
Jean-Marcel Humbert and Clifford Price (eds.), *Imhotep Today* (2003)
David Jeffreys (ed.), *Views of Ancient Egypt since Napoleon Bonaparte: Imperialism, Colonialism, and Modern Appropriations* (2003)
Sally MacDonald and Michael Rice (eds.), *Consuming Ancient Egypt* (2003)
Roger Matthews and Cornelia Roemer (eds.), *Ancient Perspectives on Egypt* (2003)
David O'Connor and Andrew Reid (eds.), *Ancient Egypt in Africa* (2003)
John Tait (ed.), *'Never had the like occurred': Egypt's View of its Past* (2003)
David O'Connor and Stephen Quirke (eds.), *Mysterious Lands* (2003)
Peter Ucko and Timothy Champion (eds.), *The Wisdom of Egypt: Changing Visions Through the Ages* (2003)

Andrew Gardner (ed.), *Agency Uncovered: Archaeological Perspectives* (2004)
Okasha El-Daly, Egyptology, *The Missing Millennium: Ancient Egypt in Medieval Arabic Writing* (2005)
Ruth Mace, Clare J. Holden, and Stephen Shennan (eds.), *Evolution of Cultural Diversity: A Phylogenetic Approach* (2005)
Arkadiusz Marciniak, *Placing Animals in the Neolithic: Social Zooarchaeology of Prehistoric Farming* (2005)
Robert Layton, Stephen Shennan, and Peter Stone (eds.), *A Future for Archaeology* (2006)
Joost Fontein, *The Silence of Great Zimbabwe: Contested Landscapes and the Power of Heritage* (2006)
Gabriele Puschnigg, *Ceramics of the Merv Oasis: Recycling the City* (2006)
James Graham-Campbell and Gareth Williams (eds.), *Silver Economy in the Viking Age* (2007)
Barbara Bender, Sue Hamilton, and Chris Tilley, *Stone Worlds: Narrative and Reflexivity in Landscape Archaeology* (2007)
Andrew Gardner, *An Archaeology of Identity: Soldiers and Society in Late Roman Britain* (2007)
Sue Hamilton, Ruth Whitehouse, and Katherine I. Wright (eds.) *Archaeology and Women* (2007)
Gustavo Politis, Nukak: *Ethnoarchaeology of an Amazonian People* (2007)
Janet Picton, Stephen Quirke, and Paul C. Roberts (eds.), *Living Images: Egyptian Funerary Portraits in the Petrie Museum* (2007)
Eleni Asouti and Dorian Q. Fuller, *Trees and Woodlands of South India: Archaeological Perspectives* (2007)
Timothy Clack and Marcus Brittain, *Archaeology and the Media* (2007)
Sue Colledge and James Conolly, *The Origins and Spread of Domestic Plants in Southwest Asia and Europe* (2007)

Dedicated to

Gordon Hillman

in honour of his contribution
to archaeobotany.

First published 2007 by Left Coast Press, Inc.

Published 2016 by Routledge
4 Park Square, Milton Park, Abingdon, Oxon OX14 4RN
605 Third Avenue, New York, NY 10017

First issued in paperback 2017

Routledge is an imprint of the Taylor & Francis Group, an informa business

Library of Congress Cataloguing-in-Publication Data

The origins and spread of domestic plants in southwest Asia and Europe / Sue Colledge and James Conolly, editors.
p. cm. — (Publications of the Institute of Archaeology, University College London)
Includes bibliographical references and index.
ISBN-13: 978-1-59874-988-5 (hardcover)
ISBN-10: 1-59874-988-9
1. Agriculture, Prehistoric—Middle East. 2. Agriculture, Prehistoric—Europe.
3. Agriculture—Origin. 4. Plants, Cultivated—Middle East. 5. Plants, Cultivated—Europe.
6. Plants, Cultivated—Origin. I. Colledge, Sue. II. Conolly, James, 1968– III. University College London. Institute of Archaeology.
GN857.O64 2007
630.93–dc22

2007005870

Typeset using $\LaTeX\ 2_\varepsilon$

ISBN 13: 978-0-8153-4749-1 (pbk)
ISBN 13: 978-1-59874-988-5 (hbk)

Contents

Acknowledgments

The papers in this volume were presented at a conference ("New Perspectives on the Origins and Spread of Farming in Southwest Asia and Europe") which was held at the Institute of Archaeology in London from 15th–17th December 2003. Funding for the conference, including travel and subsistence expenses for the invited speakers, was made available under the auspices of a three year AHRB sponsored project, "The origin and spread of neolithic plant economies in the Near East and Europe". We are thus indebted to the AHRB (now AHRC) for providing the funding for such a successful meeting.

Stephen Shennan initiated the research program and oversaw and encouraged all aspects of its implementation, including the publication of this volume, and we offer him our sincere thanks. We would also like to acknowledge the support of Peter Ucko, not least for allowing us to host the conference at the Institute of Archaeology, but also for his encouragement throughout all phases of this project. Similarly David Harris has generously offered us his expertise in too many areas to list individually, and we here recognize the important role he played in bringing this work to fruition. Lisa Usman offered invaluable help not only by overseeing arrangements prior to the conference, but also for organising everything when it was being held, and we are extremely grateful to her for her continued support. We wish also to express our appreciation to Emma Harvey and Meriel McClatchie who also provided assistance during the three days of conference meetings.

The construction of a volume of this size involves many people, and we would like in particular to thank Jeff Dillane, Rhianne McKay and Nikola Oliver from Trent University for their bibliographic and editorial assistance. At Left Coast Press, Jennifer Collier ensured that the final editorial tasks leading to publication were pleasant and efficient and Eliot Werner skillfully completed the indexes. Marion Cutting at the Institute of Archaeology provided helpful advice throughout the publication process.

We are also grateful to the anonymous reviewers who recommended changes that have greatly enhanced the final version of this volume. Most importantly we offer our sincere thanks to all the contributing authors; it has been a great pleasure to work with you all and we hope you feel this volume is a fitting tribute for your efforts.

Foreword

Stephen Shennan
University College London, UK

The origin of agriculture and its spread from Southwest Asia to Europe has been one of the most intensively discussed and debated topics in Old World archaeology for at least the last 40 years. It is all the more remarkable then that over the years so little attention has been paid in these discussions to analysis of the concrete details of the exploitation of the new resources themselves and the conclusions that can be drawn from them. This is particularly the case with the plant exploitation side of the new subsistence economy. Two connected reasons for this lack of attention spring to mind. First, it was long assumed that the only thing that needed to be established about these economies was that people exploited wheat and barley, and perhaps the odd legume. On this basis, particular cultures could easily be identified as the first farming cultures in an area and from the subsistence point of view there was little more to be said. All the attention could be focussed on aspects such as the pottery and lithics, and whether these indicated patterns of acculturation or migration. Second, because of this assumption, little effort was put into collecting archaeobotanical data, as it was perceived (wrongly) that few relevant questions could be answered by studying plant remains. Thus, the detailed studies of archaeobotanical remains that were beginning to be carried out by the first members of a new breed of specialists, on the basis of samples recovered through the application of new techniques such as flotation, were ignored by other archaeologists working on large-scale syntheses as being of little interest. The only exception to this was in Southwest Asia where the origins of domestic cereals were being sought and therefore more detailed attention was needed, for example to distinguish between wild and domestic varieties. Even non-archaeobotanists could see that this was an important issue.

In recent years, however, this situation has begun to change as it has become clear that understanding the adoption and spread of cereal agriculture is about much more than documenting the presence of a small number of domesticates. It involves the reconstruction of agricultural practices. Even more important, the work of archaeobotanists on samples collected using appropriate methods has shown that the detailed analysis of such samples, and especially of the weed assemblages, provides a basis for reconstructing changes to those practices and the social and ecological factors affecting them. In addition, genetic studies have provided new evidence on origins.

To use these new results to understand a large scale evolutionary process such as the origin and spread of cereal-based plant economies makes further demands. The samples analysed and the methods used to analyse them must produce comparable results, and those results must be shown to be at least partly a reflection of past plant exploitation practices and not simply of taphonomic processes affecting the formation of the archaeobotanical assemblages or the biases of a particular archaeobotanist. Moreover, these requirements must be met for large numbers of sites over a broad area, in this case from Southwest Asia to Scandinavia.

We have now arrived at the stage where such large-scale comparative work has begun to be possible, thanks to the accumulation of results obtained using modern methods over the last 30 years. There are two ways of carrying forward such an enterprise. One is to gather results from the whole area in a database and to analyse them together after an initial stage of source criticism. The other way is to bring together the specialists from different regions so that they can present their results to one another for comparison. This book represents the outcome of a conference designed to do precisely that and is one of the results of a project funded by the UK Arts and Humanities Research Board (as it was then) which has attempted to pursue both the paths just mentioned. Some of the results of the database work are presented elsewhere in the volume.

The book shows the enormous value of getting people together in this way. For one thing it breaks down some long-standing geographical barriers. Perhaps the most important of these is the division between those who study the origins of plant domestication in Southwest Asia and those who study early farming in Europe. They are rarely found between the same covers. Hardly less significant though is the inclusion of colleagues from the previously Soviet-dominated countries of Central and Eastern Europe, where there was very little tradition of interest in archaeobotany in general and in the reconstruction of agricultural practices in particular. It is exciting to see that the previous lack of interest on the part of archaeologists from this region in the rigorous collection of archaeobotanical data is now beginning to change. Inevitably the quality and potential of the data, as well as the sophistication of the analysis, from regions where there is a more established tradition of modern archaeobotanical work are greater. The papers in this volume present some outstanding examples of modern work that demonstrate why archaeobotanical results are of profound importance far beyond the community of archaeobotanists, and are relevant to all attempts to understand the social, cultural and economic history of the transformation that neolithisation ultimately produced.

Whether or not any of the specific studies or large-scale syntheses presented here stand the test of time, it is safe to say that this book will be a benchmark in the field and a stimulus for the future work that will supersede it. It is a particular pleasure that the conference on which it is based took place at the Institute of Archaeology, UCL, when its editors were based there. The Institute has an outstanding history of work in this field and no one has been more responsible for that reputation than Gordon Hillman, to whom the book is rightly dedicated.

Introduction: key themes in interregional approaches to the study of early crops and farming

Sue Colledge and James Conolly
University College London, UK and Trent University, Canada

This collection of papers is the outcome of a conference that was held at the Institute of Archaeology in London on 15th–17th December 2003, where many of the world's leading scholars of neolithic agriculture met to present and discuss current research on the origins and spread of farming in Southwest Asia and Europe. The purpose of the meeting was to provide a forum in which archaeobotanists could gather to review and comment on recent research on the nature of early agricultural practices. Rarely has such a complete compendium of information from such a wide geographical coverage been presented: it was in fact one of our aims to integrate archaeobotanical knowledge from Southwest Asia and Europe to dissolve the artificial barriers that have seen the study of the Neolithic carried out within regional and cultural-historical enclaves. Our belief is that this has prevented a broader interregional understanding of the reasons that might underlie local idiosyncrasies, such as some early farmers' focus on only a few of the wider range of available crops, the extensive use of wild plant foods in a few areas but not in others, and the emergence of regionally distinctive phytosociological associations.

The chapters in this volume thus comprise a wealth of new archaeobotanical data for understanding early farming practices. A number of themes emerge from such a broad regional synthesis, of which we wish to highlight five.

Regional variations in crop use Of prime importance are the detailed descriptions of the domestic species grown during the early Neolithic and, in particular, the information on the variations in the suites of crops cultivated in different regions of Southwest Asia and Europe. These variations have spatial structure and at large scales provide information about the changes in agricultural practices over time. For example, one major pattern evident from interregional assessment of archaeobotanical data is the gradual loss in diversity of cultivated crops from southeast to northwest, a pattern that has been acknowledged for some time but which becomes strikingly obvious when regional datasets are ordered spatio-temporally (Colledge et al. 2005). It is thus clear that from the earliest stage in the development of farming and throughout its subsequent establishment certain of the founder crops were adopted and grown preferentially; whether or not this was a consequence of climatic or ecological barriers, evolutionary bottlenecks, dietary or other active cultural choices remains unclear. Several of papers in this volume provide much insight into the major question of crop use and its regional variations, and the detailed taxa lists are particularly informative in this regard.

Origins and use of secondary domesticates Many of the chapters report the identification of secondary crops in early neolithic contexts. For example, *Panicum miliaceum* is included in the taxa lists for many of the sites in Central and Eastern Europe, and *Papaver somniferum* is documented on several sites in the western Mediterranean. Both species evolved outside the 'Levantine core area' where many of the primary or founder crops were domesticated and the information we have for their presence at an early date in different areas of Europe thus enables us to determine their origins much more accurately. The data on variability in the crop repertoire both within and between regions also provide important corroborative evidence about contacts and links between different cultural groups during the early Neolithic.

Use of wild taxa Taxa lists compiled by the authors include not only the domestic species that were identified at the sites but also the full range of wild species, which were found in association with the crops. The completeness of the presentation of data enables us to appreciate the breadth of the subsistence base in the different regions during the early stages of the establishment of the agricultural system. Many authors describe in detail the wild plant foods (e.g. fruits, nuts, roots, etc.) that supplemented or enhanced the harvests, and it is possible, therefore, to assess the significance of these resources in the early neolithic diet. In several chapters there are discussions about the evidence for an earlier mesolithic presence and the degree to which continued use of wild resources in the neolithic represents possible contacts with hunter-gatherers and the blending of traditions. Condiments and medicinal plants are also highlighted and commented upon in some of the papers.

Weed ecology The weeds associated with the crops provide essential information on agricultural practices, for example, sowing times, tillage methods and harvesting and processing techniques (etc.), and so contribute to our understanding of certain social aspects of the lives (i.e. the daily routines) of the early farmers. Accurate reporting of finds of weed species in the different regions allows us to monitor the development of the weed floras as farming spread across Europe. Several papers discuss the interplay between ecological and historical biogeographic processes, for example, in terms of the contrast between native weed species (apophytes) and introduced species (archaeophytes). This then adds to our knowledge about the ways in which agricultural practices evolved and were adapted as farming spread into 'uncultivated' regions. Consideration of the derivation of the introduced weed species, which were transported with grain stocks, can also inform on the routes of dispersal of the founder crops.

Impact of methodological advances Much of what is reported in the papers reflects the recent developments that there have been in archaeobotanical methods. These developments have transformed our understanding of neolithic archaeobotany. Methodologies both on-site, in terms of the recovery of plant remains, and also in the laboratory, in terms of the analysis of assemblages of seeds, fruits, etc., have been greatly improved. The quality of the resultant data is such that it allows us to make interpretations of archaeobotanical assemblages that are increasingly more insightful. On the basis that they are so fundamental to all archaeobotanical investigations, perhaps most significant of all the recent developments are the refinements in approaches to identifying many of the commonly occurring species. The implications of the use of more sophisticated diagnostic criteria to distinguish between certain crop taxa on early neolithic sites are commented upon in many of the papers. For example, several authors document the presence of a 'new glume wheat type' (Jones et al. 2000) that has been identified on the basis of the morphologies of both chaff and grains. It has yet to be formally named, but "the morphology of the grains is not inconsistent with the suggestion that the 'new' glume wheat might correspond to modern *Triticum timopheevi*"

(Kohler-Schneider 2003). Since the first accurate description of this new glume wheat there have been more reports documenting its presence on sites in different regions so that a picture of the extent of its use in Europe during the early Neolithic has gradually become clearer. This then is an example of how archaeobotanical data, which are the consequences of increasingly sophisticated analytical techniques, are so crucial to our understanding of the agricultural system at its inception. More reliable criteria for distinguishing between hexaploid and tetraploid free threshing wheats were formulated within the past 25 years (and due largely to the detailed work carried out by Hillman; see Hillman 2001) and these are now routinely used to identify grains and chaff of the two crops. Because of the enhanced accuracy with which the free threshing wheat species can be determined it has been possible, as was the case for the new glume wheat, to chart with greater precision their presence on early neolithic sites. The papers in this volume also present up to date information on the regional variability in occurrence of these two crop types and provide evidence of their dispersal during the early stages of the spread of farming. It is thus possible to chart a southern route, via the western Mediterranean and into Switzerland, of the tetraploid species, as distinct from the more northerly route, from the southeast and via Central Europe (e.g. on Linearbandkeramik settlements) of the hexaploid species.

Several other issues re-occur as common sub-themes across chapters. There are discussions of some of the more controversial issues in archaeobotany in the opening chapters; monophyletic versus polyphyletic origins for the founder crops, local domestication events and the initial routes of dispersal of neolithic farming are all debated. In the chapters covering the earliest neolithic settlements in Southeast Europe the emphasis is more on the nature of the processes underlying the spread of agriculture and possible influences on the development of the subsistence system in new territories. The distinctive nature of the Linearbandkeramik farming system is debated in chapters that cover the progression of neolithic agriculture in Central Europe. Taphonomy and the biases of the archaeobotanical record are dealt with in several papers, in particular the comparative representativeness of certain economic taxa and how these are interpreted in terms of relative 'importance' in the early neolithic. Other authors reassess theories of neolithisation in their specific regions, in particular the interaction between indigenous and exogenous peoples and modes of use of plant resources.

In our own project (which gave rise to the December conference) it was our intention to understand more fully the origins of farming and its subsequent spread to mainland Europe. Our aim was to accumulate and analyse systematically archaeobotanical data at a large spatial scale. Over the previous two years many of the archaeobotanists at this conference had contributed information towards the construction of a database of archaeobotantical records from early neolithic sites across Eurasia (overseen by the editors). The results of the analysis of this dataset has demonstrated that there are coherent trends in terms of similarities and differences of suites of crops and weeds both within and also between regionally focussed groups of early neolithic sites (Colledge et al. 2004). Some of these patterns we have interpreted as being partially environmentally determined (e.g., the decline in the range of diversity of pulses from southeast to northwest—but many seem to be culturally or historically determined (e.g., the preferred use of glume wheats in the Linearbandkeramik) (Colledge et al. 2005). In this regard our primary goal continues to be the regional-scale investigation of these changes from a historical and evolutionary perspective. We seek to understand why these differences exist in subsistence practices: for example whether the use of certain crops confer some adaptive advantage on farmers, or whether they result from other non-adaptive selective pressures such as 'founder effects' in early colonizer populations.

Some of the authors, however, question the validity of large regional scale analyses that fail to accommodate the disparities inherent in the data which derive from the smaller scale processes (i.e., involving pre- and post-depositional preservation of plant materials). While we advocate the exploration of data at a interregional level we are not unaware of the importance (and the interpretative value) of analyses carried out on a smaller scale, whereby comparisons of taxonomic composition are made on a sample by sample basis and for individual sites. Of significance in these 'smaller scale' studies are the resultant insights into routine practices involving plant resources at the 'household' level, and which are much more to do with the exploration of formation processes than investigation of characteristics that define cultural phenomena. We suggest that the value of our broad regional approach lies in the fact that similarities and differences in taxonomic composition at this level are informative about macro-scale processes with regards to how crop use changed, from the earliest adoption of the subsistence system and throughout the subsequent developmental stages.

The order of papers in this volume roughly follows the chronology of the appearance of early farming from the southeast to northwest. One of our opening chapters focuses on the origins of farming in the Euphrates Valley in Syria, and the concluding chapter deals with cereal cultivation in the British Neolithic. Our intention was to have as broad a geographic coverage as possible in this collection of papers, but we were not able to include chapters synthesizing the current state of research on early neolithic agriculture in some important regions; for example, there are glaring omissions for Turkey and France (although summary information about Turkish sites can be found in Chapters 1, 2, and 4). Our failure to involve archaeobotanists who have specific expertise in these countries is unfortunate, but this has transpired due to reasons beyond our control; every effort was made either before or subsequent to the meeting to encourage submission of papers from the missing regions.

In conclusion, this volume is an important reappraisal of the evidence for the spread of agriculture and, we hope, provides further impetus to studies of the ecology of early farming societies so as to understand better the processes, both large and small scale, of neolithisation. Rather than simply assigning plant foods to categories of 'introduced' or 'native', 'domestic' or 'wild', these papers engage properly with the complex and multifaceted role of plants and plant-use in the life of late gatherer-hunters and early farmers. The result is a more nuanced and balanced understanding of the importance of plants for understanding the social and ecological processes underlying the changes in human-plant interaction during the early Holocene.

Finally, it is appropriate to end this introduction by highlighting the enormous impact that Gordon Hillman has had on our collective understanding of early farming. His depth of knowledge of human-plant interaction, his original approaches to studying early agricultural practices by integrating ethnographic observations of smallholder farmers, and the inspiration and enthusiasm he passed to his many students (who are well-represented in this volume) are a long-lasting legacy to archaeobotanical science. It is thus with great pleasure that we dedicate this volume to him in tribute to his outstanding contribution to the study of early farming.

References

Colledge, S., J. Conolly, and S. Shennan (2004). Archaeobotanical evidence for the spread of farming in the eastern Mediterranean. *Current Anthropology 45*, S35–S58.

Colledge, S., J. Conolly, and S. Shennan (2005). The evolution of early neolithic agriculture: From SW Asian origins to NW European limits. *European Journal of Archaeology 8*(2), 137–156.

Hillman, G. (2001). Archaeology, Percival, and the problems of identifying wheat remains. In P. D. S. Caligari and P. E. Brandham (Eds.), *Wheat Taxonomy: The Legacy of John Percival*, The Linnean Special Issue No 3, pp. 27–36. Academic Press.

Jones, G., S. Valamoti, and M. Charles (2000). Early crop diversity: a 'new' glume wheat from northern Greece. *Vegetation History and Archaeobotany 9*, 133–146.

Kohler-Schneider, M. (2003). Contents of a storage pit from late Bronze Age Stillfried, Austria: another record of the 'new' glume wheat. *Vegetational History and Archaeobotany 12*(2), 105–111.

Chapter 1

Diverse origins: regional contributions to the genesis of farming

Andrew Sherratt
University of Sheffield, UK

Editors' note

Andrew Sherratt provided a lively closing address to the December meeting. From his characteristically broad, interregional, perspective he deftly brought together early farming, cultural interaction, exchange and population movement into a coherent and insightful whole. We therefore decided to have his paper open, rather than close, this volume as it provides a fitting introduction to the themes and questions that the subsequent papers engage with at local scales. Tragically, Andrew unexpectedly passed away during the latter stages of our editing of this book. The resultant loss his untimely passing brings cannot be over stated. His great depth of understanding and profound contribution to the study of Neolithic Europe was recognized by all who knew him and his work.

1.1 Introduction

This paper presents an outsider's view of the circumstances which led to the beginning and early spread of farming. It is a personal attempt to come to terms with how this historically momentous set of events came about, in light of the flood of new information which has transformed our knowledge of the subject in the last thirty years.[1] Having struggled to understand it within the changing paradigms of those argumentative decades, I feel that I have at last overcome some of the obstacles created by the language and expectations with which northwest Europeans have traditionally approached it, and which is built into the vocabulary that has been used to describe it. Insofar as this engagement has been motivated by an attempt to grasp its wider significance for prehistory as a whole, it may be a useful complement to the more professional treatment of these issues in the rest of this volume, by authors who know far more about its problems than I do. At any rate, it is motivated by a genuine desire to understand the phenomena and what they mean.

[1] The metaphor is an especially apt one, since much of the information has been retrieved by rescue-excavations in advance of dam-construction, and many of the key sites are now literally flooded.

1.2 'Neolithic': the burden of a name

A characteristic dilemma arises from using a terminology established in one set of circumstances and applying it to another—especially if it has in the meantime been given a global significance. This is the case with our basic vocabulary for dealing with this topic, that of 'neolithic origins'. 'Neolithic' was the term invented, in Northwest Europe (at a time when many neo-Greek technical terms entered the professional vocabulary, in physics and chemistry as much as in archaeology), for the evident contrasts between the earlier and later parts of the Stone Age. Another half-century elapsed before a further term, 'mesolithic', was inserted between them (on the common model of geological sequences, and in the then-fashionable mode of tripartition) to deal with the Holocene foragers who upset the neat dichotomy between Ice Age hunters and more recent stone-using farmers. The terminology achieved only a partial success, in that it was never widely applied in the Americas, and only in a typological sense in much of the rest of Eurasia; but it achieved a whole new level of meaning in Gordon Childe's unification of the language of an emerging archaeological taxonomy with that of comparative ethnography in the Enlightenment tradition, and more specifically in the classification of Lewis Henry Morgan, where pottery and polished stone axes signified the emancipation of 'Barbarism' from 'Savagery' with the appearance of farming (Childe 1936). Thus it was that Grahame Clark, as recently as 1946, could take as the title for a popular account of prehistory 'From Savagery to Civilisation', using a terminology little different from that used by Adam Smith and his contemporaries two hundred years earlier, and decidedly odd-sounding today (Clark 1946). Childe's conflation of the two vocabularies and his equation of the Neolithic with the beginnings of farming was canonized in his term the 'Neolithic Revolution', which (unlike the 'Urban Revolution'—presumably because the term 'Agricultural/Agrarian Revolution' had already been bagged by historians describing how England adopted the turnip and the Brussels sprout at the time of her eighteenth century industrialization) used an artifactual label to designate this major prehistoric change in subsistence practices. This led to inconsistencies on two levels: at a global level, where 'neolithic' continued to be used for foraging societies using pottery and polished stone axes, and at a local level as observations accumulated in Western Asia (the 'Near East' in our Eurocentric terminology) about the early village communities of that region. Hence the shock to uniformitarian expectations when Kathleen Kenyon discovered a 'Pre-Pottery Neolithic' (PPN) phase at Jericho, and Robert Braidwood an 'aceramic neolithic' phase at Jarmo—the former divided by Kenyon into A and B, and subsequently used as the basis for Levantine periodizations of the early Holocene. This has given rise to further subdivision and the emergence of such off-putting algebra as 'EPPNB' or even 'early MPPNB' (which, to make matters worse, is actually the same period in the hands of different authors). Even more confusingly, the term PPNA is used to cover both farming and foraging communities, existing contemporaneously and using the same arrowhead-types, but differing fundamentally in their modes of subsistence. Yet in fact the terminological situation is worse even than this, because of the appearance and disappearance in Western Asia of the term 'mesolithic', which was first applied by Dorothy Garrod (1957) to the Natufian, but then fell into disuse when it became evident that this group of foragers belonged to the terminal Pleistocene, and thus (on a European definition, limiting mesolithic cultures to the post-glacial) formally speaking belonging to the Palaeolithic. The creation of a special term, 'Epipalaeolithic', left them as a pendant afterthought to the older stone age, without any individuality as a historical phenomenon—unless some of them were honorifically described as 'neolithic', too. (The alternative terminology, arising from Robert Braidwood's work, 'incipient farmers', had the opposite fault—characterizing them with teleological hindsight as the precursors of an inevitable revolution.) This is not just an agreeable excursion into the antiquarian byways of our subject: it is the basis of the technical vocabulary in use today. No wonder Near Eastern prehistory (like prehistory in general, where no one can agree on a common usage of terms like eneolithic or chalcolithic, and where the onset of the Bronze Age can precede the use of tin-bronze by up to a millennium) sometimes seems arcane. The entire vocabulary is (either literally, or in its assumptions) nineteenth-century—what the Brits (or even North Americans) would call 'Victorian'.

Of course, we all know this; we explain it to our students, we apologize for its inconsistencies and the lack of anything better, but at least people know what it means. Or do they? Isn't it actually so suffused with essentialism, dualism, teleology, progressivism, uniformitarianism and unilinealism (never mind the inconsistency) that it is an active impediment to analytical thought? I realized that it was when I found myself trying to decide where the Neolithic really began (the Levant or Iran? The Jordan valley? Syria? Southeast Turkey?), and asking why domesticated cereals seemed so relatively unimportant in certain areas, which nevertheless were clearly both sedentary and complex at an early date, whereas in other areas the association between sedentism and cereal-dependence seemed to be so neat and compelling. I realized that I, too, had taken a term which made some sense at a level of general description on a global or continental scale, and used it beyond its reasonable degree of analytical resolution. Moreover, I had been guilty of making unjustified equations and assuming linear and monocausal links between phenomena whose interrelationships were far more complex and interesting: even to the extent of calorific reductionism (that complexity must result from cereal domestication) and paradigmatic over-extension ('one model fits all'). My confusion, though perhaps pardonable within the constraints of an outdated vocabulary, arose primarily from treating words as if they had some simple relationship with the real world, and as if a thing had to have a simple origin as a point in space and time. 'The Neolithic' is a slogan, not a Platonic primordium; it begins where we find it useful to begin it, not at some empirically discoverable 'point zero'. So where does this leave 'the origins of farming'?

1.3 Alternative pathways

Farming began with the cultivation of large-seeded grasses and legumes, taking annual plants and propagating them in niches to which they would not normally have access—either because of competition from perennials, or because of seasonal flooding which would destroy the seeds. (I shall argue that the latter was the original context of their cultivation.) Such plants were relatively abundant in the Fertile Crescent because of the strong seasonality of its Mediterranean climate, and their habitats existed not far from the fertile plains where they were first sustained artificially (Blumler 1996). This distinctive climate, and environmental diversity at a plate boundary, explains the unusual spatial context within which farming began. A similarly unusual context in time is provided by the events accompanying the termination of the last glaciation: increasing warmth and moisture in an interval before postglacial conditions had stabilized, and moreover punctuated by a brief stadial episode temporarily returning to colder and dryer conditions, the Younger Dryas. An unusual time in an unusual place, when the elements were shaken up and reconfigured, in the presence of behaviourally modern human populations: this is the background from which farming emerged (Sherratt 1997). It says nothing, however, about the precise circumstances of its emergence (other than that, as an 'accident waiting to happen', it could perhaps have come about in several ways other than the one it did).

Knowing that it did, indeed, happen serves to de-sensitize us to the possibility that it might not have happened, or that other alternative things were happening as well—some of which we might confuse with it, because they look very similar. Not all sedentary societies are based on farming: the ethnographic record (classically the northwest coast of North America), or the archaeological record of areas where cereal-farming was late in arriving (Jomon Japan, for instance), are evidence that foragers can be not only affluent but complex as well. Even within Europe, the late Mesolithic of the Baltic becomes more complex with each new discovery (skin boats for deep-sea fishing, cemeteries, post-built structures, permanent occupation of key sites). None of these, however, is called neolithic, in the Childean sense. Yet sites without pottery, stone axes or any evidence of cereal cultivation are routinely called 'neolithic' in the archaeological literature, just because they are more or less contemporary with ones that do, and share certain sorts of hunting equipment with them, and because they are sedentary and have houses. (I am thinking of Nemrik and Qermez Dere in Iraq, Hallan Çemi and Göbekli Tepe in Turkey, and even Nevalı Çori and early Çayönü, and no doubt others in adjacent parts of Iran: see, in general, Cauvin 1997.) These sites are clearly extremely important, and relevant to what subsequently happens (and becomes part of the 'neolithic package' as it then spreads to the Anatolian plateau and other parts of the Fertile Crescent and beyond); but is it helpful to call them 'neolithic', with all its farming associations and teleological baggage, as if they were a necessary step on the road to farming, rather than a phenomenon of global interest in their own right? I should stress that there is no dispute here about facts, and that all these points have been made forcefully by the excavators of these sites themselves (e.g., Özdoğan 1995; Hauptmann 2002); what is at issue are the conceptual categories into which we slot these observations, in our wider scheme of perception—and that I, for one, have failed to appreciate the importance of these sites because of the tendency to accommodate them to a 'neolithic' paradigm formed out of an alliance between prehistoric Europe and the southern Levant, in which 'farming' is a necessary precondition for complexity, or at least of sedentism, houses and the formation of substantial sites (e.g., Sherratt 2004, figure 1, map a).

What, then, to do? We could call such cultural phenomena 'neolithic', in a tentative sense, and await an 'achieved neolithic' at a later stage (e.g., Bischoff 2002); or we could revive the European term 'mesolithic', giving it a widened significance as a response to the new ecological opportunities of the late- and post-glacial period and perceiving it as being itself a wave-like phenomenon, beginning in low latitudes and travelling to higher latitudes as deglaciation, the range-extension of temperate species of plants and animals, and 'ecological rebound' progressed. In this conception, the classic 'Mesolithic' of Scandinavia would be a relatively late example of a more general phenomenon, beginning earlier nearer to the equator, and encompassing early seed-collectors and coastal communities, often using pottery surprisingly early (by comparison with Europe), in eastern Africa and around the Indian Ocean—with the Younger Dryas as a brief interruption of a more continuous process. The choice is essentially a rhetorical one. Using 'neolithic' is to impart a sense of teleology, but using 'mesolithic' might just confuse the issue further, by importing yet more European ideological baggage (and an implied inevitability of succession). Let us therefore be content with using the term 'forager climax', as a localised aspect of what I once (1980a) called 'the [worldwide] Postglacial Revolution'—but we have probably had enough revolutions in prehistory for now. It is sufficient to recognise that farming was a minority response to the heightened ecological productivity of the early Holocene, and only slowly took on the escalating expansionary properties which we subsequently associate with it.

1.4 The forager climax in the northern Fertile Crescent

The onset of the early Holocene brought about a bonanza. All around the world, human populations—along with those of other plants and animals—surged as a result of higher temperatures, increased precipitation, and higher levels of CO_2. Especially in the unusual conditions of adjustment—as sea levels slowly rose in response to ice-melt, and as forests slowly spread, creating vegetational associations which no longer exist today—there existed niches some of which are no longer available, and all of which have in any case been degraded by millennia of farming and intensified hunting. (We must allow, also, for a stronger monsoon system than has existed since circa 4000 cal BC, which is especially important for East Africa and Arabia: Roberts and Wright 1993.) Especially around the coasts and on coastal plains—areas subsequently submerged by rising sea levels—there were opportunities for the emergence of relatively dense forager populations. Caves in coastal regions, along the length of the Mediterranean and around the Black Sea and the Caspian, preserve a partial record of their presence which can be extrapolated to indicate the extent of mesolithic occupation more generally, as can occurrences of rock art in highland regions. As the record from Franchthi cave in the Peloponnese indicates (Runnels 2001), these communities were active in circulating materials such as obsidian (in this case from Melos) through extended networks of contact and interaction. Such webs encompassed both the Levant, supplied with obsidian from Cappadocia, and the 'hilly flanks' region of the Fertile Crescent, supplied from the area around Bingöl, long before farming began in these regions (Cauvin et al. 1998; Chataigner et al. 1998; Cauvin 2002). There is no reason to suppose that any area was 'isolated' from such small-scale flows of products and information. If for no other reason, it makes sense to view the whole perimeter of the Fertile Crescent as an interacting network of forager communities, linked in a set of interlocking circuits each with its own practical and ideological peculiarities dimly reflected in typological labels for their characeristic lithic industries (Natufian/'PPN', Trialetian, Zarzian/Zawi Chemian). What made certain parts of it peculiar (by comparison with Europe, say, or areas further east) was the degree to which a mixture of foraging (including cereal-collecting) and hunting allowed for sedentary occupation near to rivers flowing through relatively open country, with ease of movement in many directions and abundant game. Those at the exits to major routes into the mountains, where traffic-flows of resources such as obsidian were concentrated, had nodal positions and an enhanced incentive to sedentism.

These circumstances serve to make more understandable (even if they cannot fully explain) the extraordinary cultural florescence which took place at the northern end of two important routes: the 'Levantine Corridor' from the Jordan valley to the middle Euphrates (Bar-Yosef and Belfer-Cohen 1992; Bar-Yosef and Meadow 1995), and another axis of contacts (where varieties of the Kurdish language have been spoken in recent times) through the area which Robert Braidwood called the 'Hilly Flanks region' along the northern margins of the Fertile Crescent (Braidwood and Howe 1960). At the apex of these routes, where they converge in southeast Turkey, a group of major sites has been discovered in recent years (Özdoğan and Başgelen 1999; Hauptmann 1999), especially in front of the passes through the Taurus mountains, leading to obsidian and other highland resources on the other side of the mountain chain. Çayönü, on the way to the headwaters of the Tigris, lay on a major route subsequently followed by an Assyrian highway and a Roman road, and Göbekli Tepe and Nevalı Çori, between the Balikh valley and the upper Euphrates, are also positioned on critical axes of contact, linking mountain hinterlands and extensive lowland chains of onward transfer. The character of these sites in the ninth millennium BC was entirely 'mesolithic', in the sense that there is little indication of the importance of cereal cultivation, and their art depicts in three dimensions (probably with a background in wood carving before stone carving) the kinds of subjects otherwise encountered in rock art, and with an emphasis on maleness, fierce or dangerous animals, raptorial or carrion-eating birds, scorpions and snakes which contrast strikingly with the female figures and ungulates of contemporary Levantine and subsequent neolithic subjects (though they anticipate themes which will reappear on Çatalhöyük wall-paintings, and perhaps for similar reasons). They form part of a practice of male bonding in cult-houses, themselves associated both with architectural elaboration (carved pillars, terrazzo floors) and cult-equipment (decorated stone bowls, bird-head pestles) suggestive of shamanic transformation using hallucinogenic drugs (henbane?), and further associated with the transformation of natural substances into new materials ('copper metallurgy', as we reductively describe it in our utilitarian vocabulary, or lime-plastering). All this is literally fantastic, but gave rise not only to rectangular[2] houses of increasing constructional

[2]Rectangularity seems to be associated with the construction of larger buildings, in environments where timber was an important constructional element (e.g. for roofing): it is easier to roof a large rectangular building than

sophistication, and also perhaps to experiments in animal-keeping which provide the background to the first domestication of ovicaprids. (On the relation between sedentism and domestication, see Uerpmann 1996.) None of this was necessarily the result of cereal cultivation, and nor are the sites located in positions chosen for farming, even though Çayönü at any rate seems to have remained occupied at a time when cultivation seems to have become more important in this region, after 8000 cal BC.

I have phrased this description rhetorically (though by no means exaggeratedly) because it contrasts so strongly with the conventional picture of the 'origins of farming'; and nor does it in itself provide an account of how cereal cultivation might have begun. Quite the reverse: it shows how wild cereals formed a local component within a spectrum of gathered and hunted resources in supporting the elaboration of an essentially forager existence. There is no reason to suppose that this elaboration reflected an increasing reliance on cereals, or depended on their cultivation, or would necessarily have led to their cultivation. The assumption that control over carbohydrates gave rise to cultural complexity seems quite false. Nevertheless this area seems to have exerted an influence (architecturally, for instance, and perhaps in terms of animal management) on surrounding areas, and indeed to have initiated a tradition of architectural and artifactual elaboration which continued as the core of subsequent developments in the Fertile Crescent. The mesolithic origins of Near Eastern civilization?

This formula would be true only if it contained within itself the motor for expansion; and arguably this is precisely what is lacking. In showing what may be achieved by foragers, it misses the inbuilt dynamic of the conventional, calorie-based account of farming development: the feedback-loop between reliance on cultivated cereals and demographic growth. Arguably, therefore, the 'northern climax' was only one component of the neolithic revolution (to use that term with the benefit of hindsight); it was its dialectical conjunction with developments taking place in other parts of the network—with which it was increasingly combined, from 8000 BC onwards—which gave rise to the classic combination of cereal cultivation, domestic livestock and village life that was to spread with such speed in subsequent millennia, and to provide the basis for subsequent developments within the Fertile Crescent. Where, then, should we look for the origins of cultivation, and the burdens that it brought with it: 'in the sweat of thy brow shalt thou eat bread' (Genesis 3:19)?

1.5 The origins of dependence: floodwater farming and the beginning of cultivation

A characteristic class of models of farming origins may be called 'climate-driven', because they invoke climatic downturn as the forcing mechanism for intensification of subsistence practices involving higher labour-inputs than foraging. Jack Harlan long ago demonstrated how easy it was to harvest stands of wild grain (Harlan 1967), echoing the San (Bushman) informant who asked why they should perform the backbreaking labour of planting when there were so many mongongo nuts in the world (Lee and DeVore 1968)—a sentiment which may well have been held by the climax foragers of the previous section, in the northern Fertile Crescent. Both of these observations have entered the literature of farming origins, and together they convey an important truth: that the costs of cereal cultivation were a major disincentive to its adoption, which only took place in exceptional circumstances. They are reinforced by Theya Molleson's observations (1994) on the osteological pathology of female skeletons from neolithic Abu Hureyra, with their deformations and traumas resulting from the incessant grinding of grain; clearly these costs fell disproportionately on the female part of the population (who probably did much

---

a circular one, at least with systems of simple carpentry. This, rather than any specific correlation with social morphology, may have been behind the emergence of a rectangular tradition in the north.

of the cultivation as well). Cereals in general probably ranked after nuts as foods of choice, since so much labour was involved in their preparation, especially the early hulled varieties (Wright 1994, 2000). In the light of this downside of cereals, the explanation for their early cultivation has often appealed to 'Romer's rule'—the generalization from within palaeontology that innovations often arise from conservatism, in the sense that they emerged in order to permit the continuation of previous ways of life in declining circumstances but then opened up radically new alternatives (fish hopping from pool to pool as sea-levels declined, thus evolving legs from fins which then permitted the colonization of terrestrial environments, in a caricature example). The archaeological equivalent is the scenario that cultivation was an extension of foraging by other means (to adapt Clausewitz's dictum on the relation of war to diplomacy), in circumstances where previously abundant supplies of wild grain—on which human populations had become locally dependent, to sustain existing population levels—were contracting. Hence the appeal to the Younger Dryas, as a climatic forcing-mechanism leading to the deliberate cultivation of what had previously been a 'free good'. One drawback of this model is that it fails to account for the apparent fact that cultivation (to judge from the occurrence of 'domesticated' morphologies) only became common in the early Holocene, when these constraints should have been relaxed. (The climatic contexts of the earliest cultigens are not, however, entirely clear, and early Holocene cultivation could reflect practices established at an earlier stage: but nevertheless the archaeobotanical record does not at the moment clearly demonstrate an association of the first cultivation with climatic stress, and does not account for the persistence of this practice when the stress was relaxed.) Other variants of the climatic forcing model appeal to demographic displacement, and localized crowding in propitious circumstances (e.g., within the Jordan Rift: Kuijt and Goring-Morris 2002) at a time when other habitats were relatively less attractive; and such models may well capture some of the 'push' factors in a complex conjunction of conditions, as well as helping to explain why there was no going back from a dependence on cultivated grain when once it had been established.

There is, however, an alternative strand to the argument which emphasizes the 'pull' factors, and moreover one which is congruent with an expansion of cultivation in the early Holocene. This is the point of view put forward by Claudio Vita-Finzi (as a result of his early geomorphological work in the Wadi Hasa in Jordan), and succinctly summed up in his phrase 'geological opportunism', which he used as the title to his contribution to *The Domestication and Exploitation of Plants and Animals* (Ucko and Dimbleby 1969; Vita-Finzi 1969b). It is fair to say that its insights were not subsequently incorporated into the programme of work conducted by Vita-Finzi and the Higgs school (Vita-Finzi and Higgs 1970), but they stand as a far-sighted observation which may offer a clue to the circumstances we are seeking. Vita-Finzi's suggestion, derived from the characteristic locations of early farming sites in the Jordan valley, was that the increased runoff of the early Holocene created a set of spatially limited but highly productive environments where wadis or springs began to accumulate small alluvial fans, especially along the edge of the Jordan valley and more generally in the endorheic (internal drainage) basins or *khabras* behind the coastal mountain-chain of the Levant (some supported by artesian water: Wirth 1971, pp.108–114, Wirth 1998, p.15), which in the conditions of lowered evaporation during the later Pleistocene had been occupied by lakes. (This is concisely summarized in a block diagram, Fig. 23 in Vita-Finzi 1969a.) The classic example of such a location is Jericho, but its circumstances are replicated 15 km further north at Netiv Hagdud (Bar-Yosef and Gopher 1997), and perhaps also at a comparable location in the intervening Wadi 'Auja.[3] The relevance of this observation was reinforced by Bar-Yosef's suggestion (1986) that the PPN wall and tower at Jericho had been intended as a flood defense, and his definition of the small alluvial fan to the west of the site at the exit of the Wadi el Mafjah. Having noted the importance of such locations, Vita-Finzi's further observation was that these conditions were eminently suitable to

[3]Satellite images of these and other locations are available on the ArchAtlas website, at `http://archatlas.org`.

the mode of cultivation described for the traditional societies of the American Southwest, called 'floodwater farming' (Bryan 1929): a small-scale system of crop-growing taking advantage of seasonally wet ground, in which sowing took place after a small annual inundation.

Applying this model to the Jordan valley, one could envisage that the seasonal changes in water level at spring- or stream-outlets at the head of small alluvial fans along the edges of the Rift would limit the growth of other forms of vegetation in these habitats, and competition from annuals could be discouraged by small-scale clearance. Cereals would not naturally grow in such places (and even if introduced would not create a self-sustaining population), but would need to be re-sown each season. This would, in fact, create precisely the circumstances in which morphologically domesticated forms would rapidly predominate, if harvested by reaping rather than beating. Such a mode of cultivation would have been relatively light on labour, making cereals, if not a 'free good', at least a 'cheap good'. This would have been especially attractive in locations which were themselves at some distance from extensive stands of wild cereals in the hills, but which offered advantages of permanent water, concentrations of game and other gathered resources, and positions on an axis of movement along the Rift, by which exotic materials were supplied. The Jordan Rift itself was once described (by an authority on indigenous cultivation systems in Africa) as a 'natural greenhouse' (Allan 1972), emphasizing its suitability for small-scale experiments in horticulture.

Floodwater farming, which is a small-scale, horticultural version of what are more generally known as 'water-harvesting' techniques (Critchley and Siegert 1991; Prinz and Malik 2000), has been largely outgrown by most farming societies, because of the inevitably restricted conditions in which it can be practiced. Other forms of water-harvesting came into use in the agricultural margins of the southern Levant in the first millennium BC (Evenari et al. 1971), but these represent labour-intensive and spatially extensive versions of the technique at a relatively late date. A more immediately relevant analogy is provided by simple techniques of *bund* cultivation in Iran, or *sailaba* cultivation along the Indus. (The PPNA site of Gilgal, just below Netiv Hagdud, on a ridge further down the Wadi Salibiya, overlooked a natural *bund* in the early Holocene: Bar-Yosef and Gopher 1997; Tchernov 1994.) These systems merge on a larger scale with the kinds of *décrue* farming practiced around seasonally enlarged lake basins, allowing a catch-crop to be grown as the lake margin recedes. The latter are however more typical of areas with a monsoonal rainfall pattern, where—as with the Nile—these periodicities are complementary to the natural growth cycles of originally winter-grown cereals.) I discussed all these systems in 1980 (Sherratt 1980b, with references), without quite seeing how they might all fit together, beyond the observation that floodwater farming provided a plausible 'common ancestor' for systems which later differentiated into 'dry farming' and 'irrigation farming'. (Such thoughts are relevant to the origins of cereal cultivation throughout the world, especially perhaps to rice, where comparable conditions might be sought in Szechuan (Sichuan) before rice cultivation spread in more developed form down the Yangtse, to appear at sites like Jiahu and Hemudu.)

Some idea of how floodwater farming might work on the spring-fed margins of the Jordan Rift may be obtained from modern records of seasonal changes in spring discharge in these locations (e.g., Hosh 1995, fig. 1)—which of course reflect current patterns of precipitation and flow which need not be valid for the period in question—but if taken at face value indicate the possibility of a spring (three-month) growing period in the spatially limited locations affected by such discharge. A problem with this suggestion as a model for early cereal (or legume) cultivation is that although spring grown cereals were known in antiquity in the Mediterranean, it is hard (except in the case of barley) to envisage the rapid emergence in the early Holocene of such accelerated growth patterns in species genetically adapted to a longer, rainfall-dependent growing season in quite different habitats (pers. comm. Gordon Hillman, many years ago; Blumler 2002, p.105); but the suggestion is worth reviving if only because it captures something of the small-scale and experimental circumstances in which plant cultivation is likely to have emerged, and the unusual conditions of its genesis. This suggestion has a wider significance in that these conditions were replicated along the whole length of the 'Levantine Corridor' (which we might now term the 'oasis route'), in the chain of endorheic basins from the Hasa and El Jafr basins in the south, along the Rift itself (occupied in the Pleistocene by the Lisan lake) with the Dead Sea basin and the Kinneret and Huleh basins, and continuing behind the anti-Lebanon—i.e., to the east of the northern length of the Rift itself—by way of Damascus, Palmyra and El Kowm, to be within striking distance of the bend of the middle Euphrates, giving the Corridor a natural linkage to the upper Euphrates and the rainfed areas through which it flows (see maps in Kuijt and Goring-Morris 2002). As an axis of cultural contacts and communication, this existed already in late Natufian times, during the Younger Dryas, and was presumably the channel down which exotic resources such as obsidian were supplied to the Levant in the late Pleistocene and early Holocene, with Mediterranean (and, later, Red Sea) mollusc-shells exchanged in return (Bar-Yosef and Mayer 2000)—along with distinctive organic materials (drugs, feathers, bitumen) not yet identified archaeologically, from the ecologically contrasting conditions at either end of the chain. The Damascus basin occupies a pivotal position in these links, and has an extensive fan created by the Barada river (now watering the *Ghouta*, the area of orchards and gardens which sustains the modern capital city), with an extensive chain of lakes around the perimeter of the fan (now represented by the Ataibeh and adjacent small lakes, then more continuous). This lakeside location was the setting of Tell Aswad—whose situation would thus resemble, on a larger scale, that of Gilgal (in overlooking a vast natural *bund*) more than Netiv Hagdud or Jericho, in positions higher up their respective alluvial fans. (Anticipating later discussion, it is also relevant to compare these locations with that of Çatalhöyük: see the satellite images on the ArchAtlas website.) By comparison with the 'early neolithic' sites mentioned in the previous section, therefore, these oasis-like conditions, in a chain along a natural corridor of movement, provided a distinctive set of habitats in which the experimental sowing of seeds might have begun. Their distinctive locations provided a further set of incentives for cultivation, in being offset from the main stands of wild cereals, by contrast with the northern sites which were surrounded by a natural abundance of wild resources.

It is important to emphasize once again at this point the fact that all of these sites formed an interconnected network, as indicated both by the rare but diagnostic materials like obsidian which can be demonstrated to have flowed between them, and also by their membership of a common typological community in things such as hunting equipment, as reflected in their common acceptance of a sequence of arrowhead types (which are the basis for the periodization of the PPN as a whole). This does not mean, however, that they shared a common subsistence basis, or that they jointly experienced a step-like series of stages of increasing dependence on cultivation; but it does allow for the transmission of cultivated strains of crops, and the practices of their cultivation, between adjacent groups—and perhaps for a growing body of cultivators along this axis, who may initially have used different cereals (barley, emmer, einkorn) in different places and circumstances. The limited genetic variability of domesticated cereal strains, by comparison with the diversity evident in wild populations of the same species (Zohary 1996) argues for a finite number of 'domestication events' for each of the founder species, which might nevertheless have emerged in parallel in similar circumstances as 'vicariant cultigens' (Sherratt 2004, p.57). Moreover the existence of the wider community of PPN groups in the Levant and those north of the Euphrates bend suggests the possibility of a closer coupling between the postglacial 'forager florescence' in the north and the postglacial 'geological opportunism' in the south, in terms of the incentive which it gave to occupy nodal positions in a supply chain linking both ends of the Corridor, reinforcing the linear tendency in settlement distribution, and perhaps allowing us to envisage not only exchange of goods but internal migration between them and emulation of cultural achievements and practices initially restricted to particular parts of it.[4]

In some respects this reconstruction approaches Binford's (1968) model of the origins of farming, with growing populations of affluent foragers ('open systems, donor type') shedding surplus population into more marginal habitats ('open systems, recipient type') to increase pressure on resources and lead to intensification at the margin! Indeed, a fully syncretic formulation might even build in Kent Flannery's (1973) concept of initial cultivation as a 'deviation-amplifying mechanism' in the southern Levant, leading to increasing dependence on what had initially been a small-scale innovation in very particular circumstances. Almost all of the suggested models for the beginning of farming, for the last thirty years (Benz 2000), have some element of truth; the challenge is to mobilize them in their appropriate contexts, within the specific geographical circumstances of the different parts of the process, rather than treating them as competing universal explanations in a zero-sum game.

Although this discussion has focussed on the 'Levantine Corridor', it is important to bear in mind that this to some extent reflects the distribution of research opportunities in the last few decades; and it is worth noting that Çayönü, for instance, stands in the same relationship to Braidwood's 'Hilly Flanks' corridor (linking Çayönü with his earlier area of work at Jarmo in Iraqi Kurdestan) as for instance does Nevalı Çori in relation to the Balikh and the 'Levantine Corridor'—so that there may have been two axes of intensification, one down each side of the Fertile Crescent. The absence of a comparably detailed record of these periods in northeast Iraq and western Iran (and my relative unfamiliarity with their literature) gives an inevitably one-sided bias to this account, and it may well be that a more symmetrical reconstruction might be offered. On the other hand, it is equally possible—and rather more probable—that the areas along and adjacent to the Jordan Rift (the 'oasis route') did indeed provide a unique conjunction of circumstances leading to the appearance of cultivation, and that the appearance of similar phenomena along the eastern side of the Fertile Crescent represents a secondary spread.

Certainly, the appearance of substantial sites in classic farming locations in the Jordan valley such as Jericho and Netiv Hagdud (and perhaps as far up as Mureybet and Jerf el Ahmar, though their riverine locations may imply a different variety of cultivation techniques) at the very beginning of postglacial conditions, with polished axes and terracotta female figurines (those icons of the Neolithic, reflecting their characteristic tools and ideology respectively: Cauvin 1997), together with what are widely accepted as 'domesticated' plants (i.e., the outcome of artificial cultivation in the particular circumstances suggested above), do suggest that Levantine sites were practicing 'farming' when their northern neighbours were enjoying greater opportunities for 'foraging', and developing many of the characteristics which have traditionally been attributed to farming communities alone. These different responses arose from their contrasting environmental settings, within what was already a linked network of communities and which continued to be in increasing contact, as flows of products such as obsidian increased in volume to supply rising densities of population. These developments might thus be seen as a dialectical interaction, within an already established family relationship, and whose participation in a common network of exchanges was a fundamental aspect of their different forms of intensification. If we play the game of looking for a 'prime mover' of these events (a curious hangover from sixteenth-century theology), the accolade must be accorded equally to 'ecology' and 'trade'—or, put more generally, the opportunities afforded jointly by postglacial environmental changes and the geometry of webs of human communication, both set within a landscape of stark contrasts between mountains and plains, and between forests and desert, which restricted and channelled the mobilization of resources by human populations.

---

[4]It would be possible to tell this story in a slightly different way, emphasizing the centrality of the Euphrates bend as an intermediary in this chain of contacts and a focus of innovation: this would give greater prominence to the site of Jerf el Ahmar, for instance, which shares some of the characteristics of both ends of the axis, and links the Levantine corridor to the Cappadocian obsidian sources (Abbès et al. 2003).

1.6 Interaction and exchange: the synthesis and expansion of the 'neolithic package'.

At some point in the process, between 8500 and 8000 cal BC, the two areas began to interact more intimately (Harris 2002); and—especially as domestic livestock came to be integrated into the emerging package—some parts of the area began to show the kind of demographic expansion, even migrations, which were to become typical of later phases of enlargement, when the neolithic package was dispersed westwards across Europe and eastwards to the edge of the Indus valley. A key observation here is the recently discovered PPNB colonization of southern Cyprus (Peltenburg et al. 2000), at a date perhaps as early as 8700 cal BC.[5] Although it is tempting simply to draw an arrow from the nearest coastland (e.g., Sherratt 2004, fig. 1, map c), the lack of contemporary farming settlement on the northern Levantine coasts and Cilicia, by contrast with the plentiful appearance there of sites in the later Neolithic (see below), suggests that these well-watered and wooded areas—which stand out today as distinctive green strips on a satellite photograph—were only occupied by farmers at a later date, and may well have had their own hunting and foraging communities of the kinds documented for south-coastal Anatolia (at Beldibi, Belbai, Öküzini and Karain). The nearest area of the Levantine coast with contemporary evidence of PPN occupation is the Carmel region, where the Jezreel valley meets the sea, and the pattern of winds and currents suggests this as a plausible point of departure (interestingly paralleled in the late Bronze Age by the voyage of the Egyptian Wen-Amun, who was shipwrecked on the coast of Cyprus after travelling up the south Levantine coast—a suggestion I owe to Susan Sherratt). The apparently deliberate PPN colonization could thus represent a small northward movement of south-Levantine groups (albeit taking with them some features whose ancestry lies in the north), of the kind which might otherwise, but less visibly, have occurred along the Levantine corridor. In this case the Binfordian shedding of surplus forager population from north to south would have been reversed, by the early expansion of cultivators from the south. Whatever the case, it is symptomatic of expansionary tendencies in early or early middle PPNB, at the time when cultivation is plausibly suggested on sites hitherto characterized as parts of the foraging climax; and indeed it broadly coincides with a tendency to locational shift in these communities, from earlier-occupied sites to newly-founded ones at places like Cafer Höyük, Gritille or Mezraa-Teleilat—precisely the point at which it becomes appropriate to speak of an 'achieved neolithic' in these areas (Bischoff 2004). Nor was this movement one-sided, for it was apparently as part of this closer relationship between the Levantine corridor and the north that elements such as rectangular architecture and domestic livestock became increasingly evident in the south, along with an elaboration of ritual practices (like the statues of Ain Ghazal: Rollefson 1983; Salje et al. 2005). It is thus from this stage that the 'neolithic package' in its early form can be considered to have been assembled, and at which a line enclosing a distribution of 'early farmers', from the Jordan rift to southeast Anatolia, may convincingly be drawn. 'The Neolithic', in the sense that Childe conceived it, was a synthesis of this dual inheritance; and the expansion of farming settlement beyond this area was symptomatic of its potential for growth.

A second new area to be opened up to farming at around this time, in the early and middle PPNB, was the Anatolian plateau, where the occupation of Aşıklıhöyük began in the centuries just before 8000 cal BC (Esin and Harmankaya 1999). The appearance of farming settlements in this region must surely have been related to developments taking place on the other side of the mountains, reflecting both the growing demand for obsidian and perhaps also the settlement of new areas suitable for simple forms of cultivation. Unlike the Cyprus example, however, there

[5]The location of PPN Parekklisha-Shillourokambos (Guilaine and Le Brun 2003), where runoff from the southern slopes of the Troodos is concentrated behind the hills running parallel to the southern coast, offers an environment with abundant surface water.

were also indigenous population of foragers to be considered. Cappadocian obsidian sources had supplied communities on the Euphrates bend and in the Levantine corridor on a small scale in earlier periods, when they had presumably been worked (as in the terminal Pleistocene) by indigenous foraging groups in contact with related communities across the Taurus. The techniques used in the extraction-site of Kömürcü-Kaletepe P-32, however, which dates to 8300–8200 cal BC, indicate a new type of large-scale production of large, prestige types for export, of the kind which circulated within the Levantine corridor (and, indeed, Cyprus), and contrasting with contemporary microlithic industries in the caves of southern Turkey and at some sites further inland (Binder 2002). This suggests that small enclaves of farming population, directly linked to the middle Euphrates and the Levant (and perhaps attempting to bypass existing indirect modes of exchange), may have been established in the intervening area. Such groups would have avoided the forested coastal mountains surrounding the Bay of Iskenderun, and selected environments suitable for simple horticulture in moist conditions: likely places for such sites would be the intermontane basin of Elbistan, with its abundant surface water, or the eastern end of the south-central Anatolian basin south of Erciyes Dağ. The occupation of such areas would thus represent a small colonizing movement like the move to Cyprus, leading to the further proliferation of farming settlements on the Anatolian plateau—which here involved varying degrees of interaction with (and incorporation of) indigenous foraging groups.[6]

The locations occupied by early farmers in the south-central parts of the Anatolian plateau (Cappadocia and the Konya plain) bear a striking resemblance to those discussed above at an earlier phase in the Levantine corridor, typically situated on small fans at the edges of the plain or around small internal-drainage basins (some occupied by saline lakes), within relatively open environments (see especially the maps of Kuzucuoğlu 2002, cf. Cohen and Erol 1969; Cohen 1970; French 1970; Roberts et al. 1999). It is thus arguable that techniques of cultivation were transferred to the plateau environment from the Fertile Crescent without significant alteration, still at the scale of floodwater farming (though now with free-threshing wheats and complemented by some domestic livestock). A successive occupation of such locations can be postulated, from east to west in Cappadocia and the Konya plain: beginning perhaps in the Sultansazliği basin south of Erciyes Dağ (where sites such as Hacıbeyli occupy small fans from rivers entering this small endorheic basin, some 25 km in diameter), the next phase would have reached the Melendes Çay (including Aşıklıhöyük) and then extended to the small fans on the inner margins of the Taurus, hypothetically at Ereğli and demonstrably at Can Hasan III near Karaman, dating to around 7500 cal BC, and finally a century or so later to occupy the largest alluvial fan of all, that of the Çarşamba Çay in the Konya basin, on which stands Çatalhöyük (see maps in Kuzucuoğlu and Gérard 2002). The last of these represents such a major concentration of water resources in this otherwise arid basin, attractive to game as much as to human populations, as to make it likely that occupation of this major resource would have involved the absorption of a substantial number of its existing hunting and foraging occupants, to a far greater extent than at the other locations: and this, together with the concentration of farming population in a single site (perhaps as a consequence), may help to explain the unique character of the organization and art of Çatalhöyük. I see no reason, however, to doubt that it represents an unusually large-scale application of the techniques of floodwater farming which explain the pattern of locational choices back to Jericho and Netiv Hagdud (though technical discussion of the lines of evidence which lead the excavators of Çatalhöyük to question this must await publication of the evidence, now in press). In geographical terms, it represents the continuation of a pattern typical from the beginning of cultivation, in which early farming sites are closely tied to particular pockets of

[6]To some extent this may be reflected in the degree to which microlithic industries continued to be used (Binder 2002)–notably at Aşıklıhöyük (though not at Can Hasan III, in a different environmental setting). On the other hand, the agglomerated patterns of architecture at all of these sites (which together with the lack of cult-houses differentiates them from southeast Anatolian and Levantine PPNB sites) may have more to do with the more rigorous winter climate of the Anatolian plateau than with the cultural origins of their inhabitants.

well-watered terrain. This was to be the pattern for much of the early extension of cultivation systems in the following millennium, through the lowland basins of western Anatolia, Thessaly and the Balkans (Sherratt 1972, cf. van Andel et al. 1995): small enclaves of horticulturalists were established in conditions which replicated the situations where cultivation first emerged.

1.7 Subsequent spread: the dialectic of expansion

Just as the genesis of farming involved a dual origin and a dialectic between farmers and foragers, so did its subsequent spread. The story of the advance of farming across Eurasia was a two-sided affair, involving the absorption and conversion of indigenous populations as much as the demographic proliferation of farmers. Small migrations certainly occurred (and perhaps larger ones where conditions allowed), but by and large the two mechanisms of dispersal were fairly evenly matched, and certainly closely interwoven. This is the picture which is emerging from local studies in areas of abundant evidence in Europe (Bánffy 2000; Gronenborn 1998); and it is possible to suggest that a similar picture may hold true at an earlier date for Western Asia.

After 8000 cal BC, with the PPNB incursions into surrounding areas (including potentially the eastern arm of the Fertile Crescent, not followed here), the interconnected parts of the enlarged network experienced a climax in LPPNB, in the later part of the eighth millennium. This was marked by the emergence of sometimes spectacularly large sites at nodal positions throughout the network of closely interconnected communities (Çatalhöyük, Abu Hureyra, Ain Ghazal, Basta—the last two reflecting a major route down the eastern side of the Jordan valley to the Gulf of Aqaba (source of the Red Sea shells), which took over from the oases on the western side of the Rift as the main artery of communication in this area, along the line later followed by the 'King's Highway' in Biblical times. The wider linkage of this area with the eastern axis is indicated by the more general appearance in the Levant of obsidian from sources in eastern Anatolia, and the establishment of a substantial settlement at Bouqras, on the Euphrates opposite the mouth of the Khabur (Akkermans et al. 1983), which could have acted as an entrepot in exchanges across the eastern and western axes of the Crescent (in the manner of second-millennium Mari or Hellenistic Dura Europos). Such nodal centres would have arisen at critical articulation points in an arterial network.

A larger extension of the neolithic community took place during the period variously referred to as 'Final PPNB', 'PPNC', or 'early Pottery Neolithic' (since pottery was making its appearance in some parts of the area, though not others). This enlarged area, incorporating the coasts and forested mountains of the northern Levant and southeastern Anatolia mentioned earlier as initially avoided by farmers, looks like a classic case of the conversion of a pre-existing area of 'mesolithic' persistence in environments which provided ample alternatives to cereal cultivation (the mongongo nut scenario), but were ultimately absorbed in a growing network of farmers. The process is most strikingly exemplified in Cilicia (Caneva 1999), where the founding of Mersin around 7000 BC was echoed in the appearance of a multitude of later neolithic (pottery-using) sites around the margins of the bay of Iskenderun.[7] Coastal sites had begun to appear in the previous phase in the northern Levant (e.g., Ras Shamra), but now occurred at locations off the modern Carmel coast where they were subsequently covered by rising sea levels. The addition of new areas to the farming network altered the potential pattern of contacts, reducing the advantages of formerly axial routes and nodal positions. The new coastal dimension provided an alternative distribution chain for Cappadocian obsidian, for instance, which removed the monopolistic advantage of supersites along the Levantine corridor and resulted in their relative decline and a dispersal of their population (for a description of these settlement-

[7] If the radiocarbon dating of the earliest neolithic at Knossos is accurate, this is also the point at which farming arrived in Crete: can one envisage small coastal migratory movements like that which took farmers to Cyprus a millennium earlier?

pattern changes see Kuijt and Goring-Morris 2002).[8] This devolution preceded the apparently catastrophic change affecting a large area of the Levant and adjacent regions around 6200 BC which brought the whole PPN complex to an end (and which also saw the shift of location from Çatalhöyük East to Çatalhöyük West, part of a more general pattern of relocation marking the beginning of the Anatolian early Chalcolithic), which can plausibly be attributed to a brief but large-scale climatic anomaly perhaps caused (like the Younger Dryas itself) by a final pulse of glacial meltwater released into the Atlantic circulation from the remains of the Laurentide ice sheet (Klitgaard-Kristensen et al. 1998; Renssen et al. 2002).[9]

The 'Pottery Neolithic' also saw a major extension of the area within which farming occurred in western Anatolia, most clearly seen in the extension of sites in the Pisidian Lake District and other areas bordering the Konya plain (e.g., Ilıpınar) after 6500 cal BC (Kuzucuoğlu and Gérard 2002), and perhaps symptomatic of a further extension into key areas such as the plain of Eskihehir where traces are so far more enigmatic, but would provide a logical step towards Northwest Anatolia where neolithic sites are known in some numbers towards the end of the millennium. This, like the incorporation of coastal areas in the Levant, marks a scale of extension beyond simple migration of the kind seen earlier, and arguably implies the absorption of indigenous populations over a wide area—although involving the spread of a material culture closely related to that of established farming regions. Such 'block conversions' of large areas, later repeated on an even larger scale with Criş/Körös in the Danube basin, indicate how the neolithic package had taken on a new dynamic of expansion, typical of the way in which it was to spread over large areas of Europe. The collapse of 'PPN' cultures in the Levant, and the continuing spread of their spinoff-cultures into the Balkans and beyond, marked the end of the first chapter in the construction of the Neolithic.

1.8 Beyond cereals

This paper has tried to highlight two themes: the first is the parallelism in the development of forager and farmer societies, and the continuing dialectic between them; the second is the existence of large structures such as trade networks (most obvious in the case of obsidian) which lie behind both of them. The growing dependence on cereals, and the inherent dynamic which it imparted, has been less central to its concerns. This is in part because it is already an accepted element in our understanding of the Neolithic, and has been since Gordon Childe placed it as the defining characteristic of the whole concept. Yet it is salutary to look beyond it, and to see what cereals on their own fail to explain. The de-coupling of cereal domestication and social complexity (like the recognition that urban civilization emerged in the New World essentially without metallurgy or animal traction) imparts a new flexibility to concepts generated in a particular context and allows us to look for more fundamental correlations. In the case of monumental construction and the elaboration of fixed material culture, it helps us to understand why the most spectacular monuments of neolithic culture, in the form of megalithic tombs and ceremonial monuments, lay in the west and north of Europe, not in the Balkans—in areas where the 'mesolithic inheritance' was strongest, not in the areas where cereal production was most effective. It is the diversity of lifeways which came together, not the size of their calorific base, which explains their complexity. In the same spirit, we might reexamine some of the phenomena hitherto attributed to irrigation and improved cereal production in Western Asia, such as the

[8]This scenario bears a striking resemblance to that proposed for the end of the late Bronze Age by Susan Sherratt (2000), in which the traffic in exotic and scrap metal by coastal traders undercut the monopoly bulk trade in such items which was the basis for the power of palatial centres in the Aegean and Levant.

[9]Such a brief (perhaps two to three centuries) but dramatic climatic excursion may, paradoxically, have been instrumental in accelerating the spread of farming in the north Aegean and the Balkans, by destabilizing the established forager systems and allowing farmers to expand into the vacated areas... but that is another paper.

emergence of civilization in southern Mesopotamia. This, too, was an area were foragers of various kinds are likely to have persisted, and provided a diversity of lifeways in this shifting marsh. Was this diversity a factor in the rise of ceremonial centres and monumental temples, integrating fishermen and fowlers as well as farmers and herders? Such thoughts hint at factors which are as yet unexplored in conventional explanations of the rise of complexity as it was elaborated beyond the village community, and suggest deeper roots than have hitherto been sought.

In the same way, this paper has attempted to apply models of the structural relationships between sites and regions which have normally been used only in the context of more advanced economies, but can here be seen to illuminate not only neolithic supersites such as Ain Ghazal but also their forager precursors such as early Çayönü or Göbekli Tepe (cf. Jacobs 1965). Such sites only make sense within networks that extend not only between sites but between regions, at nodal points where flows of products converge and where local conditions provide the resources to sustain larger communities. These principles are independent of the specific nature both of the moving products and the sustaining resources: the analysis can apply equally to ninth millennium Çayönü or fourth millennium Susa. It is only the scale which makes the difference, not the geometry.

None of this is to downplay the crucial role of cereal cultivation in the history of humankind. Its contribution was twofold: in providing a 'calorific subsidy' comparable to that of fossil fuel in the Industrial Revolution, and in imparting a degree of instability to human populations, both through the effects of a carbohydrate diet on human demography, and through the capacity of cereal crops to respond to increasing labour inputs and to sustain unparalleled degrees of population growth. The neolithic chapter of their history was only the prelude to their expanding role throughout the Holocene, which saw the world population of the Cerealia, like that of *Homo sapiens* itself, undergo an exponential rate of proliferation. The beginnings of farming initiated an ever-accelerating pace of change, of which we (and today's genetically modified cereals) are the involuntary inheritors.

Acknowledgements

The illustrations to the original conference paper, consisting largely of satellite images of key sites, have been incorporated into the ArchAtlas website; this written version was composed during tenure of a Visiting Professorship at the University of Heidelberg, and has benefited from conversations with Harald Hauptmann and the practical help of Simone Mühl. It would not have been possible without the remarkable openness of the research community dealing with these matters, notably the members of the Central Anatolian Neolithic e-Workshop and their excellent website (`http://www.canew.org/`), and it forms a contribution to the British Institute of Archaeology at Ankara's strategic research initiative on the 'Settlement history of south-central and southeastern Anatolia'.

References

Abbès, F., L. Bellot-Gurlet, M.-C. Cauvin, S. Delerue, S. Dubernet, G. Poupeau, and D. Stordeur (2003). Provenance of the Pre-Pottery Neolithic-A Jerf el Ahmar (Middle Euphrate Valley, Syria) obsidians. *Journal of Non-Crystalline Solids 323*, 162–166.

Akkermans, P. A., J. A. K. Boerma, A. T. Clason, S. G. Hill, E. Lohof, C. Meiklejohn, M. L. Miere, G. M. F. Molgot, J. J. Roodenberg, W. Waterbolk von Rooyen, and W. van Zeist (1983). Bouqras revisited: preliminary report on a project in eastern Syria. *Proceedings of the Prehistoric Society 49*, 335–372.

Allan, W. (1972). Ecology, techniques and settlement patterns. In P. J. Ucko, R. Tringham, and G. W. Dimbleby (Eds.), *Man, Settlement and Urbanism*, pp. 211–226. London: Duckworth.

Bánffy, E. (2000). The late Starčevo and the earliest Linear Pottery groups in western Transdanubia. *Documenta Præhistorica 27*, 173–186.

Bar-Yosef, O. (1986). The walls of Jericho: an alternative interpretation. *Current Anthropology 27*, 157–162.

Bar-Yosef, O. and A. Belfer-Cohen (1992). From foraging to farming in the Mediterrannean Levant. In A. B. Gebauer and T. D. Price (Eds.), *Transitions to Agriculture in Prehistory*, pp. 21–48. Madison: Prehistory Press.

Bar-Yosef, O. and A. Gopher (Eds.) (1997). *An Early Neolithic Village in the Jordan Valley, Part I: The Archaeology of Netiv Hagdud.* American School of Prehistoric Research Bulletin 44. Cambridge, MA.: Peabody Museum of Archaeology and Ethnology.

Bar-Yosef, O. and D. E. Mayer (2000). The economic importance of molluscs in the Levant. In H. Buitenhuis (Ed.), *Archaeozoology of the Near East III: Proceedings of the Third International Symposium on the Archaeozoology of Southwestern Asia and Adjacent Areas*, pp. 218–227. Leiden: Backhuys Publishers.

Bar-Yosef, O. and R. Meadow (1995). The origins of agriculture in the Near East. In T. D. Price and A. B. Gebauer (Eds.), *Last Hunters, First Farmers: New Perspectives on the Prehistoric Transition to Agriculture*, pp. 39–94. Santa Fe: School of American Research Press.

Benz, M. (2000). *Die Neolithisierung im Vorderen Orient: Theorien, archäologische Daten und ein ethnologisches Modell.* Studies in Early Near Eastern Production, Subsistence and Environment 7. Berlin: Ex oriente.

Binder, D. (2002). Stones making sense: What obsidian could tell about the origins of the Central Anatolian Neolithic. In F. Gérard and L. Thissen (Eds.), *The Neolithic of Central Anatolia: Internal Developments and External Relations During the 9th-6th Millennia BC: Proceedings of the International CANeW Table Ronde, Istanbul, 23-24 November 2001*, pp. 79–90. Istanbul: Ege Yayınları.

Binford, L. R. (1968). Post-pleistocene adaptations. In S. R. Binford and L. R. Binford (Eds.), *New Perspectives in Archaeology*, pp. 313–341. Chicago: Aldine.

Bischoff, D. (2002). Symbolic worlds of central and southeast Anatolia in the Neolithic. In F. Gérard and L. Thissen (Eds.), *The Neolithic of Central Anatolia: Internal Developments and External Relations During the 9th-6th Millennia BC: Proceedings of the International CANeW Table Ronde, Istanbul, 23-24 November 2001*, pp. 237–252. Istanbul: Ege Yayınları.

Bischoff, D. (2004). CANeW Archaeological Sites Database, Southeastern Turkey: 10th-6th Millennia cal BC. `http://www.canew.org`.

Blumler, M. A. (1996). Ecology, evolutionary theory and agricultural origins. In D. R. Harris (Ed.), *The Origins and Spread of Agriculture and Pastoralism in Eurasia*, pp. 25–50. London: UCL Press.

Blumler, M. A. (2002). Changing paradigms, wild cereal ecology, and agricultural origins. In R. T. J. Cappers and S. Bottema (Eds.), *The Dawn of Farming in the Near East*, pp. 95–111. Berlin: Ex oriente.

Braidwood, R. and B. Howe (1960). *Prehistoric Investigations in Iraqi Kurdistan.* Studies in Ancient Oriental Civilization 31. Chicago: University of Chicago Press.

Bryan, K. (1929). Floodwater farming. *Geographical Review 19*, 444–456.

Caneva, I. (1999). Early farmers on the Cilician coast: Yumuktepe in the seventh millennium BC. In M. Özdoğan and N. Başgelen (Eds.), *Neolithic in Turkey. The Cradle of Civilization: New Discoveries*, pp. 105–114. Istanbul: Arkeoloji ve Sanat Yayınları.

Cauvin, J. (1997). *Naissance des Divinités, Naissance de l'Agriculture: La Révolution des Symboles au Néolithique, Nouvelle Édition.* Paris: C.N.R.S.

Cauvin, M.-C. (2002). L'obsidienne et sa diffusion dans le Proche-Orient néolithique. In J. Guilaine (Ed.), *Matériaux, Productions, Circulations du Néolithique à l'Age du Bronze*, pp. 13–32. Paris: Errance.

Cauvin, M.-C., A. Gourgaud, B. Gratuze, N. Arnaud, G. Poupeau, J.-L. Poidevin, and C. Chataigner (Eds.) (1998). *L'obsidienne au Proche et Moyen-Orient: du volcan à l'outil.* BAR International Series S738. Oxford: Archaeopress.

Chataigner, C., J. L. Poidevin, and N. O. Arnaud (1998). Turkish occurrences of obsidian and use by prehistoric people in the Near East from 14,000 to 6,000 bp. *Journal of Volcanology and Geothermal Research 85*, 515–537.

Childe, V. G. (1936). *Man Makes Himself.* London: Watts and Co.

Clark, J. G. D. (1946). *From Savagery to Civilisation.* London: Cobbett Press.

Cohen, H. R. (1970). The palaeoecology of south central Anatolia. *Anatolian Studies 20*, 119–137.

Cohen, H. R. and O. Erol (1969). Aspects of the palaeogeography of central Anatolia. *Geographical Journal 135*, 388–397.

Critchley, W. and K. Siegert (1991). *Water Harvesting.* Rome: Food and Agriculture Organization of the United Nations. `http://www.fao.org/docrep/U3160E/U3160E00.htm`.

Esin, U. and S. Harmankaya (1999). Aşıklı. In M. Özdoğan and N. Başgelen (Eds.), *Neolithic in Turkey. The Cradle of Civilization: New Discoveries*, pp. 115–132. Istanbul: Arkeoloji ve Sanat Yayınları.

Evenari, M., L. Shanan, and N. H. Tadmor (1971). *The Negev: The Challange of a Desert.* Cambridge, MA.: Harvard University Press.

Flannery, K. V. (1973). The origin of agriculture. *Annual Review of Anthropology 2*, 271–310.

French, D. (1970). Notes on site distribution in the Çumra area. *Anatolian Studies 20*, 139–148.

Garrod, D. (1957). The Natufian culture: the life and economy of a Mesolithic people in the Near East. *Proceedings of the Prehistoric Society 43*, 211–227.

Gronenborn, D. (1998). Ältestbandkeramische Kultur, la Hougette, Limburg, and... what else? Contemplating the Mesolithic–Neolithic transition in southern Central Europe. *Documenta Præhistorica 25*, 189–202.

Guilaine, J. and A. Le Brun (Eds.) (2003). *Le Néolithique de Chypre: Actes du Colloque International Organisé par le Département des Antiquités de Chypre et l'Ècole Française d'Athènes, Nicosie 17-19 Mai.* Athens: Ècole Française d'Athène.

Harlan, J. R. (1967). A wild wheat harvest in Turkey. *Archaeology 20*, 199–201.

Harris, D. R. (2002). Development of the agro-pastoral economy in the Fertile Crescent during the Pre-Pottery Neolithic period. In R. T. J. Cappers and S. Bottema (Eds.), *The Dawn of Farming in the Near East*, pp. 67–84. Berlin: Ex oriente.

Hauptmann, H. (1999). The Urfa region. In M. Özdoğan and N. Başgelen (Eds.), *Neolithic in Turkey. The Cradle of Civilization: New Discoveries*, pp. 65–86. Istanbul: Arkeoloji ve Sanat Yayınları.

Hauptmann, H. (2002). Upper Mesopotamia in its regional context during the early Neolithic. In F. Gérard and L. Thissen (Eds.), *The Neolithic of Central Anatolia: Internal Developments and External Relations During the 9th-6th Millennia BC: Proceedings of the International CANeW Table Ronde, Istanbul, 23-24 November 2001*, pp. 263–271. Istanbul: Ege Yayınları.

Hosh, L. M. (1995). Preliminary Evaluation of the Aquaculture Potential in Palestine. Master's thesis, University of California at Davis. `http://www.arij.org/pub/aquacul/index.htm`.

Jacobs, J. (1965). *The Economy of Cities.* Harmondsworth: Penguin Books.

Klitgaard-Kristensen, D., H. P. Sejrup, H. Haflidason, S. Johnsen, and M. Spurk (1998). A regional 8200 cal. yr B.P. cooling event in Northwest Europe, induced by final stages of the Laurentide ice-sheet deglaciation? *Journal of Quaternary Science 13*(2), 165–169.

Kuijt, I. and N. Goring-Morris (2002). Foraging, farming, and social complexity in the Pre-Pottery Neolithic of the southern Levant: a review and synthesis. *Journal of World Prehistory 16*(4), 361–440.

Kuzucuoğlu, C. (2002). The environmental frame in central Anatolia from the 9th to the 6th millennia cal BC. In F. Gérard and L. Thissen (Eds.), *The Neolithic of Central Anatolia: Internal Developments and External Relations During the 9th-6th Millennia BC: Proceedings of the International CANeW Table Ronde, Istanbul, 23-24 November 2001*, pp. 33–58. Istanbul: Ege Yayınları.

Kuzucuoğlu, C. and F. Gérard (2002). Geoarchaeological Maps: Central Anatolia, 9th–6th millennia cal BC. `http://www.canew.org/data.html`.

Lee, R. B. and I. DeVore (Eds.) (1968). *Man the Hunter.* Chicago: Aldine de Gruyter.

Molleson, T. (1994). The eloquent bones of Abu Hureyra. *Scientific American 271*(2), 70–75.

Özdoğan, M. (1995). Neolithization of Europe: a view from Anatolia. Part 1: The problem and the evidence of east Anatolia. *Porocilo o Raziskovanju Paleolita, Neolita in Eneolita v Sloveniji 22*, 25–61.

Özdoğan, M. and N. Başgelen (Eds.) (1999). *Neolithic in Turkey. The Cradle of Civilization: New Discoveries.* Istanbul: Arkeoloji ve Sanat Yayınları.

Peltenburg, E., S. Colledge, P. Croft, A. Jackson, C. McCartney, and M. A. Murray (2000). Agro-pastoral colonization of Cyprus in the 10th millennium BP: initial assessments. *Antiquity 74*, 844–853.

Prinz, D. and A. H. Malik (2000). Runoff Farming. `http://www.plantstress.com/Articles/drought_m/runoff_farming.pdf`.

Renssen, H., H. Goosse, and T. Fichefet (2002). Modeling the effect of freshwater pulses on the early Holocene climate: the influence of high-frequency climate variability. *Paleoceanography 17*(2), 10.1029/2001PA000649.

Roberts, N., M. Kuzucuoglu, and M. Karabiyikoglu (1999). The Late Quaternary in the eastern Mediterranean region. *Quaternary Science Reviews 18*(4-5), 497–716.

Roberts, N. and H. E. Wright (1993). Vegetational, lake-level and climatic history of the Near East and Southwest Asia. In H. E. Wright, J. E. Kutzbach, T. Webb, W. F. Ruddimann, F. A. Street-Perrott, and P. J. Bartlein (Eds.), *Global Climates Since the Last Glacial Maximum*, pp. 194–220. Minneapolis: University of Minnestota Press.

Rollefson, G. O. (1983). Ritual and ceremony at neolithic 'Ain Ghazal (Jordan). *Paléorient 9*(2), 29–38.

Runnels, C. (2001). The stone age of Greece from the palaeolithic to the advent of the neolithic. In *Aegean Prehistory. A Review*, American Journal of Archaeology Supplement I, pp. 225–254. Boston: Archaeological Institute of America.

Salje, B., N. Riedl, and G. Schauerte (2005). *Gesichter des Orients. 10000 Jahre Kunst und Kultur aus Jordanien.* Begleitband zur Ausstellung der Kunst- und Ausstellungshalle der Bundesrepublik Deutschland, Bonn in Kooperation mit dem Vorderasiatischen Museum, Staatliche Museeen zu Berlin – Stiftung Preußischer Kulturbesitz, Mainz 2004. Mainz: Zabern.

Sherratt, A. G. (1972). Socioeconomic and demographic models for the Neolithic and Bronze Ages of Europe. In D. L. Clarke (Ed.), *Models in Archaeology*, pp. 477–542. London: Methuen.

Sherratt, A. G. (Ed.) (1980a). *The Cambridge Encyclopaedia of Archaeology.* Cambridge: Cambridge University Press.

Sherratt, A. G. (1980b). Water, soil and seasonality in early cereal cultivation. *World Archaeology 11*, 313–330.

Sherratt, A. G. (1997). Climatic cycles and behavioural revolutions: the emergence of modern humans and the beginning of farming. *Antiquity 71*, 271–287.

Sherratt, A. G. (2004). Fractal farmers: patterns of neolithic origin and dispersal. In J. Cherry, C. Scarre, and S. Shennan (Eds.), *Explaining Social Change: Studies in Honour of Colin Renfrew*, pp. 53–63. Cambridge: McDonald Institute for Archaeological Research.

Sherratt, S. (2000). Circulation of metals and the end of the bronze age in the eastern Mediterranean. In F. E. Pare (Ed.), *Metals Make the World Go Round: The Supply and Circulation of Metals in Bronze Age Europe*, pp. 82–98. Oxford: Oxbow Books.

Tchernov, E. (1994). *An Early Neolithic Village in the Jordan Valley Part II: The Fauna of Netiv Hagdud.* Cambridge, MA: Peabody Museum of Archaeology and Ethnology.

Ucko, P. J. and G. Dimbleby (Eds.) (1969). *The Domestication and Exploitation of Plants and Animals.* London: Duckworth.

Uerpmann, H.-P. (1996). Animal domestication: accident or intention? In D. R. Harris (Ed.), *The Origins and Spread of Agriculture and Pastoralism in Eurasia*, pp. 227–237. London: UCL Press.

van Andel, T., K. Gallis, and G. Toufexis (1995). Early neolithic farming in a Thessalian river landscape, Greece. In M. G. M. J. Lewin and J. C. Woodward (Eds.), *Mediterranean Quaternary River Environments*, pp. 131–143. Rotterdam: Balkema.

Vita-Finzi, C. (1969a). Fluvial geology. In D. Brothwell and E. S. Higgs (Eds.), *Science in Archaeology*, pp. 135–150. London: Thames and Hudson.

Vita-Finzi, C. (1969b). Geological opportunism. In P. J. Ucko and G. Dimbleby (Eds.), *The Domestication and Exploitation of Plants and Animals*, pp. 31–34. London: Duckworth.

Vita-Finzi, C. and E. Higgs (1970). Prehistoric economy in the Mount Carmel area of Palestine: site catchment analysis. *Proceedings of the Prehistoric Society 36*, 1–37.

Wirth, E. (1971). *Syrien, eine geographische Landeskunde.* Darmstadt: Wissenschaftliche Buchgesellschaft.

Wirth, E. (1998). Die natürlichen Ressourcen Vorderasiens als Handlungsrahmen der holozänen Kulturen und Hochkulturen. *Baghdader Mitteilungen 29*, 9–26.

Wright, K. (1994). Ground stone tools and hunter-gatherer subsistence in Southwest Asia: implications for the transition to farming. *American Antiquity 59*(2), 238–263.

Wright, K. (2000). The social origins of cooking and dining in early villages of Western Asia. *Proceedings of the Prehistoric Society 66*, 89–121.

Zohary, D. (1996). The mode of domestication of the founder crops of Southwest Asian agriculture. In D. R. Harris (Ed.), *The Origins and Spread of Agriculture and Pastoralism in Eurasia*, pp. 142–158. London: UCL Press.

Chapter 2

The adoption of farming and the beginnings of the Neolithic in the Euphrates valley: cereal exploitation between the 12th and 8th millennia cal BC

George Willcox
CNRS Jalès, France

2.1 Introduction

An exceptional body of archaeological and archaeobotanical data has been amassed from several early farming sites along the Euphrates valley over the last thirty years. These sites were the focus of a number of archaeological rescue operations prior to the construction of a series of dams (figure 2.1). At the earliest site, Natufian Abu Hureyra, small pit dwellings were uncovered; during the three millennia that followed there was a progressive increase in both the size of the sites and the complexity of the architecture, culminating in large sites, e.g., Halula, which is dated to the first half of the 8th millennium cal BC (Molist and Stordeur 1999; Moore et al. 2000; Stordeur 2000). The seven sites along the Euphrates belonging to the late 10th and early 9th millennium cal BC that are referred to in this paper exhibit considerable cultural coherence in terms of artefacts, architecture and symbolic representations, which implies that these societies likely shared similar beliefs and spoke the same language. The inhabitants of the sites were some of the earliest farmers in the Near East. Of the seven sites, five are in Syria and are located at regular intervals along the Euphrates, in some cases at distances of less than 25 kms. There is convincing evidence for contact between the sites as indicated by the movement of raw materials: for example, obsidian, chloritic stone, and cedar wood were imported to the Syrian sites from sources further upstream, in Anatolia.

Within the regional context of the Near East, the early Neolithic Euphrates sites in Syria differ from sites in the southern Levant and Anatolia because they are situated at regular intervals on a major river and are also a long way from present-day wild wheat stands. Archaeobotanical data from 17 sites across the Near East clearly show that other early agriculture sites are located in or near areas that are coincidental with present-day distributions of the cereals that have been identified in the archaeobotanical assemblages found on each site. So for other areas, for example the southern Levant and Anatolia, the past cereal distribution appears to have

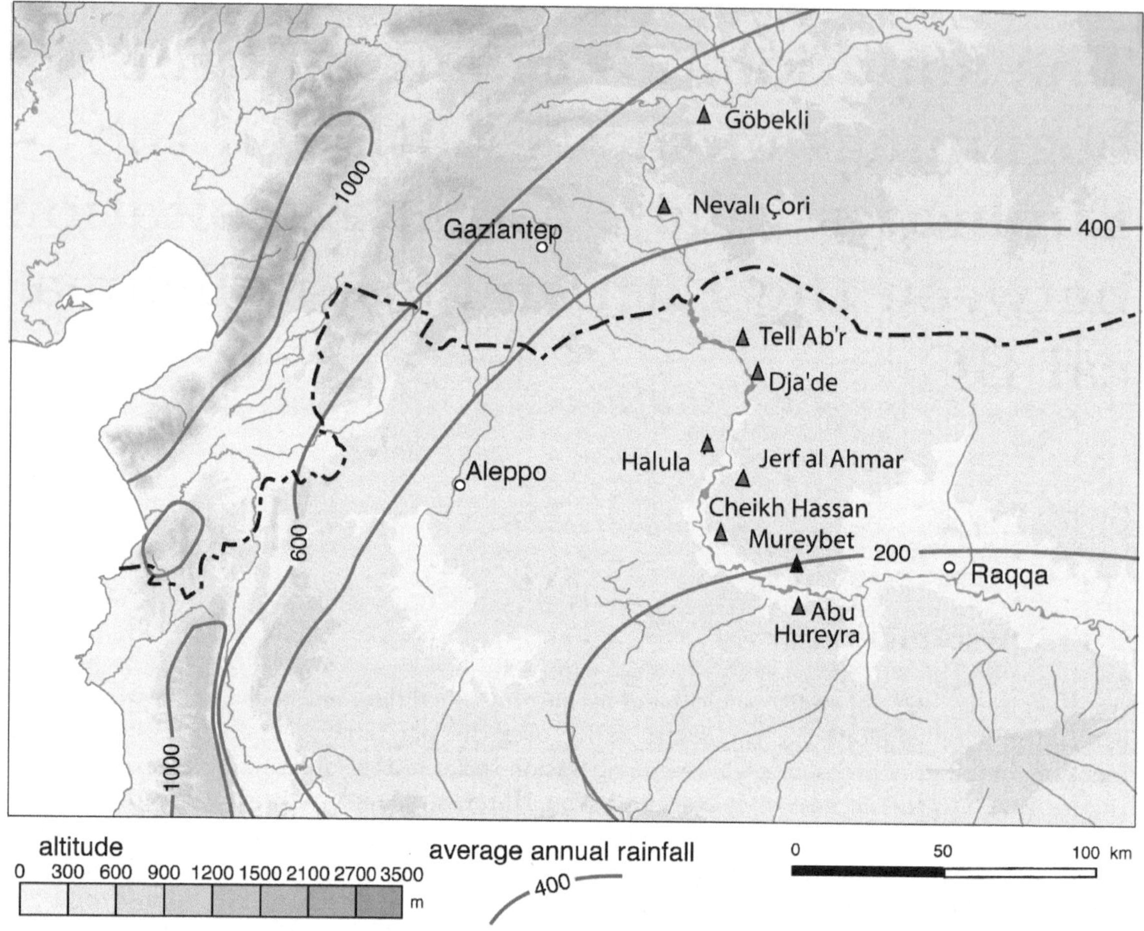

Figure 2.1: Map showing early farming sites on the Euphrates mentioned in the text with rainfall isohyets. So far no archaeobotanical remains have been published from Cheikh Hassan and Tell Ab'r. The site of Mureybet is situated at the very limit of the dry farming zone, yet at this site cereals were present during the Younger Dryas, which is a period considered to have been less humid than at present (irrigation would not have been possible due to the high flood levels that coincide with harvest time; see Willcox and Roitel 1998).

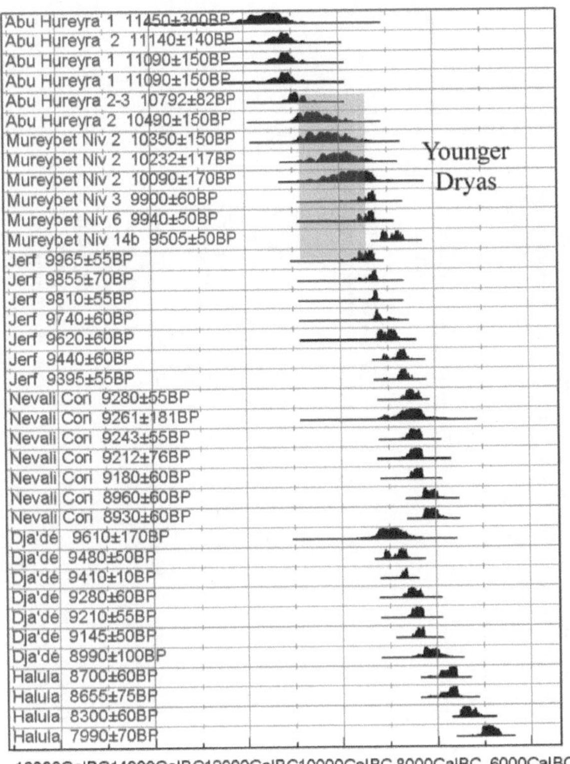

Figure 2.2: Range of selected radiocarbon dates from the Euphrates early neolithic sites using the Oxford calibration curve.

been similar to that seen today. The regional differences in ancient cereal assemblages provide evidence for independent domestication events. Thus in the southern Levant there is evidence for the domestication of barley and emmer, the only wild cereals found growing in the region; and rye and einkorn, which are absent in the south, were domesticated in the northern Levant. Barley and emmer occur in both areas and may have been domesticated more than once (table 2.1) (Willcox 2002c). These conclusions are based on archaeobotanical data and are supported by evidence from DNA analyses of modern cereal populations. (Badr et al. 2000; Özkan et al. 2002; Salamini et al. 2002; Heun et al. 1997; Ishtii et al. 2001; Tanno et al. 2002; Thuillet et al. 2002).

The charred plant remains from the Euphrates sites present a wealth of information about the adoption of agriculture by these early neolithic societies. In this paper I will examine the evidence from the seven sites in the Euphrates valley (see figure 2.1 for site location and 2.2 for dating) using data from published reports (Hillman 2000; Hillman et al. 2001; Neef 2003; Pasternak 1998; van Zeist and Bakker-Heeres 1986; Willcox 2002a, 1996; Willcox and Fornite 1999; Stordeur 2000) and also from my as yet unpublished work.

2.2 Past and present wild cereal habitats in the middle Euphrates and their availability in the past

Knowledge of the ecology of wild cereals is essential for an understanding of their distribution and availability of these plants in relation to archaeological sites; however information in the

Site	Units	Date bp	Einkorn 1g	Einkorn 2g	Emmer	Barley	Rye	cal BC
Southern Levant								
Ohalo II	1	19000	**0**	**0**	21	629	**0**	20500
Wadi Hammeh 27		12200-11920	**0**	**0**	0	p	**0**	12200-11900
Netiv Hagdud	8	9900-9400	**0**	**0**	27 (4)	541 (8)	**0**	9500-8800
Zad 2		9800-9500	**0**	**0**	P	P	**0**	9400-8900
Iraq ed-Dubb		9900	?	**0**	?	P		9500
Aswad I	9	9300-9000***	**0**	**0**	**23(6)**	**32(6)**	**0**	8600-8250
Middle Euphrates and Iraq								
Abu Hureyra	24	1150-10000	**0**	>5000(21)	**0**	**0**	**p**	11000-7000
Qermez Dere		10100-9700	**0**	0	**0**	p		10100-9300
Mureybet I-II	33	10200-9900	**0**	19	**0**	5	p	10200-9900
Mureybet III	27	9800-9400	**0**	>2000	**0**	164	p	9300-8800
Jerf el Ahmar*	430	9800-9400	67	2539	**0**	9641	p	9300-8800
Dja'de*	225	9500-9000	265	1120	191	3763		8800-8000
Eastern Anatolia								
Göbekli Tepe	1	9550-9450	0	5	0	16		8900-8800
Cafer Höyük XII XIII		9400-8800	p	p	**p**	0	p	8800-8000
Çayönü g, bp, ch		9250-8500	p	p	**p**	p		8200-7500
Nevalı Çori	1	9250-9000	**661**	p	129	89		8200-8000
West Syria								
Tell el-Kerkh**		9250-9000	**p**		p			8200-8000
Cyprus								
Mylouthkia/Shillourokambos		9250-9000	**p**	**p**	**p**	**p**		8200-8000

Units=number of samples; 1g=single grained; 2g=two grained; numbers not in brackets=absolute numbers of grains identified; numbers in brackets=number of samples in which taxa is present; **0**=absence significant; 0=absence not significant; other **bold** entries=possible domestication;*=preliminary results from author's unpublished data; **=Ken-Itchi Tanno pers. com.; ***=based on new AMS dates (GrA-25913, GrA-25915, GrA-25916, GrA-25917) of emmer grains from phase Ia and Ib of the 1973 excavations.

Table 2.1: Cereal assemblages from different regions of the Near East. Presence/absence of cereal finds indicate that ancient regional distributions were similar to present-day patterns, but changes may have occurred on a local level. The differences also indicate that local cereals were first taken into cultivation in each region. The Syrian Euphrates sites are an exception because of the greater distance from wild stands of the cereals found in assemblages (einkorn and rye) compared to the equivalent distance for other sites.

literature is sparse and even recent publications (for example, Blumler 2002) are not based on first-hand field observations. Numerous field trips to Turkey, Jordan, Armenia and Syria over the past 30 years have enabled me to examine the ecology of the wild cereals. In southeastern Anatolia and northern Syria, for example, I have observed that the five wild cereals (*Secale* spp., *Triticum urartu*, *Triticum boeoticum*, *Triticum dicoccoides* and *Hordeum spontaneum*) growing in this region have different and distinct habitat requirements, particularly in terms of rainfall and soil tolerance. This is particularly clear in the Euphrates valley region as a result of the strong climatic gradient, where rainfall varies from >900 mm per annum in the north to <150 mm south (see isohyets figure 2.1 and Willcox and Roitel 1998; Helmer et al. 1998; van Zeist and Bakker-Heeres 1986). The distribution of the five cereals extends across the region, with those least tolerant of aridity occurring in the north and those most tolerant in the south. Ryes and wheats are calcifuge plants, thus today rye is restricted to volcanic soils in the Taurus mountains above 900 m and emmer extends to the lower slopes, in areas with a minimum of 400 mm of rain per year. Two-grained einkorns (*Triticum urartu* and *Triticum boeoticum thaoudar*), which are more aridity-tolerant than emmer (or single-grained einkorn) reach as far as the Syrian/Turkish border and are found just east of Ain al Arab on basalt soils in an area with 300–350 mm of rain per year. Barley, the most drought-resistant cereal, is tolerant of poor calcareous soils and occurs further south into the Syrian steppe, where it grows on the poor chalk soils of the middle Euphrates in areas with 200–250 mm of rain per annum. Wild barley grows in rich stands near Halula and Jerf el Ahmar where grazing is restricted, and it is commonly found even further to the south.

To understand the early history of the six wild cereal taxa (with the addition of single-grained einkorn), one has to take into account that in the area where they were found there was a strong climatic gradient, and also that the sites at which they were present spanned a period of climate change. What do we know about the effects of climatic change on the vegetation in this area? The only information we have is local in focus, based on the charred macro-remains that represent gathered plants. The Younger Dryas, which is interpreted as a period of climatic deterioration at the end of the Pleistocene, has been recognised in many parts of the world. In the Near East its effects on the vegetation can be detected in lakebed pollen diagrams from the Mediterranean zone at Ghab (Yasuda et al. 2000) and at Hula (Baruch and Bottema 1999), but they are not as obvious in pollen diagrams from lakes situated in the more continental regions. Indeed, Bottema states that during the Younger Dryas in these areas there appears to be little evidence for vegetation change (Bottema 1995), but notes a relative increase in cereal type pollen (Bottema 2002), thus contradicting the popularly held view that cereals became sparse during this period. The relative frequencies of selected taxa from archaeological sites along the Euphrates are compared in figure 2.3.

The reduction in *Stipa*, *Amygdalus* and *Pistacia* during the Younger Dryas may be the result of worsening conditions; however, it is significant that they are present before, during and after this period (figure 2.3) (Hillman 2000; Willcox 2002b; Roitel and Willcox 2000). This continuity would appear to indicate that the climatic deterioration was not a catastrophic event and may merely represent a period of cooler, less stable conditions, which did not result in radical changes in the vegetation cover. One important question to then ask is whether or not these late Pleistocene climatic conditions would have resulted in a more southerly distribution of the wild cereals; lower sea levels combined with lower temperatures at the end of this period could have led to a lowering of the altitudinal limits of the vegetation zones and to their subsequent southward extension. Indeed the presence of *Pistacia* and *Amygdalus* and even some of the gallery forest species, such as *Fraxinus* and *Platanus,* which are absent today, may be a sign of lower-altitude vegetation zones.

If past climates permitted a more southerly distribution of wild cereals, how important a factor was soil tolerance? Wild rye is rarely found growing on soils that aren't volcanic, while the

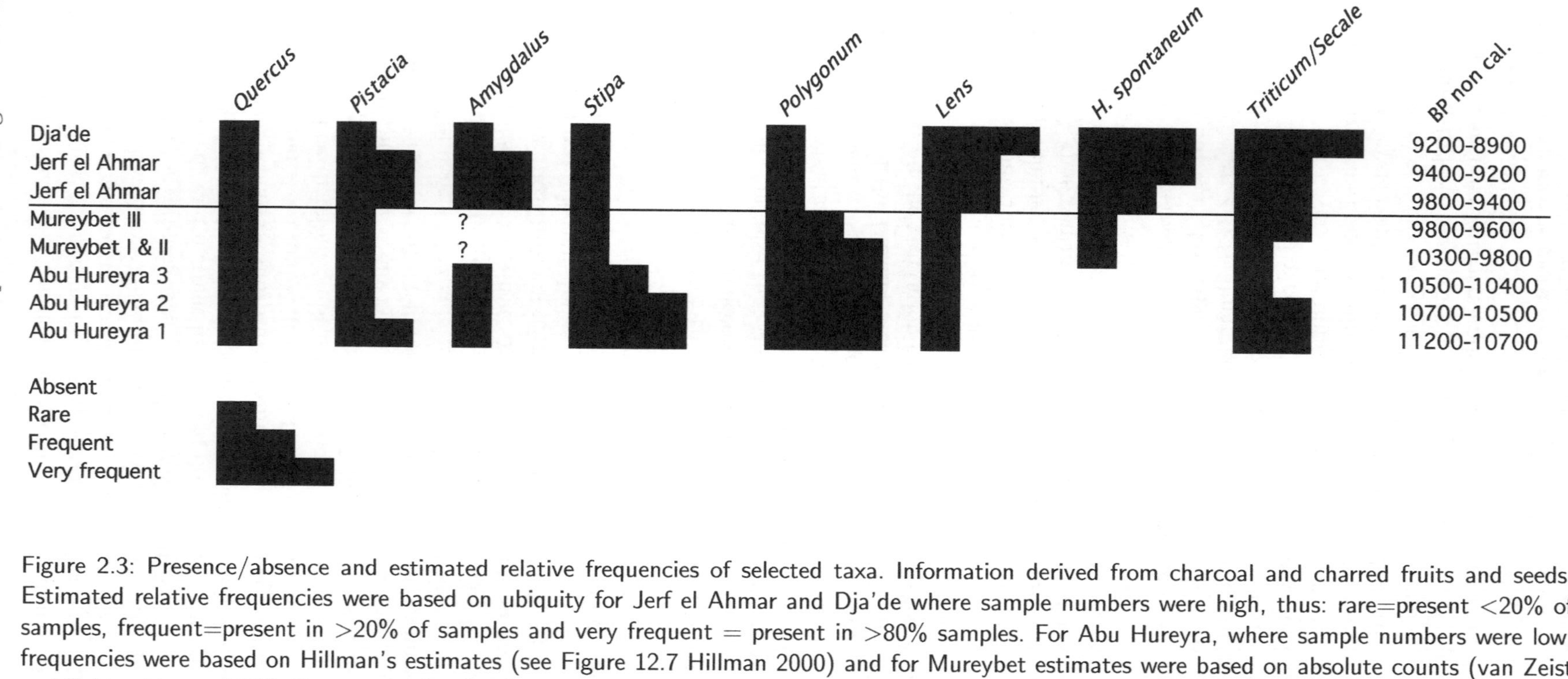

Figure 2.3: Presence/absence and estimated relative frequencies of selected taxa. Information derived from charcoal and charred fruits and seeds. Estimated relative frequencies were based on ubiquity for Jerf el Ahmar and Dja'de where sample numbers were high, thus: rare=present <20% of samples, frequent=present in >20% of samples and very frequent = present in >80% samples. For Abu Hureyra, where sample numbers were low, frequencies were based on Hillman's estimates (see Figure 12.7 Hillman 2000) and for Mureybet estimates were based on absolute counts (van Zeist and Bakker-Heeres 1986). Frequencies for the three tree species were based on fruit and charcoal identifications. Note that *Pistacia* persists throughout the sequence. *Stipa* and *Polygonum* tend to diminish while *Lens* increases. The major change occurs at the beginning of the Holocene (marked with horizontal line) and may coincide with increasing use of cultivation.

two wheat species do not tolerate calcareous soils but may be found on deep, decalcified terra rossa soils that occur in hard limestone areas (these are rare in the middle Euphrates region). Given the cooler conditions and a more southerly distribution of vegetation zones, the nearest potential habitat for wild rye may have been at Qara Perguel Dah, which is situated on the left bank of the Euphrates about 15 km south of the Turkish border. This basalt massif, rising to 694 metres, is a likely habitat for late Pleistocene stands of wild rye and wild einkorn. Further south, between Dja'dé and Jerf el Ahmar there is a small basalt lava flow near the village of Serine and this might also have been a favourable, if more restricted, location. In general the Pliocene chalks that cover most of the area give rise to very thin poor soils with high salt levels, and so it is highly improbable that they would have been suitable habitats for wild wheat or wild rye, and neither would the river terraces because of their limited surface area and regular inundation. This leads us to the inevitable conclusion that the cereals used at the Natufian sites of Mureybet and Abu Hureyra may not have been growing locally, which is why researchers such as van Zeist and Bakker-Heeres (1986), Cauvin (1994), Salamini (2000) and Willcox (2002c) have suggested it is likely that they were imported or introduced from further north. Trace element analysis is also being carried out to investigate whether rye was growing on volcanic soils to the north; at the time of writing, preliminary results show a marked difference in trace elements between barley and *Triticum/Secale*, which suggests they may have been growing under different edaphic conditions. However, before making conclusion we are awaiting further results.

2.3 Materials and methods

The data set available for the six cereal taxa from the area is given in table 2.2. Charred material was collected at the sites using standard flotation techniques (the number of samples and volume of sediment are also shown in table 2.2). Final reports have not yet been published for Halula, Dja'dé and Jerf el Ahmar but a sufficient number of samples have been examined for the purposes of this study.

Presence/absence of taxa is the most reliable indicator because frequencies may be biased by sampling methods and the presence of storage structures. The total number of individual identifications for all samples combined is given in table 2.2. Absence is considered to be real in the cases where the cells with zeros are in bold text, and this interpretation is justified given the large-scale sampling that these sites underwent. The presence/absence scores and frequencies show a coherent pattern from period to period that adds weight to the reliability of the data.

2.4 Results

2.4.1 Rye

Wild rye has been identified at the two Natufian sites of Mureybet and Abu Hureyra. The grains of wild rye and wild two-grained einkorn are difficult to distinguish from each other, however, evidence from spikelet bases suggests that rye was dominant during this period at Mureybet (figures 2.4 and 2.5) (Willcox and Fornite 1999). No spikelet bases were found at Abu Hureyra. Wild ryes may have had a more extensive distribution during both the last glacial period and the Younger Dryas, but the middle Euphrates with its dry chalk hills would have been an unlikely habitat even in the cooler conditions. Rye continues to be found in this area up to the early Holocene at Jerf el Ahmar, where spikelet bases are common in the early levels, but they become progressively less frequent in the later levels (Willcox and Fornite 1999; Willcox 2004) before disappearing altogether.

Taxon	Abu Hureyra 1	Mureybet I and II	Mureybet III	Jerf al Ahmar	Göbekli	Dja'dé	Nevalı Çori	Halula
Einkorn 2 grained/*Secale* grains	>5000	19	>2000	2539	5	1120	0	0
Triticum boeoticum/urartu spikelet bases	0	0	0	5	0	9	P	0
Secale sp. spikelet bases	0	p	p	145	0	16	0	0
Hordeum spontaneum/sativum grains	**0**	5	164	9461	16	3763	89	151
Hordeum spontaneum/sativum rachis	**0**	**0**	6	3326	0	152	P	p
Einkorn 1 grained	**0**	**0**	**0**	67	0	265	661	P
Triticum dicoccum/dicoccoides grains	**0**	**0**	**0**	0	0	191	129	47
Naked wheat grains	**0**	**0**	**0**	**0**	**0**	**0**	**0**	23
Total volume of sediment in litres	>10000	c.310	c.310	16600		>6000		>200
Number of samples	21	31	31	430	1	>225	267	>20

Table 2.2: Data from charred cereal remains from seven sites in the Euphrates region. The number of cereal taxa increases progressively with time (from left, Natufian through Khamian, PPNA, early PPNB to middle PPNB). This change is interpreted as due to increasing use of cultivation which may have resulted in the appearance of taxa previously absent such as barley, one-grained einkorn, emmer and finally, naked wheat. The bold 0 represents genuine absence; the 0 may not represent true absence and could result from shortcomings of sampling techniques.

Figure 2.4: Digital photographs of (above) spikelet bases of rye and grains recovered from Jerf al Ahmar, and (below) rye grains recovered from pisé at Mureybet, which were associated with spikelet fragments shown in figure 2.5. I have included these photos because agronomists have questioned the presence of rye in this area on ecological grounds.

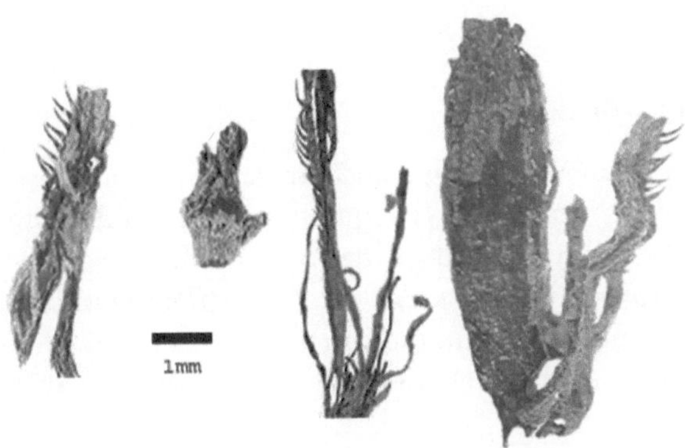

Figure 2.5: Digital photos of rye spikelets recovered from the pisé at PPNA Mureybet. These exceptionally fragile remains were preserved in building-earth that was baked when one of the houses was destroyed by fire.

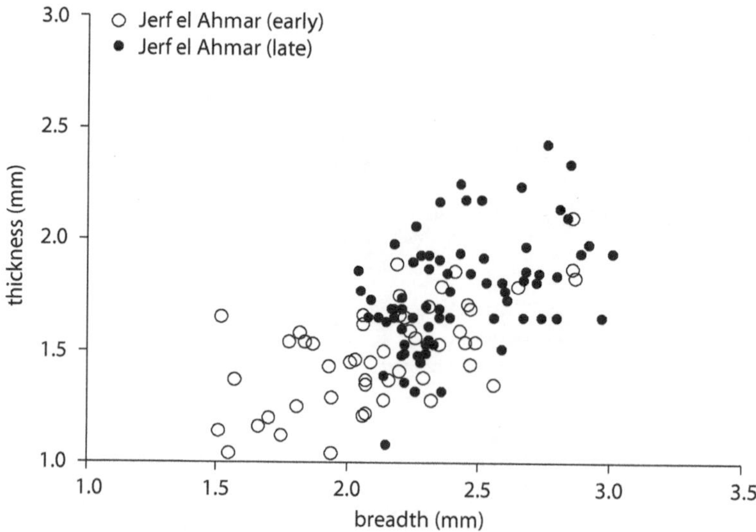

Figure 2.6: Scatter diagram of measurements taken from barley grains (*Hordeum spontaneum*) at Jerf el Ahmar, which show an increase in size between early and late levels. This may not be the result of domestication since rachis fragments are of the wild type, but cultivation using optimum growing conditions may have produced better-formed grains. Another explanation is the introduction of large-grained varieties. Some of the smallest grains may come from *Hordeum murinum* complex or *Hordeum bulbosum*.

2.4.2 Two-grained einkorn

The two species of two-grained einkorn (i.e., *Triticum urartu* and *Triticum boeoticum thaoudar*) cannot be distinguished if found in charred remains and their grains are difficult to differentiate from those of wild rye, although the latter are generally smaller. Two-grained einkorn was identified at Natufian Abu Hureyra and Mureybet from charred grains; at Jerf el Ahmar, where spikelet bases and grains appear to become common only in the upper levels; at Dj'adé, where spikelet bases are relatively more frequent than grains; and possibly at Nevalı Çori and Halula in small quantities.

2.4.3 Single-grained einkorn

The only site in our area where single-grained einkorn is dominant is Nevalı Çori. This sub-species (*Triticum boeoticum aegilopoides*) requires more rainfall than the other einkorn species, and so it is not surprising to find that it occurs only at the most northerly site where rainfall is higher. Both *Triticum urartu* and *Triticum boeoticum thaoudar* occasionally produce spikelets with one grain, thus it is only when the majority of the einkorn grains in an assemblage have a strongly convex ventral face that the single grained sub-species can be identified with certainty.

2.4.4 Barley

Wild barley (*Hordeum spontaneum*) is absent at Abu Hureyra and rare at Mureybet during the Natufian. It becomes progressively more frequent during the mid to late 10th millennium cal BC at Mureybet and Jerf el Ahmar, and is the dominant cereal at Dj'adé and Halula. An increase in grain size has been recorded at Jerf el Ahmar (figure 2.6), which is similar to the size changes noted at other sites in the Levant by Colledge (2001). This increase is probably due to the effects of cultivation but is not necessarily a consequence of the domestication process. During the 8th millennium cal BC domestic two-rowed barley is the most common cereal at middle PPNB sites (as for example at Halula). Wild barley is the only wild cereal that today extends well into Syria (see above) and domestic barley is the only dry-farming cereal in this area, i.e., in the Euphrates valley. Further north in Anatolia at Nevalı Çori, barley is less common than wheat and this regional trend is confirmed by the results from Cafer Hüyök and Çayönü (van Zeist and de Roller 1994; de Moulins 1997).

2.4.5 Emmer

Emmer is absent from Natufian sites and from Jerf el Ahmar; it appears for the first time on the Syrian sites at Dja'dé and is present at Nevalı Çori. Four grains were found among the thousands of other cereal grains at Jerf el Ahmar, while at Dja'dé emmer is more frequent albeit still rare. Emmer is also present at Nevalı Çori. On middle PPNB sites on the Euphrates, for example at Halula, domestic emmer is the most common wheat.

2.4.6 Naked wheat

This wheat, found only as a domesticate, appears for the first time during the 8th millennium cal BC at the middle PPNB sites of Halula and Abu Hureyra.

2.4.7 Morphological signs of domestication

A few plump rye grains from Abu Hureyra have been identified as domestic (Hillman et al. 2001), however, the vast majority of grains are of the wild type and no spikelet material was recovered. At Mureybet only wild rye, einkorn and barley have been found and, similarly, at Jerf el Ahmar all spikelet remains were identified as wild types. Grain size was measured for einkorn/rye and for barley and an increase between the early and late levels was observed at Jerf el Ahmar (figure 2.6); this increase in size is not necessarily due to changes in morphology that are the result of domestication (Willcox 2004), but may merely be a consequence of cultivation under optimum growing conditions. The introduction to the site of large grained varieties is an alternative but less plausible hypothesis. The earliest incontestable evidence for morphological domestication on the Euphrates sites at the time of writing is demonstrated by the rough abscission scars occurring on einkorn spikelet bases recovered from Nevalı Çori, dated to end of the mid-9th millennium cal BC. The solid rachises of barley and semi-solid rachises of emmer, and naked wheat dated to the early 8th millennium cal BC at Halula, are clear signs of domestication.

2.5 Sickle blades, querns and harvesting techniques

Glossed flint sickle blades and querns are present from the Natufian levels at Abu Hureyra and Mureybet but are rare (M.-C. Cauvin, pers. com.). By the mid-10th millennium (that is, on PPNA sites) there is a significant increase in both size and numbers of the glossed flint blades. The use of sickles for harvesting was also suggested by the presence of basal spikelets which occurred among the archaeobotanical remains. At Jerf el Ahmar and Dja'dé future work on grain size, wild/weed assemblages and the chaff remains may give us more insight into the evolution of harvesting techniques. The apparent absence of chaff on the Natufian sites and the abundance at Jerf el Ahmar may well be linked to different methods of harvesting and/or crop processing.

2.6 Conclusions

The questions most frequently asked concerning the origins of farming are how and where were cereals first taken into cultivation. The explanation frequently expounded over the last fifteen years (Moore and Hillman 1992; Bar-Yosef and Belfer-Cohen 2002) is that diminishing wild cereal resources during the Younger Dryas climatic deterioration (10700–9600 cal BC) favoured the adoption of cultivation by hunter/gatherers in core areas. Within these areas the cultivation of wild cereals led to rapid domestication and subsequent diffusion of better adapted cultivars. In the light of recent evidence from across the Near East this explanation appears too simplistic. It is now clear that agriculture developed very gradually among societies dispersed over a wide region with diverse climatic and environmental conditions. Depending on the specific area, a different set of local cereals was taken into cultivation quite independently. This leads us to the conclusion that the process by which societies abandoned gathering in favour of cultivation may have varied from one region to another.

The Syrian Euphrates sites appear to have been situated outside the areas of natural distribution of the wild progenitors of rye and probably also the wheats. Finds of emmer at Zad 2 (Edwards et al. 2001) in the lower Jordan valley are an example from the southern Levant, but in this case the distance from potential wild habitats is far less than that for the sites along the Euphrates. It would appear, therefore, that cereals were transported down the Euphrates valley into areas where they could not have grown naturally because of unsuitable edaphic and climatic conditions. The appearance in the archaeobotanical record of barley, which was well-adapted to local conditions, may mark the point when the inhabitants started to rely on resources in the immediate vicinity of their sites and this may coincide with greater reliance on cultivation. Cultivation of the less well-adapted wheats could have taken place in favourable micro-habitats.

By examining the presence and, to some extent, the frequencies of the six cereal taxa on sites within the Euphrates valley over a period of some 3,500 years we can attempt to trace the developments that turned simple gatherers into fully-fledged cultivators. But we must accept that given the time span involved and the coverage of samples, clearly what we are left with is a very discontinuous chronological sequence from a few widely dispersed sites. The cereal remains alone provide a one-sided picture of the sequence of events, however, if we combine the archaeobotanical and archaeological data certain patterns emerge. Figure 2.7 gives a schematic representation of how cereals came into and went out of use in the Euphrates valley and also how this relates to the periods during which the different sites were occupied.

The earliest inhabitants of Abu Hureyra were undoubtedly gatherers. In the summer they relied on seeds collected from the flood plain and in the late spring they gathered cereals, mainly rye, from wild stands, possibly at some considerable distance to the north. There was global climatic change during the occupation of the site. We do not know how this area of the Near East was affected, but it is probable that the eastern Mediterranean coast with the elevated coastal range provided a buffer zone, which may have moderated the worsening conditions in the interior. The persistence of *Pistacia atlantica* throughout the Younger Dryas in the Euphrates basin would appear to be evidence of this (figure 2.3). The first two phases at Mureybet show remarkable similarities with Abu Hureyra; however, the assemblages differ in that at the former site barley is present (but rare) and *Stipa* is absent. Both sites have sickle blades and querns.

At the beginning of the Holocene (that is, during the PPNA), data from both Mureybet phase III and Jerf el Ahmar show an increase in barley, lentils and in some of the weed taxa (Colledge 1998), but a decrease in grains gathered from the flood plain (for example, *Polygonum* [figure 2.3], and also *Scirpus*). These changes coincide with an increase in the number of sickle blades, saddle querns, storage structures and the appearance of more uniform well-developed cereal grains. These combined changes appear to be the first indication of cultivation, yet there are no morphological signs of domestication, perhaps because gathering continued alongside cultivation

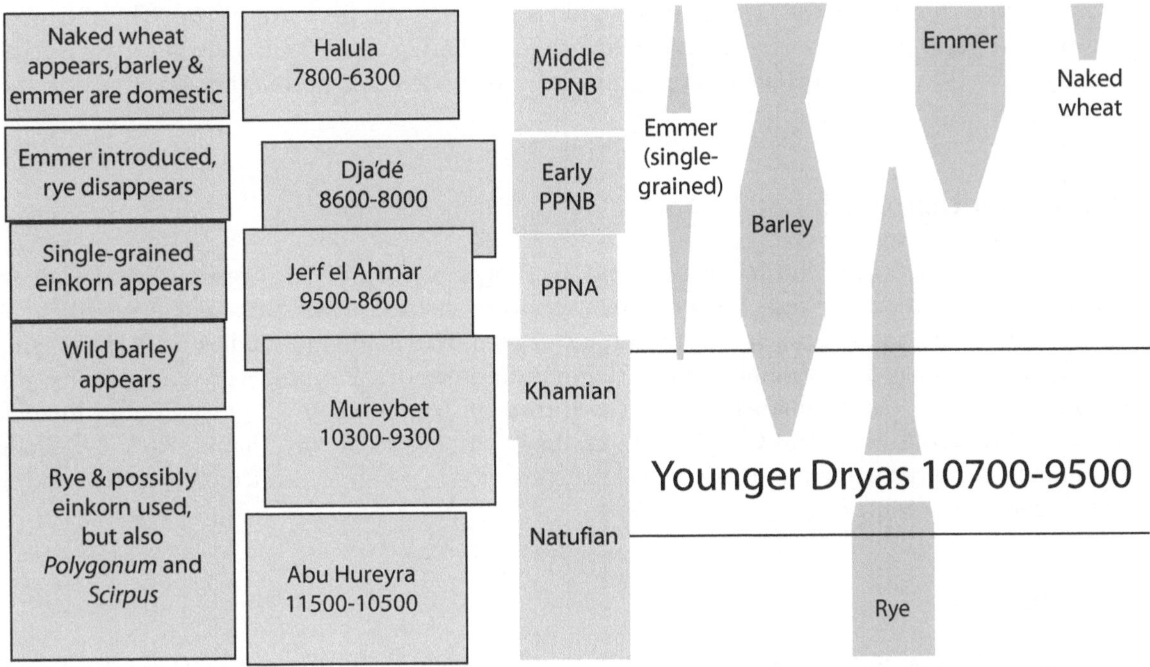

Figure 2.7: A schematic diagram giving changes in cereal use during a 3,500 year period at sites on the Syrian Euphrates (dates cal BC). Barley, emmer and naked wheat are absent from the earliest levels, and then increase to become the dominant cereal crops at the expense of einkorn and rye, which were progressively abandoned.

and reduced the probability of a domestic population becoming established. There may well have been cultivation before this date, which did not leave any detectable signs in the archaeobotanical record, however, the evidence does indicate that there was increasing management of crops. This is exactly at the point when we see the onset of more stable conditions, which were more favourable for growing cereals than the unstable conditions of the Younger Dryas.

During the 10th millennium at Jerf el Ahmar, the frequency of barley increases progressively, while that of rye and einkorn decreases. There is an increase in size of barley grains (figure 2.6) and possibly also einkorn grains, which is considered to be yet another indication of cultivation but not necessarily of domestication (Willcox 2004). Emmer is introduced at Dja'dé during the early part of the 9th millennium cal BC (figure 2.7), whereas the situation at this time further north is different and at Nevalı Çori, which is well within the natural habitat of einkorn, single-grained einkorn was dominant. Another important difference is that there is the first appearance of spikelet bases with rough abscission scars characteristic of the domestic species. Barley and emmer were also present but at low frequencies.

There is a significant development in architectural styles towards a more complex architecture from the 12th millennium to the end of the 10th millennium cal BC, for example, from the small pit dwellings at Abu Hureyra to the monumental structures at Nevalı Çori, together with a progressive increase in the surface area of each site (Molist and Stordeur 1999). These developments culminate during the first half of the 8th millennium cal BC with the standardized dwellings, as shown by eight excavated houses which have identical plans at Halula (M. Molist pers. com. 2004)—a site which, like the contemporary levels at Abu Hureyra, covers several hectares. At these sites there were domestic species of cereals (including naked wheat), sheep, goats and cattle. Although we cannot estimate the number of inhabitants at these sites we can safely assume that there was a population explosion during the first half of the 8th millennium cal BC and it seems reasonable to postulate that gathering must have become a very minor activity, and that these societies were completely dependent on production farming for their subsistence.

Acknowledgments

My thanks go first to Gordon Hillman, friend and colleague, whose enthusiasm, knowledge and immense contribution to Near Eastern archaeobotany helped make this article possible. My thanks also to the many French and Syrian colleagues who made my studies at Jerf el Ahmar and Dja'dé possible, in particular Linda Herveux and Sandra Fornite, who are working with me on the charred plant remains, and to the European Commission who contributed financial support under contract ICA3-CT-2002-10022. Last but not least, my thanks to Sue Colledge and James Conolly for making many helpful suggestions to the first draft of this article and for a most rewarding conference.

References

Badr, A., K. Müller, R. Schäfer-Pregl, H. El Rabey, S. Effgen, H. H. Ibrahim, C. Pozzi, W. Rohde, and F. Salamini (2000). On the origin and domestication history of barley (*Hordeum vulgare*). *Molecular Biology and Evolution 17*, 499–510.

Bar-Yosef, O. and A. Belfer-Cohen (2002). Facing environmental crisis. societal and cultural changes at the transition from the Younger Dryas to the Holocene. In R. T. J. Cappers and S. Bottema (Eds.), *The Dawn of Farming in the Near East*, Studies in Near Eastern Production, Subsistence and Environment 6, pp. 55–66. Berlin: Ex oriente.

Baruch, U. and S. Bottema (1999). A new pollen diagram from Lake Hula. In H. Kawanabe, G. W. Coulter, and A. C. Roosevelt (Eds.), *Ancient Lakes: Their Cultural and Biological Diversity*, pp. 75–86. Ghent, Belgium: Kenobi Productions.

Blumler, M. A. (2002). Changing paradigms, wild cereal ecology, and agricultural origins. In R. T. J. Cappers and S. Bottema (Eds.), *The Dawn of Farming in the Near East*, pp. 95–111. Berlin: Ex oriente.

Bottema, S. (1995). The Younger Dryas in the eastern Mediterranean. *Quaternary Science Reviews 14*, 883–891.

Bottema, S. (2002). The use of palynology in tracing early agriculture. In R. T. J. Cappers and S. Bottema (Eds.), *The Dawn of Farming in the Near East*, Studies in Near Eastern Production, Subsistence and Environment 6, pp. 27–38. Berlin: Ex oriente.

Cauvin, J. (1994). *Naissance des Divinités, Naissance de l'Agriculture: La Révolution des Symboles au Néolithique.* Paris: C.N.R.S.

Colledge, S. (1998). Identifying pre-domestic cultivation using multivariate analysis. In A. Damania, J. Valkoun, G. Willcox, and C. Qualset (Eds.), *The Origins of Agriculture and Crop Domestication*, pp. 121–131. Aleppo, Syria: ICARDA.

Colledge, S. (2001). *Plant Exploitation on Epipalaeolithic and Early Neolithic Sites in the Levant.* BAR International Series 986. Oxford: John and Erica Hedges Ltd.

de Moulins, D. (1997). *Agricultural Changes at Euphrates and Steppe Sites in the Mid-8th to the 6th Millennium B.C.* Oxford: British Archaeological Reports.

Edwards, P., S. Falconer, P. Fall, I. Berelov, C. Davies, J. Meadows, C. Meegan, M. Metzger, and G. Sayej (2001). Archaeology and environment of the Dead Sea Plain: preliminary results of the first season of investigations by the joint La Trobe University/Arizona State University project. *Annual of the Department of Antiquities of Jordan 45*, 135–157.

Helmer, D., V. Roitel, M. Sana, and G. Willcox (1998). Interprétations environnementales des données archéozoologiques et archéobotaniques en Syrie du nord de 16000 bp à 7000 bp, et les débuts de la domestication des plantes et des animaux. *Bulletin of the Canadian Society for Mesopotamian Studies 33*, 9–34.

Heun, M., R. Schäfer-Pregl, D. Klawan, R. Castagna, M. Accerbi, B. Borghi, and F. Salamini (1997). Site of einkorn wheat domestication identified by DNA fingerprinting. *Science 278*(5341), 1312–1314.

Hillman, G., R. Hedges, A. Moore, S. Colledge, and P. Pettitt (2001). New evidence of Late Glacial cereal cultivation at Abu Hureyra on the Euphrates. *Holocene 11*(4), 383–393.

Hillman, G. C. (2000). Plant food economy of Abu Hureyra. In A. M. T. Moore, G. C. Hillman, and T. Legge (Eds.), *Village on the Euphrates: From Foraging to Farming at Abu Hureyra*, pp. 372–392. Oxford: Oxford University Press.

Ishtii, T., N. Mori, and Y. Ogihara (2001). Evaluation of allelic diversity at chloroplast microsatellite loci among common wheat and its ancestral species. *Theoretical and Applied Genetics 103*, 896–904.

Molist, M. and D. Stordeur (1999). Le moyen Euphrate Syrien et son rôle dans la neolithisation: specificité et évolution des architectures. In G. del Olmo Lete and J.-L. Montero Fenollos (Eds.), *Archaeology of the Upper Syrian Euphrates in the Tishrin Dam Area*, pp. 395–412. Barcelona: AUSA.

Moore, A. M. T. and G. C. Hillman (1992). The Pleistocene to Holocene transition and human economy in Southwest Asia: the impact of the Younger Dryas. *American Antiquity 57*, 482–494.

Moore, A. M. T., G. C. Hillman, and T. Legge (2000). *Village on the Euphrates. From Foraging to Farming at Abu Hureyra.* Oxford: Oxford University Press.

Neef, R. (2003). Overlooking the steppe forest: preliminary report on the botanical remains from early neolithic Göbekli Tepe (southern Turkey). *Neolithics 2*, 13–15.

Özkan, H., A. Brandolini, R. Schäfer-Pregl, and F. Salamini (2002). AFLP analysis of a collection of tetraploid wheats indicates the origin of emmer and hard wheat domestication in southeast Turkey. *Molecular Biology and Evolution 19*, 1797–1801.

Pasternak, R. (1998). Investigations of botanical remains from Nevalı Çori PPNB, Turkey. In A. Damania, J. Valkoun, G. Willcox, and C. Qualse (Eds.), *The Origins of Agriculture and Crop Domestication*, pp. 170–177. Syria: Aleppo.

Roitel, V. and G. Willcox (2000). Analysis of charcoal from Abu Hureyra. In A. M. T. Moore, G. C. Hillman, and T. Legge (Eds.), *Village on the Euphrates: From Foraging to Farming at Abu Hureyra*, pp. 544–547. Oxford: Oxford University Press.

Salamini, F. (2000). La première céréale cultivée. *Pour la Science 274*, 58–63.

Salamini, F., A. B. H. Özkan, R. Schäfer-Pregl, and W. Martin (2002). Genetics and geography of wild cereal domestication in the near east. *Nature Reviews Genetics 3*, 429–441.

Stordeur, D. (2000). Jerf el Ahmar et l'émergence du néolithique au Proche Orient. In J. Guilaine (Ed.), *Premiers Paysans du Monde: Naissance des Agricultures*, pp. 33–60. Paris: Errance.

Tanno, K., S. Takeda, K. Takeda, and T. Komatsuda (2002). A DNA marker closely linked to the *vrs 1* locus (row-type gene) indicates multiple origins of six-rowed cultivated barley (*Hordeum vulgare L.*). *Theoretical and Applied Genetics 104*, 54–60.

Thuillet, A.-C., D. Bru, J. David, P. Roumet, S. Santoni, P. Sourdille, and T. Bataillon (2002). Direct estimation of mutation rate for 10 microsatellite loci in durum wheat, *Triticum turgidum* (L.) Thell. ssp durum desf. *Molecular Biology and Evolution 19*(1), 122–125.

van Zeist, W. and J. Bakker-Heeres (1986). Archaeobotanical studies in the Levant 3. Late Palaeolithic Mureybet. *Palaeohistoria 26*, 171–199.

van Zeist, W. and G. de Roller (1994). The plant husbandry of Aceramic Çayönü, E. Turkey. *Palaeohistoria 33/34*, 65–96.

Willcox, G. (1996). Evidence for plant exploitation and vegetation history from three early neolithic pre-pottery sites on the Euphrates (Syria). *Vegetational history and archaeobotany 5*, 143–152.

Willcox, G. (2002a). Charred plant remains from a late tenth millennium kitchen at Jerf el Ahmar (Syria). *Vegetational History and Archaeobotany 11*, 55–60.

Willcox, G. (2002b). Evidence for ancient forest cover and deforestation from charcoal analysis of ten archaeological sites on the Euphrates. In S. Thiébault (Ed.), *Charcoal Analysis: Methodological Approaches, Palaeoecological Results and Wood Uses. Proceedings of the Second International Meeting of Anthracology, Paris, September 2000*, BAR International Series 1063, pp. 141–145. Oxford: Archaeopress.

Willcox, G. (2002c). Geographical variation in major cereal components and evidence for independent domestication events in western Asia. In R. T. J. Cappers and S. Bottema (Eds.), *The Dawn of Farming in the Near East*, Studies in Near Eastern Production, Subsistence and Environment 6, pp. 133–140. Berlin: Ex oriente.

Willcox, G. (2004). Measuring grain size and identifying Near Eastern cereal domestication: evidence from the Euphrates valley. *Journal of Archaeological Science 31*, 145–150.

Willcox, G. and S. Fornite (1999). Impressions of wild cereal chaff in pisé from the tenth millennium at Jerf el Ahmar and Mureybet: northern Syria. *Vegetational History and Archaeobotany 8*, 21–24.

Willcox, G. and V. Roitel (1998). Rapport archéobotanique préliminaire de trois sites préceramiques du Moyen Euphrate (Syria). *Cahiers de l'Euphrate 8*, 65–84.

Yasuda, Y., H. Kitagawa, and T. Nakagawa (2000). The earliest record of major anthropogenic deforestation in the Ghab valley, northwest Syria: a palynological study. *Quaternary International 73/74*, 127–136.

Chapter 3

East of Eden? A consideration of neolithic crop spectra in the eastern Fertile Crescent and beyond

Michael Charles
University of Sheffield, UK

> Humans have a choice and it's choice that makes them human. (After John Steinbeck, *East of Eden*, 1952.)

3.1 Introduction

While the timing and mode of the spread of crops westwards from the Fertile Crescent into Europe has been the subject of considerable debate (Barker 1985; Bogucki 1996; Halstead 1989; Price 2000; Zvelebil 2000), the easterly movement of the neolithic crop package has received much less attention. This bias towards western developments applies also to the dispersal of crops within the region. Currently there is a consensus that crop domestication had taken place in the Levant by the end of the 10th millennium cal BC, that domesticated crops had dispersed to Cyprus by the beginning of the 9th millennium cal BC (Peltenburg et al. 2001; Peltenburg 2004), and that farming reached Europe some 1500 years later. Early archaeobotanical work at the sites of Jarmo and Ali Kosh by the Danish scientist Hans Helbaek (1959; 1960; 1969) still represents a major source of information about early agriculture at the eastern end of the Fertile Crescent (figure 3.1). This work has subsequently been augmented by the results from a series of excavations in this region and from others further to the east in Central Asia. The conference has provided an opportunity to consider the evidence available for the eastward spread of crops.

The 20th century can broadly be characterised as a period of data collection and methodological development in archaeobotany. In contrast, data compilation and synthesis have characterised the last decade of the old millennium and the first decade of the new millennium. The dataset that will be discussed in this paper is in part drawn from a new database of bioarchaeological evidence from Iraq (the Mesopotamian Environmental Archaeology Database, Charles and Dobney nd). This database is just one example of a growing tendency in archaeobotany to compile and synthesise assemblages that are published in disparate locations and in different formats; indeed, the conference provided an example of this current approach (Colledge et al. this volume). It is appropriate, therefore, to consider briefly the nature of databases and their use or abuse. In order to understand any archaeobotanical assemblage it is necessary to identify the individual 'behavioural episodes' (Jones 1991) from which it derived; this means that archaeological deposits must be sampled and analysed separately in terms of their taphonomy

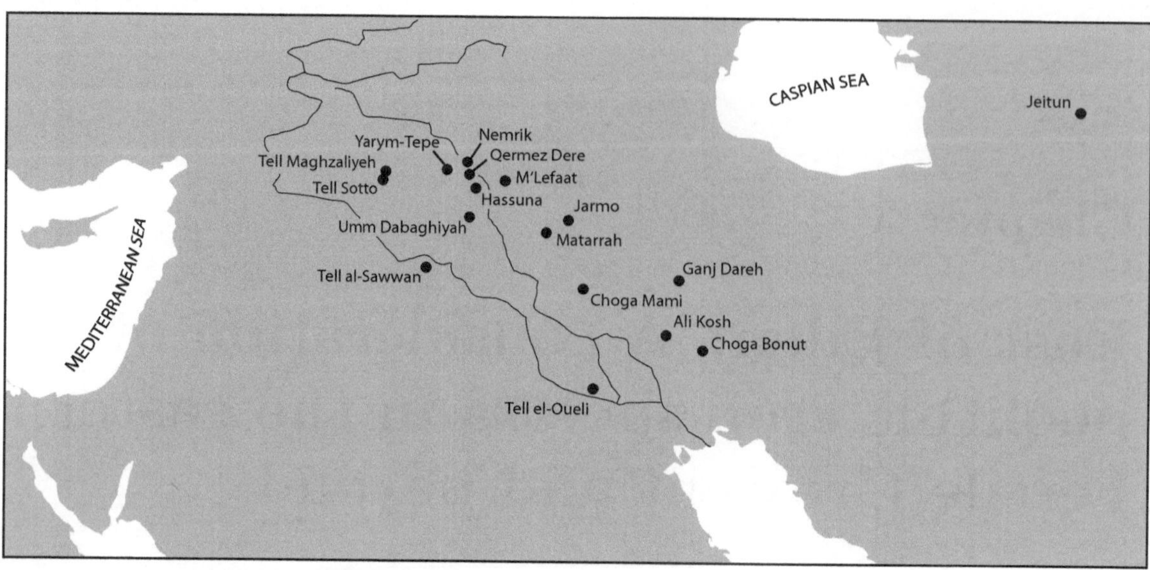

Figure 3.1: The location of sites referred to in the text.

and composition. The structure of databases encourages users (if not compilers) to amalgamate data at an early stage of analysis, and this can involve amalgamation of both identification categories and (perhaps more critically) of individual samples. The result is a tendency to neglect sample-level investigations in favour of those focused at the regional or period levels. As this paper sets out to show, however, general trends must ultimately be understood in terms of the episodes and activities that generated the archaeobotanical assemblages.

3.2 Methods

The data table constructed (table 3.1) comprises all neolithic sites (10000–6500 bp / 10000–5500 cal BC) in Iraq, Iran and Turkmenistan for which archaeobotanical records could be found. The abundance of the major crops (wild and domesticated) is recorded on a 1–3 scale (see the table for further explanation). The way in which the original data are recorded (sample-by-sample, per phase, or for the whole site) is also noted. There are only a small number of sites and so the chronological groupings used are relatively broad.

It is not always easy to establish the domesticated status of crops; as far as possible that given in the tables is based on the assessment of the original author, but where there is ambiguity the frequency value for the taxon is underlined.

Although it was tempting to apply some minimum criteria for the inclusion of sites (e.g., in terms of sample number and quantity of items) this was not done, as by doing so many of the sites would have been excluded. The subsequent discussion, however, is based on crops that occur with a frequency of 2 or more (table 3.1).

One striking feature of the data under study is the overall scarcity of plant remains, which is because of the small number of sites represented and the small number of samples studied at these sites; there are only 27 sites and just over 250 samples in total (table 3.1). Three sites (Ganj Dareh, Ali Kosh and Jeitun) account for more than 50% of the samples and will be dealt with in greater detail.

Site date bp	Site name and sample phase	No. samples		Sample method	No. of wild crop types (>1)	No. of crop types (>1)		Wild barley	Wild einkorn	Wild emmer	Wild flax	Wild lentil	Wild pea	(Wild) bitter vetch	Barley	Einkorn	Emmer	Flax	Lentil	Pea	Grass pea
9800–9600	M'Lefaat	1		s	1	0		2	1	.	.	1	.	1	.	.	.	.	.	.	.
10100–9700	Qermez Dere	1		s	2	0		1	.	.	.	2	.	2	.	.	.	.	.	.	.
10000	Nemrik	1		s	2	0		.	.	.	.	2	2	.	.	.	.	.	.	.	.
10000–9600	**PPNA**	**3**	sum		**2**	**0**	mean	**1**	**0**	**0**	**0**	**2**	**1**	**1**	**0**	**0**	**0**	**0**	**0**	**0**	**0**
8900	Ganj Dareh	122		s	2	1		3	.	.	.	2	1	.	3	.	.	.	.	.	.
8800–8000	Ali Kosh (Bus Mordeh)	21		p	0	2		1	1	.	1	.	.	.	3	1	3	.	.	.	.
7600	Ali Kosh (Ali Kosh)	13		p	0	2		.	1	.	1	.	.	.	3	1	3	.	1	.	.
8600–7800	Tell Maghzaliyeh	2		p	1	5		.	.	.	2	.	.	.	2	2	2	2	2	.	.
9000–8400	Tepe Abdul Hosein	1		si	0	3		.	.	.	.	.	.	.	2	.	2	.	2	.	.
8500	Jarmo	?		si	0	6		1	1	1	.	.	.	.	2	2	2	.	2	2	2
8600–8000	Choga Bonut	24		s	0	4		.	.	.	.	.	.	.	2	2	3	.	2	.	.
9600–8000	**PPNB(/C)**	**184**	sum		**0**	**4**	mean	**1**	**0**	**0**	**1**	**1**	**0**	**0**	**7**	**3**	**6**	**1**	**4**	**1**	**1**
8000–6500	Umm Dabaghiyah	3		s	0	2		.	.	.	.	.	.	.	2	.	2	.	.	.	.
8000–6500	Yarim Tepe I	5		s	0	2		.	.	.	.	.	.	.	2	.	2	.	.	.	.
7000	Tell Sotto	1		s	0	2		.	.	.	.	.	.	.	2	.	2	.	.	.	.
7270–7100	Jeitun	35		s	0	1		.	.	.	.	.	.	.	1	3	1	.	.	.	.
7100	Tell al-Sawwan	3		s	0	4		.	.	.	.	.	.	.	3	2	2	2	.	.	.
8000–6500	Choga Mami (Samarra)	7		s	0	5		.	.	.	.	.	.	.	3	2	2	2	3	.	.
6100	Hassuna	1		s	0	1		.	.	.	.	.	.	.	2	.	.	.	.	.	.
7820	Ali Kosh (Mohammed Jaffar)	13		p	0	2		.	.	.	1	.	.	.	3	.	2	.	1	.	.
6600	Matarrah	1		si	0	1		.	.	.	.	.	.	.	2	.	.	.	.	.	.
8000–6500	Tell el-Oueli	5		s	0	2		.	.	.	.	.	.	.	3	1	2	1	.	.	.
8000–6500	**Late Neolithic**	**74**	sum		**0**	**2**	mean	**0**	**0**	**0**	**0**	**0**	**0**	**0**	**9**	**3**	**7**	**2**	**1**	**0**	**0**

Data compiled from: Bakheyev and Yanushevich (1980); Charles and Bogaard (nd); Helbaek (1959, 1960, 1964, 1969, 1972a, b); Hubbard (1990); Jacobsen (1982); Kozlowski (1989); Lisicyna (1983); Miller (2003); Neef (1991); Nesbitt (1998, 2002); Watkins et al. (1989); Watson (1983); van Zeist et al. (1984).

Table 3.1: The abundance of major crops (wild and domesticated) in samples from sites in the eastern Fertile Crescent and beyond, scored on a 3-point scale (1=rare, 2=occasional, 3=frequent). Underlined values denote uncertainty of identification.

3.3 The spread of domesticated cereals and pulses in the western part of the Fertile Crescent

In order to evaluate the nature of crop patterns in the eastern part of the Fertile Crescent, it is necessary first to outline briefly developments in the Levant. The earliest evidence for domesticated cereal morphology dates to the PPNA (ca. 10000–8800 cal BC) and derives from the southern Levantine sites of Aswad and Iraq ed-Dubb (Colledge 2001; Colledge et al. 2004), but the number of specimens involved is low and their preservation is generally poor (Willcox 1998; Garrard 1999; Colledge 2004). Futhermore, recent excavations at Aswad appear to indicate that the site is PPNB in date Stordeur (2003).

Otherwise, the earliest sites with evidence of domesticated cereals and pulses date to the second half of the 10th millennium bp and the first half of the 9th millennium bp (early/middle PPNB, ca. 8800–7500 cal BC) and are concentrated in the northern Levant; for example, Abu Hureyra in the semi-desert region of Syria, and a cluster of sites in southeastern Turkey (Nevalı Çori, Cafer Höyük and Çayönü) (Nesbitt and Samuel 1996; Willcox 1998; Nesbitt 2002). Each of these sites contains a wide range of 'crops' (wild and domesticated) that includes both cereals and pulses. There are no sites of this date further east than Çayönü with domesticated crops; thus, the current evidence points towards the western half of the Fertile Crescent as the area where domestic crops originated.

By the end of the 9th millennium bp (late PPNB period, ca. 7000 cal BC), the western part of the Fertile Crescent appears to have become more densely populated and there is evidence of over 30 sites in the area. A wide range of crops has been identified at a number of the sites; these are the so-called neolithic 'founder crops' (emmer, barley, einkorn, lentil, pea, chickpea, bitter vetch and flax; Zohary 1996). Free-threshing wheat has also been found at many of the sites. By this time there had been an expansion of domesticated crops further west, into central Anatolia (e.g., as the evidence from the sites of Asikli Höyük and Çatalhöyük shows). In addition, there is a notable expansion into the drier, semi-desert margins (Garrard et al. 1996).

The final phase of the aceramic Neolithic (final PPNB or PPNC) and the onset of the ceramic Neolithic sees a continuation of relatively dense occupation within the rain-fed agricultural zone of the Fertile Crescent, although many large middle-late PPNB sites had been abandoned in the first part of the 7th millennium cal BC (Banning 1998).

3.4 The spread of domesticated cereals and pulses in the eastern part of the Fertile Crescent and beyond

3.4.1 PPNA: 10000–8800 cal BC

In this early phase, it appears that a limited range of wild crops (i.e., the wild progenitor species of the domestic crops) was being used (table 3.1). The three sites dating to the PPNA are situated close to each other (within 100 km) in the steppic area of northern Iraq (figure 3.1). Wild lentil occurs at two of the three sites and is the best represented crop type; otherwise, the wild crops are site-specific in their distribution, a pattern that matches contemporary trends in wild cereal use in the western part of the Fertile Crescent (Willcox 2002).

3.4.2 PPNB: 8800–7000 cal BC

The predominant crops of this and subsequent phases are emmer and barley, both of which are well attested at five of the six PPNB sites. Lentil and einkorn each occur at four or more sites, while pea and grass pea are only common in the exceptionally rich deposits at Jarmo. Thus, five of the 'founder crops' together with grass pea occur at the PPNB sites; the presence of at least three of these (with the exception of Ganj Dareh) at each site is remarkable, given that the sites are distributed across the rain-fed farming area of the Zagros foothills and cover several vegetation zones. This would seem to indicate the arrival of a 'package' rather than the local domestication of wild crops (cf. Willcox 1999). It should be stressed, however, that the concentration of research in the western part of the Fertile Crescent (Wilkinson 2000) for political reasons has inevitably biased perspectives on the 'origins' of cereal domestication, but it is perhaps significant that the primacy of crop domestication in the southern Levant is now in some doubt (Nesbitt 2002). For example, little work has been possible in northeastern Iraq since the pioneering work of Braidwood and his colleagues in the middle of the 1900s. The replacement during this phase of wild crops by domesticated species is notable, though wild types are also attested at Ganj Dareh, Jarmo and Ali Kosh. The persistence of wild crop species in the PPNB has also been noted in the western region of the Fertile Crescent (Willcox 2002).

3.4.3 Late Neolithic: 7000–5500 cal BC

By ca. 7000 cal BC, there is evidence in the eastern region of the Fertile Crescent for most of the major crops that are found in the west, but only the glume wheats and barley occur at two or more sites during this long period; Choga Mami has the broadest crop spectrum (with five of the founder crops). Flax is a crop with a relatively high water requirement and is common only at Choga Mami and Tell es-Sawwan, both of which are located out in the dry steppe, away from the foothills of the Zagros. Rainfall levels here are too low for dry farming and there are claims that there was some form of supplementary watering at these sites (Helbaek 1964; Oates and Oates 1976). Jeitun, which is located at the edge of the Kara Kum desert in Central Asia, is another marginal site; claims have again been made for supplementary watering at the site (Charles and Hillman 1992; Harris et al. 1993; Harris and Gosden 1996), however, in contrast to Choga Mami the crop range is very narrow and only einkorn occurs frequently.[1]

3.4.4 Summary

Even though little work has been carried out on PPNA sites in the eastern part of the Fertile Crescent, wild crop use is fairly well established at this time in the region, and seems to have involved the exploitation of a different set of wild crop types at each of the sites. In the PPNB these wild crops were replaced or supplemented by their domesticated equivalents and the general trend is for the range of crops (both wild and domesticated) to increase. The later Neolithic sees a return to narrower crop ranges; however, the remains from Choga Mami are a notable exception, despite the concerns expressed by the site archaeobotanist (Helbaek 1972a) about the methods of recovery used.

The restricted range of crops commonly found at the later sites, in comparison to the crop diversity of the PPNB, is worth noting. Concentration on the cultivation of such a narrow range would seem surprising given the breadth of the founder crop package, which was widely attested by the later aceramic Neolithic. The role of crop diversity in buffering against inherent agricultural risks is well known (e.g., Halstead 1987), but a narrowing of the crop spectrum has also been noted in the westward spread of crops beyond the Fertile Crescent, into Central Europe (Halstead 1989). However, here there appears to have been widespread use of at least two cereal and pulse crops (Jacomet and Kreuz 1999, pp.294–299).

[1]Early analyses at Jeitun reported the presence of two-row barley, bread wheat and club wheat (Masson 1971, p.79). Recent investigations by Charles and Bogaard (nd) have confirmed the presence of barley but the presence of free-threshing wheat is uncertain. Two-row barley and a form of free-threshing wheat were reported from Chagylli-depe (Masson and Sarianidi 1972, pp.33, 42).

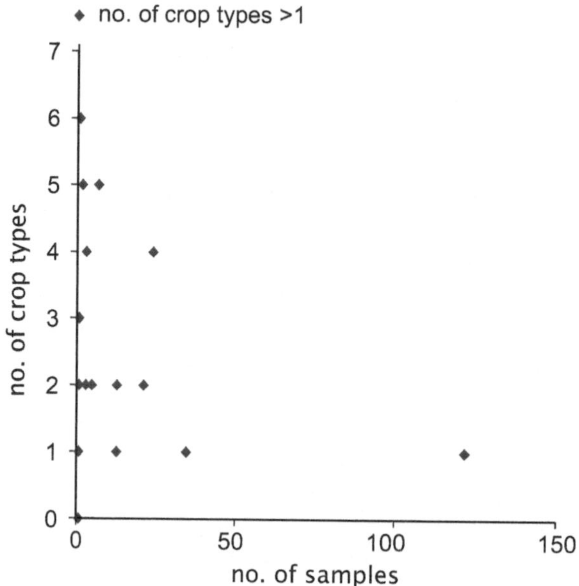

Figure 3.2: Scatter plot of the number of samples against number of well-attested crop types (>1).

3.5 Narrow crop spectra: possible explanations

Three explanations can be put forward to account for the narrow crop spectra recovered at some of the eastern sites: inadequate sampling, taphonomic bias, or human selection of a limited number of crops on the basis either of their tolerance of certain conditions (e.g., crops suitable for marginal environments), and/or of their cultural preference. The role of human selection can only be considered once the first two alternative explanations have been investigated. Thus, sampling and taphonomy will be considered first.

3.5.1 Inadequate sampling

The small number of samples (<30) from most of the sites is unlikely to reveal the full spectrum of plant use; more representative assemblages are likely to be recovered from thoroughly sampled sites than from those with only a few samples. Concentration in this study on the better-sampled sites and on the commonest crops is an attempt to minimise this problem. A scatter plot of sample number against number of well attested crop types (table 3.1) failed to show any clear relationship (figure 3.2); for example, Ganj Dareh, which is one of the most thoroughly sampled sites, has 121 samples but only three crop types, whereas the much smaller assemblage from Choga Mami (seven samples) has a more diverse spectrum comprising five crop types. While limited sampling may be a contributing factor, therefore, it does not account fully for the narrow crop range at some of the sites.

3.5.2 Taphonomic bias

A typical source of charred plant remains on Western Asian sites is material burnt as fuel in small quantities on domestic fires, in the form of animal dung or kindling (Miller 1996; Charles 1998). At the sites that were relatively well sampled, such as Ganj Dareh, Ali Kosh and Jeitun, there may be an emphasis on this 'background noise' of fuel-derived plants. Given the probable importance of animal dung as a source of fuel in the Neolithic, it could be argued that the breadth of animal, rather than human, diets is typically represented, the exception being in rare cases where stored crops for (potential) human consumption are accidentally destroyed by fire.

A second form of possible taphonomic bias relates to crop processing. Glume wheats can be stored as spikelets (providing protection from attack by insects etc.; Sigaut 1988), which means that late-stage processing (grain dehusking) often occurs on a daily basis on-site and, as such, the waste from processing is likely to provide the principal source of cereal material in charred plant assemblages (Hillman 1984; Jones 1987). In contrast, free-threshing cereal processing regularly takes place away from sources of fire (e.g., at threshing floors), so the chaff tends to be under-represented (unless it is collected and used systematically as fodder or fuel—van der Veen 1999).

In order to assess possible biases in archaeobotanical remains it is necessary, at least initially, to consider each sample or context (each behavioural or depositional event) individually. While some processes (e.g. crop processing, animal consumption etc.) influence sample composition before deposition, potentially varying from one deposit to the next, others (e.g. soil freezing/thawing, wetting/drying etc.) operate after deposition at a larger scale (part or all of a site). Identification of the effects of taphonomic filters can be made on a sample-by-sample basis using the composition of the crops and wild plants, the condition of the remains and so on. Once taphonomy has been assessed, it may be useful to group samples together to investigate broader trends, but this grouping should be based around samples with a common taphonomic history (i.e. comparing 'like-with-like'). To evaluate the role of taphonomic bias, therefore, the evidence from three relatively well-sampled sites dating to the PPNB and late Neolithic is reviewed in detail.

3.6 Case studies

Ganj Dareh At Ganj Dareh, a small mound site in the forest-steppe zone of the Zagros mountains in western Iran (1400 m), a total of 121 samples were processed from a range of ashy contexts (e.g., hearths, firepits) (van Zeist et al. 1984). Most of the samples contained small numbers of identifiable items (<50), however, the volume of soil processed is not recorded and so the density of items cannot be calculated.

Eight samples contained at least 50 items and the contents of these are briefly assessed here. Barley occurs mainly in the form of grain, though a small amount of rachis material (both wild and domesticated) is present in more than half of these samples. Two samples contain more than 50 barley items (grains) together with the seeds of wild plants. One of these samples (no. 70) comprises seeds of a range of wild taxa, which include those of a species of feather grass (*Stipa* indet.) (10 seeds). The other sample (no. 136) is dominated by small legumes (50 seeds) but also contains 19 seeds of the sea clubrush (*Scirpus maritimus*).

It is unlikely that the combination of barley grain and these wild plant seeds represents a harvested crop and its segetal flora. Grain-rich samples in association with seeds of species that mimic the size and behaviour of cereals as they are processed (Hillman 1984; Jones 1984) are likely to represent a gathered crop and its weeds, whereas the two Ganj Dareh samples do not conform to this model. It is possible instead that at least some of the wild plant seeds in the two samples were eaten by animals, *Scirpus* and small legumes have both been associated with animal dung at many sites (Miller 1996, 2002; Charles 1998; Fairbairn et al. 2002).

Three of the six remaining samples that contain more than 50 identifiable items are dominated by small legumes (one of which also has a concentration of *Scirpus maritimus*); two are dominated by a tiny-seeded grass (*Agrostis* type); and one sample has more or less equal amounts of pistachio (*Pistacia*), *Agrostis* and *Scirpus maritimus*. These samples appear to contain varying combinations of dung-derived plant remains (e.g., small-seeded legumes, *Scirpus*, possibly also *Agrostis*) and other discard (*Pistacia* nutshell).

Miller and Smart (1984) and (Charles 1998) have proposed a number of criteria for the recognition of dung-derived plant remains in archaeobotanical assemblages. Dung-producing livestock, namely, early herded goats, were certainly present in and around the site, as indicated by the faunal remains (Zeder and Hesse 2000; Zeder 2001) as well as by hoof prints preserved in mud brick (Smith 1990). This complies with one criterion described by Miller and Smart (1984). The presence in the assemblage of late-maturing taxa such as *Scirpus*, which is unlikely to be harvested in fruit with winter crops, fulfils one of the four criteria suggested by Charles (1998), however it should be stressed that the frequent occurrence of a single crop type (barley) makes it impossible to apply two of the other criteria.

To summarise, the samples from Ganj Dareh are equally likely to represent both animal and human diets. The predominance of barley, a free-threshing cereal, cannot be explained simply as a taphonomic bias related to crop processing (which would tend to lead to an over-representation of glume wheats—see above). The archaeobotanical assemblage may shed additional light on the early husbandry of goats, as inferred from mortality data (Zeder and Hesse 2000). It is tempting to argue that the use of barley as a fodder crop can be traced back from recent times to the PPNB (cf. van Zeist et al. 1984; Miller 2002); ethnographic and experimental work suggests that hulled barley grain can survive caprine digestion (Charles and Valamoti nd).

Ali Kosh Ali Kosh is located in the small alluvial Deh Luran plain of Khuzistan in Iran; the area is characterised by dry steppe, which is marginal for rain-fed cultivation. Unfortunately, sample-by-sample data are not available for Ali Kosh (Helbaek 1969), but the site merits attention from a taphonomic point of view. The major crop types and parts (e.g., chaff and/or grain) present during the three phases of occupation are similar. The commonest crops throughout are emmer, which occurs primarily as chaff, and hulled (six-row) barley, which is represented primarily by grain. Naked barley (probably six-row) is present at lower frequencies, and there are occasional seeds of einkorn, linseed and lentil. The dominant feature of the assemblage, however, is the amount of 'endemic legumes'; these comprise a range of small-seeded taxa that make up more than 50% of all the plant material recovered and more than 90% of the remains from the earliest (Bus Mordeh) phase. The samples also contain a range of potential crop weeds: goat-faced grass (*Aegilops*), ryegrass (*Lolium*), wild oats (*Avena*), mallow (*Malva*) and plantain (*Plantago*). Another taxon of note is *Prosopis*, which is a deep-rooted perennial plant that grows and matures after the harvesting of wheat and barley; this taxon increases in frequency through time and constitutes almost 5% of the remains in the last (Mohammad Jafar) phase.

Helbaek's (1969) interpretation of the plant remains from Ali Kosh is couched entirely in terms of human consumption; what people were eating, what proportion of their dietary requirements were being met by which plant groups, and the relative importance of the food plants through time (including a number of wild plants such as caper, not mentioned above, and *Prosopis*). The apparent increase in the number of *Prosopis* seeds is discussed in considerable detail and the author tentatively puts forward the notion that the plant was not available during the earliest phases of occupation but was collected in the later phases after it had been 'introduced' or its growth had been stimulated by human land use. Crop husbandry practices are also considered and there is some discussion of the use of irrigation, which is implied by the presence of crops with high water-requirements such as six-row barley. The possibility that at least some of the plant material recovered from the site could have been collected for purposes other than human consumption is not considered.

An alternative interpretation for the remains is provided by Miller (1996), who in a detailed reconsideration concluded that the remains "fit all the criteria for the use of dung fuel suggested by Miller and Smart (1984)"—namely, the location of the site in a marginal environment, the presence of dung-producing animals and the occurrence of recognisable dung pellets in the archaeological deposits (e.g., hearth contexts). In addition, the presence of late-maturing taxa, such as *Prosopis*, which would not be harvested in fruit with winter cereals, meant that they must have entered the archaeological record by some other route; and the fact that *Prosopis* is palatable for livestock but not for humans (other than in times of famine) indicates the likelihood that this was via dung fuel (Charles 1998). The application of further criteria suggested by Charles (such as a the presence in an assemblage of a mixture of crops and crop parts incompatible with a single stage of crop processing [Charles 1998]) requires sample-by-sample analysis. The predominance of emmer chaff versus barley grain may reflect the frequent dehusking of emmer on site. The combination of glume wheat chaff and barley grain has been interpreted elsewhere as a likely fodder/dung mixture (Charles 1998), but the lack of sample-by-sample data for Ali Kosh means that the coincidence of these components in the same deposits cannot be confirmed.

Jeitun The low-lying mound of Jeitun sits at the edge of the Kara Kum desert some 30 km from the Kopet Dag mountains in an area of low rainfall (ca. 200–250 mm per annum). Occupation at the site lasted approximately 500 years between ca. 6100–5600 cal BC and comprises several phases characterised by small rectangular buildings (Harris et al. 1993; Harris and Gosden 1996). The buildings are often associated with yards that are commonly covered by layers of dark ashy deposits. All sealed deposits (e.g., courtyard fills and mudbricks) were systematically sampled for the recovery of charred plant remains. Approximately 220 samples were taken from over 150 different contexts. Of these, 39 samples were chosen for analysis and these were estimated to contain at least 100 identifiable botanical items (based on scans of all of the flots).

Glume wheat chaff (consisting of spikelet forks and glume bases) is present in all but one of the samples and is the commonest type of crop material. Einkorn was by far the most frequent type of glume wheat identified (both grain and chaff) but there were also small quantities of chaff resembling emmer and the 'new' type (Jones et al. 2000). Other cereal crops present are barley (*Hordeum sativum*), both naked and hulled, and a few rachis internodes tentatively identified as free-threshing wheat. Whole or fragmentary charred dung pellets of sheep or goat occurred in the majority of samples.

The seeds of wild plants found in the Jeitun samples were identified as having derived from two sources: 1. crop weeds harvested with the cereals, e.g., goat-face grass (*Aegilops*) and *Eremopyrum*; and 2. other wild plants, a majority of which fruit after the crop harvest and were probably brought onto the site in animal dung (Charles and Bogaard nd), e.g., small-seeded grasses such as *Aeluropus* and sea blite (*Suaeda*), both of which were observed in the matrix of fragmented dung pellets.

The Jeitun samples were found to fulfil three of the four criteria proposed by Charles (1998) for recognising dung-derived material in archaeobotanical assemblages: the presence of identifiable animal dung (and especially the presence of charred plant remains within dung); the dissimilarity between the composition of the samples and crop processing products/by-products; and the presence of late-maturing wild plants unlikely to have been harvested in fruit with crops. A fourth and final criterion is the occurrence in assemblages of mixtures of crops that are unlikely to have been processed together for practical reasons (e.g., glume wheat and free-threshing cereals). The predominant cereal components in the Jeitun assemblage are the glume wheat glume bases (chaff), and their presence can be related to frequent dehusking of spikelets. Whether this by-product material was fed to livestock, mixed with dung to form dung cakes or discarded directly into fires is as yet unclear (Charles and Bogaard nd). It can be argued, therefore, that the taphonomic bias towards presence in the samples of glume wheats at Jeitun relates primarily to the way in which they are processed and not to animal diet per se. There is no firm evidence for deliberate and systematic mixing of crops as fodder.

3.7 Discussion

A closer site-by-site and sample-by-sample examination of the broad changes in plant use that were tentatively identified in the first half of the paper has revealed a more complex situation, whereby the taphonomy or 'origin' of the plant remains is critical for the interpretation of crop spectra. Two of the sites that were described in detail, Ali Kosh and Jeitun, have revealed evidence of dung-derived plant material. Evidence from Ganj Dareh is less clear but the possibility also exists that the remains originated from dung. These sites were selected on the grounds that they were well-sampled and that the samples were likely to provide information on plant use. The fact that these remains have been identified (or at least partially so) as the result of animal feeding rather than produce (or the by-products from its preparation) for human consumption highlights the need for caution when dealing with archaeobotanical databases. Regional scale consideration of databases does not 'remove' the problem of taphonomy; the danger is that it simply 'buries it deeper' in the interpretation process. The assumption that plant remains reflect human consumption is no longer tenable; instead there is a need to determine the source(s) of plant remains per sample and per site before any attempt is made to interpret in terms of crop selection.

With regard to the apparent reduction of crops as neolithic farming spread eastwards, the question arises whether it is possible to assess if this trend represents a real shift towards the selection of certain crops from the 'neolithic package' or is a result of taphonomic biases. At Jeitun, for example, the issue is whether the predominance of a single crop (einkorn) indicates crop selection or is the result of taphonomic biases relating to crop processing and/or animal feeding. The criteria used to identify dung-derived material at the site, relate to wild plants and not to the crops. Jeitun does not contain a mixture of crop types and crop parts suggestive of fodder/dung, as was the case for example at the bronze age site of Abu Salabikh in southern Iraq (Charles 1998). The dominance of glume wheats (as chaff) could reflect the bias towards the crops or crop parts that are most likely come into contact with fires, whereas any link between the crops and animal fodder is as yet unclear. A means of determining whether or not crop remains have passed through the digestive system of livestock would greatly assist in this interpretation; for example, animal feeding experiments that have been carried out suggest that glume bases can survive digestion (Charles and Valamoti in press).

For Jeitun it appears that the predominance of einkorn is at least partially due to taphonomic bias (crop processing), but this does not explain the prevalence of einkorn per se rather than emmer. Einkorn is known to be more tolerant of poor growing conditions than emmer (Percival 1974, p.171), and so its dominance over other cereal species may reflect the selection of a hardier crop for the marginal environment of Jeitun. The presence of einkorn at Jeitun has been contrasted with the situation in the late PPNB of the Fertile Crescent, where it has been found in lower frequencies, possibly indicative of the fact that it was weed rather than a crop in its own right (Willcox 1999).

The interpretation of the assemblages at Ali Kosh is even less clear due to the lack of sample-by-sample data. The crop spectrum is broader than that at Jeitun, but only barley and emmer are well attested. The predominance of emmer(/einkorn) glume bases in the assemblage as a whole (Helbaek 1969, table 3) could also be related to a taphonomic bias favouring the preservation of glume wheat chaff. Mixtures of glume wheat chaff and free-threshing cereal grain in individual samples (which has not been proven at Ali Kosh) may indicate the deliberate amalgamation of crops for the preparation of fodder and/or dung cakes (cf. Charles 1998).

At Ganj Dareh the single well-attested crop (barley) is free-threshing and is not likely, therefore, to be over-represented in the assemblage due to crop processing bias; for this reason the suggestion of van Zeist et al. (1984) that 'barley was grown primarily for animal food' is not so implausible.

3.8 Conclusion

An often unstated assumption underlying much of the writing on the emergence and spread of domesticated crops is that patterns in the data through time and across regions reflects similarities or differences in human diet. The assumption has been questioned at individual sites and this paper suggests that animal rather than human consumption (i.e., fodder and/or dung) is as likely a reason for the patterning. The argument is not that early crop-growing was only for the purposes of either fodder or food production; ethnographic studies have shown that crops may play both roles (Jones and Halstead 1995). The important point to be made here is that the use of a crop component as animal feed, and its subsequent charring in dung fuel, may lead to an (over-)representation of that crop in the archaeobotanical record.

There are indications at sites such as Jeitun that crops may have been chosen for their ability to tolerate marginal conditions beyond the Fertile Crescent. Similar narrowing of crop spectra has been noted in the westward spread of crops in climatic zones where the burning of animal dung fuel is less likely (cf. Charles and Halstead 2001). Future work could profitably explore the ecological and cultural underpinnings of such changes in cuisine, but the fundamental role of taphonomy in shaping archaeobotanical assemblages must always be considered first.

This paper started with a consideration of the 'nature of databases'; it is now worth returning to that theme by asking what the role of databases is in the investigation of trends in archaeobotanical data. The London conference provided a forum for the evaluation of the considerable body of data on the emergence and spread of crops during the Neolithic of Western Asia and Europe. A number of broad trends in the data were highlighted which will fuel and direct research and debate. Two issues, however, are inescapable: firstly, the datasets for many regions (e.g., the eastern half of the Fertile Crescent) are too sparse to sustain anything but a superficial survey of major events, in this respect databases are only as good as the data they contain; and secondly, as this paper has demonstrated, the fundamental role of taphonomy in shaping archaeobotanical assemblages should always be considered before grand theories are constructed.

Acknowledgements

The author would like to thank: Amy Bogaard for helpful comments on the manuscript; Naomi Miller for bibliographic references; Sue Colledge for her invitation to the London conference (and her patience!); and the British School of Archaeology in Iraq for funding to develop the bioarchaeological database.

References

Bakheyev, F. and Z. Yanushevich (1980). Discoveries of cultivated plants in the early farming settlements of Yarym-Tepe I and Yarym-Tepe II in Northern Iraq. *Journal of Archaeological Science 7*, 167–178.

Banning, E. (1998). The Neolithic period: triumphs of architecture, agriculture and art. *Near Eastern Archaeology 61*, 188–237.

Barker, G. (1985). *Prehistoric Farming in Europe.* Cambridge: Cambridge University Press.

Bogucki, P. (1996). The spread of early farming in Europe. *American Scientist 84*, 242–253.

Charles, M. (1998). Fodder from dung: the recognition and interpretation of dung-derived plant material from archaeological sites. *Environmental Archaeology 1*, 111–122.

Charles, M. and A. Bogaard (n.d.). The archaeobotany of Jeitun: implications for early cultivation and herding practices in western Central Asia. In D. R. Harris (Ed.), *Origins of Agriculture in Western Central Asia: Archaeological-Environmental Investigations in Southern Turkmenistan.* [In Preparation].

Charles, M. and K. Dobney (n.d.). Mesopotamian environmental archaeology database: Phase I Iraq. [In Prep].

Charles, M. and P. Halstead (2001). Biological resource exploitation: problems of theory and method. In D. Brothwell and A. Pollard (Eds.), *Handbook of Archaeological Sciences*, pp. 365–378. Chichester: John Wiley and Sons.

Charles, M. and G. Hillman (1992). Crop husbandry in a desert environment: evidence from the charred plant macro-remains. In V. Masson (Ed.), *New Research at the Jeitun Settlement (Preliminary Reports on the Work of the Soviet-British Expedition)*, pp. 83–84. Ashkabad: Academy of Sciences of Turkmenistan. [In Russian].

Charles, M. and S. Valamoti (n.d.). Distinguishing food from fodder through the study of charred plant remains. [In Prep for *Vegetation History and Archaeobotany*].

Colledge, S. (2001). *Plant Exploitation on Epipalaeolithic and Early Neolithic Sites in the Levant.* BAR International Series 986. Oxford: John and Erica Hedges Ltd.

Colledge, S. (2004). Reappraisal of the archaeobotanical evidence for the emergence and dispersal of the 'founder crops'. In E. Peltenburg and A. Wasse (Eds.), *Neolithic Revolution: New Perspectives on Southwest Asia in Light of Recent Discoveries on Cyprus*, Levant Supplemantary Series I, pp. 49–60. Oxford: Oxbow Books.

Colledge, S., J. Conolly, and S. Shennan (2004). Archaeobotanical evidence for the spread of farming in the eastern Mediterranean. *Current Anthropology 45*, S35–S58.

Fairbairn, A., A. Asouti, J. Near, and D. Martinoli (2002). Macro-botanical evidence for plant use at Neolithic Çatalhöyük. *Vegetational History and Archaeobotany 11*, 41–54.

Garrard, A. (1999). Charting the emergence of cereal and pulse domestication in south-west Asia. *Environmental Archaeology 4*, 67–87.

Garrard, A., S. Colledge, and L. Martin (1996). The emergence of crop cultivation and caprine herding in the "marginal zone" of the southern Levant. In D. R. Harris (Ed.), *The Origins and Spread of Agriculture and Pastoralism in Eurasia*, pp. 204–226. London: UCL Press.

Halstead, P. (1987). Traditional and ancient rural economy in Mediterranean Europe: plus ça change? *Journal of Hellinistic Studies 107*, 77–87.

Halstead, P. (1989). Like rising damp? an ecological approach to the spread of farming in Southeast and Central Europe. In A. Milles, D. Williams, and N. Gardner (Eds.), *The Beginnings of Agriculture*, Symposia of the Association for Environmental Archaeology No. 8. BAR International Series 496, pp. 23–53. Oxford: British Archaeological Reports.

Harris, D. and C. Gosden (1996). The beginnings of agriculture in western Central Asia. In D. R. Harris (Ed.), *The Origins and Spread of Agriculture and Pastoralism in Eurasia*, pp. 370–389. London: UCL Press.

Harris, D., V. Masson, Y. Berezkin, M. Charles, G. Korobkova, K. Kurbansakhatov, A. Legge, and S. Limbrey (1993). Investigating early agriculture in Central Asia: new research at Jeitun, Turkmenistan. *Antiquity 67*, 324–338.

Helbaek, H. (1959). Domestication of food plants in the Old World. *Science 130*, 365–372.

Helbaek, H. (1960). The palaeoethnobotany of the Near East and Europe. In R. Braidwood and B. Howe (Eds.), *Prehistoric Investigations in Iraqi Kurdistan*, pp. 99–118. Chicago: University of Chicago Press.

Helbaek, H. (1964). Early Hassunan vegetable remains at es-Sawwan near Samarra. *Sumer 20*, 45–48.

Helbaek, H. (1969). Plant collecting, dry farming, and irrigation agriculture in prehistoric Deh Luran. In K. V. Flannery and J. A. Neely (Eds.), *Prehistory and Human Ecology of the Deh Luran Plain*, pp. 383–426. Ann Arbor: University of Michigan.

Helbaek, H. (1972a). Samarran irrigation agriculture at Choga Mami in Iraq. *Iraq 34*, 35–48.

Helbaek, H. (1972b). Traces of plants in the Early Ceramic site of Umm Dabaghiyah. *Iraq 34*, 17–20.

Hillman, G. C. (1984). Interpretation of archaeological plant remains: the application of ethnographic models from Turkey. In W. van Zeist and W. A. Casparie (Eds.), *Plants and Ancient Man*, Studies in Palaeoethnobotany. Proceedings of the Sixth Symposium of the International Work Group for Palaeoethnobotany, Groningen, 30 May–3 June 1983, pp. 1–42. Rotterdam: A. A. Balkema.

Hubbard, R. (1990). Archaeobotany of Abdul Hosein. The carbonised seeds from Tepe Abdul Hosein: results of preliminary analyses. In J. Pullar (Ed.), *Tepe Abdul Hosein: a Neolithic site in western Iran, excavations 1978*, BAR International Series 563, pp. 217–221. Oxford: British Archaeological Reports.

Jacobsen, T. (1982). *Salinity and Irrigation Agriculture in Antiquity: Diyala Basin Archaeological Projects. Report on Essential Results 1957–58.* Bibliotheca Mesopotamica 14. Malibu: Undena.

Jacomet, S. and A. Kreuz (1999). *Archäobotanik. Aufgaben, Methoden und Ergebnisse vegetations—und agrargeschichtlicher Forschung.* UTB für Wissenschaft 8158. Stuttgart: Verlag Eugen Ulmer. With contributions by M. Rösch.

Jones, G. (1984). Interpretation of archaeological plant remains: ethnographic models from Greece. In W. van Zeist and W. A. Casparie (Eds.), *Plants and Ancient Man*, Studies in Palaeoethnobotany. Proceedings of the Sixth Symposium of the International Work Group for Palaeoethnobotany, Groningen, 30 May–3 June 1983, pp. 43–61. Rotterdam: A. A. Balkema.

Jones, G. (1987). A statistical approach to the archaeological identification of crop processing. *Journal of Archaeological Science 14*, 311–323.

Jones, G. (1991). Numerical analysis in archaeobotany. In W. van Zeist, K. Wasylikowa, and K.-E. Behre (Eds.), *Progress in Old World Palaeoethnobotany: A Retrospective View on the Occassion of 20 Years of the International Work Group for Palaeoethnobotany*, pp. 63–80. Rotterdam: A. A. Balkema.

Jones, G. and P. Halstead (1995). Maslins, mixtures and monocrops: on the interpretation of archaeobotanical crop samples of heterogeneous composition. *Journal of Archaeological Science 22*, 103–114.

Jones, G., S. Valamoti, and M. Charles (2000). Early crop diversity: a 'new' glume wheat from northern Greece. *Vegetation History and Archaeobotany 9*, 133–146.

Kozlowski, K. (1989). Nemrik 9, a PPN neolithic site in northern Iraq. *Paléorient 15*, 25–31.

Lisicyna, G. (1983). Die ältesten paläoethnobotanischen Funde in Nordmesopotamien. *Zeitschrift für Archäologie 17*, 31–38.

Masson, V. M. (1971). *The Jeitun Settlement: The Emergence of a Productive Economy [In Russian]*. Materials and Research on the Archaeology of the U.S.S.R. 180. Moscow: Nauka.

Masson, V. M. and V. I. Sarianidi (1972). *Central Asia: Turkmenia Before the Achaemenids*. London: Thames and Hudson.

Miller, N. (1996). Seed eaters of the ancient Near East: human or herbivore? *Current Anthropology 37*, 521–528.

Miller, N. (2002). Tracing the development of the agropastoral economy in southeastern Anatolia and northern Syria. In R. Cappers and S. Bottema (Eds.), *The Dawn of Farming in the Near East*, Studies in Near Eastern Production, Subsistence and Environment 6, pp. 85–94. Berlin: Ex oriente.

Miller, N. (2003). Plant remains from the 1996 excavation. In A. Alizadeh (Ed.), *Excavations at the Prehistoric Mound of Chogha Bonut, Khuzestan, Iran, Seasons 1976/77, 1977/78, and 1996*, Oriental Institute Publications 120, pp. 123–128. Chicago: The Oriental Institute.

Miller, N. and T. Smart (1984). Intentional burning of dung as fuel: a mechanism for the incorporation of charred seeds into the archaeological record. *Journal of Ethnobiology 4*, 15–28.

Neef, R. (1991). Plant remains from archaeological sites in lowland Iraq: Tell el 'Oueili. In J.-L. Huot (Ed.), *'Oueili, Travaux de 1985*, pp. 321–329. Paris: Editions Recherche sur les Civilisations.

Nesbitt, M. (1998). Preliminary report on the plant remains from M'lefaat. *Cahiers de l'Euphrate 8*, 232–233, 272.

Nesbitt, M. (2002). When and where did domesticated cereals first occur in Southwest Asia? In R. T. J. Cappers and S. Bottema (Eds.), *The Dawn of Farming in the Near East*, Studies in early Near Eastern Production, Subsistence and Environment 6/1999, pp. 113–132. Berlin: Ex oriente.

Nesbitt, M. and D. Samuel (1996). From staple crop to extinction? The archaeology and history of the hulled wheats. In S. Padulosi, K. Hammer, and J. Heller (Eds.), *Hulled wheats. Promoting the Conservation and Use of Underutilized and Neglected Crops*, pp. 41–100. Rome: IPGRI.

Oates, D. and J. Oates (1976). Early irrigation agriculture in Mesopotamia. In G. Sieveking and I. Longworth (Eds.), *Problems in Economic and Social Archaeology*, pp. 109–135. London: Duckworth.

Peltenburg, E. (2004). Introduction: a revised Cypriot prehistory and some implications for the study of the Neolithic. In E. Peltenburg and A. Wasse (Eds.), *Neolithic Revolution: New Perspectives on Southwest Asia in Light of Recent Discoveries on Cyprus*, Levant Supplemantary Series I, pp. xi–xx. Oxford: Oxbow Books.

Peltenburg, E., S. Colledge, P. Croft, A. Jackson, C. McCartney, and M. Murray (2001). Neolithic dispersals from the Levantine Corridor: a Mediterranean perspective. *Levant 33*, 35–64.

Percival, J. (1974). *The Wheat Plant.* London: Duckworth.

Price, T. (2000). The introduction of farming in northern Europe. In T. Price (Ed.), *Europe's First Farmers*, pp. 260–300. Cambridge: Cambridge University Press.

Sigaut, F. (1988). A method for identifying grain storage techniques and its application for European agricultural history. *Tools and Tillage 6*, 3–32.

Smith, P. E. L. (1990). Architectural innovation and experimentation at Ganj Dareh, Iran. *World Archaeology 21*, 323–335.

Stordeur, D. (2003). Des cranes surmodelés à Tell Aswad de Damascène (PPNB—Syrie). *Paléorient 29*(2), 109–116.

van der Veen, M. (1999). The economic value of chaff and straw in arid and temperate zones. *Vegetation History and Archaeobotany 8*, 211–224.

van Zeist, W., P. E. L. Smith, R. M. Palfenier-Vegter, M. Suwijn, and W. Casparie (1984). An archaeobotanical study of Ganj Dareh Tepe, Iran. *Palaeohistoria 26*, 201–224.

Watkins, T., D. Baird, and A. Betts (1989). Qermez Dere and the early aceramic of N. Iraq. *Paléorient 15*, 19–24.

Watson, P. J. (1983). Note on the Jarmo plant remains. In L. S. Braidwood, R. Braidwood, B. Howe, C. Reed, and P. Watson (Eds.), *Prehistoric Archaeology Along the Zagros Flanks*, Oriental Institute Publications 105, pp. 501–503. Chicago: The Oriental Institute.

Wilkinson, T. (2000). Regional approaches to Mesopotamian archaeology: the contribution of archaeological surveys. *Journal of Archaeological Science 8*, 219–267.

Willcox, G. (1998). Archaeobotanical Evidence for the Beginnings of Agriculture in Southwest Asia. In A. B. Damania, J. Valkoun, J. Willcox, and C. O. Qualset (Eds.), *The Origins of Agriculture and Crop Domestication. Proceedings of the Harlan Symposium 10-14 May 1997*, pp. 25–38. Aleppo: ICARDA.

Willcox, G. (1999). Agrarian change and the beginnings of cultivation in the Near East: evidence from wild progenitors, experimental cultivation and archaeobotanical data. In C. Gosden and J. Hather (Eds.), *The Prehistory of Food: Appetites for Change*, pp. 478–500. London: Routledge.

Willcox, G. (2002). Geographical variation in major cereal components and evidence for independent domestication events in the Western Asia. In R. T. J. Cappers and S. Bottema (Eds.), *The Dawn of Farming in the Near East*, pp. 133–140. Berlin: Ex oriente.

Zeder, M. (2001). A metrical analysis of a collection of modern goats (*Capra hircus aegargus* and *C. h. hircus*) from Iran and Iraq: implications for the study of caprine domestication. *Journal of Archaeological Science 28*, 61–79.

Zeder, M. and B. Hesse (2000). The inital domestication of goats (*Capra hircus*) in the Zagros mountains 10,000 years ago. *Science 287*, 2254–2257.

Zohary, D. (1996). The mode of domestication of the founder crops of Southwest Asian agriculture. In D. R. Harris (Ed.), *The Origins and Spread of Agriculture and Pastoralism in Eurasia*, pp. 142–158. London: UCL Press.

Zvelebil, M. (2000). The social context of the agricultural transition in Europe. In C. Renfrew and K. Boyle (Eds.), *Archaeogenetics: DNA and the Population Prehistory of Europe*, pp. 57–79. Cambridge: McDonald Institute for Archaeological Research.

Chapter 4

A review and synthesis of the evidence for the origins of farming on Cyprus and Crete

Sue Colledge and James Conolly
University College London, UK and Trent University, Canada

4.1 Introduction

Much recent work that has considered the phenomenon of human migration has emphasised the complexity of the *process* of population movement, its impact on population structures and its role in social transformation (e.g., Anthony 1990; Burmeister 2000; Goldstein and Chikhi 2002; Härke 1998). In addition, we maintain that large-scale population movement and, in particular, the movement of farmers into previously uncultivated areas, will have ecological impacts on the receiving environments that may be detected in archaeobotantical remains recovered from 'first' (or early) coloniser settlements.

Migration is rarely, if ever, a single *event* and even the sudden (in archaeological terms) appearance of settlements in a previously uninhabited territory is typically preceded by earlier essential phases of exploration and 'landscape learning' (e.g., Anderson 2003; Rockman 2003; cf. Schwartz 1970). It is not unexpected that the first permanently inhabited settlements on islands were preceded by earlier, archaeologically invisible, visitations for the purpose of short-term exploitation of island resources and, synchronously, obtaining information about the nature and structure of the landscape. This information, fed into the information networks that typify small-scale mobile communities, provides the requisite level of prior knowledge and confidence of success that underlies the directional, long-distance, colonisation of new landscapes.

As we review in this paper, the evidence from the earliest agropastoralist settlement of Cyprus and Crete suggests that some prior knowledge of the target landscape and its suitability for settlement was held by the colonists (Broodbank and Strasser 1991; Peltenburg et al. 2000). We here show that the nature of the crop system also informs on the origins and connections between the inhabitants of Crete and Cyprus and mainland neolithic communities.

4.2 Cyprus

The shortest distance between the Cyprus and the Anatolian mainland (at current sea levels—distances would be reduced at time of first occupation) is about 70 km from Anamur Burnu to Cape Kormakiti, with an average of about 90 km from coast to coast. Distances are greater to the Levant, with the shortest just of 100 km from Latakia in Syria to Cape Apostolos Andreas, with an average to the mid-Levantine coast in the region of 170 km. These distances place Cyprus within view of both coasts (Broodbank 2001, pp.40–41) and within approximately two to nine days travel on the basis of estimates of 20 km a day for a small canoe and 40–50 km a day for longboat (although with favourable winds, and slightly reduced distances, possibly reachable within a day) (Broodbank 2001, p.102).

Prior to the 1980s there was very little unequivocal proof of habitation on Cyprus dating earlier than an aceramic neolithic phase of ca. 6000 cal BC (Peltenburg 2004, pp.xi–xiv, table 1). However, by the end of that decade there was evidence from the excavation at the site of Akrotiri-*Aetokremnos* of a hunter-gatherer presence from the early 10th millennium cal BC (Simmons 2004).[1] Links with the Levantine mainland are inferred from the chipped stone artefacts, which "would fit within a late natufian or early neolithic mainland assemblage" (Simmons 2001, p.10). During the 1990s further significant contributions were made to the understanding of the early prehistory of the island by the excavations of newly discovered aceramic sites that were older by about 2000 years than those recorded in the previous decade. On the basis of lithic technology, imports/trade, building styles, settlement organisation, artefact types, and ideology and symbolism (Peltenburg 2003a, p.87; McCartney 2004, p.119), these settlements are considered to be part of the 'Pre-Pottery Neolithic interaction sphere' (Bar-Yosef and Belfer-Cohen 1989; Gopher 1989). Peltenburg argues that the architectural, artefactual and artistic parallels with sites in the mid-upper Euphrates basin provide *overwhelming* evidence of connections with Cyprus (Peltenburg 2003a, p.96, 102). Simmons, however, suggests that migration from the southern Levant was as likely at this time, and he cites the similarities between the mainland and Cypriot Pre-Pottery Neolithic (PPN) settlement types to support his theory (Simmons 2004, p.11; see also McCartney 2005, pp.219–221). Our own work has also indicated links between Cypriot and southern Levantine sites in terms of the overall similarity in composition of archaeobotanical samples of crops and associated weed taxa (Colledge et al. 2004, p.s47). On the basis of characteristic tool types in lithic scatters found during recent surveys at several locations it now appears that there were probably temporary occupations which post-dated the Akrotiri-*Aetokremnos* phase but pre-dated the hitherto earliest aceramic sites (McCartney 2005), thus closing the 'gap' in the habitation record for the island between the early 10th and mid 9th millennia cal BC (Peltenburg et al. 2000).

Peltenburg et al. (2000) have argued that coastally-adapted epipalaeolithic and early neolithic communities along the Syrian Levantine coast would have had knowledge of seafaring and maritime resources. For reasons explained in the introduction it is likely that periodic visits were made to Cyprus prior to colonisation that provided the necessary knowledge for a successful colonisation. Tantalisingly, evidence for coastal Levantine PPN sites, which may have been the origin of the Cypriot neolithic farmers, is extremely limited.[2] One possible reason is that these coastal regions may have subsequently been inundated (as a result of sea level rise) so that the sites became 'invisible' to archaeological surveys (Peltenburg 2003a, p.97). Peltenburg suggests if there was flooding of cultivable land that this would have led to a considerable reduction of resources (at a time when the harvests had already become a necessary part to the yearly subsistence regime) and would consequently have prompted attempts to compensate for the loss by exploring new territories; by this time Cyprus was known to the inhabitants of the mainland and so would have been an attractive focus for relocation.

[1] The fact that the island was inhabited—at least temporarily—at this time is undisputed; there has been much debate amongst archaeologists, however, about whether there is adequate evidence of direct human intervention in the killing and butchering of the pygmy hippos, the bones of which were found in large numbers at the site (see, for example, Reese 2001).

[2] At present, the earliest Levantine coastal settlements where there is evidence of a substantial dependence on marine resources date to the end of the 9th millennium uncal bp (Galili et al. 2004, p.99), approximately 1500 years later than Cypro-EPPNB sites.

4.2.1 Archaeobotanical data

Archaeobotanical investigations at the Cypriot aceramic neolithic sites began in the 1970s, shortly after the technique of flotation was first described and applied by Helbaek at the site of Ali Kosh in Iran.[3] There are seven aceramic neolithic sites on Cyprus for which there is evidence of plant materials in the form of either charred remains retrieved by flotation or impressions of plant parts in pisé/mudbrick/plaster (Kissonerga-*Mylouthkia*, Perekklisha-*Shillourokambos*, Kalavassos-*Tenta*, Cape Andreas-*Kastros*, Khirokitia, Dhali-*Agridhi*[4] and Kholetria-*Ortos*[5]). The sites cover the time periods from the Cypro-EPPNB (Cypro-Early Pre-Pottery Neolithic B) to the late Aceramic Neolithic/Khirokitian and range in date from the mid 9th to the 6th millennia cal BC (Peltenburg 2003a, p.87, table 11.3).

Methods have improved since the 1970s and more recent recovery procedures have been standardised to ensure maximum comparative potential. Interpretations of data from sites where processing techniques (and also preservation conditions) were less than ideal are to some extent limited in their scope. These limitations apply in varying degrees to all archaeobotanical investigations but are perhaps of greater consequence for those that concern the dispersal of the earliest domestic crop species immediately after their evolution and subsequent emergence into new territories. For example, archaeobotanical finds at early aceramic neolithic sites in the eastern Mediterranean are crucial to the investigation of the routes and timings of the initial spread of the 'founder crop' species (Zohary 1992, p.82; 1996, pp.143–144) from their origins in Southwest Asia (see Colledge et al. 2004), and for this reason details of on-site methodology, depositional conditions and also identification criteria are relevant in instances where claims have been made that data represent the first evidence of domestic crops (i.e., and by implication the earliest crop-based agriculture) at certain sites and in certain localities. Below we describe (or cite publications with details of) factors that may have influenced the representation of taxa on the Cypriot sites.

At the excavations of the neolithic settlement at Dhali-*Agridhi*, which were initiated in 1972 (Lehavy 1974), flotation using the 'Streuver' method was undertaken (Stewart 1974). Out of 99 samples from neolithic contexts[6] (total volumes sampled for specific contexts were not given) only 18 different taxa are recorded. It is unclear what size of sieve was used but the fact that there are few very small seeded species suggests that the mesh diameter may have been larger than the standard size frequently used today. In this instance, therefore, the limited spectrum of plant taxa represented at the site could be due to differential recovery on the basis of size of charred items. In contrast, at *Mylouthkia* (excavated from the early 1990s; Peltenburg 2003a) where flotation was undertaken using the standard 1.0mm and 0.25mm sieves, 25 taxa are recorded from five phase IA samples (total volume 250 litres) and 34 are listed in the seven samples (total volume 630 litres) from phase IB (Murray 2003). At *Shillourokambos*, which was also excavated in the 1990s, 2,446 litres of sediment from 19 samples were processed by flotation using a sieve with a mesh size of 0.5mm and only nine taxa are recorded (including those taxa listed in the 10 samples that were from mixed contexts; Willcox 2001). In this instance ineffective

[3] "Therefore, when called in on the excavation I made up my mind to transfer to the field, for the first time, the laboratory technique for segregating plant remains from mineral samples by means of buoyancy. This process in its most primitive form, is carried out by drying the soil or ash sample and then pouring it into a basin with water." (Helbaek 1969, p.385).

[4] Kissonerga-*Mylouthkia*: Murray 2003; Perekklisha-*Shillourokambos*: (Willcox 2001); Kalavassos-*Tenta*: Hansen 1978; Cape Andreas-*Kastros*: van Zeist 1981; Khirokitia: Hansen 1989, 1994; Miller 1984; Waines and Stanley-Price 1977; Dhali-*Agridhi*: Lehavy 1974

[5] Kholetria-*Ortos*: A. Simmons pers. comm.; only a summary of the results from this site could be included here as the full report is not yet published. Archaeobotanical investigations of material recovered from the recent excavations at the Cypro-MPPNB site of Kritou Marottou-*'Ais Yiorkis* had not been completed at the time of writing (L. Espinda pers. comm.) and so the data could not be incorporated in the comparisons.

[6] In the published report the neolithic samples are listed according to locus numbers (cf. Lehavy 1974, p.102, appx. a and b); locus numbers above 28 (which were not included in the appendices) were assumed to be neolithic.

recovery methods are less likely to be the reason why there are so few remains; instead Willcox notes that the charred plant material had been fragmented by the precipitation and subsequent crystallisation of calcium carbonate in the porous structure of the grains (Willcox 2003, p.234). The low numbers of taxa and identifiable items are probably a result of taphonomic processes and consequent 'invisibility' in the archaeobotanical record. At *Shillourokambos* there was slightly better representation of taxa as imprints in pisé (11 taxa in total) including several that were not present in the charred material.

Presence of taxa at the aceramic neolithic sites excavated in the 1990s

Kissonerga-*Mylouthkia* is located in the southwest of the island (see Peltenburg 2003a for details of location and dates) and is one of the earliest aceramic neolithic sites. Charred plant remains were recovered from well 116[7] at the site (period IA), which has been assigned to the Cypro-EPPNB (equivalent to the Levantine early PPNB; Peltenburg 2003a, p.86) and dated to the mid 9th millennium cal BC.[8] This well and the others at the site had been dug in order to gain access to underground streams that flowed between impervious strata (i.e., aquicludes: confined or closed aquifers that are sandwiched in between dense impermeable layers), and appear to have acted as repositories for waste while in use and after abandonment. Peltenburg (2003a, p.90) records three different discard groups in well 116 that relate to the sequences of disposal; the refuse includes both natural infill and domestic waste, which comprise bones (both animal and human), marine and land molluscs, groundstone artefacts, bone tools, fragments of painted daub and charred plant material. The environmental and artefactual evidence indicates that there was an associated settlement, the permanency of which is suggested by the presence of the remains of house mice, (i.e., [human] commensal species; see Cucchi et al. 2002), and whose inhabitants were engaged in a suite of varied economic activities (Peltenburg 2003a, p.90). Acquaintance with and exploitation of the sea is attested at this time by the finds of marine resources, a pig tusk fishing hook and mainland imports (e.g., obsidian).

The plant taxa found in the five samples taken from the fill of well 116 at Mylouthkia are recorded in table 4.1 (Murray 2003, pp.59–71). Of prime interest at this early date are the economic taxa and much effort was devoted to the identification of the cereal grains and pulses found in the assemblages, in particular to establish their wild or domestic status (Peltenburg et al. 2001; Colledge 2004, pp.53–54). On the basis of the assessment of morphological and metrical analyses it was concluded that the three founder crop cereals: einkorn (*Triticum monococcum*), emmer (*Triticum dicoccum*) and hulled barley (*Hordeum sativum*) are present in Cypro-EPPNB contexts in well 116. The status of the pulses (e.g., *Lens* sp.) is less clear and only genus-level identification was assigned. Murray records that glume wheat chaff is present in four of the five period IA samples and hulled barley chaff is present in three of the samples (Murray 2003, pp.64–65). Of significance is the fact that there are high proportions of weed taxa in association with the domestic crops of which rye-grass (*Lolium* sp.) and canary grass (*Phalaris* sp.) are the most common (found in 80% of period IA samples according to Murray 2003, pp.68). In terms of overall characterization it appears that the samples are typical of fine sieving residues that derive from crop cleaning. The archaeobotanical evidence is highly indicative, therefore, that domestic cereal (and possibly also pulse) crops were being cultivated and processed in the vicinity of the site.

The later structures at *Mylouthkia* (period IB), which are dated to the late 8th–early 7th millennia cal BC and assigned to the Cypro-LPPNB (Cypro-Late Pre-Pottery Neolithic B; equivalent to the Levantine late PPNB), comprised a well (133), a pit and an eroded curvilinear building (B340; Peltenburg 2003a, p.85). The artefacts and structures from this later phase were

[7]To date, this is one of the earliest wells known (Peltenburg 2003a, p.84).

[8]At the time of writing several more wells had been discovered and excavated but are not yet published and the environmental materials have not been analysed (Peltenburg 2003b, p.20).

		Cypro-PPNB						Khirokitian				
		Early		Middle/Late								
Taxon	Plant part	KMIA	PSI	PSII	PSIII	KMIB	Kal-T	KhE	KhTh	KhW	CAK	DA
Cereals, pulses and flax												
Hordeum spontaneum	G	.	+	+	.	.	.	.	.	.	.	.
	R	.	+	.	+	.	.	.	.	.	.	.
Hordeum spontaneum/sativum	G	.	+	.	+	.	.	.	.	.	.	.
Hordeum sativum	G	+	.	.	+	+	+	+	+	+	+	+
	R	+	.	.	.	+	.	+	.	+	.	.
Hordeum sativum	G	.	.	.	.	.	.	.	.	.	.	+
Triticum monococcum	G	+	.	.	.	+	+	+	+	+	+	.
	GB/SF	.	.	.	.	.	.	+	.	+	.	.
Triticum monococcum/dicoccum	G	+	.	.	.	+	.	.	.	.	.	.
	GB/SF	+	.	.	.	+	.	.	.	.	+	.
Triticum dicoccum/dicoccoides	G	.	+	+	.	.	.	.	.	.	.	.
	SF	.	.	.	+	.	.	.	.	.	.	.
Triticum dicoccum	G	+	.	.	.	+	+	+	+	+	+	+
	GB/SF	.	.	.	.	.	.	+	.	+	.	.
Triticum spp.	G	.	.	.	.	.	.	+	.	.	.	+
Lathyrus sativus	.	.	.	.	.	.	.	.	+	.	.	.
Lens culinaris	.	.	.	.	.	.	+	.	.	.	+	.
Lens sp.	.	+	.	.	.	+	+	+	+	+	.	+
Pisum sativum	.	.	.	.	.	.	+	.	.	.	+	.
Pisum sp.	.	.	.	.	.	.	.	+	+	.	.	.
Vicia ervilia	.	.	.	.	.	.	+	+	.	.	.	.
Vicia faba/narbonensis	.	.	.	.	.	.	.	.	+	.	+	.
Linum sp.	.	+	.	.	.	+	.	.	.	.	+	.
Trees and shrubs												
Pistacia spp.	.	+	.	.	.	+	+	+	+	+	+	+
Capparis spinosa	.	.	+	.	.	.	.	.	.	+	.	.
Capparis sp.	.	.	.	.	.	.	.	+	.	.	.	.
Ficus carica	.	.	.	.	.	.	.	.	.	+	.	.
Ficus sp.	.	.	.	.	.	+	+	+	.	.	+	+
Olea europaea	.	.	.	.	.	.	.	+	.	+	.	.
Olea sp.	.	.	.	.	.	.	.	.	.	.	+	+
Amygdalus sp.	.	.	.	.	.	.	.	.	.	.	.	+
Prunus sp.	.	.	+	+	.	.	.	+	+	.	.	.
Pyrus sp.	.	.	.	.	.	.	+	.	.	.	.	.
Celtis sp.	.	.	.	.	.	.	.	+	.	+	.	+
Vitis sp.	.	.	.	.	.	.	.	.	.	.	.	+

Table 4.1: Summary of the presence of taxa identified at aceramic neolithic sites on Cyprus (taxa ordered alphabetically by family). Plant part codes: G=grain, R=rachis, GB=glume base, SF=spikelet fork. Site codes: KMIA/B= Kissonerga-*Mylouthkia* Phase IA/IB, PSI-III= Perekklisha-*Shillourokambos* Phases I–III, Kal-T= Kalavassos-*Tenta*, KhE/KhTh/KhW=Khirokitia Trench E/small sounding/Trench W, CAK= Cape Andreas-*Kastros*, DA= Dhali-*Agridhi*.

		Cypro-PPNB						Khirokitian				
		Early	Middle/Late									
Taxon	Plant part	KMIA	PSI	PSII	PSIII	KMIB	Kal-T	KhE	KhTh	KhW	CAK	DA
Wild herbaceous taxa												
Amaranthus retroflexus	.	.	.	.	.	.	+	.	.	.	.	.
Anchusa sp.	.	.	.	.	.	.	.	.	.	.	.	+
Buglossoides arvensis	.	.	.	.	.	.	+	+	.	.	.	.
Buglossoides sp.	.	.	.	.	.	.	.	+	.	+	.	.
Buglossoides tenuiflora	.	+	.	.	.	.	.	.	.	.	.	.
Echium sp.	.	+	.	.	.	.	.	.	.	+	.	.
Lithospermum officinale	.	.	.	.	.	.	+	.	.	.	.	.
Spergularia arvensis	.	.	.	.	.	.	+	.	.	.	.	.
Beta sp.	.	.	.	.	.	+	.	.	.	.	.	.
Carex sp.	.	.	.	.	.	.	.	+	.	+	.	.
Schoenus nigricans	.	.	.	.	.	.	+	.	.	+	.	.
Fumaria sp.	.	.	+	.	.	+	.	.	.	+	+	.
Agropyron sp.	.	.	.	.	.	.	+	.	.	.	.	.
Avena sp.	.	+	.	.	.	+	.	+	.	+	+	+
Bromus sp.	.	.	.	.	.	.	.	+	.	+	.	.
Hordeum murinum	.	.	.	.	.	.	.	.	.	+	.	.
Hordeum spp. (weed)	.	.	.	.	.	+	.	.	.	.	.	.
Lolium rigidum/perenne	.	.	.	.	.	.	.	.	.	.	+	.
Lolium sp.	.	+	.	.	.	+	+	+	.	+	.	.
Phalaris sp.	.	+	.	.	.	+	.	+	.	.	+	.
Setaria sp.	.	.	.	.	.	.	.	.	.	+	.	.
Astragalus sp.	.	.	.	.	.	.	+	+	.	.	+	.
Genista sp.	.	.	.	.	.	.	+	.	.	.	.	.
Lathyrus sp.	.	+	.	+	.	.	.	.	.	.	.	.
Medicago sp.	.	.	.	.	.	.	+	+	.	+	+	+
Scorpiurus sp.	.	.	.	.	.	+	.	.	.	.	.	.
Trifolium resupinatum	.	.	.	.	.	.	+	.	.	.	.	.
Trigonella sp.	.	.	.	.	.	.	.	+	.	.	.	.
Vicia sp.	.	.	.	.	.	.	+	+	.	+	+	+
Vicia/Lathyrus spp.	.	+	.	.	.	+	.	.	.	.	.	+
Asphodelus sp.	.	.	.	.	.	.	.	+	.	.	.	.
Muscari sp.	.	.	.	.	.	.	.	+	.	.	.	.
Malva sylvestris/nicaensis	.	.	.	.	.	.	.	.	.	.	+	.
Malva sp.	.	.	.	.	.	+	.	+	.	+	.	.
Polygonum sp.	.	.	.	.	.	.	+	.	.	+	.	.
Rumex sp.	.	.	.	.	.	+	.	.	.	.	.	.
Adonis dentata	.	.	.	.	.	.	.	.	.	.	+	.
Adonis sp.	.	.	.	.	.	+	.	.	.	.	.	.
Rubus sp.	.	.	.	.	.	.	+	.	.	.	.	.
Galium sp.	.	+	.	.	+	+	+	.	.	.	.	.
Pimpinella sp.	.	.	.	.	.	.	.	.	.	+	.	.

Table 4.1, cont.

also considered to be evidence of an *in situ* settlement. Two archaeobotanical samples were taken from the pit, one was taken from building B340, and four samples were taken from well 133 in which there was a greater density of artefacts than in the earlier well and an equally rich array of faunal and botanical remains (Peltenburg 2003a, p.91). The composition of these samples is similar to those from the earlier phase (table 4.1), i.e., predominantly cereal-rich assemblages with high numbers of wild/weed taxa (Murray 2003, p.61, table 7.2), and the same conclusion was reached regarding the local cultivation of the domestic crops. Peltenburg comments that the evidence suggests the agricultural economy at *Mylouthkia* had remained broadly the same for a millennium, which he states was a measure of the success of the early neolithic farmers (Peltenburg 2003a, p.91).

The aceramic neolithic site of *Shillourokambos* is located about 70 kilometres to the east of *Mylouthkia* and further inland. Four aceramic neolithic levels have been defined at the site: the Early Phase A (the earliest level represented at the site) is considered to be broadly contemporary with *Mylouthkia* IA (for ^{14}C dates see Guilaine 2003); the Early Phase B is slightly later and is assigned to the Cypro-MPPNB (Cypro-Middle Pre-Pottery Neolithic B; equivalent to the Levantine middle PPNB); the Middle Phase has dates that span the middle and late Cypro-PPNB; and the Late Phase (the latest aceramic level at the site), which is contemporary with *Mylouthkia* IB, corresponds to the Cypro-LPPNB (Peltenburg 2003a, pp.84–87). There is substantial evidence of settlement architecture in the earliest phase at *Shillourokambos*, which includes a trapezoidal enclosure (total area 76m^2) bounded by palisade trenches, and pits, post-hole alignments and also several wells (Guilaine et al. 2000, pp.75–76; Guilaine 2003; Peltenburg 2004, p.xiv). Willcox (2001, 2003) records that the only archaeobotanical evidence for the earliest phase (Early Phase A: PSI) is plant impressions in pisé, and he identifies wild barley (*Hordeum spontaneum*) grains and rachis, and indeterminate glume wheat grains and chaff (i.e., of uncertain wild/domestic status). Imprints of large seeded/stoned wild taxa were also found in pisé fragments dated to this phase (table 4.1). Circular stone structures, hearths, pits, wells, cobble layers and other associated features have been assigned to the later aceramic phases (Guilaine et al. 2000, pp.75–76; Guilaine 2003). Samples taken from these contexts (dated to Cypro-MPPNB/LPPNB: PSII/III) produced a few charred remains that were recovered by flotation and these complemented the range of taxa present in the form of impressions. Wild barley and glume wheat(s) of indeterminate wild/domestic status were also identified in both the Middle and Late Phases, but of significance are the impressions of domestic hulled barley grains in the latter.

The information provided by the two contemporary sites on the island with the earliest archaeobotanical evidence is very different and poses somewhat of a dilemma about the interpretation of the status of neolithic crop-based agriculture at this early date. The *Mylouthkia* IA data indicate that there was cultivation of a subset of the neolithic 'crop package', whereas at an equivalent date at *Shillourokambos* it appears that the only unambiguous evidence is of wild taxa (including wild cereals). As previously stated, the preservation conditions at the latter site were poor, and in comparison with *Mylouthkia* it is perhaps less likely that as full a set of taxa that were originally transported to the settlement are represented either as impressions or, in later phases, as charred remains. The range and/or types of taxa represented in the earliest phase at *Shillourokambos* were possibly also limited because preservation of the plant materials was solely in the form of impressions. For example, impressions in the pisé might comprise only those taxa considered more valuable as temper (relative to their usefulness for other purposes), and those of wider economic importance, therefore, may be 'missing' from the archaeobotanical record (c.f. Hillman 1989, p.218). If during future excavations larger quantities of better-preserved charred remains are recovered at *Shillourokambos* it may be possible to shed light on the reasons underlying the contradictory nature of the evidence from the two sites, and to determine whether the differences in the plant spectra are a result of actual differences in plant-based subsistence, or

		Cypro-PPNB						Khirokitian				
		Early		Middle/Late								
Taxon	Plant part	KMIA	PSI	PSII	PSIII	KMIB	KaI-T	KhE	KhTh	KhW	CAK	DA
wild barley	grains	.	+	+	.	.	.	.	.	.	.	.
wild barley	chaff	.	+	.	+	.	.	.	.	.	.	.
wild/dom barley	grains	.	+	.	+	.	.	.	.	.	.	.
wild/dom emmer	grains	.	+	+	.	.	.	.	.	.	.	.
wild/dom emmer	chaff	.	.	.	+	.	.	.	.	.	.	.
hulled barley	grains	+	.	.	+	+	+	+	+	+	+	+
hulled barley	chaff	+	.	.	.	+	.	+	.	+	.	.
naked barley	grains	.	.	.	.	.	.	.	.	.	.	+
einkorn	grains	+	.	.	.	+	+	+	+	+	+	.
einkorn	chaff	.	.	.	.	.	.	+	.	+	.	.
einkorn/emmer	grains	+	.	.	.	+	.	.	.	.	.	.
einkorn/emmer	chaff	+	.	.	.	+	.	.	.	.	+	.
emmer	grains	+	.	.	.	+	+	+	+	+	+	+
emmer	chaff	.	.	.	.	.	.	+	.	+	.	.
free threshing	grains	.	.	.	.	.	.	+	.	.	.	+
rye	.	.	.	.	.	.	.	.	.	.	.	.
lentil*	.	+	.	.	.	+	+	+	+	+	+	+
pea*	.	.	.	.	.	.	+	+	+	.	+	.
chick pea*	.	.	.	.	.	.	.	.	.	.	.	.
bitter vetch*	.	.	.	.	.	.	+	+	.	.	.	.
grass pea*	.	.	.	.	.	.	.	.	+	.	.	.
faba bean*	.	.	.	.	.	.	.	.	+	.	+	.
flax*	.	+	.	.	.	+	.	.	.	.	+	.

Table 4.2: Summary of the presence of domestic taxa at the aceramic neolithic sites. The domestic status for the pulses and flax (marked with *) has not been confirmed in all cases. Site codes: KMIA/B= Kissonerga-*Mylouthkia* Phase IA/IB, PSI-III= Perekklisha-*Shillourokambos* Phases I–III, Kal-T= Kalavassos-*Tenta*, KhE/KhTh/KhW=Khirokitia Trench E/small sounding/Trench W, CAK= Cape Andreas-*Kastros*, DA= Dhali-*Agridhi*.

are merely artefacts of taphonomy and/or preservation.

Presence of taxa at the other aceramic neolithic sites

Table 4.1 is a summary of the taxa recorded at the two sites mentioned above and at the other aceramic sites on Cyprus (the order left to right corresponds approximately to their chronological order from earliest to latest although there is considerable overlap in dates, particularly for the Khirokitian sites) and table 4.2 summarises the presence of domestic crop species (it should be noted, however, that for some records the domestic status of the pulses and flax is uncertain).

With the exception of *Shillourokambos* (where hulled barley only is recorded for the latest phase) and Dhali-*Agridhi* (where no einkorn was identified), grains and/or chaff of the three founder crop cereals are present at all the sites and in the different trenches or areas of the sites and similarly, but again with the exception again of *Shillourokambos*, lentils (*Lens culinaris/Lens* sp.) are ubiquitous; these four founder crops were also recorded at Kholetria-*Ortos* (A. Simmons pers. comm). The earliest records of pea (*Pisum sativum*) and bitter vetch (*Vicia ervilia*) are at Kalavassos-*Tenta*; bitter vetch, grass pea (*Lathyrus sativus*) and faba bean (*Vicia faba/narbonensis*) are present at Khirokitia, and the last-named species was also identified at

Cape Andreas-*Kastros*. Flax (*Linum* sp.) is recorded at *Mylouthkia* (IA and IB) and at Cape Andreas-*Kastros* (at the latter site van Zeist identified *Linum usitatissimum/bienne*; van Zeist 1981, pp.96, 98, table 1). The five founder crops: einkorn, emmer, hulled barley, lentil and pea, are the most commonly occurring domestic species during the aceramic Neolithic on Cyprus, and four of these are present from the earliest phase, the Cypro-EPPNB, at *Mylouthkia*. Free threshing wheats were identified in one sample (out of 131) at Khirokitia (East trench; Hansen 1994), and in two samples (out of 99) at Dhali-*Agridhi* (Stewart 1974). The implications of the presence (albeit tentative) of these secondary domesticates at such an early date on Cyprus are considered below.

4.2.2 Changes in the proportions of taxa with time: the development of neolithic agriculture

Figure 4.1 presents a comparison of the numbers of domestic crops and wild taxa represented at the aceramic neolithic sites. Data from Ayios Epiktitos-*Vrysi*, a ceramic/late neolithic site (5th millennium cal BC; Kyllo 1982) in the north of the island, have been included to contrast the earlier results with those from a site at which there is evidence of a mature agricultural economy. There is an overall trend in the data; for example, the numbers of domestic taxa increase from the earliest aceramic phase to the ceramic Neolithic, with the notable exceptions of *Shillourokambos*, Dhali-*Agridhi* (see previous section) and Cape Andreas-*Kastros*[9]. This can be explained by the addition over time of imported species: free threshing wheat, rye and several of the pulses. Horwitz et al. (2004, p.38) note that the timing of the introductions of the wild animals at *Shillourokambos* (see section 4.4) appears to have been staggered and they suggest there is evidence of at least five separate importation events; it seems reasonable to propose that in the same way different crops were added to the repertoire of domestic plant taxa at the neolithic sites. It appears, therefore, that agriculture thrived once it was established on the island and that the new crops adapted well to the new environment.

The wild taxa are divided into trees or shrubs and herbaceous plants; the former category comprises many fruit-bearing species, which are common on the island today and that no doubt would have been exploited to supplement the cereal and pulse harvests in the Neolithic; the latter category includes many taxa that could be classified as weeds or 'potential' weeds, which would have been growing with the crops in the fields. As with the domestic taxa, there is a gradual rise over time in the numbers of wild herbaceous taxa, from the earliest aceramic phase at *Mylouthkia* to the ceramic Neolithic at Ayios Epiktitos-*Vrysi*, but once again with the exception of *Shillourokambos*, Dhali-*Agridhi* and Cape-Andreas-*Kastros*. We have suggested elsewhere that the relatively low taxon diversity of wild or weed species at aceramic neolithic sites could in part be explained by the fact that in the initial stages of colonisation of the island there would have been a greater investment in the maintenance of the cultivated fields to sustain stocks of grain, and that this would have been essential once the farmers and their crops had been isolated from the mainland (Colledge et al. 2004, p.47).

Our model is in concordance with the conclusions drawn from more detailed quantitative analyses that were designed to explore the development of neolithic farming systems in Central Europe, and in which the numbers of weed species found in association with crops in archaeobotanical samples were found to be lowest at the initial stages of colonization (i.e., in the early Neolithic), with a significant increase by the late Neolithic (Willerding 1986; Rösch 1998). By the later periods on Cyprus, when agriculture was fully established, there would have been cultivation of larger areas of land, with inclusion in the fields of greater number of species from previously undisturbed habitats, hence the greater diversity of wild taxa at some of the later

[9]Very little information is given for the site in the 1981 report about the recovery procedures or preservation conditions.

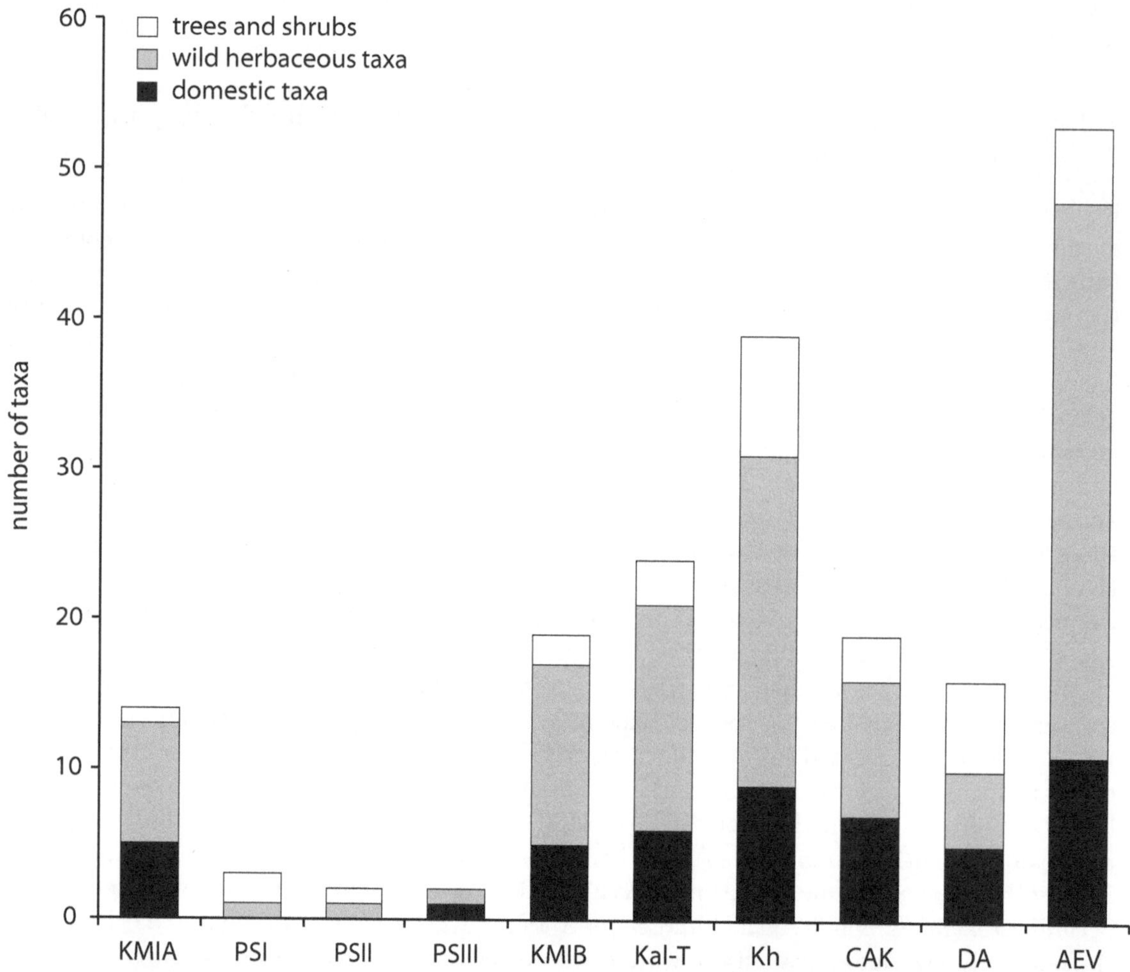

Figure 4.1: Comparison of the numbers of domestic and wild taxa identified on the aceramic neolithic sites and at one ceramic neolithic site (AEV= Ayios Epiktitos-*Vrysi*; the taxa lists from the different areas of excavation at Khirokitia (Kh) have been amalgamated; see table 4.1 for other site codes). Totals were calculated after taxa identified only to family level had been omitted from the lists and there are no other exclusions from the datasets.

aceramic and the ceramic neolithic sites. In the Central European study it was noted that the numbers of immigrant species (anthropochores) increased over time and for these it was possible to suggest origins and thus to define the direction from which neolithic farming spread into the region (Rösch 1998, p.121–122). Non-native weeds would have been transported to Cyprus with the new crops and although it is beyond the scope of this paper, a similar exploration of the provenance of introduced species (e.g., at Ayios Epiktitos-*Vrysi* there are 25 taxa not previously recorded in the earlier aceramic period) may provide evidence of links with the mainland via which essential supplies were imported.

4.3 Crete

Knossos is one of a small handful of sites that provide direct evidence for the origins and earliest stages of farming in Greece. The current consensus is that Knossos was founded by a small (ca. 50 people) community of agropastoralists from coastal region of Anatolia who island-hopped in canoes or rafts to Crete (Broodbank and Strasser 1991; Broodbank 1992; van Andel and Runnels 1995). It is the next earliest example, after the Cypriot aceramic sites, of a successful colonisation of an island by agropastoralists, with ^{14}C dates of the earliest (aceramic) level clustering around the beginning of the 7th millennium cal BC (Perlès 2001, table 5.3). Crete is well over 200 km from the southwestern Anatolian coast but because of the many intervening islands is not within visible range of the mainland, although in good conditions can be seen from the nearby island of Karpathos. Travel time from the mainland would have been perhaps in the region of four to ten days by canoe although, unlike Cyprus, the intervening islands reduce the extent, and thus the risk, of open water voyaging.

It is probably correct to describe the colonisation of Crete as the tail end of the extensive population movements of the PPNB, and thus a later part of the same migration phenomenon of the earlier migrations to Cyprus. This model is far from definitive, however, and variability in the subsistence base and forms of material culture in the earliest Greek farming communities suggest that there were several separate pulses of population movement from Southwest Asia to Southeast Europe (Colledge et al. 2004; Perlès 2001). Explaining the origins of Greek farming is thus a difficult task and requires a comparative regional approach. One body of evidence that has previously been used to examine possible origins are archaeobotantical data (e.g., Helbaek 1970; Perlès 2001, pp.74–75, 154–166), which we here review to provide additional insight into crop dispersal processes (Colledge et al. 2004).

4.3.1 Archaeobotanical evidence from an aceramic neolithic level at Knossos

At present very little has been published about the archaeobotany from Knossos; however, a report on samples (including those taken from aceramic, early, middle and late neolithic levels) from a rescue excavation undertaken in 1997 will soon be made publicly available (A. Sarpaki pers. comm.). The only results that can be described here are those from Hans Helbaek's (unpublished) analysis of charred material taken from a sounding during John Evans' earlier excavations in the 1950s and 60s. Helbaek examined a cache of grains from an aceramic neolithic level (level X) in area AC at the site (a date of 7740±130 bp [BM-436] was obtained on a sample of the grains).

Evans states that no remains of buildings were found in area AC, "but there was plentiful evidence of other activities, including threshing, corn grinding, digging pits and interring the bodies of a number of children of various ages" (Evans 1994, p.2). He suggests that this area was where the activities of the inhabitants of the aceramic settlement (located further to the south on the mound) "spilled over on the northern, and possibly the western slope" (Evans 1994, p.4). A number of pits and hollows dug into bedrock are recorded in AC, some of which contained charcoal and charred grain, and layers of ash are described, which were interpreted as the debris from small fires. Helbaek sampled where Evans records that "a field of breadwheat had apparently been threshed", this area was bounded by a row of stake-holes; the remains of one of the charred stakes were recovered and two dates were obtained: 8050±180 bp (BM-124) and 7910±140 bp (BM-278).

The list of taxa that Helbaek recorded in the sample from level X is given in table 4.3. The four founder crops of neolithic agriculture (einkorn, emmer, hulled barley and lentils) that were identified at the earliest aceramic sites on Cyprus are also present in this early context on Crete. The status of the lentils is uncertain, as was the case for the early Cypriot finds, but the presence of 350 seeds is evidence of deliberate harvesting and some authors have suggested that large numbers of pulses are indicative of the cultivation of a domestic crop (e.g., Garfinkel et al. 1988). Helbaek records 15 grains of naked barley. Grains of this variety of barley were also identified at Dhali-*Agridhi* on Cyprus and there are many finds on other aceramic sites in the Levant and Anatolia (see table 4.4). The presence at Knossos of naked barley has not aroused as much attention amongst archaeobotanists (and archaeologists) as the finds of bread wheat (*Triticum aestivum*), which are recorded in high numbers in the level X cache in area AC. Helbaek identified both grains and rachis fragments of hexaploid free-threshing wheat, and of the latter he apparently stated that they were short and that the ears would have been dense but not compact.

The presence of bread wheat on the island at this early date is intriguing; it is not one of the founder crops and so did not evolve as a result of the primary domestication events. Hexaploid free-threshing wheat is a secondary domesticate, it evolved under cultivation as a result of the crossing of already domesticated emmer with a diploid wild grass *Aegilops squarrosa* (synonym: *Aegilops tauschii*; Zohary and Hopf 2000, p.51). It would not, therefore, have been a component of the neolithic crop package that formed the foundation of the first farming communities in Southwest Asia and the eastern Mediterranean. Hexaploid free-threshing wheat could only have evolved once domestic emmer had spread to the north and east of its place of origin, to where the centre of distribution of *Aegilops squarrosa* was (e.g., in continental or temperate Central Asia); it has been suggested that the domestication took place somewhere in the southwest corner of the Caspian Belt (Zohary and Hopf 2000, p.54). The time lapse between the primary and secondary domestication events is unknown, as is the means by which the secondary domesticates became amalgamated in the early neolithic crop repertoire.

4.4 Conclusion: implications for the spread of founder crops and neolithic agriculture from Southwest Asia

The earliest domestic crops have been found on PPNA sites in the Levant and are dated to the 10th millennium cal BC (see Colledge et al. 2004). They are first recorded in the late 10th/early 9th millennium cal BC in eastern Anatolia, the mid 9th millennium cal BC on Cyprus, and the mid-late 9th millennium cal BC in central Anatolia (table 4.5). The appearance of four of the founder crops of neolithic agriculture on Cyprus at such an early date has inevitably prompted speculation about the possibility of indigenous domestication.

The debate concerning monophyletic versus polyphyletic origins of the founder crops is ongoing. Zohary is a proponent of the former and he argues from the (albeit incomplete) genetic evidence that "the development of grain agriculture in Southwest Asia was triggered (in each crop) by a single domestication event or at most by very few such events" (Zohary 1996, 1999; see also Heun et al. 1997). On the basis of the available data he doubts if the issue of whether these domestication events occurred in one location or in several locations can be resolved and is equally uncertain, therefore, whether the process by which the eight founder crops became 'packaged' as the constituent elements of neolithic agriculture can be elucidated. However, Zo-

Taxon	Common name	Plant part	No. of specimens
Hordeum sativum	domestic barley (hulled)	grains	17
		rachis frags (6-row)	3
		rachis frags (2-row)	9
Hordeum sativum	domestic barley (naked)	grains	15
Triticum dicoccum	emmer	grains	10
Triticum monococcum	einkorn	grains	10
		spikelet forks	1
Triticum monococcum/dicoccum	einkorn/emmer	spikelet forks	191
Triticum spp.	free threshing wheat	grains (hexaploid)	5400
		rachis frags (hexaploid)	3
Lens sp.	lentil		350
Avena sp.	oats	awn fragment	1
Lolium sp.	rye-grass		1
Malva sp.	mallow		52
Plantago sp.	plantain		1

Table 4.3: Helbaek's list of taxa from level X at Knossos.

Site	Period	Phase	Plant part	No. of specimens
Turkey				
Aşıklıhöyük	MPPNB	phase 2	grains	6
	MPPNB	phase 2	rachis fragments	9
Can Hasan III	LPPNB	trench 49L	grains	presence only
	LPPNB	trench 49L	rachis fragments	presence only
Hacılar	MPPNB	PPN levels	grains	presence only
	MPPNB	PPN levels	rachis fragments	1
Çatalhöyük	L/FPPNB	pre-level XIIA	grains	193
	L/FPPNB	pre-level XIIA	rachis fragments	6
	L/FPPNB	pre-level XIIB	grains	40
	L/FPPNB	pre-level XIIB	rachis fragments	9
	L/FPPNB	pre-level XIIC/D	grains	41
Syria				
Tell Abu Hureyra	L/FPPNB	level 2B	grains	6
Tell Aswad	MPPNB	level II	grains	15
Bouqras	LPPNB	levels 3/8/9	grains	374
El Kowm I	LPPNB	phase A, level IX	grains	2
	FPPNB	El Kowm 2 Caracol	grains	4
Tell Ghoraifé	LPPNB	level II	grains	3
Tell Ramad	LPPNB	level I	grains	1
	FPPNB	level II	grains	58
	FPPNB	level II	rachis fragments	1
Jordan				
Beidha	MPPNB	phase B	grains	impressions
Palestinian Territories				
Jericho	MPPNB	trench FI	grains	8
Cyprus				
Dhali-Agridhi	Khirokitian	layer III	grains	5

Table 4.4: Records of naked barley (grains and chaff) for Pre-Pottery Neolithic B (PPNB) sites in Turkey, Syria, Jordan (and the Palestinian Territories) and Cyprus. Period codes: MPPNB=middle PPNB, LPPNB=late PPNB, L/FPPNB=late/final PPNB, F=final PPNB.

Site/Region	^{14}C date	cal BC	Lab code	Context	Material	Notes
Southeast Turkey						
Cafer Höyük	9560±190	9250–8700	Ly-4436	Early phase level XII: east area layer J2a	emmer chaff; emmer/einkorn grains and chaff	samples taken from hearths 126 and 127, and burnt layer
Cafer Höyük	8990±160	8340–7830	Ly-2182	Early phase level XII: 1978 base of sondage, on virgin soil	emmer chaff; emmer/einkorn grains and chaff	samples taken from hearths 126 and 127, and burnt layer
Cafer Höyük	8950±80	8270–7970	Ly-4437	Early phase level IX: east area layer J1a, hearth 124	einkorn grains	sample 88 from hearth 124
Çayönü	9320±55	8700–8470	GrN-6243	Basal Pits sub–phase	emmer grains; emmer/einkorn chaff	samples taken from basal pits (unspecified)
Çayönü	9275±95	8630–8340	GrN-6241	Chanelled Building sub–phase: ch1–4	emmer/einkorn chaff	samples taken from chanelled building (unspecified)
Çayönü	9250±60	8570–8340	GrN-8079	Basal Pits sub–phase: hearth	emmer grains; emmer/einkorn chaff	samples taken from basal pits (unspecified)
Nevalı Çori	9261±181	8740–8280	Hd-16781-835	Level I/II	glume wheat spikelet forks (assigned 'indeterminate' category)*	samples taken from hearths and other contexts, and from the fill between stones in 'cult building'
Nevalı Çori	9243±55	8560–8340	Hd-16782-351	Level I/II	glume wheat spikelet forks (assigned 'indeterminate' category)*	samples taken from hearths and other contexts, and from the fill between stones in 'cult building'
Nevalı Çori	9212±76	8540–8310	Hd-16783-769	Level I/II	glume wheat spikelet forks (assigned 'indeterminate' category)*	samples taken from hearths and other contexts, and from the fill between stones in 'cult building'
Central Turkey						
Aşıklıhöyük	8958±130	8300–7910	P–1240	Base of site: burnt layer NW cut	einkorn grains; emmer grains; emmer/einkorn chaff	sample provenance unspecified
Aşıklıhöyük	8920±50	8240–7980	GrN-19116	Phase 2C–A: square 2J, room FF	einkorn grains; emmer grains; emmer/einkorn chaff	sample provenance unspecified
Aşıklıhöyük	8880±70	8230–7950	GrN-19865	Phase 2E/2D: area JY, dump/workshop	einkorn grains; emmer grains; emmer/einkorn chaff	sample provenance unspecified
Can Hasan III	8584±65	7660–7540	HU–11	Trench 49L, near basal levels	einkorn grains; emmer grains	samples taken from trench 49L
Can Hasan III	8543±66	7600–7525	HU–12	Trench 49L, basal levels	einkorn grains; emmer grains	samples taken from trench 49L
Can Hasan III	8470±140	7650–7320	BM–1664R	Trench 49L, sample 156F	einkorn grains; emmer grains	samples taken from trench 49L
Çatalhöyük	8240±55	7360–7140	OxA-9778	Pre XII.D	einkorn grains and chaff; emmer grains and chaff [wheat grain dated]	samples taken from level pre XII.C/D
Çatalhöyük	8160±50	7250–7060	OxA-9777	Pre XII.C	einkorn grains and chaff; emmer grains and chaff	samples taken from level pre XII.C/D
Çatalhöyük	8155±50	7250–7060	OxA-9893	Pre XII.D	einkorn grains and chaff; emmer grains and chaff [wheat grain dated]	samples taken from level pre XII.C/D
Southwest Turkey						
Hacılar	8700±180	8200–7550	BM127	level V: area Q, courtyard floor hearth	emmer grains and chaff	sample taken from ashy layer at base of aceramic levels

*Nesbitt 2002, after examination of the Nevalı Çori glume wheat chaff, stated that it was all of the domestic type, but to date full results have not been published by the primary analyst.

Table 4.5: Dates of earliest domestic cereals in Turkey (dates from canew.org).

hary's bias in favour of single (or at most very few) origins for the majority of the founder crops is substantiated by the rapid spread of agriculture, as there was thus little time for additional domestication events to occur elsewhere in Southwest Asia or Europe prior to the diffusion of domestic species from their centre of origin (Zohary 1996, p.156).

Other authors take an opposing view and argue in favour of multiple independent domestication events for individual species (Allaby e.g., 2000; Allaby and Brown e.g., 2003; Jones et al. e.g., 1998; Willcox e.g., 2002, this volume). Willcox, for example, has proposed that barley was domesticated in the southern Levant and perhaps also in the Zagros, and that domestic emmer possibly evolved both in the northern and the southern Levant (Willcox 2002). In support of his theories of multiple origins he refers to DNA data and to the geographical correlation between areas where wild progenitors grow and where there are early sites with evidence of the founder crops (Willcox 2004, s53; see also Willcox in this volume). He suggests the early finds at *Mylouthkia* and *Shillourokambos* indicate that there could have been local domestication of cereals on Cyprus (Willcox 2002, p.137). For einkorn and emmer this would only have been possible, however, if there had been earlier import of wild species[10], because their progenitors have not been recorded as being native to the island (Meikle 1985, p.1834; Murray 2003, p.63; Holmboe 1914; Christodoulou 1959). We thus conclude that there is insubstantial corroborative evidence in terms of the distributional concurrence of wild and domestic species to support the theory of indigenous domestication events. Moreover, local domestication would seem less plausible than the introduction of domestic cereal and pulse crops at a time when there was contact between neolithic sites on the mainland where agriculture was already established.

The evidence from the faunal remains in the early phases of the two earliest aceramic neolithic sites (*Mylouthkia* IA and *Shillourokambos* Early Phase A) indicates that there was introduction of several wild species (e.g., sheep, goat, fallow deer, pig, cattle and mouse; Horwitz et al. 2004, p.38, but unlike the crops, this was at a time when there were no domestic equivalents (i.e., of the species that were eventually domesticated) on the mainland (Horwitz et al. 2004). The faunal data are of relevance to discussions about the status of crop-based agriculture on Cyprus during the Cypro-EPPNB because they provide unequivocal evidence of large-scale transportation at this early date of animals, presumably with the intention of providing sustenance for long-term residence on the island. By comparison the equivalent strategy of transporting stocks of grain to ensure fresh supplies of crops would seem logistically less complex and altogether more manageable. The long tradition of links between Cyprus and the mainland would have meant that the immigrant farmers had some knowledge of the landscape and of the available natural resources (or the lack of; cf. Peltenburg who suggests that at the time there would have been "[an] impoverished island ecology" 2003a, p.95) and would have been aware of the nutritional requirements to support communities on the island.

At Knossos the remains of ovicaprines, pigs and cattle in the earliest occupation levels underscore the degree of organisation required to colonise a new landscape successfully. The transport of grain stock would have been comparatively easy, although considerable effort would have then needed to be expended to prepare new fields for first planting in the autumn.

Helbaek apparently compared the Knossos bread wheat with that from Hacılar (Helbaek 1970) and commented that they were 'the same race', thus presumably implying links between southwest Turkey and Crete. But the free-threshing grains and rachis at Hacılar were found in pottery neolithic contexts, which post-date the aceramic level where the Cretan bread wheat grains were found, and are not likely, therefore, to have been the source for the crop at Knossos. Aceramic sites in southewest Asia and the eastern Mediterranean with records of free-threshing

[10]Not as has been suggested, however, with the deliberate intention of 'domesticating' the wild species; the following statement, which implies the domestication process could have been a conscious act is misleading: "In this case, independence means that islanders took the initiative to enhance their subsistence requirements [by importing wild species] from abroad and to engage in the complex business of domestication" (Peltenburg 2003a, p.94).

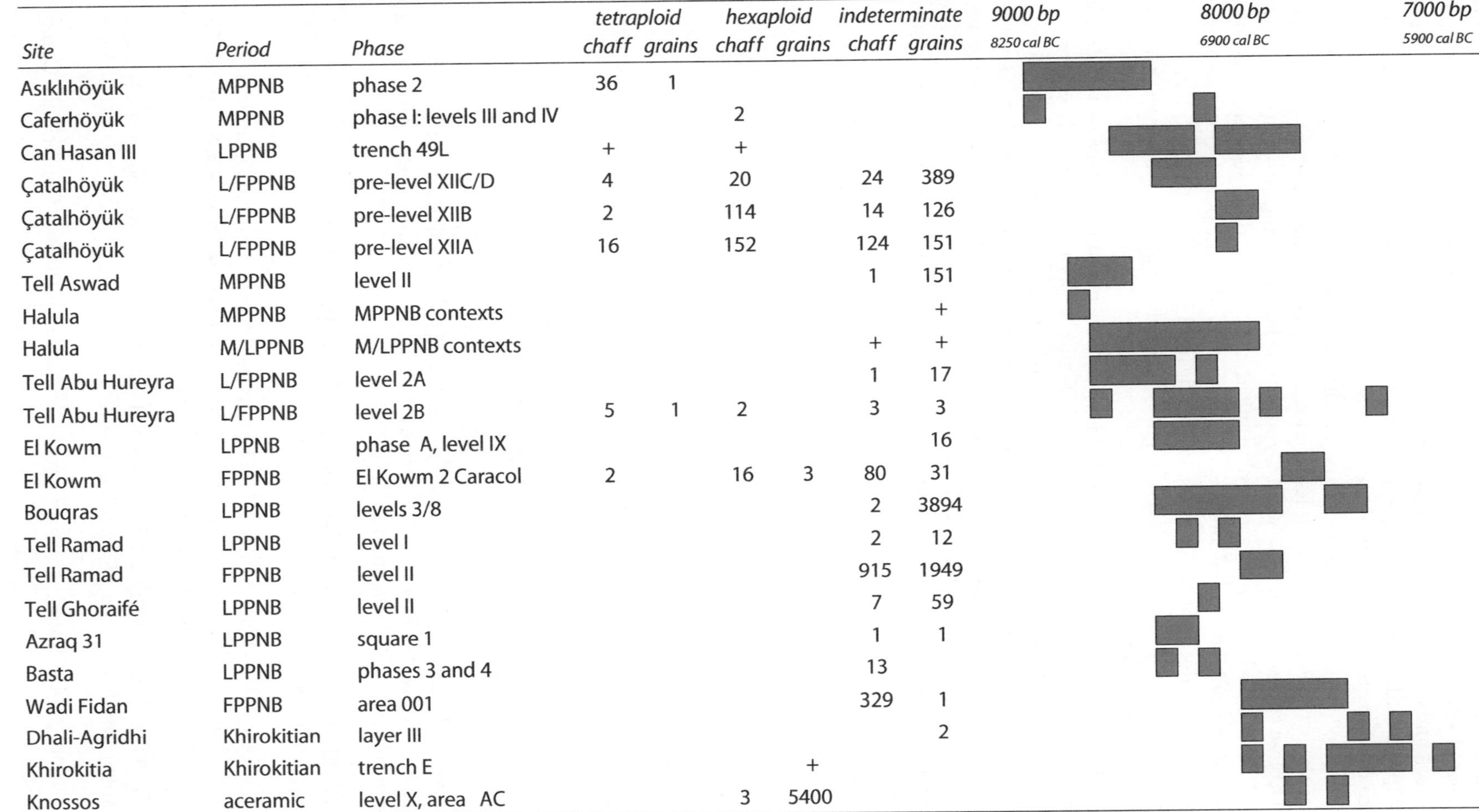

Site	Period	Phase	tetraploid chaff	tetraploid grains	hexaploid chaff	hexaploid grains	indeterminate chaff	indeterminate grains
Asıklıhöyük	MPPNB	phase 2	36	1				
Caferhöyük	MPPNB	phase I: levels III and IV			2			
Can Hasan III	LPPNB	trench 49L	+		+			
Çatalhöyük	L/FPPNB	pre-level XIIC/D	4		20		24	389
Çatalhöyük	L/FPPNB	pre-level XIIB	2		114		14	126
Çatalhöyük	L/FPPNB	pre-level XIIA	16		152		124	151
Tell Aswad	MPPNB	level II					1	151
Halula	MPPNB	MPPNB contexts						+
Halula	M/LPPNB	M/LPPNB contexts					+	+
Tell Abu Hureyra	L/FPPNB	level 2A					1	17
Tell Abu Hureyra	L/FPPNB	level 2B	5	1	2		3	3
El Kowm	LPPNB	phase A, level IX						16
El Kowm	FPPNB	El Kowm 2 Caracol	2		16	3	80	31
Bouqras	LPPNB	levels 3/8					2	3894
Tell Ramad	LPPNB	level I					2	12
Tell Ramad	FPPNB	level II					915	1949
Tell Ghoraifé	LPPNB	level II					7	59
Azraq 31	LPPNB	square 1					1	1
Basta	LPPNB	phases 3 and 4					13	
Wadi Fidan	FPPNB	area 001					329	1
Dhali-Agridhi	Khirokitian	layer III						2
Khirokitia	Khirokitian	trench E				+		
Knossos	aceramic	level X, area AC			3	5400		

Figure 4.2: Presence of free-threshing wheats on pre-pottery neolithic sites in Turkey, Syria, Jordan, Cyprus and Crete, showing dates for the phases in which the grains/chaff were identified. Numbers of grains and rachis fragments are the totals per phase.

wheat are listed in figure 4.2, and dates for the phases in which the crops were identified are also shown. Several sites in Turkey, Syria and Cyprus have evidence of bread wheat that pre-dates the Knossos finds, and other indeterminate finds of free-threshing wheats (e.g., that could represent tetraploid and/or hexaploid species) are recorded on many more sites, and so links could be inferred between Crete and all these regions.

In the same way that our previous study highlighted the similarity between the overall composition of the archaeobotanical samples from sites in the southern Levant and Cyprus, we also identified a close correspondence between Cyprus and the assemblage from aceramic Knossos (Colledge et al. 2004, p.s47). While archaeobotanical data alone does not provide sufficient information about the spheres of influence within which goods and populations were moving, our analysis nevertheless does not support the van Andel and Runnels (1995) model for a central Anatolian origin. It thus provides additional impetus to explore alternative routes, particularly coastal, for the origins of the Greek Neolithic.

In conclusion, therefore, it is clear that archaeobotanical remains do have a role to play in the interpretation of population movements of early farmers—as papers in this volume clearly demonstrate—as any crops and weeds introduced into new areas provide clues about farming practices and the origins of migrants. Although not unambiguous, such data nevertheless provides additional (and much needed) evidence to examine existing models and to explore alternatives. The use of archaeobotantical data in this way, although not novel, reinforces its importance and the need to include more fully such information when considering the processes of neolithisation.

Acknowledgments

We thank Anaya Sarpaki for making available the unpublished report by Hans Helbaek.

References

Allaby, R. (2000). Wheat domestication. In C. Renfrew and K. Boyle (Eds.), *Archaeogenetics: DNA and the Population Prehistory of Europe*, pp. 321–32. Cambridge: McDonald Institute for Archaeological Research.

Allaby, R. G. and T. A. Brown (2003). AFLP data and the origins of domesticated crops. *Genome 46*(3), 448–453.

Anderson, A. (2003). Entering uncharted waters: Models of initial colonization in Polynesia. In M. Rockman and J. Steele (Eds.), *Colonization of Unfamiliar Landscapes: The Archaeology of Adaptation*, pp. 169–189. London: Taylor and Francis.

Anthony, D. W. (1990). Migration in archaeology: the baby and the bathwater. *American Anthropologist 92*(4), 895–914.

Bar-Yosef, O. and A. Belfer-Cohen (1989). The Levantine 'PPNB' interaction sphere. In I. Hershkovitz (Ed.), *People and Culture in Change*, BAR International Series 508, pp. 59–72. Oxford: British Archaeological Reports.

Broodbank, C. (1992). The neolithic labyrinth: social change at Knossos before the Bronze Age. *Journal of Mediterranean Archaeology 5*(1), 39–75.

Broodbank, C. (2001). *An Island Archaeology of the Early Cyclades.* Cambridge: Cambridge University Press.

Broodbank, C. and T. Strasser (1991). Migrant farmers and the neolithic colonization of Crete. *Antiquity 65*, 233–245.

Burmeister, S. (2000). Archaeology and migration. *Current Anthropology 41*(4), 539–567.

Christodoulou, D. (1959). *The Evolution of Rural Land Use Patterns in Cyprus.* Bude: Geographical Publications Limited.

Colledge, S. (2004). Reappraisal of the archaeobotanical evidence for the emergence and dispersal of the 'founder crops'. In E. Peltenburg and A. Wasse (Eds.), *Neolithic Revolution: New Perspectives on Southwest Asia in Light of Recent Discoveries on Cyprus*, Levant Supplemantary Series I, pp. 49–60. Oxford: Oxbow Books.

Colledge, S., J. Conolly, and S. Shennan (2004). Archaeobotanical evidence for the spread of farming in the eastern Mediterranean. *Current Anthropology 45*, S35–S58.

Cucchi, T., J.-D. Vigne, J.-C. Auffray, P. Croft, and E. Peltenburg (2002). Introduction involontaire de la souris domistique (*Mus musculus domesticus*) à Chypre dès le Néolithique précéramique ancien (fin IXe et VIIIe millénaires av. J.-C.). *Compte Rendus Palevol 1*, 235–241.

Evans, J. D. (1994). The early millennia: continuity and change in a farming settlement. In D. Evely, H. Hughes-Brock, and N. Momigliano (Eds.), *Knossos: A Labyrinth of History. Papers Presented in Honour of Sinclair Hood*, Chapter 1, pp. 1–20. London: The British School at Athens.

Galili, E., A. Gopher, B. Rosern, and L. K. Horwitz (2004). The emergence of the Mediterranean fishing village in the Levant and the anomaly of neolithic Cyprus. In E. Peltenburg and A. Wasse (Eds.), *Neolithic Revolution: New Perspectives on Southwest Asia in Light of Recent Discoveries on Cyprus*, Levant Supplementary Series I, pp. 91–102. Oxford: Oxbow Books.

Garfinkel, Y., M. E. Kislev, and D. Zohary (1988). Lentil in the Pre-Pottery Neolithic B Yiftah'el: Additional evidence of its early domestication. *Israel Journal of Botany 37*, 49–51.

Goldstein, D. B. and L. Chikhi (2002). Human migrations and population structure: what we know and why it matters. *Annual Review of Genomics and Human Genetics 3*, 129–152.

Gopher, A. (1989). Diffusion process in the Pre-pottery Neolithic Levant: The case of the Helwan Point. In I. Hershkovitz (Ed.), *People and Culture in Change*, BAR International Series 508, pp. 91–105. Oxford: British Archaeological Reports.

Guilaine, J. (2003). Parekklisha-*Shillourokambos* périodisation et aménagements domestiques. In J. Guilaine and A. Le Brun (Eds.), *Le Néolithique de Chypre. Actes du Colloque International Organisé par le Départment des Antiquités de Chypre et l'Ecole Française d'Athènes, Nicosie 17-18 Mai 2001*, Bulletin de Correspondance Hellénique, pp. 3–14. Athènes: École Française d'Athènes.

Guilaine, J., F. Briois, J.-D. Vigne, and I. Carrère (2000). Découverte d'un Néolithique précéramique ancien chypriote (fin 9e, début 8e millénaires cal. BC), apparenté au PPNB ancien/moyen du Levant nord. *Earth and Planetary Sciences 330*, 75–82.

Hansen, J. (1978). Botanical remains, I.A. Todd, ed. "Vasilikos Valley Project: Second Preliminary Report, 1977". *Journal of Field Archaeology 5*, 161–195.

Hansen, J. (1989). Khirokitia plant remains preliminary report (1980–1981, 1983). In A. Le Brun (Ed.), *Fouilles récentes à Khirokitia (Chypre) 1983–1986*, pp. 235–250. Paris: Éditions Recherche sur les Civilisations.

Hansen, J. (1994). Khirokitia plant remains preliminary report. In A. Le Brun (Ed.), *Fouilles récentes à Khirokitia (Chypre) 1988–1991*, pp. 393–409. Paris: Éditions Recherche sur les Civilisations.

Härke, H. (1998). Archaeologists and migrations: a problem of attitude? *Current Anthropology 39*(1), 19–44.

Helbaek, H. (1969). Plant collecting, dry farming, and irrigation agriculture in prehistoric Deh Luran. In K. V. Flannery and J. A. Neely (Eds.), *Prehistory and Human Ecology of the Deh Luran Plain*, pp. 383–426. Ann Arbor: University of Michigan.

Helbaek, H. (1970). The plant husbandry of Hacilar. In J. Mellaart (Ed.), *Excavations at Hacilar*, pp. 189–244. Edinburgh: Edinburgh University Press.

Heun, M., R. Schäfer-Pregl, D. Klawan, R. Castagna, M. Accerbi, B. Borghi, and F. Salamini (1997). Site of einkorn wheat domestication identified by DNA fingerprinting. *Science 278*(5341), 1312–1314.

Hillman, G. C. (1989). Late Palaeolithic plant foods from Wadi Kubbaniya in Upper Egypt: dietary diversity, infant weaning, and seasonality in a riverine environment. In D. R. Harris and G. C. Hillman (Eds.), *Foraging and Farming: The Evolution of Plant Exploitation*, pp. 207–239. London: Unwin Hyman.

Holmboe, J. (1914). *Studies on the Vegetation of Cyprus*. Bergen: John Griegs Boktrykkeri.

Horwitz, L. K., E. Tchernov, and H. Hongo (2004). The domestic status of the early neolithic fauna of Cyprus: a view from the mainland. In E. J. Peltenburg (Ed.), *Neolithic Revolution: New Perspectives on Southwest Asia in Light of Recent Discoveries on Cyprus*, pp. 35–48. Oxford: Oxbow Books.

Jones, M., R. G. Allaby, and T. A. Brown (1998). Wheat domestication. *Science 279*, 302–303.

Kyllo, M. (1982). The botanical remains. In E. J. Peltenburg (Ed.), *Vrysi. A Subterranean Settlement in Cyprus. Excavations at Prehistoric Ayios Epiktitos Vrysi 1969-1973*, pp. 90–93. Warminster: Aris & Phillips Ltd.

Lehavy, Y. M. (1974). Excavations at neolithic Dhali-Agridhi. In L. E. Stager, A. Walker, and G. E. Wright (Eds.), *American Expedition to Idalion, Cyprus. First Preliminary Report: Seasons of 1971 and 1972.*, pp. 94–103. Massachusetts: The American School of Oriental Research.

McCartney, C. (2004). Cypriot neolithic chipped stone industries and the progress of regionalization. In E. Peltenburg and A. Wasse (Eds.), *Neolithic Revolution: New Perspectives on Southwest Asia in Light of Recent Discoveries on Cyprus*, Levant Supplemantary Series I, pp. 103–122. Oxford: Oxbow Books.

McCartney, C. (2005). Chipped stone. In I. A. Todd (Ed.), *Vasilikos Valley Project 7: Excavations at Kalavassos-Tenta*, Volume II of *Studies in Mediterranean Archaeology 71/7*, pp. 176–228. Sävedalen: Paul Åströms Verlag.

McCartney, C. (2005). Preliminary report on the re-survey of three early neolithic sites in Cyprus. *Report of Department of Antiquities, Cyprus 2005*, 1–21.

Meikle, R. D. (1985). *Flora of Cyprus*, Volume 2. London: The Bentham-Moxon Trust, Royal Botanic Gardens, Kew.

Miller, N. F. (1984). Some plant remains from Khirokitia, Cyprus: 1977 and 1978 excavations. In A. Le Brun (Ed.), *Fouilles récentes à Khirokitia (Chypre) 1977–1981*, pp. 183–188. Paris: Éditions Recherche sur les Civilisations.

Murray, M. A. (2003). The plant remains. In E. Peltenburg (Ed.), *The Colonisation and Settlment of Cyprus. Investigations at Kissonerga-Mylouthkia, 1976–1996. Lemba Archaeological Project, Cyprus III.1*, Studies in Mediterranean Archaeology 70/4, pp. 59–71. Sävedalen: Åströms Förlag.

Nesbitt, M. (2002). When and where did domesticated cereals first occur in Southwest Asia? In R. T. J. Cappers and S. Bottema (Eds.), *The Dawn of Farming in the Near East*, Studies in early Near Eastern Production, Subsistence and Environment 6/1999, pp. 113–132. Berlin: Ex oriente.

Peltenburg, E. (2003a). Conclusions: Mylouthkia 1 and the early colonists of Cyprus. In E. Peltenburg (Ed.), *The Colonisation and Settlement of Cyprus. Investigations at Kissonerga-Mylouthkia, 1976-1996*, Lemba Archaeological Project, Cyprus III.1/Studies in Mediterranean Archaeology LXX:4, pp. 83–103. Sävedalen: Paul Åströms Verlag.

Peltenburg, E. (2003b). Identifying settlement of the X–IXth millennium BP in Cyprus from the contents of Kissonerga-*Mylouthkia* wells. In J. Guilaine and A. Le Brun (Eds.), *Le Néolithique de Chypre. Actes du Colloque International Organisé par le Départment des Antiquités de Chypre et l'Ecole Française d'Athènes, Nicosie 17-18 Mai 2001*, Bulletin de Correspondance Hellénique, pp. 15–33. Athènes: École Française d'Athènes.

Peltenburg, E. (2004). Introduction: a revised Cypriot prehistory and some implications for the study of the Neolithic. In E. Peltenburg and A. Wasse (Eds.), *Neolithic Revolution: New Perspectives on Southwest Asia in Light of Recent Discoveries on Cyprus*, Levant Supplemantary Series I, pp. xi–xx. Oxford: Oxbow Books.

Peltenburg, E., S. Colledge, P. Croft, A. Jackson, C. McCartney, and M. Murray (2001). Neolithic dispersals from the Levantine Corridor: a Mediterranean perspective. *Levant 33*, 35–64.

Peltenburg, E., S. Colledge, P. Croft, A. Jackson, C. McCartney, and M. A. Murray (2000). Agropastoral colonization of Cyprus in the 10th millennium BP: initial assessments. *Antiquity 74*, 844–853.

Perlès, C. (2001). *The Early Neolithic of Greece.* Cambridge: Cambridge University Press.

Rockman, M. (2003). Knowledge and learning in the archaeology of colonization. In M. Rockman and J. Steele (Eds.), *Colonization of Unfamiliar Landscapes: The Archaeology of Adaptation*, pp. 3–24. London: Taylor and Francis.

Rösch, M. (1998). The history of crops and crop weeds in south-western Germany from the Neolithic period to modern times, as shown by archaeobotanical evidence. *Vegetational History and Archaeobotany 7*, 109–125.

Schwartz, D. (1970). The postmigration culture: a base of archaeological inference. In W. Longacre (Ed.), *Reconstructing Prehistoric Pueblo Societies*, pp. 175–193. Albuquerque: University of New Mexico Press.

Simmons, A. (2001). The first humans and last pigmy hippopotami of Cyprus. In S. Swiny (Ed.), *The Earliest Prehistory of Cyprus: From Colonization to Exploitation*, Cyprus American Archaeological Research Institute Monography Series 12, pp. 1–18. Boston: American School of Oriental Research.

Simmons, A. (2004). Bitter hippos of Cyprus: the island's first occupants and the last endemic animals—setting the stage for colonisation. In E. Peltenburg and A. Wasse (Eds.), *Neolithic Revolution: New Perspectives on Southwest Asia in Light of Recent Discoveries on Cyprus*, Levant Supplemantary Series I, pp. 1–14. Oxford: Oxbow Books.

Stewart, R. T. (1974). Palaeobotanic investigation: 1972 season. In L. E. Stager, A. Walker, and G. E. Wright (Eds.), *American Expedition to Idalion, Cyprus. First Preliminary Report: Seasons of 1971 and 1972*, pp. 123–129. Massachusetts: The American School of Oriental Research.

van Andel, T. and C. Runnels (1995). The earliest farmers in Europe. *Antiquity 69*, 481–500.

van Zeist, W. (1981). Plant remains from Cape Andreas-Kastros (Cyprus). In *Un site néolithique précéramique en Chypre: Cap Andreas Kastros*, Recherche sur les grandes civilisations. Memoirs 5, pp. 95–99. Paris: Éditions ADPF.

Waines, J. G. and N. P. Stanley-Price (1977). Plant remains from Khirokitia in Cyprus. *Paléorient 3*, 281–284.

Willcox, G. (2001). Présence des céréales dans le Néolithique précéramique de Shillourokambos à Chypre: résultats de la campagne 1999. *Paléorient 26*(1), 129–135.

Willcox, G. (2002). Geographical variation in major cereal components and evidence for independent domestication events in the Western Asia. In R. T. J. Cappers and S. Bottema (Eds.), *The Dawn of Farming in the Near East*, pp. 133–140. Berlin: Ex oriente.

Willcox, G. (2003). The origins of Cypriot farming. In J. Guilaine and A. Le Brun (Eds.), *Le Néolithique de Chypre. Actes du Colloque International Organisé par le Départment des Antiquités de Chypre et l'Ecole Française d'Athènes, Nicosie 17-18 Mai 2001*, Bulletin de Correspondance Hellénique, pp. 231–238. Athènes: École Française d'Athènes.

Willerding, U. (1986). *Zur Geschichte der Unkräuter Mitteleuropas.* Gttinger Schriften zur Vor-und Frühgeschichte, Band 22. Neumunster: Karl Wachholtz.

Zohary, D. (1992). Domestication of the neolithic Near Eastern crop asssemblage. In P. Anderson-Gerfaud (Ed.), *Préhistoire de l'agriculture: nouvelles approches expérimentales et ethnographiques*, Monographie du CRA 6, pp. 81–86. Paris: C.N.R.S.

Zohary, D. (1996). The mode of domestication of the founder crops of Southwest Asian agriculture. In D. R. Harris (Ed.), *The Origins and Spread of Agriculture and Pastoralism in Eurasia*, pp. 142–158. London: UCL Press.

Zohary, D. (1999). Monophyletic vs. polyphyletic origin of the crops on which agriculture was founded in the Near East. *Genetic Resources and Crop Evolution 46*(2), 133–142.

Zohary, D. and M. Hopf (2000). *Domestication of Plants in the Old World: The Origin and Spread of Cultivated Plants in West Asia, Europe and the Nile Valley* (3rd ed.). Oxford: Oxford University Press.

Chapter 5

Transitions to agriculture in the Aegean: the archaeobotanical evidence

Soultana-Maria Valamoti and Kostas Kotsakis
Aristotle University of Thessaloniki, Greece

5.1 Introduction

The exploitation of plants for food production as part of an emerging neolithic way of life at the onset of the Holocene in the Aegean is subject to multiple definitions. This makes the investigation of the processes leading to the adoption of agriculture in the region a challenging subject that merits closer study. We here present a brief review of the ideas put forward in relation to the appearance of agriculture in the Aegean, with an emphasis on the available archaeobotanical data and the ways they have been interpreted by various researchers. Our point is that transitions to agriculture in the Aegean, rather than representing a package of domesticates, permanent settlements and 'new gods' (Cauvin 2000), whether established because of an influx of newcomers or not, represented a continuous process during which the elements that constituted what we call 'neolithic' were subject to negotiation, definition and redefinition through daily practice. Plants formed part of this process and the archaeobotanical record from Greece is also subject to alternative interpretations.

5.2 Background to the Mesolithic/Neolithic transition

The archaeological record informs us that, at the onset of the Holocene on mainland Greece and in the Aegean islands, human settlement existed both in caves and open sites on the coast and also inland (Jacobsen 1976; Kyparissi-Apostolika 2000; Perlès 2001; Runnels 2001; Galanidou and Perlès 2003) (figure 5.1). In addition, it informs us that wherever the tool industry has been studied it is exceptional and different to contemporary industries in other parts of Europe or the Near East. Mesolithic assemblages are characterised by end-scrapers, notches and denticulates of the same form and manufacture (flaking method) as their Upper Palaeolithic counterparts (Perlès 2001; Perlès 2003, p.80). It has also been suggested that the Greek Mesolithic lithics resemble those from the east Mediterranean (Runnels 2001, p.252). Subsistence evidence shows that people were gathering plants (Hansen 1991), hunting (Payne 1982), and using marine resources (Shackleton 1988). Towards the beginning of the 7th millenium BC in Thessaly and other parts of Greece, e.g., Crete and the Peloponnese, 'permanent' villages were established

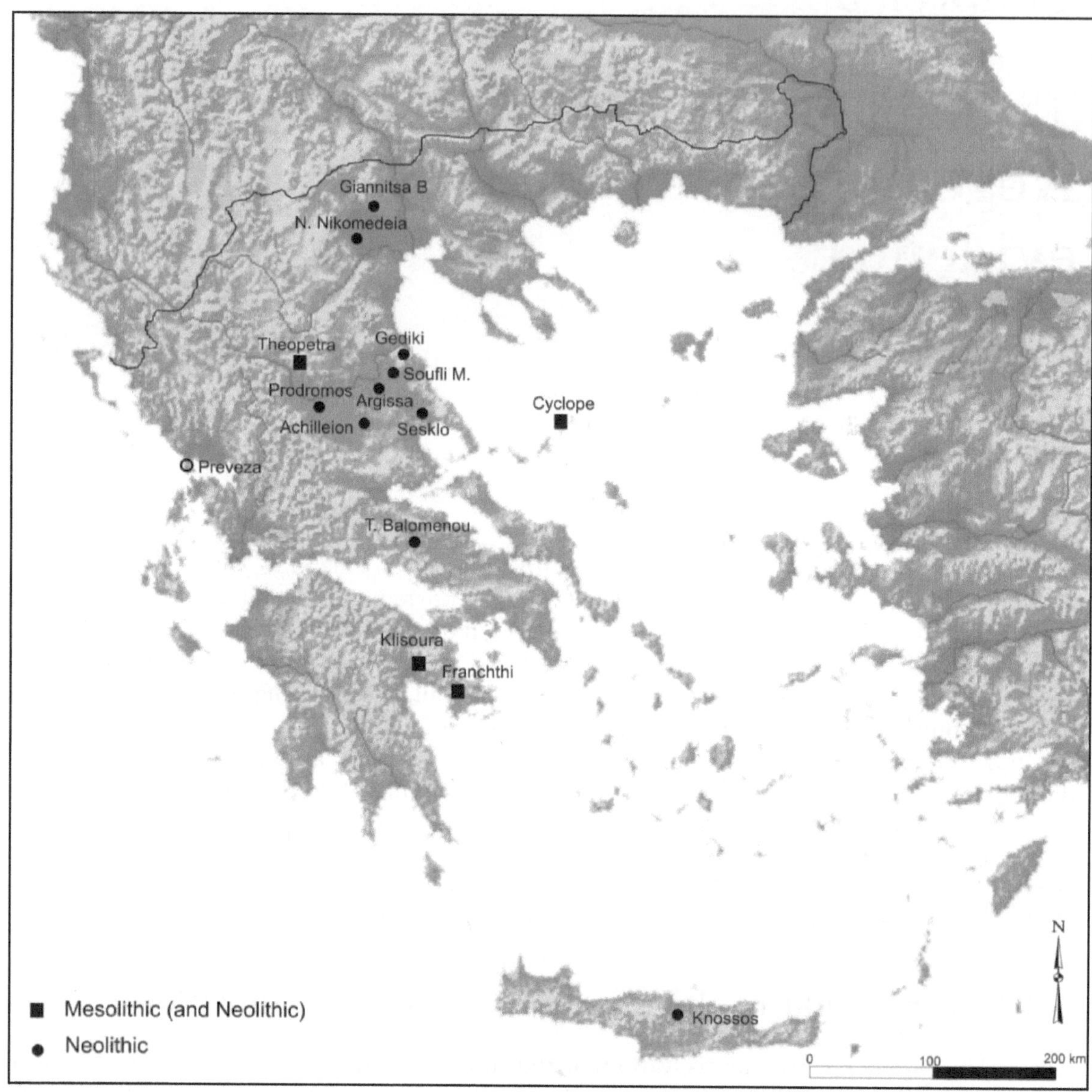

Figure 5.1: Map of Greece with mesolithic and early neolithic sites mentioned in the text (after Perlès 2001; Runnels 2001; Galanidou and Perlès 2003).

whose inhabitants had a different tool industry (characterised by blades; Perlès 2001, 2003), and who made pots, cultivated plants and kept animals, with apparently minimal reliance on hunting wild game. These very general descriptions form the basis on which various views about the causes underlying the transition from the Mesolithic to the Neolithic have been elaborated.

The flimsy nature of the mesolithic archaeological record, which appears to be specific to Greece, has led some to argue for a sparse (to the point of being almost invisible) mesolithic culture, such that "large areas were uninhabited during the Mesolithic" (Hansen 1992, p.242), subsequent to which neolithic groups migrated to Greece from the coast of Asia Minor (Runnels 2001, p.251; Runnels 2003, p.123, 128). The poor representation of anthropogenic deposits during the millennia following the onset of the Holocene in the region, may partly be due to inadequate research. Recent finds from Theopetra cave (Kyparissi-Apostolika 2000, 2003), Yioura (Sampson et al. 2003), Klissoura (Koumouzelis et al. 2003) and the Preveza region in Epirus (Runnels 2001, 2003) have collectively increased the evidence for mesolithic presence. Moreover, as already suggested (e.g., Chapman 1994), the apparent near-absence of evidence of human settlement may well be the outcome of considerable changes of the relief caused by erosion and sea level changes. The nature of the mesolithic background has played an important part in the formation of ideas concerning the origins of the Greek Neolithic and thus of agriculture in Greece, as the two have been viewed as closely interwoven. In examining these ideas on the issue of origins, we see that the emphasis has been placed on different elements of this process and various interpretations of the data.

Several archaeologists acknowledge the presence of a mesolithic population, and stress the contribution of indigenous elements in the process of cultural change; their ideas can range from those that view the Neolithic as a complex process in which the indigenous elements played a significant, but not necessarily independent part (Theocharis 1967, 1973; Dennell 1983; Barker 1985; Kotsakis 1992, 2000, 2001, 2003; Seferiadis 1993; Halstead 1996), to those in which the Neolithic is seen as a local development (Kyparissi-Apostolika 2000, 2003). Others view the emergence of neolithic societies in Greece as the adoption or imposition of a way of life that originated elsewhere in a fully fledged form, either through colonization (Ammerman and Cavalli-Sforza 1971, 1973, 1984; Hansen 1992; Perlès 2001; Runnels 2003) or by means of acculturation, or by a combination of both (Dennell 1992; Runnels and van Andel 1988). The important question in the debate seems to be why the transformation of the Mesolithic, otherwise termed neolithisation, took place in this region; i.e., what were the causes and by what processes did a new way of life become established? The key gaps in the discussion will remain open until substantial data become available and, quite clearly, the need to investigate the mesolithic background of the region is imperative (Efstratiou 1996; Kotsakis 2003).

5.3 Neolithic Agriculture in the Aegean

5.3.1 Definitions

The exploitation of plants as a basic component of food-production is a key element of our definitions of the Neolithic. The term 'agriculture' refers unambiguously to the cultivation of domesticated plants (e.g., Harris 1989, p.19), however, the forms that people-plant relationships can take have been defined in various ways (e.g., Jarman 1972; Jarman et al. 1982; Harris 1989, 1996). Harris (1996) summarises the transitional stages from the gathering of wild plants to the cultivation of domestic plants as follows:

- Harvesting of wild plants: This can range from gathering from wild stands to tending the wild stands, with the latter relationship is referred to as plant husbandry (Jarman 1972) or plant nursing (Hastorf 1998).

- Cultivation of wild plants: This involves the gathering and subsequent sowing of the seeds of morphologically wild plants with small or larger scale disturbance of the cultivated land (e.g., clearance, tillage). In this situation people become involved in the reproduction of the plant without altering its genetic characteristics and physiology.

- Cultivation of domesticated plants—namely, agriculture: In this relationship the plant has become dependent on people for its reproduction, and its physiological, phenotypic and genetic characteristics are transformed accordingly (Harlan 1992).

It becomes clear from the above categories that agriculture is the outcome (but not a necessary or obligatory one) of a long-term process of people-plant interaction and, as ethnographic examples show, this represents a continuum of diverging, rather than a specific set of practices (Harris 1989; Rindos 1984; Hastorf 1998). Moreover, the crucial change in the people-plant relationship is from a reliance on naturally occurring wild stands to controlled reproduction in the wild (i.e., by saving and sowing seeds), and this practice could have also taken place in areas where the wild plants were not locally abundant (Harlan and Zohary 1966, p.1079; Jones et al. 1998). The traits signifying domestication are a by-product of this relationship, and although of crucial importance in generating new species with great potential in terms of yield, the morphological traits are probably more important to us for pin-pointing early agriculture than they were for those pioneers experimenting with the cultivation of wild stands: they were essentially an unforeseen outcome of long-term plant-people interaction (cf. Bender 1978, p.203). The archaeobotanical evidence from the Near East suggests that during the course of the Neolithic this plant-people relationship underwent different phases, which is also supported by ethnographic and experimental observations (Harris 1996; Hillman and Davies 1992; Kislev 1989b; Willcox 2000).

Archaeological recovery of morphologically wild plants could signify the harvesting of either wild or cultivated stands of wild plants. Domesticated plants are not required for defining a 'neolithic' way of life as demonstrated, for example, by the lack of domesticates in Levantine Pre-Pottery Neolithic A contexts (Willcox 2000; Kislev 1992; Nesbitt 2002). Moreover, no archaeobotanical evidence exists for plants that exhibit morphological features that are transitional between wild and domesticated that would indicate they were under 'incipient' domestication (Hillman and Davies 1992; Hansen 1992, p.238; however, see Willcox, this volume, for evidence of grain size increase in wild cereals). One other line of evidence would be a co-existence of wild and domesticated plants in the same assemblage in proportions different to those characteristic of natural wild stands (Hillman and Davies 1992, pp.144–145, 151; Nesbitt 2002, p.117). The ecology of the wild flora accompanying plants that were later domesticated provides an alternative route to the investigation of people-plant relationships (though in itself not devoid of problems) as the case study of Abu Hureyra has demonstrated (Hillman et al. 1989; Hillman 2003).

5.3.2 A review of the available archaeobotanical data from Greece

Neolithic

The practice of agriculture as a means of food production is evident in the Aegean at the end of the 8th millenium cal BC (Perlès 2001, pp.90–92). Plant remains from deposits dated to the earliest Neolithic, termed by some authorities as aceramic, come mainly from five sites in Thessaly: Gedhiki, Sesklo, Achilleion, Soufli Magoula and Argissa (Milojcic et al. 1962; Renfrew 1966; Kroll 1983). Unfortunately the number of samples with seeds of cultivated plants from these sites is small (out of a total of four samples: one sample/71 seeds from Ghediki, one sample/seven seeds from Achilleion and two samples/10 seeds from Sesklo; Renfrew 1966), and the data are difficult to evaluate from summary tables, a problem that has also been encountered for early neolithic finds (e.g., Kroll 1983; Valamoti 1995). There is also evidence available for this period from Franchthi (Hansen 1991, 1992), and Knossos (identified by Helbaek in Evans 1964; recent finds from Knossos are currently being studied by Anaya Sarpaki). The species identified from these earliest levels are einkorn, emmer, free-threshing wheat, hulled and naked two-row barley, and hulled six-row barley.

There is more evidence from the ceramic Neolithic but the archaeobotanical data set is still relatively poor: only 10 sites have published results, a majority of which are not quantifiable. The body of available data comprise samples from Franchthi (Hansen 1991), Toumba Balomenou (Sarpaki 1995), Sesklo, Otzaki, Argissa, Achilleion, Soufli (Milojcic et al. 1962; Renfrew 1966; Kroll 1981, 1983), Prodromos (Halstead and Jones 1980), Nea Nikomedia (van Zeist and Bottema 1971) and Giannitsa B (Valamoti 1995). Remains dated to the 7th millennium cal BC comprise a variety of cultivated cereals, including einkorn, emmer, a free-threshing wheat (for which the ploidy level has not yet been determined) and both two-row and six-row barley, and pulses, including lentils, bitter vetch, pea, grass pea and chickpea (table 5.1). There were few samples for the vast majority of sites (e.g., one to 10 samples for the Thessalian sites) and these were relatively poor in plant remains (Franchthi, N. Nikomedeia and Toumba Balomenou having both the greatest numbers of samples and seeds). It is evident, therefore, that the picture obtained from the archaeobotanical record is very fragmentary. Thus, although the data set is adequate for establishing that the cultivation of domesticated plants was part of early neolithic life in Greece, it is too small to permit the investigation of either the co-occurrence and dominance of different species at each site or the full range of people-plant relationships.

Mesolithic

The evidence for what preceded the Neolithic is even scarcer. Despite the key position Greece holds in the debate of the 'hows' and 'whys' agriculture emerged in Europe, the archaeobotanical evidence is limited. Two of the 13 mesolithic sites on the map of Greece (Runnels 2001, p.247; Runnels and van Andel 2003, pp.117–123; Galanidou and Perlès 2003, p.30) have yielded archaeobotanical remains, Franchthi and Theopetra, and published remains are only available from the former (Hansen 1991, 1992). The archaeobotanical finds presented in the Theopetra preliminary publication (Mangafa 2000) should be treated with caution because six-row barley is listed among the mesolithic finds, whereas there is no evidence for wild or two-row barley in the preceding finds. Wild einkorn is also identified; but on the basis of the presence of grains, unlike the spikelet forks, these do not provide a reliable basis for the distinction between wild and domesticated forms (Nesbitt 2002, p.116). Moreover, no quantitative data have yet been made available, which renders the interpretation of the results difficult for the time being. In addition, the presence of a single specimen of a spikelet fork of *Triticum dicoccum* among the Upper Palaeolithic layers should be considered as intrusive, given the absence of *Triticum dicoccoides* from the upper palaeolithic and mesolithic finds at the site, as well as from the flora of Greece.

On the other hand, there is evidence for the use of wild barley, lentils and oats, as well as fruits and nuts, during the Mesolithic at Franchthi. The near absence of stone mortars and pestles at the site, as well as the absence of blades, all of which are implements associated with cultivation and plant processing and that occur in abundance in Near Eastern sites from the Natufian onwards, has been used as an argument against a possible formation of a close people-plant relationship (Perlès 2001). The only relevant evidence consists of a small number of quern fragments (Runnels 2001, p.247) and a flake with use-wear traces, which suggests its use for cutting fresh soft plant stems that could include grasses (Vaughn 1990). Such differences in artefact assemblages between populations in the Near East and Greece are characteristic of the whole range of material culture and not only that relevant to dealing with plants. The archaeobotanical finds from Franchthi suggest that food was acquired partly through gathering

Taxon	Franchthi (Meso)	Franchthi (Neo)	Gediki	Achilleion	Argissa	Soufli	Sesklo	Otzaki	Prodromos	Nea Nikomedeia	Giannitsa B	Knossos	Toumba Balomenou
T. monococcum	.	D	D	.	D	D	D	D	D	D	D	.	D
T. dicoccum	.	D	D	D	D	D	D	D	D	D	D	D	.
Triticum sp. 'new type'	.	.	.	.	.	.	.	.	.	.	D	.	.
T. aestivum/durum	.	.	.	.	.	.	.	D	.	.	D	D	.
H. vulgare	W	D	D	.	D	D	D	D	D	D	D	D	D
Avena sp.	W	.	.	?	.	.	.	.	.	.	.	.	?
L. culinaris	W	D	D	.	D	D	D	D	D	D	D	.	D
V. ervilia	.	.	.	.	D	.	D	D	.	D	.	.	D
Vicia sp.	.	.	D	.	.	.	.	.	.	.	.	.	.
P. sativum	.	.	D	.	D	D	D	D	D	D	.	.	D
L. sativus	.	.	.	.	.	.	.	.	D	.	.	.	D
C. arietinum	.	.	.	.	.	.	.	.	.	.	.	.	.

Table 5.1: Cereal and pulse species present at mesolithic and early neolithic sites in Greece (data from Theopetra not included due to incomplete study; millet from Argissa not included because it is likely to be intrusive). Pulses from neolithic levels are assumed to be domesticated. Table based on: Milojcic et al. (1962); Evans (1964); Renfrew (1966); van Zeist and Bottema (1971); Halstead and Jones (1980); Kroll (1983); Hansen (1991); Sarpaki (1995); Valamoti (1995). W: wild, D: domesticated, ? unknown status.

wild plants and that some form of people-plant interaction had been initiated. However, the question of how to define the nature of the relationship represented by the archaeobotanical material at Franchthi is unresolved and it is unclear whether the plants were harvested from wild stands or from cultivated plots of morphologically wild plants. This issue is addressed by Hansen and remains open as the evidence from the plant morphology is equivocal (Hansen 1992, p.258); nevertheless she draws the conclusion that there is no indication in the archaeobotanical record for plant cultivation (Hansen 1992, p.241).

5.3.3 From the Mesolithic to the Neolithic: the archaeobotanical data revisited

On the basis of the available evidence, therefore, the transition from mesolithic plant exploitation to neolithic agriculture can only be studied at Franchthi. After several meters of deposits succeeding the mesolithic levels (recently recognised as representing a hiatus of 600 years; Farrand 2003), which are characterised by a significant reduction in plant remains (Hansen 1992, p.241), the wild barley, lentils and oats found in abundance in the Mesolithic are completely replaced in the early Neolithic by emmer, barley and lentils and later by einkorn and bitter vetch. Thus, there are records in the later period of the domestic equivalents of some species that were harvested earlier from the wild. Although Hansen considers the evidence for the local domestication or introduction of lentil *per se* as inconclusive (1991, p.59), she interprets the neolithic finds as having been imported from elsewhere (i.e., the Near East) as fully domesticated crops. The main arguments for non-local domestication of plants that were locally available in the wild during the Mesolithic are: (a) the indications of an abandonment of the cave prior to the Neolithic, and; (b) the 'sudden' appearance of fully domesticated crops (together with a different material culture) in the Neolithic (Hansen 1992). For plants that were not locally available in the wild (for example, emmer, for which the wild progenitor is not reported from Greece), their introduction from elsewhere (namely from the regions where today they grow, and presumably in antiquity also grew, in the wild) needs no further justification. Although it is obvious that neolithic agriculture in Greece involved the cultivation of plants that were both locally and non-locally available in a wild form during the Mesolithic, it does not necessarily follow that all the domesticated crops were introduced from elsewhere as part of a fully-fledged agricultural system. This is one interpretation of the currently available archaeobotanical data set, which is greatly influenced by the concept of a 'neolithic package'.

Consideration of the modern-day geographical distributions of the wild progenitor species has been important in addressing the origins of agriculture, as these are regarded as the areas where the domestication events occurred. A lot of emphasis has been placed on the importance of emmer, for example, the restricted distribution of the wild species and its monophyletic domestication, hence its spread to other parts of Asia and to Europe (Zohary 1996). Recent DNA studies of modern wild and domesticated einkorn populations from an area covering the Balkans, Anatolia, and the Near and Middle East suggest that it was probably domesticated in the area of the Karacadağ mountains (Heun et al. 1997), and that the wild populations currently reported from Greece (Zamanis et al. 1988) are feral, their closest link being with the Karacadağ wild populations. This could be taken to support a monophyletic origin for domesticated einkorn; however, we must take into consideration a point emphasized by various authorities, who have pointed out that current-day domesticates do not necessarily represent the whole range of domesticated populations, as some of them could have become extinct through time (Willcox 1999, p.480, Willcox 2002, p.133, Nesbitt 2001, p.50) and that multiple domestications could have taken place (e.g., Willcox 1999, 2002). For example, recent evidence from genetic analysis shows at least two centres of domestication for emmer (Allaby et al. 1999). Moreover, the modern distribution of the wild relatives of domesticated plants does not necessarily correspond to their distribution during the onset of the Holocene (Dennell 1983; Theocharis 1973) and it is not necessarily related to the place where domestication took place (Jones et al. 1998). The recent recognition in archaeobotanical samples of a type of wheat that in morphology closely resembles *Triticum timopheevi* (Jones et al. 2000) does indicate the likelihood of domestication having taken place more than once, or having involved more wild species than had previously been thought. For example, *Triticum araraticum* or different populations of *Triticum dicoccoides* may have been domesticated and subsequently spread to Europe (in this regard, note that the modern distribution of *Triticum araraticum* overlaps with the northern part of the Fertile Crescent).

The Aegean could have been the area where some wild plants were domesticated, as the case of lentil and grass pea might suggest. In the context of Franchthi, the main argument that Hansen presents for the adoption of domesticated lentils that came from the Near East, rather than the local domestication of wild populations of lentils that existed in the Upper Palaeolithic and Mesolithic (12,500–8,000 bp or ca. 12,650–7,000 cal BC), is that *Lens orientalis* (the wild progenitor of *Lens culinaris*) could not have grown in the vicinity as it prefers a higher altitude (by 50 metres) than that encountered at the site. It is questionable whether this difference in altitude of 50 metres is an important criterion for eliminating the possibility of local domestication, or whether it was necessarily the case that the early neolithic lentils were *Lens culinaris*-type and not another extinct form, as suggested by Ladizinsky's work on *Lens nigricans* (a species with a circum-Mediterranean distribution; Ladizinsky et al. 1983; Ladizinsky 1989). Grass pea, the wild progenitor species of which occurs in the Mediterranean basin and Southwest Asia, is rarely present in assemblages from the Near East (Zohary and Hopf 2000) compared to the frequent neolithic finds from Greece and Bulgaria (Halstead and Jones 1980; Zohary and Hopf 2000; Marinova, this volume), and has not been considered a component of the 'crop package' that spread from the Near East into Europe (Zohary and Hopf 2000). This has lead Kislev to suggest that grass pea "may perhaps be the first crop domesticated in Europe" (Kislev 1989a, p.262). Thus, multiple domestications could have taken place in various parts of the Fertile Crescent and beyond (cf. Jones et al. 1998).

The appearance of domesticates is 'proof' of agriculture in the very specific sense of cultivating domesticated plants, however, as regards the transition a whole range of people-plant relationships can be detected from morphologically wild cereals and pulses. Thus, the presence of wild plants from mesolithic levels at Franchthi could have resulted from harvesting wild stands by gathering, or from sowing and harvesting wild plants in the same place or at locations across the landscape. One possible indication for the process of domestication being under way would be a co-occurrence of wild and domesticated forms in the archaeobotanical assemblages (Hillman and Davies 1992, pp.144–145; Nesbitt 2002, p.117). Although wild and domesticated forms of barley don't occur in the same samples at Franchthi, there is other evidence that leaves open the possibility of a local contribution. The discontinuous occupation of Franchthi prior to the early Neolithic, which has been used in support of the introduction of a 'crop package' from the Near East, is estimated to have lasted approximately 600 years (Farrand 2003, pp.77–78). The domesticated crops in the early Neolithic deposits were undoubtedly introduced to the cave from elsewhere, but from where, however, is a difficult question to answer.

The possibility of local origins could be supported by a continuity in material culture and subsistence practices between the Mesolithic and early Neolithic in the cave as highlighted, for example, by the continuous use of the same molluscan food source, *Cerinthium vulgatum*, during the Palaeolithic, Mesolithic and Neolithic (Dennell 1983; Halstead 1996; Dilcock 2001) and the persistence of mesolithic tool types (notches, end-scrapers and denticulates) into the initial Neolithic (Perlès 2001, p.46). The cave was a place that was constantly visited by human groups for millennia; it was a landmark and a point of reference, as indicated by the presence of burials (Cullen 1995; Dilcock 2001). Thus, as we cannot grasp what went on in terms of the relationship between people and wild cereals and pulses during the long period when the cave remained unoccupied, the question that needs to be addressed is why do some of the domesticated plants that appear at Franchthi in the early Neolithic (i.e., lentils and barley), which were previously gathered from the wild, have to be introduced from far away rather than nearby?

Of course, none could argue that the full suite of Greek Neolithic crops was domesticated at Franchthi; the presence of emmer in the neolithic levels is sufficient to exclude this possibility. The points we are trying to make are that: (a) there are many alternative interpretations of the available archaeobotanical evidence, and; (b) the most prevalent of these are influenced by the well-established perception of the 'crop package'. In our veiw, the picture that emerges from this review is that the neolithisation of Greece is more complex and interesting than proposed by the simple model of colonists who introduced a 'neolithic package' consisting of domesticates and the requisite know-how for their reproduction.

5.4 Crops as culture-specific

The emphasis that has been placed on the origins and domestication of crops, and the spread of a neolithic 'package', has perhaps downplayed the significance of other lines of evidence available in the archaeobotanical record that underline the subtle nuances of the interaction between plants and people. The example of barley is a case in point. The earliest occurrence of domesticated barley (based on a positive identification of barley rachises rather than grain) appears as early as 9100–8550 bp (ca. 8300–7500 cal BC) at Beidha (after Nesbitt 2002, table 1), at approximately the time of the mesolithic occupation at Franchthi (9500–8000 bp, ca. 8800–7000 cal BC). This means that while certain people in the Near East used domesticated barley, others on the Greek mainland were using morphologically wild plants (if the Franchthi evidence represents a generalised situation for Greece as a whole). This, however, does not necessarily imply that Franchthi was a place where the 'wave of colonists' carrying the new crops had not yet arrived; there was also continued use of morphologically wild plants as late as 8100–7600 bp (ca. 7200–6400 cal BC) at sites in Syria and Jordan (e.g., Wadi el-Jilat, Ghoraifé, Halula and Ras Shamra) that were near to those where domesticates were grown (after Nesbitt 2002, table 1), which was about the time of early neolithic settlements in Greece. It is therefore legitimate to ask whether domesticated crops had not yet arrived at Franchthi between 9500-8000 bp (ca. 9000–7000 cal BC) or whether, despite an awareness of their availability over a long period of time, they were not adopted by the mesolithic inhabitants. A distinction between wild and domestic plants and animals is encountered among different cultures and can be seen clearly expressed as a dichotomy between civilised grain growers and barbaric eaters of raw and wild food in the ancient Greek world (Vidal-Naquet 1983).[1] Such a distinction has been used to interpret neolithic material culture (e.g., Hodder 1990) and following this one could argue that in early neolithic Greece the 'new foodstuffs', which resulted from cultivated crops, represented a different worldview and were used to define 'self' and the 'other(s)' (cf. Perlès 2001, pp.171–172). We should bear in mind, however, that the distinction between the concepts of 'wild' and 'domestic' varies in place and time and that the cultivation and use of domesticates does not necessarily imply such a distinction (Ingold 1996). The adoption of domesticated plants according to the archaeobotanical, literary and ethnographic evidence could not have been a straightforward, unidirectional process whereby domesticates, by virtue of their properties and lifestyle requirements, spread and dominated the world; moreover, it need not have represented a radical change in existing ways of acquiring food or of relating to plants.

[1]There is evidence for the co-existence of cultivators and food collectors in the circum-Mediterranean world: in Homer's verses, written several millennia after the establishment of the first neolithic villages in Greece, it is written that harvesting cereals and grapes from abundant wild stands was practised by Cyclops and his clan at a time when Ulysses and Greeks were cultivating corn (Odyssey I, v. 109-112).

This dynamism of the adoption of new foodstuffs and a new way of life that we group under the term neolithic is further emphasised by the examination of the plant components of the so called 'package' that appears in Greece. The origins of crops and the case for monophyletic domestication in a restricted area have been closely linked to the notion of the neolithic 'crop package' comprising the following plants: emmer, einkorn, barley, pea, lentil, chickpea, bitter vetch and flax (Zohary 1996, p.144); however, the composition of this package in terms of the different plant components seems to vary in the Near East and in parts of Europe (a point clearly demonstrated during the course of this workshop). Close examination of the archaeobotanical records from the early and late neolithic sites on mainland Greece indicates the subtle nuances in diversity. For example, chickpea is included in the 'neolithic package' although it does not seem to become established in northern Greece; it is only reported from Otzaki Magoula in the early Neolithic (in one sample with a total of 57 identified charred items (Kroll 1983), in a small cache of 32 seeds from Dimini during the late Neolithic (Kroll 1979) and from Sitagroi where one seed is recorded (Renfrew 2003). Grass pea, on the other hand, which does not seem to have been part of the package, is present during the course of the Neolithic and is occasionally found in rich caches (Halstead and Jones 1980; Hubbard 1979; Valamoti 2004).

In the context of early neolithic communities of the Near East, it seems that the role of einkorn has been unfairly downplayed in comparison with the significance that has been attributed to emmer wheat (Zohary 1996). In Nesbitt's tabulation of Near/Middle Eastern archaeobotantical idenfications (Nesbitt 2001, 2002) wild einkorn is more common than wild emmer and for the domestic species einkorn is only slightly less frequent than emmer (domestic einkorn is represented at 19 sites and domestic emmer at 24; Nesbitt 2001). A similar situation is observed within the context of Aegean neolithic economies and, in northern Greece at least, it appears that einkorn wheat may have been the dominant crop in assemblages from a large number of late neolithic sites (Valamoti 2004), a trend that apparently continued into the Bronze Age (Kroll 1983, Jones *et al.* 1986, Halstead 1994). Einkorn wheat dominates the early neolithic assemblage from Toumba Balomenou, which according to Sarpaki could represent a 'cultural traditionalism' (Sarpaki 1995: 294). On this basis it is tempting to suggest that the long tradition of einkorn cultivation in northern Greece (the early and late neolithic were separated by approximately 2,000 years) may have something to do with origins, given that einkorn was the staple crop of the ancestors, and as such culturally defined the people who occupied parts of the Aegean. Einkorn could have either originated from wild stands in Greece, or been brought in as wild or domesticated crops from elsewhere (e.g., Anatolia). Emmer, on the other hand, was preferred in other parts of the Mediterranean from the onset of the Neolithic, as demonstrated by the evidence from Italy and Spain (see Peñ-Chocarro; Zapata Peña; Rottoli and Pessina, this volume), an observation that further undermines the concept of the existence of a 'package' in the Fertile Crescent.

The co-occurrence of the type of '*timopheevi*-like' wheat in some of the einkorn-rich samples (Jones et al. 2000; Valamoti 2004) could be an indication of different contact pathways via which crops may have been circulated, thus generating different combinations of cultigens in the broad region of the Fertile Crescent (cf. Perlès 2001, p.62). Following Hastorf's views, developed from ethnographic observations, wild stands of cereals or pulses could have been domesticated through a daily process of experimenting and 'nursing' plants, and their diffusion into neighbouring communities could have taken place through mating networks as people carried their crops with them (Hastorf 1998). By tracing the crops cultivated on early settlements in both regions on either side of the Aegean, it may be possible to establish the nature of the networks of relationships between the early Neolithic groups of Anatolia and Greece. Such networks continued to exist in the Bronze Age, as seems to be apparent from the later appearance of spelt (compared to regions further north) and an oil-producing crop (with an Anatolian/Transcaucasian modern distribution) found in northern Greek early bronze age contexts (Valamoti et al. nd;

Jones and Valamoti 2005). We believe that an understanding of the range of crops represented at each site from the beginning of the Neolithic together with a better knowledge of the mesolithic background could throw some light on the processes of change.

5.5 Conclusion

The preceding discussion leaves the following propositions open to further archaeobotanical investigation:

- That different combinations of domesticates, whether regionally and/or culturally specific, were available during the early Neolithic in the Near East, Anatolia and Greece.

- That different routes via which the domesticated crops could have spread were followed.

- That some of the plants growing wild in Greece today could have been independently domesticated and incorporated in a new way of life that we call the Neolithic.

The adoption of domesticated crops by the inhabitants of Greece was only part of a process of a changing way of life we call neolithic, and as we have already stated, it was not a necessary part of that process. This way of life may have had different manifestations, involving combinations of attributes that constitute the Neolithic. In this regard, the process would have involved the exploitation of different mixes of plants, both native and imported, exchanged or transported, together with the negotiation and assignment of ancestral and extrinsic meaning to plants. It is imperative that current and future research of the Mesolithic and Neolithic of Greece includes the systematic retrieval of archaeobotanical material. Last but not least, we would like to emphasize that we are not arguing from an 'indigenist' standpoint, but are in agreement with Pluciennik (1998), who suggests that the adoption of agriculture involved a complex process of interactions between local people and those from elsewhere and between local plants and plants imported from elsewhere. As regards the theme of this workshop ('New Perspectives on the Origins and Spread of Farming in Southwest Asia and Europe'), we believe that during the course of the 'journey' from Southwest Asia to Europe[2] agriculture was transformed, enriched and modified through the contribution of dynamic local traditions that had been in existence for well over a millennium.

Acknowledgements

S.M. Valamoti would like to thank Stephen Shennan, Sue Colledge and James Conolly for inviting her to participate in the workshop 'New Perspectives on the Origins and Spread of Farming in Southwest Asia and Europe', and the following friends and colleages for providing references, sharing unpublished data and ideas on the subject dealt with in our paper: Stelios Andreou, Nikos Efstratiou, Mark Nesbitt, Anaya Sarpaki and George Willcox. None, however, should be held responsible for the views expressed here. She also wishes to express her gratitude to Gordon Hillman, to whom the volume is dedicated, for the influence he has had on her since they first met at the World Archaeological Conference in Southampton in 1986 (where as an undergraduate student of archaeology she happened to be having breakfast next to Gordon). Since then, he has been a constant source of help, advice and inspiration and has been responsible for focusing her interests on archaeobotany.

[2]In Greek mythology Europe represents a Levantine Princess abducted by the Greek god Zeus; this may have some bearing on current perceptions of space and identity, though it is highly questionable that they existed 10,000 years ago (cf. Tringham 2000).

References

Allaby, R. G., M. Banerjee, and T. A. Brown (1999). Evolution of the high molecular weight glutenin loci of the A, B, D and G genomes of wheat. *Genome 42*, 296–307.

Ammerman, A. J. and L. L. Cavalli-Sforza (1971). Measuring the rate of spread of early farming in Europe. *Man 6*, 674–688.

Ammerman, A. J. and L. L. Cavalli-Sforza (1973). A population model for the diffusion of early farming in Europe. In C. Renfrew (Ed.), *The Explanation of Culture Change*, pp. 343–357. London: Duckworth.

Ammerman, A. J. and L. L. Cavalli-Sforza (1984). *The Neolithic Transition and the Genetics of Population in Europe.* Princeton: Princeton Univeristy Press.

Barker, G. (1985). *Prehistoric Farming in Europe.* Cambridge: Cambridge University Press.

Bender, B. (1978). Gatherer-hunter to farmer: a social perspective. *World Archaeology 10*, 203–221.

Cauvin, J. (2000). *The Birth of the Gods and the Origins of Agriculture.* Cambridge: Cambridge University Press.

Chapman, J. (1994). The origins of farming in south east Europe. *Préhistoire Européenne 6*, 133–156.

Cullen, T. (1995). Mesolithic mortuary ritual at Franchthi Cave, Greece. *Antiquity 69*, 270–289.

Dennell, R. W. (1983). *European Economic Prehistory: A New Approach.* London: Academic Press.

Dennell, R. W. (1992). The origins of crop agriculture in Europe. In C. Wesley Cowan and P. J. Watson (Eds.), *The Origins of Agriculture: An International Perspective*, pp. 71–100. Washington: Smithsonian Institution Press.

Dilcock, J. (2001). Returning to the ancestors? Interpretation of the Mesolithic/Neolithic transition at Franchthi Cave, Greece. In K. J. Fewster and M. Zvelebil (Eds.), *Ethnoarchaeology and Hunter-Gatherers: Pictures at an Exhibition*, BAR International Series 955, pp. 73–80. Oxford: British Archaeological Reports.

Efstratiou, N. (1996). The neolithisation of Europe: the Greek experience. *Documenta Præhistorica 22*, 63–82.

Evans, J. D. (1964). Excavations at the neolithic settlement of Knossos 1957–60. Part I. *The Annual of the British School at Athens 59*, 132–240.

Farrand, W. R. (2003). Depositional environments and site formation during the Mesolithic occupations of Franchthi Cave, Peloponnesos, Greece. In N. Galanidou and C. Perlès (Eds.), *The Greek Mesolithic: Problems and Perspectives*, British School at Athens Studies 10, pp. 69–78. London: British School at Athens.

Galanidou, N. and C. Perlès (Eds.) (2003). *The Greek Mesolithic: Problems and Perspectives.* British School at Athens Studies 10. London: British School at Athens.

Halstead, P. (1996). The development of agriculture and pastoralism in Greece: when, how, who and what? In D. R. Harris (Ed.), *The Origins and Spread of Agriculture and Pastoralism in Eurasia*, pp. 297–309. London: UCL Press.

Halstead, P. and G. Jones (1980). Early neolithic economy in Thessaly: some evidence from excavations at Prodromos. *Anthropologika 1*, 93–117.

Hansen, J. (1991). *The Palaeoethnobotany of Franchthi Cave.* Excavations at Franchthi Cave, Greece. Fascicle 7. Bloomington: Indiana University Press.

Hansen, J. (1992). Franchthi cave and the beginnings of agriculture in Greece and the Aegean. In P. Anderson-Gerfaud (Ed.), *Préhistoire de l'Agriculture: Nouvelles Approches Expérimentales et Ethnographiques*, pp. 231–247. Paris: C.N.R.S.

Harlan, J. (1992). Wild grass seed harvesting and implications for domestication. In P. Anderson-Gerfaud (Ed.), *Préhistoire de l'Agriculture: Nouvelles Approches Expérimentales et Ethnographiques*, pp. 21–28. Paris: C.N.R.S.

Harlan, J. R. and D. Zohary (1966). Distribution of wild wheats and barley. *Science 153*, 1074–1080.

Harris, D. R. (1989). An evolutionary continuum of people-plant interaction. In D. R. Harris and G. C. Hillman (Eds.), *Foraging and Farming: The Evolution of Plant Exploitation*, pp. 12–26. London: Unwin Hyman.

Harris, D. R. (1996). Introduction: themes and concepts in the study of early agriculture. In D. R. Harris (Ed.), *The Origins of Agriculture: An International Perspective*, pp. 1–9. London: UCL Press.

Hastorf, C. A. (1998). The cultural life of early domestic plant use. *Antiquity 72*, 773–782.

Heun, M., R. Schäfer-Pregl, D. Klawan, R. Castagna, M. Accerbi, B. Borghi, and F. Salamini (1997). Site of einkorn wheat domestication identified by DNA fingerprinting. *Science 278*(5341), 1312–1314.

Hillman, G. (2003). Investigating the start of cultivation in western Eurasia: studies of plant remains from Abu Hureyra on the Euphrates. In A. J. Ammerman and P. Biagi (Eds.), *The Widening Harvest. The Neolithic Transition in Europe: Looking Back, Looking Forward*, pp. 75–97. Boston: Archaeological Institute of America.

Hillman, G. and M. Davies (1992). Domestication rate in wild wheats and barley under primitive cultivation: preliminary results and archaeological implications of field measurements of selection co-eficient. In P. Anderson-Gerfaud (Ed.), *Préhistoire de l'Agriculture: Nouvelles Approches Expérimentales et Ethnographiques*, pp. 113–158. Paris: C.N.R.S.

Hillman, G. C., S. M. Colledge, and D. R. Harris (1989). Plant food economy during the Epipalaeolithic period at Tell Abu Hureyra, Syria: dietary diversity, seasonality and modes of exploitation. In D. R. Harris and G. C. Hillman (Eds.), *Foraging and Farming: The Evolution of Plant Exploitation*, pp. 240–268. London: Unwin Hyman.

Hodder, I. (1990). *The Domestication of Europe. Structure and Contingency in Neolithic Societies.* Oxford: Blackwell.

Hubbard, R. N. L. B. (1979). Ancient agriculture and ecology at Servia. *Annual of the British School at Athens 74*, 226–228.

Ingold, T. (1996). Growing plants and raising animals: an anthropological perspective on domestication. In D. R. Harris (Ed.), *The Origins and Spread of Agriculture and Pastoralism in Eurasia*, pp. 12–24. London: UCL Press.

Jacobsen, T. W. (1976). 17,000 years of Greek prehistory. *Scientific American 234*, 76–87.

Jarman, H. N. (1972). The origins of wheat and barley cultivation. In E. S. Higgs (Ed.), *Papers in Economic Prehistory*, pp. 15–26. Cambridge: Cambridge University Press.

Jarman, M. R., G. N. Bailey, and H. N. Jarman (1982). *Early European Agriculture: Its Foundations and Developments.* Cambridge: Cambridge University Press.

Jones, G. and S. Valamoti (2005). *Lallemantia*, an imported oil plant in Bronze Age northern Greece? *Vegetational History and Archaeobotany 14*(4), 571–577. In Press.

Jones, G., S. Valamoti, and M. Charles (2000). Early crop diversity: a 'new' glume wheat from northern Greece. *Vegetation History and Archaeobotany 9*, 133–146.

Jones, M., R. G. Allaby, and T. A. Brown (1998). Wheat domestication. *Science 279*, 302–303.

Kislev, M. E. (1989a). Origins of the cultivation of *Lathyrus sativus* and *L. cicera* (Fabaceae). *Economic Botany 43*, 262–270.

Kislev, M. E. (1989b). Pre-domesticated cereals in the Pre-Pottery Neolithic A period. In I. Hershkovitz (Ed.), *People and Culture in Change*, BAR International Series 508, pp. 147–151. Oxford: British Archaeological Reports.

Kislev, M. E. (1992). Agriculture in the Near East in the VIIth millennium bc. In P. Anderson-Gerfaud (Ed.), *Préhistoire de l'Agriculture: Nouvelles Approches Expérimentales et Ethnographiques*, pp. 87–93. Paris: C.N.R.S.

Kotsakis, K. (1992). The neolithic way of production: indigenous or immigrant? [in Greek]. In *Diethnes Synedrio gia tin Archaia Thessalia sti Mnimi tou Dimitri P. Theochari*, pp. 120–135. Athens: Tameio Archaiologikon Poron kai Apalotrioseon.

Kotsakis, K. (2000). The beginning of the Neolithic in Greece [In Greek]. In N. Kyparissi-Apostolika (Ed.), *Theopetra Cave: Twelve Years of Excavations and Research 1987-1998. Proceedings of the International Conference, Trikala 6-7 November 1998*, pp. 173–180. Athens: Institute for Aegean Prehistory.

Kotsakis, K. (2001). Mesolithic to neolithic in Greece. Continuity, discontinuity or change of course? *Documenta Præhistorica 28*, 63–78.

Kotsakis, K. (2003). From the neolithic side: the Mesolithic/Neolithic interface in Greece. In N. Galanidou and C. Perlès (Eds.), *The Greek Mesolithic: Problems and Perspectives*, pp. 217–221. London: British School at Athens.

Koumouzelis, M., J. K. Kozlowski, and B. Ginter (2003). Mesolithic finds from Cave I in the Klisoura Gorge, Argolid. In N. Galanidou and C. Perlès (Eds.), *The Greek Mesolithic: Problems and Perspectives*, pp. 113–122. London: British School at Athens.

Kroll, H. (1981). Thessalische Kulturpflanzen. *Zeitschrift für Archäologie 15*, 97–103.

Kroll, H. J. (1979). Kulturpflanzen aus Dimini. *Archaeophysika 8*, 173–189.

Kroll, H. J. (1983). *Die Pflanzenfunde. Kastanas: Ausgrabungen in einem Siedlungshugel der Bronze und Eisenzeit Makedoniens 1975-1979* . Prähistorische Archaeologie in Südosteuropa 2. Berlin: Verlag Volker Spiess.

Kyparissi-Apostolika, N. (Ed.) (2000). *Theopetra Cave: Twelve Years of Excavation and Research 1987-1998. Proceedings of the International Conference, Trikala 6-7 November 1998.* Athens: Institute for Aegean Prehistory.

Kyparissi-Apostolika, N. (2003). The Mesolithic in Theopetra Cave: new data on a debated period of Greek prehistory. In N. Galanidou and C. Perlès (Eds.), *The Greek Mesolithic: Problems and Perspectives*, British School at Athens Studies 10, pp. 189–198. London: British School at Athens.

Ladizinsky, G. (1989). Origin and domestication of the Southwest Asian grain legumes. In D. R. Harris and G. C. Hillman (Eds.), *Foraging and Farming: The Evolution of Plant Exploitation*, pp. 374–388. Unwin Hyman.

Ladizinsky, G., D. Braun, and F. J. Muehlbauer (1983). Evidence for domestication of *Lens nigricans* (M. Bieb.) Godron in S Europe. *Botanical Journal of the Linnean Society 87*, 169–176.

Mangafa, M. (2000). I ekmetalefsi ton fyton apo ti mesi palaeolithiki eos ti neolithiki periodo apo tin karposyllogi stin kalliergeia. In N. Kyparissi-Apostolika (Ed.), *Theopetra Cave: Twelve Years of Excavations and Research 1987-1998. Proceedings of the International Conference, Trikala 6-7 November 1998*, pp. 135–137. Athens: Institute for Aegean Prehistory.

Milojcic, V., J. Boessneck, and M. Hopf (1962). *Die Deutschen Ausgrabungen auf der Argissa-Magula in Thessaliken I: das Praekeramische Neolithicum sowie die Tier und Pflanzenreste.* Bonn: Dr. Rudolf Habelt Verlag.

Nesbitt, M. (2001). Wheat evolution: integrating archaeological and biological evidence. In P. D. S. Caligari and P. E. Brandham (Eds.), *Wheat Taxonomy: The Legacy of John Percival*, The Linnean Special Issue No 3, pp. 37–59. London: Academic Press.

Nesbitt, M. (2002). When and where did domesticated cereals first occur in Southwest Asia? In R. T. J. Cappers and S. Bottema (Eds.), *The Dawn of Farming in the Near East*, Studies in early Near Eastern Production, Subsistence and Environment 6/1999, pp. 113–132. Berlin: Ex oriente.

Payne, S. (1982). Faunal evidence for environmental climatic change at Franchthi cave (southern Argolid) 25.000 BP-5000 BP: preliminary results. In J. L. Bintliff and W. van Zeist (Eds.), *Palaeoclimates, Palaeoenvironments, and Human Communities in the Eastern Mediterranean Region in Later Prehistory*, BAR International Series 133, pp. 133–136. Oxford: British Archaeological Reports.

Perlès, C. (2001). *The Early Neolithic of Greece.* Cambridge: Cambridge University Press.

Perlès, C. (2003). The Mesolithic at Franchthi: an overview of the data and problems. In N. Galanidou and C. Perlès (Eds.), *The Greek Mesolithic: Problems and Perspectives*, British School at Athens Studies 10, pp. 79–88. London: British School at Athens.

Pluciennik, M. (1998). Deconstructing 'the Neolithic' in the Mesolithic-Neolithic transition. In M. Edmonds and C. Richards (Eds.), *Understanding the Neolithic of North-Western Europe*, pp. 61–83. Glasgow: Cruithne Press.

Renfrew, J. M. (1966). A report on recent finds of carbonized cereal grains and seeds from prehistoric Thessaly. *Thessalika 5*, 31–36.

Renfrew, J. M. (2003). Grains, seeds and fruits from prehistoric Sitagroi. In E. S. Elster and C. Renfrew (Eds.), *Prehistoric Sitagroi: Excavations in Northeast Greece, 1968-1970, Volume 2: The Final Report*, pp. 1–28. Los Angeles: Cotsen Institute of Archaeology.

Rindos, D. (1984). *The Origins of Agriculture: An Evolutionary Perspective.* New York: Academic Press.

Runnels, C. (2001). The stone age of Greece from the palaeolithic to the advent of the neolithic. In *Aegean Prehistory. A Review*, American Journal of Archaeology Supplement I, pp. 225–254. Boston: Archaeological Institute of America.

Runnels, C. (2003). The origins of the Greek Neolithic: a personal view. In A. J. Ammerman and P. Biagi (Eds.), *The Widening Harvest. The Neolithic Transition in Europe: Looking Back, Looking Forward*, pp. 121–132. Boston: Archaeological Institute of America.

Runnels, C. and T. van Andel (1988). Trade and the origins of agriculture in the eastern Mediterranean. *Journal of Mediterranean Archaeology 1*, 83–109.

Runnels, C. and T. van Andel (2003). The Early Stone Age of the Nomos of Preveza: landscape and settlement. In J. Wiseman and K. Zachos (Eds.), *Landscape Archaeology in Southern Epirus, Greece I*, Hesperia Supplement 32, pp. 47–134. Princeton: The American School of Classical Studies in Athens.

Sampson, A., J. K. Kozlowski, and M. Kaczanowska (2003). Mesolithic chipped stone industries from the Cave of Cyclope on the island of Youra (Northern Sporades). In N. Galanidou and C. Perlès (Eds.), *The Greek Mesolithic: Problems and Perspectives*, British School at Athens Studies 10, pp. 123–130. London: British School at Athens.

Sarpaki, A. (1995). Toumba Balomenou, Chaeronia: plant remains from the early and middle neolithic levels. In H. Kroll and R. Pasternak (Eds.), *Res Archaeobotanicae: 9th Symposium IWGP*, pp. 281–300. Kiel: Oetker-Voges Verlag.

Seferiadis, M. L. (1993). The European neolithisation process. *Documenta Præhistorica 21*, 137–162.

Shackleton, J. C. (1988). *Marine Molluscan Remains from Franchthi Cave.* Excavations at Franchthi Cave, Greece. Fascicle 4. Bloomington: Indiana University Press.

Theocharis, D. (1967). *I Avgi tis Thessalikis Proistorias.* Volos: Filarhaios Etaireia Volou.

Theocharis, D. (1973). *Neolithic Greece.* Athens: National Bank of Greece.

Tringham, R. (2000). Southeastern Europe in the transition to agriculture in Europe: bridge, buffer or mosaic. In T. Douglas-Price (Ed.), *Europe's First Farmers*, pp. 19–56. Cambridge: Cambridge University Press.

Valamoti, S. M. (1995). Georgika proionda apo to neolithico oikismo Giannitsa B: mia prokatarktiki proseggisi meso ton archeovotanikon dedomenon. *AEMTHΘ 6*, 177–184.

Valamoti, S. M. (2004). *Plants and People in Late Neolithic and Early Bronze Age Northern Greece: An Archaeobotanical Investigation.* BAR S1258. Oxford: British Archaeological Reports.

Valamoti, S. M., A. Papanthimou, and A. Pilali (n.d.). Cooking ingredients from EBA Archondiko: the archaeobotanical evidence. In Y. Facorellis, N. Zacharias, K. Polikreti, T. Vakoulis, Y. Bassiakos, V. Kiriatzi, and E. Aloupi (Eds.), *Proceedings of the 4th Symposium on Archaeometry*, BAR International Series. Oxford: Archaeopress. In Press.

van Zeist, W. and S. Bottema (1971). Plant husbandry in early neolithic Nea Nikomedeia, Greece. *Acta Botanica Neerlandika 20*, 542–538.

Vaughn, P. C. (1990). Use-wear analysis of Mesolithic chipped-stone artifacts from Franchthi Cave. In C. Perlès (Ed.), *Les Industries lithiques taillées de Franchthi (Argolide, Grèce). Tome II: Les Industries du Mésolithique et du Néolithique initial*, Excavations at Franchthi Cave, Greece. Fascicle 5, pp. 239–253. Bloomington: Indiana University Press.

Vidal-Naquet, P. (1983). *Le Chasseur Noir.* Greek Translation. Athens: Livanis-Nea Synora Publications.

Willcox, G. (1999). Agrarian change and the beginnings of cultivation in the Near East: evidence from wild progenitors, experimental cultivation and archaeobotanical data. In C. Gosden and J. Hather (Eds.), *The Prehistory of Food: Appetites for Change*, pp. 478–500. London: Routledge.

Willcox, G. (2000). Nouvelles données sur l'origine de la domestication des plantes au Proche-Orient. In J. Guilaine (Ed.), *Premiers Paysans du Monde*, pp. 121–140. Paris: Editions Errance.

Willcox, G. (2002). Geographical variation in major cereal components and evidence for independent domestication events in the Western Asia. In R. T. J. Cappers and S. Bottema (Eds.), *The Dawn of Farming in the Near East*, pp. 133–140. Berlin: Ex oriente.

Zamanis, A., S. Samaras, N. Stavropoulos, and J. E. Dille (1988). *Report of an Expedition to Rescue Germplasm of Wild Species of Wheat and Relatives in Greece, June 1988.* Greek Gene Bank Scientific Bulletin 5. Thessaloniki: Agricultural Research Centre.

Zohary, D. (1996). The mode of domestication of the founder crops of Southwest Asian agriculture. In D. R. Harris (Ed.), *The Origins and Spread of Agriculture and Pastoralism in Eurasia*, pp. 142–158. London: UCL Press.

Zohary, D. and M. Hopf (2000). *Domestication of Plants in the Old World: The Origin and Spread of Cultivated Plants in West Asia, Europe and the Nile Valley* (3rd ed.). Oxford: Oxford University Press.

Chapter 6

Archaeobotanical data from the early Neolithic of Bulgaria

Elena Marinova
Sofia University 'St Kliment Ohridsky', Bulgaria

6.1 Introduction

The territory of modern Bulgaria is situated on one of the routes of distribution of early neolithic agriculture from the Near East to Europe. One of the sources of information about the dispersal processes is the archaeobotanical studies carried out on neolithic sites in the area. The current study presents some recent archaeobotanical results for the early neolithic layers of three sites, Kovačevo, Slatina and Kapitan Dimitrievo, situated in the modern territory of Bulgaria (figure 6.1). Archaeobotanical data available from 12 other sites will be compared with these results.

6.2 Archaeological and chronological settings

Prehistorians generally agree that the origins of the Neolithic in the central and eastern Balkans are connected with Anatolia, although this is still debated. In this context, different Balkan-Anatolian interaction models have been developed. According to a recently presented model (Nikolov 2004), the early Neolithic in south and west Bulgaria included four successive stages. Period I involves the initial development phase of the early neolithic Karanovo I culture, a major characteristic of which is the white-painted pottery. At that time, the culture (i.e., its classical period) comprised only the southwest parts of the territory; it was established in the Mesta Valley, the eastern parts of the Sofia basin and the westernmost parts of Thrace. During the classical period of the Karanovo I culture, which ended with the emergence of the Karanovo II culture at Tell Karanovo, isolated dark-painted vessels appeared in addition to the white-painted pottery. At that time the Karanovo I culture reached its maximum area and extended from the western parts of Thrace to the Tundzha and from the sub-Balkan plains to the northern foothills of the Rhodope Mountains. This period (II) involves the Karanovo I layer of the eponymous Tell Karanovo, the early layers of Tell Azmak, Stara Zagora and Tell Kazanlak, the early phase II layers of the early neolithic deposits of Tell Kapitan Dimitrievo, phase II of Kovačevo, as well as the early layer of Aşağı Pınar near Kırklareli in eastern Thrace. During Period III, the early neolithic Karanovo II culture evolved in the northeast parts of Thrace; the development of Karanovo I culture continued in the other parts of this area. In the marginal zones, this culture is represented by its Azmak variant. During Period IV, which involved the middle neolithic cultural phenomenon Proto-Karanovo III, the northeast parts of Thrace and the Kazanlak basin were

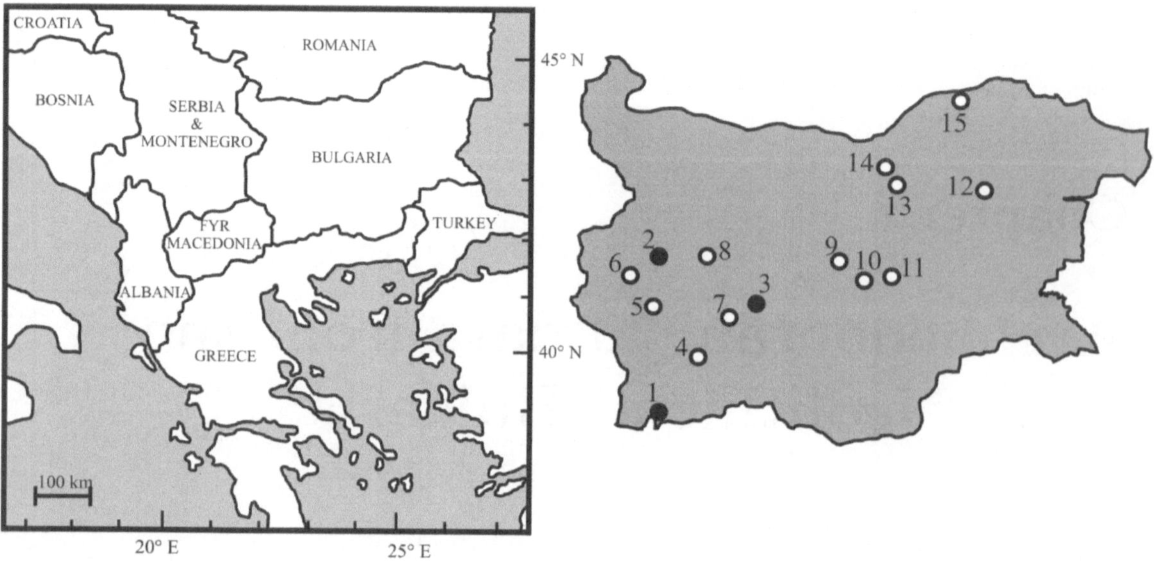

Figure 6.1: Location of sites referred to in the text (black circles=author's studies; open circle=others' studies). 1 Kovačevo; 2 Slatina; 3 Kapitan Dimitrievo; 4 Elešnica, 5 Kremenik (Spareva Banja); 6 Gălăbnik; 7 Rakitovo; 8 Čavdar; 9 Azmak; 10 Okrăžna Bolnica, Stara Zagora; 11 Karanovo; 12 Poljanica Platoto; 13 Orlovec; 14 Koprivec; 15 Malăk Preslavec.

occupied; this period is represented by two territorial variants: Karanovo II–III and Karanovo I–III. The development of the Karanovo I continued in the other parts of Thrace.

The earliest neolithic developments in the central Balkans are even more complicated. Different cultural phenomena are registered here, but due to the lack of well-stratified materials and the confusingly varied local designations, they have been consolidated in the Carpatho-Danubian culture of the early Neolithic (Nikolov 2002). There were probably two stages: the first characterized by the monochrome red-slipped pottery, and the second by the sporadic use of specific and rather scarce white ornamentation. It has been difficult to say anything until now about mesolithic occupation in the area, since no sites are known. Some stone artefacts without stratigraphic context, which were found in the area of the Bulgarian northern Black Sea coast, are the only evidence for this period. We exclude this subject from the current discussion, therefore, due to the lack of data on any connections with mesolithic populations at the sites considered.

The archaeobotanical information presented here will use the former chronological framework for the early Neolithic of Bulgaria, which divides this period into two halves (Nikolov 2000), the first corresponding to Period I and the second half corresponding to Periods II and III defined most recently by Nikolov (2004). According to the radiocarbon evidence from Bulgarian prehistoric sites (Görsdorf and Bojadziev 1996), these two halves of the early Neolithic belong to the time span 6000–5700 cal BC (early Neolithic I) and 5700–5450 cal BC (early Neolithic II).

The lowest layers in the site of Kovačevo (Kovačevo Ia and Ib) have evidence of direct connections with northern Greece and the early Neolithic of FYR Macedonia. The layers above (Kovačevo Ic and Id) contain ceramics that have some parallels in the Karanovo I culture (Lichardus-Itten et al. 2002). The site is considered to be a multilayer 'flat' site (or 'open' settlement), with a surface area of 5–6 ha and a neolithic cultural layer that is 2-m thick. The site of Slatina has four building horizons that belong to the first half of the early Neolithic and two building horizons that belong to the second half of the early Neolithic, the latter show influence from the developed Starčevo (II–III) culture (Nikolov 1992). The site was distributed over an area of 8 ha and in some locations the early neolithic cultural layer was 4-m thick. To date, this is one of the largest sites found in west Bulgaria. The cultural layer of tell Kapitan Dimitrievo has a thickness of 13 m (with occupation during the Neolithic, Chalcolithic, Bronze Age and Iron Age) and its diameter east-west is 140 m and north-south is 110 m. The early Neolithic occupation of the site covers both halves of the Bulgarian early Neolithic. During this period, this Thracian site shows quite clear connections with southwestern Bulgaria.

6.3 Environmental setting

About 60% of the territory of Bulgaria consists of hilly and mountainous areas. Until now most of the earliest neolithic sites that have been found are in an area with an altitude of about 300–500 m above sea level (asl); that is, in the transitional zone between the lowlands and the mountains. The main reason for this is probably the fact that many neolithic sites in the lowlands and the river valleys are covered with thick sediments and are thus not easily accessible for investigation.

The three sites considered belong to the ecotone between the lowlands and mountains, which is particularly rich in various natural resources. Kovačevo is situated at an altitude of 450 m asl, Slatina is at 550 m asl and tell Kapitan Dimitrievo is 350 m above sea level. All were located near small rivers and springs. The deep river valleys in southern Bulgaria (Struma, Mesta, Maritza, etc.) provide connections with the Aegean area and northern Greece, via which early agriculture probably spread from Anatolia. It is likely that these geographical factors played a decisive role in the establishment and development of the neolithic cultures in Bulgaria.

The reconstruction of the temperatures during the period around 6000 cal BC in Europe show that in the southeast the mean winter temperatures were almost at today's levels and the summer temperatures were slightly lower (Davis et al. 2003). The climate in the territory of Bulgaria is continental, but a strong Mediterranean influence exists especially in the southern parts of the country. The mean annual temperature today is about 10–12° C and the mean annual precipitation in the lowlands is between 550–700 mm. In comparison to the eastern Mediterranean, the area studied represents a region with a colder climate, where there are frosts in the winters and long periods of below-zero temperatures. Olives, which are a crop that can be used as an indicator of a Mediterranean climate, do not survive in the Bulgarian winters and this is proof that the area under consideration represents a different climatic and vegetational gradient. Therefore, the crops and agriculture introduced from the Mediterranean area and Near East had to adapt to these conditions, which were transitional between the eastern Mediterranean and Europe.

The natural vegetation in Bulgaria belongs to the province of European deciduous forests that have a prevalence of plants with mainly European and Euro-Asiatic distributions. Today, the area in which the sites are located belongs to the vegetation belt of the xerothermic oak forests. The data from pollen analyses covering the period concerned are numerous, but well-dated diagrams are still scarce. Due to the climatic conditions the sites most suitable for pollen analysis are located in the mountains; there is almost no information from the lowlands and the mid and low altitudes. Most of the studies show that in the southern parts of the country the development of the deciduous oak forests started at ca. 7000 cal BC (Huttunen et al. 1992; Stefanova and Ammann 2003), which is similar to the data from northern Greece, however, in the Balkan Mountains, the expansion of the forests probably started at 6500 cal BC (Filipovitch and Stefanova 1998). These forests were rich in mesophilous species like maple/sycamore (*Acer*), elm (*Ulmus*), common/European ash (*Fraxinus excelsior*), lime (*Tilia*), ivy (*Hedera*), hops (*Humulus*), etc. The uppermost limit of the oak forests during the period considered was higher than today.

If we adhere to the widely accepted assumption of the researchers of Bulgarian prehistory, the process of neolithisation started around 6100–5900 cal BC. At this time, therefore, the oak forests were probably already developed in the hilly areas when the first farmers arrived. The palaeoecological data from the site of Tschokljovo marsh, located at 850 m asl in western Bulgaria, show that oak forests with hazel as a pioneering element were present at ca. 6000 cal BC (Tonkov and Bozilova 1992) and, according to Willis and Bennet (1994), there was a similar composition to the vegetation in the whole Balkan area. The authors noticed an increase of hazel (*Corylus* sp.) and oriental hornbeam/hop hornbeam (*Carpinus orientalis*/*Ostrya*) type pollen during the neolithisation period of the Balkan area between 6000–5000 cal BC. The latter is a sub-Mediterranean floristic element that grows on thin soils and has a greater distribution today as a result of anthropogenic impact.

6.4 The evidence

The archaeobotanical material from the sites of Kovačevo, Slatina and Kapitan Dimitrievo was collected systematically during two excavation seasons. It was processed by means of manual flotation. The archaeological contexts studied from the three sites and volumes floated are given in table 6.1. Due to the preservation conditions, a majority of the material found was in a charred state and mineralised remains were present in very small quantities. Most of the information from the 12 earlier archaeobotanical studies of the Bulgarian early Neolithic originated from occasional finds of storage contexts. Systematic sampling and flotation was undertaken at Čavdar (Dennell 1978) and at Karanovo (Thanheiser 1997). A preliminary study was carried out by the author at Orlovec and Koprivec and flotation of about 80–100 litres distributed over five to eight samples/contexts was applied at each of the sites. About 100 litres were wet sieved at the site of Gălăbnik. The chronological attribution of the 15 sites is given in table 6.2 and the presence of the plant remains is given in table 6.3. Approximately 60 taxa were found at Kovačevo, Slatina and Kapitan Dimitrievo (for details see Marinova 2006) but here only those relevant to the discussion will be mentioned.

6.4.1 Cultivated plants

The main trends in the cultivated plants can be observed in the flotation samples from the three sites studied (Kovačevo, Slatina, Kapitan Dimitrievo, figure 6.2) in combination with the data already published from the 12 other sites (table 6.3). In many cases, the dominance of one crop or another is statistically not significant because of the small size of samples studied. It is difficult to derive a clear regional pattern from the small archaeobotanical data set that is presently available for the Bulgarian early Neolithic (figure 6.1, tables 6.2 and 6.3). In general the main crops were cultivated cereals, including hulled wheat (einkorn and emmer), barley and naked wheat in smaller quantities, various pulses (lentils, grass pea, pea, bitter vetch and chickpea) and flax. All of these plants are known from Bulgarian early neolithic storage contexts. This combination of crops corresponds fully to the Near Eastern crop assemblage defined by Zohary and Hopf (Zohary and Hopf 1993).

Archaeobotanical information is available for nine sites from the first half of the early Neolithic (table 6.3). In many of the sites during this period the predominant wheat is einkorn (e.g., at Slatina, Elešnica, Kapitan Dimitrievo, Poljanica platoto, Koprivec and Orlovec). Most of these sites are situated in the mountainous or northern parts of the country. Hulled and naked barley is present. Hulled barley is prevalent in the floated material from Kovačevo, Kapitan Dimitrievo, Elešnica, Karanovo and Orlovec and naked barley was identified in the storage contexts in a house from this period at Slatina. Of the pulses, lentil and grass pea are quite important.

Kovačevo		
Sector	Context	Volume (l)
E	Posthole	15
I	Ditch	30
K	House 2199	20
N	House 2071	70
E	House 1714	20
A, K, N	Cultural layer	286
Total		441

Slatina		
Building horizon	Context	Volume (l)
I	House	20
II	House	36
IV	Ditch	48
IV	House	141
Total		245

Kapitan Dimitrievo		
Sondage	Context	Volume (l)
I	House 2	40
I	Pit	20
I	Cultural layer	25
II	Cultural layer	20
II	House 1, floor	95
II	House 1, oven	18
II	House 2, floor	40
III	Depression, waste place	160
Total		318

Table 6.1: Archaeological contexts and sample volumes from Kovačevo, Slatina and Kapitan Dimitrievo.

SW Bulgaria	Thrace	North Bulgaria
First half of the Neolithic (6000–5650 cal BC)		
Kovačevo (Marinova 2006)	Karanovo (Thanheiser 1997)	Poljanica Platoto (Hopf 1988)
Gălăbnik (Marinova et al. 2002)	Kapitan Dimitrievo (Marinova 2006)	Koprivec (Marinova, unpubl.)
Slatina (Dontscheva 1990)		Orlovec (Marinova, unpubl.)
Elešnica (Dontscheva, unpubl.)		
Second half of the Neolithic (5650–5400 cal BC)		
Kovačevo (Marinova, unpubl.)	Karanovo (Hopf 1973; Thanheiser 1997)	Malăk Preslavec (Panayotov et al. 1992)
Slatina (Marinova, unpubl.)	Kapitan Dimitrievo (Arnaudov 1939; Marinova 2006)	
Kremenik (Cakalova and Sarbinska 1986)	Okrăžna Bolnica (Lisitzina and Filipovitch 1980)	
Rakitovo (Tschakalova and Bozilova 2002)	Azmak (Hopf 1973)	
Čavdar (Hopf 1973; Dennell 1978)		

Table 6.2: Chronological and regional settings of the sites discussed.

Site	Cultivated plants												Collected plants									
	T. monococcum	*T. dicoccum*	*T. aestivum/durum*	*Hordeum vulgare*	hulled barley	naked barley	*Lens culinaris*	*Pisum sativum*	*Latyrus sativus/cicera*	*Vicia ervilia*	*Cicer arietinum*	*Linum* sp.	*Cornus mas*	*Vitis vinifera* ssp. *sylvestris*	*Prunus* sp.	*Pyrus/ Malus/ Sorbus*	*Corylus avellana*	*Sambucus* sp.	*Rubus* sp.	*Fragaria vesca*	*Physalis alkekengi*	*Pistacia terebinthus*
Early Neolithic I																						
Kovačevo	++	+	+	+	++	.	++	+	++	+	.	.	+	+	+	.	+	+	.	.	.	+
Gălăbnik	+	++	+	+	.	+	+	+	+	+	*+*	.	+	.	+	.	.	.	.	.	.	.
Slatina	*++*	*+*	*+*	.	.	*++*	*+*	+	*++*	+	.	+	.	.	.	.	.	.	.	.	.	.
Elešnica	++	+	+	+	+	+	+	+	+	+	.	.	.	+	.	.	.	.	.	.	.	.
Kap. Dimitrievo	*++*	*+*	+	+	+	+	+	+	*++*	.	.	.	+	.	+	.	+	+	+	.	.	+
Karanovo	+	++	+	+	+	.	++	+	+	++	.	.	+	+	.	.	.	+	+	.	.	.
Poljanica Platoto	++	+	.	+	.	.	.	.	.	.	.	.	.	.	.	.	.	.	.	.	.	.
Koprivec	++	+	.	+	+	.	+	+	.	.	.	.	+	.	+	.	.	.	.	.	+	.
Orlovec	++	+	.	+	+	.	+	+	.	.	.	.	+	+	+	.	.	.	.	.	+	.
Early Neolithic II																						
Kovačevo	+	++	+	+	++	.	+	+	+	.	.	.	+	+	+	.	+	+	+	+	+	.
Slatina	+	++	.	+	++	.	++	+	++	+	.	+	+	+	+	.	+	+	+	+	.	.
Kremenik	*++*	*+*	+	+	.	+	+	+	+	+	.	.	+	.	+	.	.	.	.	.	.	.
Čavdar	*+*	++	+	*+*	.	.	*+*	*+*	.	.	.	+	+	.	.	+	+	+	+	.	.	.
Rakitovo	*+*	*++*	.	+	*++*	*+*	.	.	.	.	.	.	.	.	.	.	+	.	.	.	.	.
Kap. Dimitrievo	*++*	*+*	+	.	*++*	+	+	*++*	+	+	+	+	.	+	.	+	.	.	.	.	.	+
Azmak	*+*	*++*	*+*	+	.	.	+	*++*	*+*	.	.	.	.	.	.	+	.	.	.	.	.	.
Okrăžna Bolnica	*+*	*++*	.	.	.	*+*	.	.	.	.	.	.	.	.	.	.	.	.	.	.	.	.
Karanovo	+	++	+	+	+	.	+	+	+	+	.	.	+	+	.	.	.	+	+	.	.	.
M. Preslavec	++	.	.	.	.	+	+	+	.	.	.	.	.	+	.	.	.	.	.	.	.	.

Table 6.3: The most frequent plants found at Bulgarian Neolithic sites. +=present; ++=predominant; *+* or *++*=found as storage.

Site	*Ajuga chamaepitys*	*Bromus* sp.	*Chenopodium* sp.	*Galium aparine/spurium*	*Galium* sp.	*Lithospermum arvense*	*Malva* sp.	*Medicago* sp.	*Polygonum convolvulus*	*Polygonum* sp.	*Rumex* sp.	*Setaria verticillata/viridis*	*Teucrium chamaedrys*	*Thymelaea passerina*
	Weeds													
Early Neolithic I														
Kovačevo	+	+	+	+	+	+	·	·	+	+	+	+	·	+
Gălăbnik	·	·	·	·	·	·	·	·	+	·	·	·	·	·
Slatina	·	·	·	+	·	·	·	·	+	+	·	·	·	·
Elešnica	·	·	·	+	·	·	·	·	·	+	·	·	·	·
Kap. Dimitrievo	+	+	+	+	+	+	·	·	+	+	+	+	+	+
Karanovo	·	+	+	·	+	+	+	+	+	+	+	·	·	·
Poljanica platoto	·	·	·	·	·	·	·	·	·	·	·	·	·	·
Koprivec	·	·	+	·	·	+	·	·	·	·	·	·	·	·
Orlovec	+	+	+	+	+	+	·	·	+	+	+	·	+	·
Early Neolithic II														
Kovačevo	+	+	+	+	+	+	·	·	+	+	+	+	·	+
Slatina	+	+	+	+	+	+	·	·	+	+	·	+	+	+
Kremenik	·	+	·	+	·	·	·	·	·	·	·	·	·	·
Čavdar	·	+	+	+	+	+	+	·	+	+	+	·	·	·
Rakitovo	·	+	·	·	·	·	·	·	·	·	·	·	·	·
Kap. Dimitrievo	+	+	+	+	+	+	+	·	+	+	+	+	+	+
Azmak	·	·	·	·	·	·	·	·	·	·	·	·	·	·
Okrăžna Bolnica	·	·	+	·	·	·	·	·	·	+	+	·	·	·
Karanovo	·	+	+	+	+	+	+	·	+	+	+	·	·	·
M. Preslavec	·	·	·	·	·	·	·	·	·	·	·	·	·	·

Table 6.3, cont.

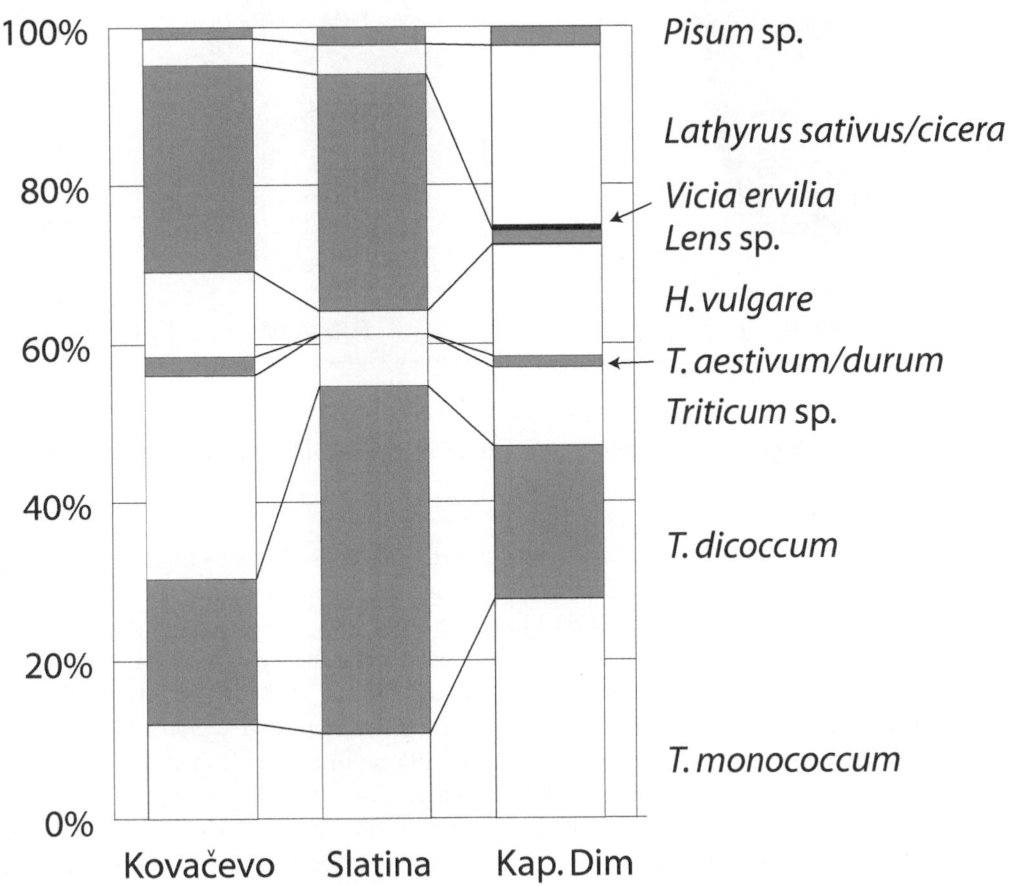

Figure 6.2: Ratios of cultivated plants recovered from Kovačevo, Slatina and Kapitan Dimitrievo (excluding storage remains). Note that the quantification of *T. aestivum/durum* is based on rachis fragments.

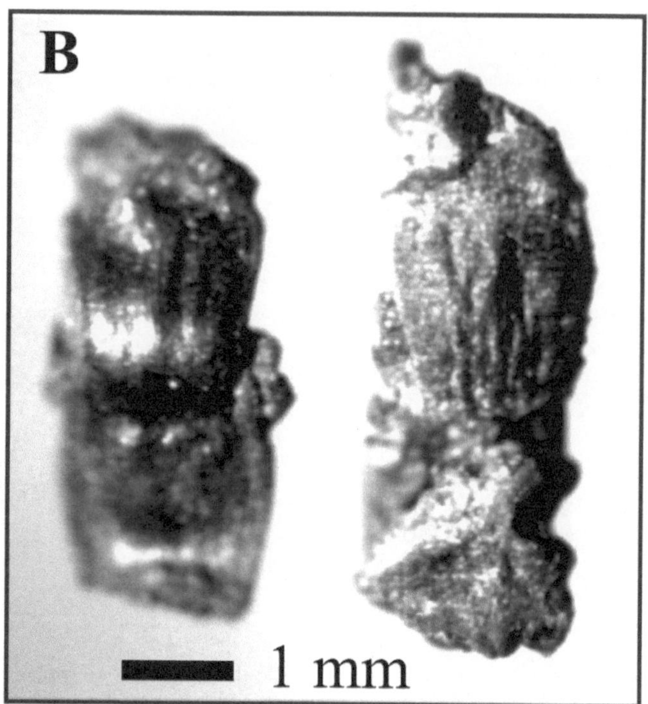

Figure 6.3: Rachis fragments of naked wheat (possibly hexaploid).

Ten sites (but not all the same as in the previous period) were also taken into consideration for the second half of the early Neolithic (table 6.3). Emmer is more common than einkorn at almost all of the sites in this period. There are only two exceptions, the sites of Kremenik and Kapitan Dimitrievo, and at these einkorn is predominant. For the former site this is probably connected with its location at an altitude of about 700 m, and for the latter this could be related to microclimatic or edaphic conditions. In the stores of hulled wheats found in three different sondages from Kapitan Dimitrievo there are many underdeveloped thin emmer grains. The frequency of occurrence of naked barley increases during the second half of the early Neolithic and it becomes more common than hulled barley in the late Neolithic. Lentil was as important in the second half of the early Neolithic, but numerous stores of pea and grass pea are also recorded for this period (table 6.3). Rachis fragments of naked wheats (figure 6.3) were identified in the archaeobotanical material from the sites of Kovačevo and Kapitan Dimitrievo. The upper parts of some of these rachis fragments were wider than the lower parts. These morphological features correspond at least partly to those of the hexaploid naked wheats and it is likely that some of the rachis fragments found belonged to this type of wheat. Most, however, do not exhibit clear morphological characteristics (e.g., figure 6.3). Similar rachis fragments were found on neolithic sites in Anatolia (G. Hillman, personal communication). Grains of mainly naked wheat were found in the Bulgarian early Neolithic (e.g., at Karanovo, Azmak, Čavdar and Slatina). Naked wheat was even identified in stores in the large house in Slatina dated to the first half of the early Neolithic (Dontscheva 1990). In the samples from Slatina, studied by the author of the present paper, there were numerous flat and broad grains of emmer wheat that resembled naked wheat. Similar grains were described in other Bulgarian prehistoric material, for example, from the chalcolithic site of Slatino, and in previous studies they were confused with naked wheat and spelt (Marinova et al. 2002). Similar observations of 'naked wheat-like' emmer grains were made for neolithic material from the region of Southeastern Europe at the site of Nea Nikomedeia in

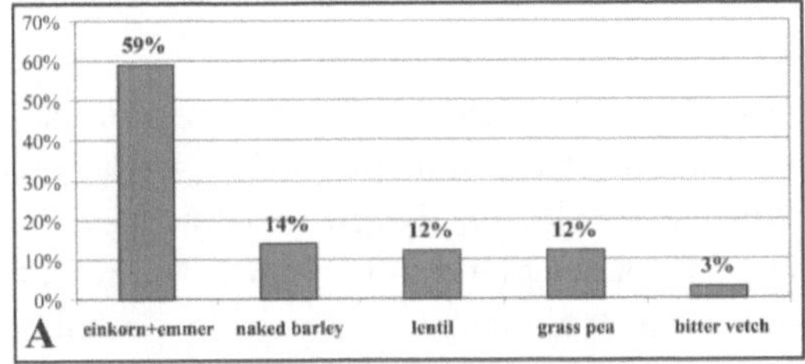

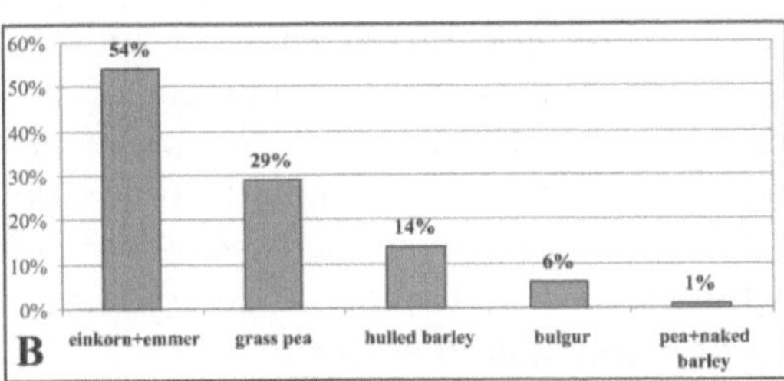

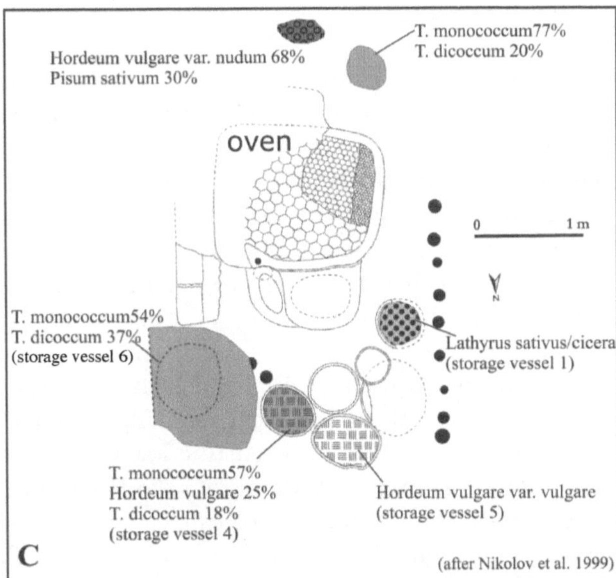

Figure 6.4: Comparison of the crop stores found in houses, as percentages. A. Slatina (after Dontscheva 1990); B. Kapitan Dimitrievo; C. Location of the stores in an early neolithic house in Kapitan Dimitrievo.

northern Greece (van Zeist and Bottema 1971). Considering these observations, the significance of naked wheats in the Neolithic in this region is to some extent overstated although their presence is important for comparative studies.

The storage finds should also be mentioned in connection with the cultivated plants. In two of the sites, Slatina in first half of the early Neolithic (Dontscheva 1990) and Kapitan Dimitrievo in the second half of the early Neolithic (the author's own study), almost entire stores were recovered and, in both cases, einkorn was the predominant stored wheat crop. Two main types of wheat stores were found in the houses, those in storage vessels and those comprising very large quantities outside well-defined containers; the crops present were not threshed in either. The hulled wheats in the vessels were usually found in the form of spikelets, that is, partially processed. In the wheat stores at Kapitan Dimitrievo almost whole ears and fragments of cereal stems could be observed, as well as mineralised fragments of wheat straw and leaves (although it is not clear whether the latter belonged to the stored ears or were used as insulating material for the wheat store). This evidence could be used to suggest that in some cases the hulled wheats (emmer and einkorn) were harvested and stored as sheaves and this is also indicated by some of the weeds that were found (see below). In one of the storage vessels at Kapitan Dimitrievo, roughly ground and broken grains of wheat and barley were found (for example, in fragments of 2–4 mm) and these probably resulted from processing prior to cooking. Such cracked grains (bulgur) could be more easily boiled and prepared for eating even if only soaked in hot water. About 20–30% of the volume of stored crops were leguminous plants (figure 6.4). They were found in storage vessels, as well as in the upper areas of the houses, for example, under or on the ceilings in baskets, bags, etc.. Storage in the upper areas of the houses has been assumed, because in some cases large quantities of charred leguminous crops were found in the highest destruction levels, above the floor. From the composition of the crop stores it is possible to suggest that the hulled wheats were grown together and this was probably to secure the harvest, as einkorn is more resistant to unfavourable conditions. All other crops were found unmixed, so it is likely that they were grown separately.

6.4.2 Weeds

Some weeds (and plants that could be weeds) were also found together with the crop plants in the archaeobotanical material. In this paper only the finds of weeds and potential weeds in the three sites that were the main focus of discussion (Kovačevo, Slatina, Kapitan Dimitrievo) are considered, and those that have been identified to species level are given in table 6.4. Many of the weeds are archaeophytes; that is, they were included in the Bulgarian flora during prehistoric times. Numerous species indicate winter crops; the climatic conditions in the area are also favourable for cereal winter crops and wheats sown in autumn give higher yields than those sown as summer crops. Barley and pulses on the other hand could be summer crops, but the weeds found in these stores provide mixed evidence, and given the current state of knowledge it is difficult to assess this definitively. The maximum growing height of a number of weeds reaches about 30–40 cm and this could be considered as an indication that the wheat crops were probably harvested by cutting low to the ground, a method that is usually done by sickles. The flint assemblages studied from the Bulgarian early Neolithic show an increase of flint sickle elements in the second part of the early Neolithic (Gurova 2001). The habitat requirements of weed and potential weed taxa indicate that the fields probably had light, fertile soils.

Setaria verticillata/viridis is one of the most numerous 'weeds' at the site of Kovačevo and it appears in 98% of the samples. The high percentage presence could be explained by the fact that this plant was collected for use. There are also some finds of larger quantities of *Setaria* stored in sheaves from later periods in the region (Tschakalova and Bozilova 1984). An alternative explanation could be that these small seeds are derived from dung; there are numerous and frequent finds of sheep/goat coprolites in the three sites considered. However, for the site Kovačevo it would be difficult to prove either of the two hypotheses because of the very bad preservation conditions at the site. Some of the mineralised material probably originates from coprolites. A number of the chaff and awn fragments are mineralised because of their high silica (SiO_2) content. In general the spikelet forks found at the sites look robust, possibly due to the accumulated silica.

6.4.3 Collected plants

A great range of collected plants was found at the sites, which indicates a good knowledge and rational use of wild plant resources. These plants originated from varied habitats in the surrounding area, including riverine and mixed oak forests, their borders, rocky areas, etc. Some of the light-demanding species (e.g., *Cornus mas*, *Rubus* sp., *Sambucus* sp., *Pistacia terebinthis*) probably indicate the forest borders or rocky slopes, but it should be not forgotten that usually their wider distribution is also a result from opening of the forests through human activities.

In the material studied the most abundant and numerous collected plant was the cornelian cherry (*Cornus mas*); it was present in a majority of contexts in Kovačevo (92%), Kapitan Dimitrievo (88%) and in Slatina (60%). The cornel is probably to some extent over-represented, because even when there are only very small fragments in the samples they are easily identifiable. The wood charcoal analysis from Kovačevo also shows a very high ubiquity for the wood of the cornel, probably due to the fact that it was used as a building material. There is evidence that the fruits of plum (*Prunus* sp.), raspberry (*Rubus idaeus*), blackberry (*Rubus caesius*), strawberry (*Fragaria vesca*), physalis/winter or bladder cherrry (*Physalis alkekengi*), hazel (*Corylus avellana*), elder (*Sambucus* sp.), mountain ash (*Sorbus aucuparia*), and apple/pear (*Malus/Pyrus* sp.) were collected and consumed either immediately or later, in dry state. Such finds are especially numerous in the houses. Pips of wild wine (*Vitis vinifera* ssp. *sylvestris*) were found in seven of the sites distributed throughout the whole of the territory of Bulgaria. In the sites of Kovačevo and Kapitan Dimitrievo, were the sub-Mediterranean element in the vegetation is

Sites	Taxon	Apophytes	Archaeophytes	On calcerous or basic soils	Xerophytes	Oriental	Mediterranean	On sandy, light soils	Eutrophic	Thermophilous	Winter crop	Weak competitor	Perennial	Growing height
3	*Adonis flammea*		+	+	+	.	.	+	+	+	+	+	.	10–40 cm
3	*Ajuga chamaepitys*	+	.	+	+	.	+	+	+	.	+	+	.	20–40 cm
1,3	*Anagallis arvensis*		+	.	.	.	+	.	.	+	+	.	.	15–40 cm
1,3	*Asperula arvensis*	+	.	+	.	.	.	.	.	.	+	+	.	15–30 cm
2,3	*Bromus sterilis/tectorum*	+	.	.	.	.	+	.	.	.	+	.	.	80–140 cm
1,2,3	*Bromus arvensis*	+	.	.	+	.	+	.	+	.	+	.	.	80–140 cm
3	*Convolvulus arvensis*		+	.	+	+	+	.	+	.	.	.	.	50–140 cm
3	*Coronilla* cf. *scorpioides*	+	.	+	+	.	+	.	.	.	.	+	+	60–90 cm
2,3	*Fumaria officinalis*	+	.	.	.	.	.	+	.	.	(+)	.	.	10–40 cm
2,3	*Galega officinalis*	+	.	.	.	.	.	+	.	.	+	+	+	20–60 cm
2,3	*Galium* cf. *mollugo*	+	.	.	.	.	.	.	+	.	+	.	.	30–100 cm
1,2,3	*Galium spurium*	+	.	+	.	.	+	.	+	.	+	.	.	30–100 cm
3	*Heliotropium europaeum*		+	.	+	.	+	+	+	+	.	+	.	10–30 cm
3	*Lapsana communis*	+	.	.	.	.	.	+	+	.	+	+	.	30–120 cm
1,2,3	*Lithospermum arvense*		+	+	.	.	+	.	+	.	.	.	.	10–50 cm
1.3	*Polycnemum arvense*		+	.	+	+	+	+	+	+	+	+	.	10–30 cm
1,2,3	*Polygonum aviculare*		+	.	.	+	.	.	.	.	.	.	.	10–30 cm
1,2,3	*Polygonum convolvulus*		+	.	.	.	.	.	.	.	(+)	.	.	50–180 cm
1,2,3	*Portulaca oleracea*		+	+	.	.	+	+	+	+	.	+	.	10–40 cm
3	*Plantago* cf. *lanceolata*	+	.	.	.	.	.	.	+	+	.	+	+	15–50 cm
1,2	*Scleranthus annuus*	+	(+)	.	.	.	.	+	.	.	.	+	.	10–20 cm
1,2,3	*Setaria verticillata*		+	.	.	.	+	.	.	.	.	.	.	15–50 cm
1,2,3	*Setaria viridis*		+	.	.	.	+	+	.	.	.	.	.	20–60 cm
1,2,3	*Sherardia arvensis*		+	(+)	.	.	+	.	.	+	.	+	.	10–20 cm
2,3	*Teucrium chamaedrys*	+	.	+	+	.	.	.	.	.	.	+	.	15–30 cm
1,2,3	*Thymelaea passerina*		+	+	+	.	+	.	+	+	.	.	.	15–40 cm
2,3	*Valerianella dentata*		(+)	+	.	.	.	.	.	.	.	.	.	10–30 cm
1,2,3	*Verbena officinalis*		+	.	.	.	+	.	+	+	+	.	.	10–30 cm
1,2,3	*Vicia hirsuta*		+	.	+	.	+	.	+	+	+	.	.	30–90 cm
1,2,3	*V. tetrasperma*	+	.	.	.	.	+	.	+	+	+	.	.	20–70 cm

Table 6.4: Documented weeds (and potential weeds) and some of their characteristics relevant to the discussion. Sites: 1= Kovacevo (EN-1st half); 2= Slatina (EN-2nd half); 3= Kapitan Dimitrievo (EN-2nd half).

stronger, terebinth (*Pistacia terebinthus*) also appears in the archaeobotanical record and was obviously collected.

6.5 Connections with the neighbouring regions

In general the predominant components of the archaeobotanical material at the Bulgarian sites are the cultivated plants, which provide evidence of connections with the Near East and the Mediterranean area, as is the case in the all neighbouring areas in the region—for example, the territory of northwest Turkey, northern Greece, FYR Macedonia and Serbia. These cultivated plants correspond to the Near Eastern crop assemblage. A crop that should also be mentioned in connection with this is chickpea (*Cicer arietinum*), which is rare in the Bulgarian Neolithic. It appears in crop stores at the site of Gălăbnik and in the flotation samples from tell Kapitan Dimitrievo. Chickpea is clearly a Mediterranean plant commonly found at neolithic sites in Anatolia (Nesbitt 1995). A find of chickpea is also known from the early Neolithic of Thessaly (Kroll 1981). Chickpea only appears in the southern part of the area in question during the early Neolithic and at sites that have evidence of clear connections with northern Greece and Anatolia. It did not become an established crop in the territory of Bulgaria probably because of the climatic conditions. The distribution of many of the weed species is concentrated in the Mediterranean and especially in the east of this region. Of particular interest in this connection are certain taxa, for example, *Heliotropium europaeum*, *Thymelaea passerina* and *Adonis* sp.; these are considered as archaeophytes, which do not belong to the natural vegetation of Bulgaria and were brought in with neolithic agriculture. The connections between northern Greece (west Macedonia and Thessaly) and Anatolia, and the similarities in the development of the Neolithic in these regions as observed in the pottery and other archaeological evidence (Nikolov 1999) could be confirmed from the archaeobotanical data. The composition of the crop plants at Nea Nikomedeia (van Zeist and Bottema 1971) corresponds to that found at Gălăbnik (Marinova et al. 2002) and in the first half of the early Neolithic at Slatina (Dontscheva 1990). One of the characteristic crops for the Bulgarian early Neolithic, *Lathyrus sativus/cicera*, was found in storage contexts in the early Neolithic at Prodromus in Thessaly (Halstead and Jones 1980). As a whole, the suite of the crop plants established in Neolithic Servia (Hubbard and Housley 2001) seems to be quite similar to that founded in southern Bulgaria. The recently published archaeobotanical studies of neolithic sites from northern Greece (Valamoti 2004) also provide good parallels for early neolithic Bulgarian plant assemblages.

6.6 Conclusions

Almost all of the crop plants recorded for the entire Neolithic and Chalcolithic periods were present from the earliest stages of the Neolithic (6000-5605 cal BC). The cultivated plants correspond to the so called Near Eastern crop assemblage. The natural distributions of many of the weeds discovered at the sites are in the Mediterranean area and this also confirms the Near Eastern connections. The crop stores found in the houses show that the cereals were kept in an un-threshed state, and the high concentration of chaff suggests that they were threshed and cleaned on site. The prevailing sowing time was autumn and the fields were on light, sandy soils with high fertility. The indication from some of the weeds is that the cereals were harvested low to the ground.

Differences between the regions of Bulgaria and the various time periods seem to be more apparent in terms of the quantitative proportions of the crops rather than in the principal composition of the cultivated plants; in some cases these differences might reflect limited data rather than real discrepancies.

Acknowledgements

I am very greatful to Dr. Krum Băčvarov, Institute of Archaeology and Museum, Sofia, for his valuable help with this chapter.

References

Arnaudov, N. (1939). Pflanzenreste aus prähistorischen Siedlungen in Südbulgarien. *Österreichische Botansche Zeitschrift 88*(1), 54–57.

Cakalova, E. and E. Sarbinska (1986). Pflanzenreste aus der neolithischen Siedlung Kremenik bei Sapareva Banja. *Studia Praehistorica 8*, 48–56.

Davis, B., S. Brewerb, A. Stevensona, J. Guiotc, and Data Contributors (2003). The temperature of Europe during the Holocene reconstructed from pollen data. *Quaternary Science Reviews 22*, 1701–1716.

Dennell, R. (1978). *Early Farming in South Bulgaria from the 4th to the 3rd Millennia BC.* BAR International Series 45. Oxford: British Archaeological Reports.

Dontscheva, E. (1990). Plant macrorest research of early neolithic dwelling in Slatina. *Studia Praehistorica 10*, 86–89.

Filipovitch, L. and I. Stefanova (1998). Anthropogenic changes in the vegetation of the Balkan Range according to data obtained from polen and macrofossil analyses. *Phytologia Balcanica 4*, 37–44.

Görsdorf, J. and J. Bojadziev (1996). Zur Absoluten Chronologie der bulgarischen Urgeschichte. Berliner C14 Dattierungen von bulgarischen archäologischen Fundplätzen. *Eurasia Antiqua 2*, 105–173.

Gurova, M. (2001). Eléments de tribulum de Bugarie—références ethnographiques et contexte préhistorique. *Archaeologia Bulgarica 5*(1), 1–19.

Halstead, P. and G. Jones (1980). Early neolithic economy in Thessaly: some evidence from excavations at Prodromos. *Anthropologika 1*, 93–117.

Hopf, M. (1973). Frühe Kulturpflanzen aus Bulgarien. *Jahrbuch des Römisch-Germanischen Zentralmuseums Mainz 20*, 1–47.

Hopf, M. (1988). Frühneolithische Kulturpflanzen aus Polajnica-Palteau bei Targoviste (Bulgarien). *Studia Praehistorica 8*, 34–36.

Hubbard, R. and R. Housley (2001). The agriculture in prehistoric Servia. In C. Ridley, K. Wardle, and C. Mould (Eds.), *Servia I: Anglo-Hellenic Rescue Excavations 1971-73*, British School at Athens Supplementary Volume 32, pp. 330–336. Oxford: Oxbow Books.

Huttunen, A., R.-L. Huttunen, Y. Vasary, H. Panovska, and E. Bozilova (1992). Late glacial and Holocene history of flora and vegetation in the Western Rhodopes Mountains, Bulgaria. *Acta Botanica Fennica 144*, 63–80.

Kroll, H. (1981). Thessalische Kulturpflanzen. *Zeitschrift für Archäologie 15*, 97–103.

Lichardus-Itten, M., J.-P. Demoule, L. Perniceva, M. Grebska-Kulova, and I. Kulov (2002). The site Kovačevo and the beginnings of the neolithic period in southwestern Bulgaria. In M. Lichardus-Itten, J. Lichardus, and V. Nikolov (Eds.), *Beiträge zu jungsteinzeitlichen Forschungen in Bulgarien*, Volume 74 of *Saarbrücker Beiträge zur Altertumskunde*. Bonn: Dr. Rudolf Habelt Verlag.

Lisitzina, G. and L. Filipovitch (1980). Palaeoethnobotanical findings in the Balkan Peninsula. *Studia Praehistorica 4*, 5–90.

Marinova, E. (2006). *Vergleichende paläoethnobotanische Untersuchung zur Vegetationsgeschichte und zur Entwicklung der prähistorischen Landnutzung in Bulgarien.* Dissertationes Botanicae 401. Stuttgart: Gebr. Borntraeger Verlagsbuchhandlung, Science Publishers.

Marinova, E., E. Tchakalova, D. Stoyanova, S. Grozeva, and E. Doceva (2002). Ergebnisse archäobotanischer Untersuchungen aus dem Neolithikum und Chalcolithikum in Südwestbulgarien. *Archaeologia Bulgarica 4*, 1–11.

Nesbitt, M. (1995). Plants and people in ancient Anatolia. *Biblical Archaeologist 58*, 68–81.

Nikolov, V. (1992). Frühneolithisches Haus in Slatina (Sofia). In V. Nikolov (Ed.), *Ausgrabungen und Forschungen 25*, pp. 163. Sofia: Verlag der Bulgarischen Akademie der Wissenschaften.

Nikolov, V. (1999). The Neolithic culture in the Bulgarian lands in the context of Anatolia and Balkan. *Annual of Department of Archaeology New Bulgarian University 2/3*, 133–144.

Nikolov, V. (2000). Neolithische Keramikkomplexe in Thrakien. In S. Hiller and V. Nikolov (Eds.), *Karanovo: Band III, Beiträge zum Neolithikum in Südosteuropa*, pp. 11–19. Wien: Phoibos Verlag.

Nikolov, V. (2002). Problems of cultural development and chronology of the early Neolithic in west Bulgaria. In M. Lichardus-Itten, J. Lichardus, and V. Nikolov (Eds.), *Beiträge zu jungsteinzeitlichen Forschungen in Bulgarien*, Volume 74 of *Saarbrücker Beiträge zur Altertumskunde*, pp. 159–164. Bonn: Dr. Rudolf Habelt Verlag.

Nikolov, V. (2004). Dynamics of the cultural processes in neolithic Thrace. In V. Nikolov, K. Băčvaro, and P. Kalchev (Eds.), *Prehistoric Thrace*, pp. 18–25. Sofia: Stara Zagora.

Panayotov, I., I. Gatsov, and T. Popova (1992). "Pompena stancia" bliz s. Malk Preslavec - rannoneoliticheskoe poselene s intaramuralnmi pogrebenijami. *Studia Praehistorica 11/12*, 51–61. In Russian. ??Check translation??

Stefanova, I. and B. Ammann (2003). Lateglacial and Holocene vegetation belts in the Pirin Mountains (southwestern Bulgaria). *The Holocene 13*, 97–107.

Thanheiser, U. (1997). Botanische Funde. In S. Hiller and V. Nikolov (Eds.), *Karanovo. Bild I, Die Ausgrabungen im Südsektor 1984-1992*, Österreichisch-bulgarische Ausgrabungen und Forschungen in Karanovo, pp. 429–454. Horn/Wien: Berger & Söhne.

Tonkov, S. and E. Bozilova (1992). Paleoecological investigation of Tschokljovo marsh (Konjavska Mountain). *Annuaire de l'universite de Sofia "St. Kliment Ohridski", Faculte de Biologie, Botanique 2*, 5–16.

Tschakalova, E. and E. Bozilova (1984). Rastitelni ostanki ot rannobronzovata epoha. *Annuaire de l'universite de Sofia "St. Kliment Ohridski", Faculte de Biologie, Livre 2–Botanique 74*, 18–27.

Tschakalova, E. and E. Bozilova (2002). Paleoecologichni i paleoetnobotanicni materiali ot selishtanat mogila do gr. Rakitovo. In A. Raduncheva, V. mit Beiträgen von Macanova., I. Catsov, G. Kovacev, G. Georgiev, E. Tschakalova, and E. Bozilova (Eds.), *Neolitnoto selishte do grad Rakitovo*, Ausgrabungen und Forschungen 29, pp. 192–201. Sofia: Gal-Iko.

Valamoti, S. M. (2004). *Plants and People in Late Neolithic and Early Bronze Age Northern Greece: An Archaeobotanical Investigation.* BAR S1258. Oxford: British Archaeological Reports.

van Zeist, W. and S. Bottema (1971). Plant husbandry in early neolithic Nea Nikomedeia, Greece. *Acta Botanica Neerlandika 20*, 542–538.

Willis, K. and K. Bennet (1994). The neolithic transition—fact or fiction? *The Holocene 4*, 326–330.

Zohary, D. and M. Hopf (1993). *Domestication of Plants in the Old World* (2nd ed.). Oxford: Claredon Press.

Chapter 7

The spread of cultivated plants in the region between the Carpathians and Dniester, 6th–4th millennia cal BC

Felicia Monah
Institute of Archaeology, Iaşi, Romania

7.1 Geographic borders and natural environment

The region we are concerned with borders on the mountain range of the eastern Carpathians in the west, the Danube and the Black Sea in the south, and the Dniester in the east and northeast (figure 7.1). Most of this region is covered by a vast system of sub-Carpathian hills and plateaux. The northern area, which comprises the depressions of Jijia and Bălţi, has the characteristics of a forest steppe, while in the south of the region there are a few genuine steppes.

The hydrographical system is made up of the rivers Siret, Prut and Dniester, as well as a number of smaller rivers. While the plateaux and the sub-Carpathian hills are covered with brown diluvial clay and brown podzol-like forest soils, the plains are characterised by the presence of chernozems, which had already formed by the beginning of the neolithisation process, although at that time they were probably quite shallow. The climate of the region was of the temperate-continental type. The chronological sequence I refer to overlaps to a great extent with the climatic period known as the Atlantic, between approximately 6000–3000 cal BC, which had higher mean temperatures and greater rainfall than the present day (Kremeneckij 1991, p.159).

7.2 Neolithisation

As far as the neolithisation of the area between the Carpathians and the Dniester is concerned, there are two hypotheses: (i) that the appearance of the Neolithic in the area between the Dniester and the south Bug was the result of the evolution of a local mesolithic culture—i.e., the so-called Bug-Dniester culture (Markevič 1974; Larina 1994, pp.42–43), or; (ii) that the neolithisation of the area between the Carpathians and the Dniester was a consequence of the migration of the Starčevo-Criş communities, which came from the western and northwestern part of the territory.

The Bug-Dniester culture is, broadly speaking, contemporary with the Starčevo-Criş culture and the LBK. However, for the early phases of Bug-Dniester there are only two questionable

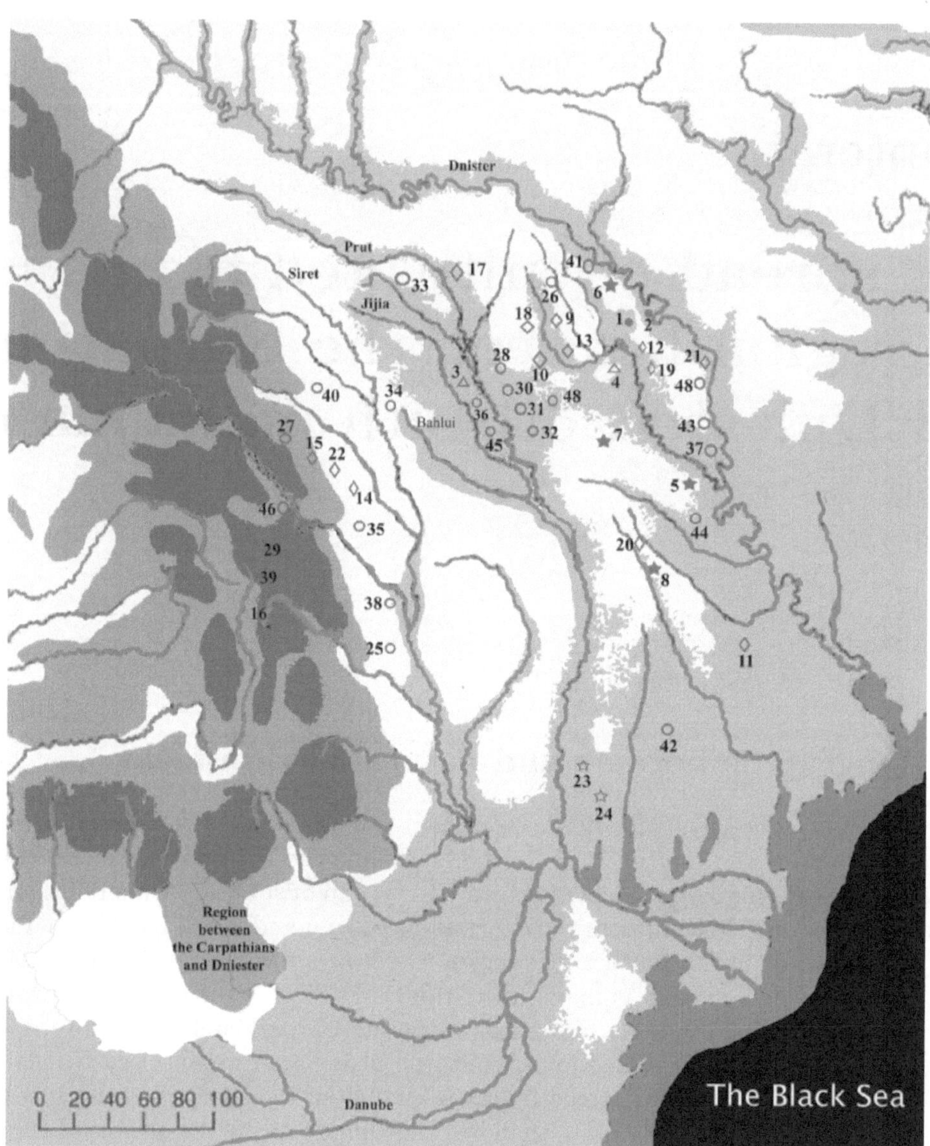

Figure 7.1: The study area and sites referred to in the text where plant remains have been recovered. *Bug-Dniester culture (?)*: 1—Ruptura, 2—Soroca. *Starčevo-Criş culture*: 3—Glăvăneştii Vechi, 4—Sacarovca. *Linearbandkeramik (LBK)*: 5—Brăneşti, 6—Gura Camenca, 7—Mândreşti, 8—Ruseştii Noi I. *Pre-Cucuteni culture*: 9—Alexandrovca, 10—Bahrineşti, 11—Cărbuna, 12—Coşerniţa, 13—Floreşti, 14—Ghigoieşti, 15—Izvoare, 16—Poduri, 17—Pererâta I, 18—Putineşti I, 19—Rogojeni I, 20—Ruseştii Noi I, 21—Solonceni I, 22—Târpeşti. *Stoicani-Aldeni-Bolgrad culture*: 23—Lopăţica, 24—Vulcăneşti II. *Cucuteni culture*: 25—Bălăneasa, 26—Bolboci, 27—Bodecsti-Frumuşica, 28—Brânzeni, 29—Calu-Piatra Şoimului, 30—Costeşti, 31—Cuconeştii Vechi, 32—Cuban, 33—Drăguşeni, 34—Hăbăşeşti, 35—Izvoare, 36—Cucuteni, 37—Jora de Sus, 38—Mărgineni, 39—Poduri, 40—Preuteşti, 41—Racoveţ, 42—Ruseştii Noi, 43—Solonceni, 44—Ţipordei, 45—Valea Lupului, 46—Văleni, 47—Varvarovka, 48—Vâhvatinţi.

dates from the Soroca settlements (see figure 7.1) of 7520±120 bp and 7420±120 bp[1] (Markevič 1974, pp.128–129, 138; Larina 1994, p.43).

The first hypothesis, autochthonous development, is not very convincing for several reasons. At Soroca there are several settlements assigned to the Bug-Dniester culture which have led researchers to claim local evolution from the Mesolithic to the Neolithic, but the excavations were undertaken over a very small area (between 70 and 120 square metres) and there is insufficient evidence to sustain this argument. In contrast, V. I. Markevič and O. Larina assert that the inhabitants of these settlements were hunters and fishers. The ceramics (fragments of five vessels) are of good quality and Larina considers that they were imported—which could explain the impressions of caryopses. No pollen analysis has been carried out and the only argument for agriculture is the discovery of several polished pieces of flint (Larina 1994, pp.44–45). Moreover, most of the settlements described as Bug-Dniester culture were categorised as such on the basis of only surface material. Similar settlements have been reexamined after excavation and are now described as Starčevo-Criş (Larina 1994, p.43). A good example of this is the settlement of Sacarovca I, initially labeled as Bug-Dniester but after excavation attributed to the Starčevo-Criş culture (Kuzminova et al. 1998, pp.166–170).

On balance, therefore, the second hypothesis of exogenous origins is more convincing because the settlements in the area between the Carpathians and the Dniester belong to the late stages (i.e., the 3rd and 4th) of the Starčevo-Criş culture and are thus a developed, rather than formative, neolithic culture.

7.3 Overview

There is an outstanding tradition of archaeobotanical studies in the region we are dealing with. In the first half of the 20th century renowned archaeobotanists, for example, Fritz Netolilitzky, Radu Popovici and, more recently, Traian Săvulescu, Zoia Januševič, Marin Cârciumaru and Natalia Kuzminova, have been actively involved in research. There have been palynological and anthracological analyses and studies of plant macro-remains. During the last few decades numerous plant macro-remains from chalcolithic and neolithic sites in the above-mentioned area have been analysed and it seems likely that the majority of the species cultivated in the period of interest have been identified, although certain surprises can still occur.

7.3.1 Archaeobotantical material that has been studied

There are a few pollen analyses (Marinescu-Bîlcu et al. 1981; Marinescu-Bîlcu et al. 1984; Kremeneckij 1991; Volontir 1989) and a number of anthracological studies (Popovici 1932, 1934), but most of the archaeobotanical information is provided by the identifications of plant remains. In Bessarabia studies of the impressions left by the seeds and the fragments of other plant parts on pottery and in the adobe of houses were carried out on a large scale. This method has seldomly been used in Romania where, more commonly, the rich carbonised material discovered mainly on chalcolithic settlements has been investigated.

7.4 Plants cultivated in the early Neolithic (ca. 5600–5400 cal BC)

The Starčevo-Criş tribes brought with them the primary cultigens as outlined in tables 7.1 and 7.2. The most interesting information regarding the cultivation of plants by the Starčevo-Criş communities comes from the Sacarovca I settlement. Zoia Januševič, for instance, has studied

[1]The laboratories are unspecified.

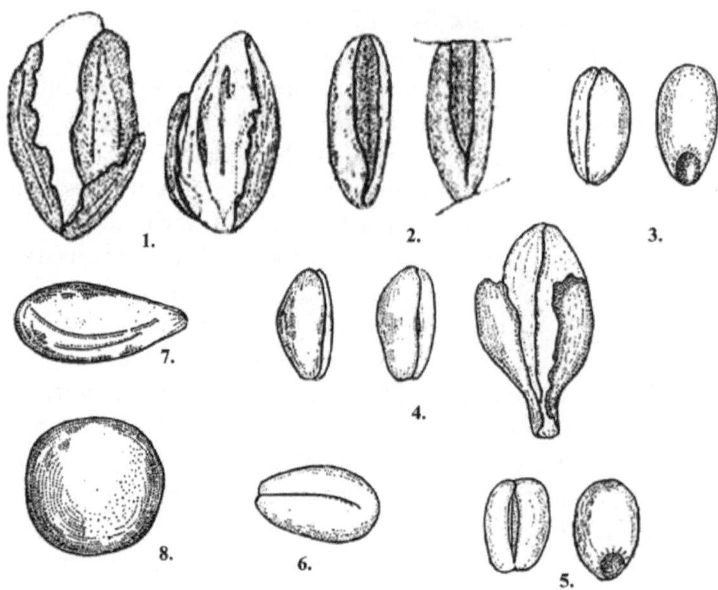

Figure 7.2: Impressions of plant remains from the Starčevo-Criş site of Sacarovca (after Kuzminova et al. 1998): 1 *Hordeum vulgare*, 2 *Avena* sp., 3 *Triticum monococcum*, 4 *Triticum dicoccum*, 5 *Triticum aestivum* ssp. *compactum*, 6 *Triticum spelta*, 7 *Malus*/ *Pyrus*, 8 *Pisum* sp.

an important set of impressions (Januševič 1986, pp.4–14). Natalia Kuzminova identified the material recovered in the 1989–1991 and 1995 campaigns. In this recent material there is also a quantity of carbonised remains. The burned adobe frequently retained impressions of plant organs, for example, cereal spikelet fragments, paleae and rachises; but the impressions of fruit kernels and seeds from tree species and of seeds of leguminous plants were found more rarely (Kuzminova et al. 1998).

Zoia Januševič noticed that on most adobe fragments there were impressions of paleae and spikelet fragments of *Triticum dicoccum*. Clear impressions of *Hordeum vulgare* var. *coeleste* and *Triticum spelta* were identified on the burnt adobe (Januševič 1986, pp.5–8) and, in addition, two narrow impressions appeared to indicate the presence of oats (*Avena* sp.). Impressions possibly of *Pisum* sp. were discovered in the first set of archaeobotanical materials at Sacarovca I (Januševič 1986, p.8). More distinct impressions of kernels of *Prunus insititia* and *Prunus spinosa* were also identified, whereas for those attributed to the genus *Malus* it was not possible to specify the species or determine whether or not they were cultivated (Januševič 1986, pp.4–14). A few paleae remains, which were attributed to *Triticum dicoccum*, were discovered in the less intensely burnt adobe. In the same adobe a few incomplete caryopses of *Triticum spelta* were found, together with carbonised seeds of segetal species (Kuzminova et al. 1998, p.167).

Based on the frequency of the identifications of impressions, Zoia Januševič concluded that the inhabitants of Sacarovca I cultivated *Triticum dicoccum* and *Hordeum vulgare* var. *coeleste*, while *Triticum spelta* and *Avena* sp., which appear less frequently (figure 7.2), seem not to have been grown deliberately (Januševič 1986, pp.13–14).

Natalia Kuzminova succeeded in identifying 878 impressions of plant organs, caryopses and seeds. Besides the impressions of plants, my colleague from Chişinău also analysed a considerable quantity of carbonised remains discovered in a mud hut. Kuzminova identified impressions of the following taxa: *Triticum monococcum*, *Triticum dicoccum*, *Triticum spelta*, *Triticum compactum*, *Hordeum vulgare* (hulled and naked), *Avena* sp., *Panicum miliaceum*, *Pisum sativum*, *Prunus*

Taxon	Neolithic			Chalcolithic		
	Starčevo Criş	Bug-Dniester(?)	LBK	Pre-Cucuteni	Stoicani-Aldeni-Bolgrad	Cucuteni
Triticum aestivum	+	.	.	+	.	+
Triticum compactum	+	.	.	+	.	+
Triticum dicoccum	+	+	+	+	+	+
Triticum monococcum	+	+	.	+	+	+
Triticum cf. *durum*	.	.	.	+	.	+
Triticum spelta	+	+	+	+	+	+
Triticum sp.	.	.	+	.	.	.
Hordeum vulgare	+	+?	.	+	.	+
Hordeum vulgare var *coeleste*	+	.	.	+	+	.
Hordeum distichum	.	.	.	.	.	+
Avena sativa	.	.	.	+	.	+
Avena sp.	+	.	+	.	+	.
Secale cereale	.	.	.	+	.	+
Panicum miliaceum	+	+?	+	+	+	+
Lathyrus sativus	+	.	.	+	.	+
Pisum sativum	+	.	.	+	.	+
Vicia ervilia	+	.	.	+	.	+
Vicia sativa	.	.	.	.	.	+
Coriandrum sativum	.	.	.	+	.	+
Fagopyrum esculentum	.	.	.	+	.	.
Camelina sativa	.	.	.	.	.	+
Sinapis alba	.	.	.	.	.	+
Papaver somniferum	.	.	.	.	.	+
Malus sp.	.	.	.	.	.	+
Pyrus sp.	.	.	.	+	.	.
Malus/ Pyrus	+	.	.	.	.	+
Cerasus avium	.	.	.	+	.	.
Armeniaca vulgaris	.	.	.	+	.	.
Prunus cerasifera	+	.	.	+	.	.
Prunus domestica	+	.	.	.	.	+
Prunus insititia	+	.	.	.	.	+
Vitis vinifera	.	.	.	.	.	+

Table 7.1: Plants cultivated in the region between the Carpathians and the Dniester, 6th–4th millennia cal BC.

	Neolithic			Chalcolithic		
Taxon	Starčevo Criş	Bug-Dniester(?)	LBK	Pre-Cucuteni	Stoicani-Aldeni-Bolgrad	Cucuteni
Agrostemma githago	.	.	.	+	.	+
Agrostemma sp.	+	.	.	.	.	.
Alyssum sp.	+	.	.	.	.	+
Amaranthus retroflexus	.	.	.	+	.	+
Atriplex sp.		.	.	+	.	.
Brassica rapa	.	.	.	+	.	.
Brassica nigra		.	.	.	.	+
Bromus secalinus	.	.	.	.	.	+
Bromus sp.		.	.	.	.	+
Chenopodium album		.	.	+	.	+
Chenopodium hybridum	.	.	.	.	.	+
Convolvulus arvensis		.	.	.	.	+
Cuscuta sp.		.	.	.	.	+
Datura stramonium	.	.	.	.	.	+
Echinochloa crus-galli		.	.	.	.	+
Galium sp.	+	.	.	.	.	.
Galium spurium	.	.	.	+	.	+
Glaucium corniculatus		.	.	.	.	+
Leonurus cardiaca	.	.	.	+	.	.
Lolium perenne		.	.	.	.	+
Lolium temulentum	.	.	.	.	.	+
Medicago lupulina	.	.	.	.	.	+
Medicago falcata	.	.	.	.	.	+
Polygonum sp.	+	.	.	.	.	.
Polygonum aviculare	.	.	.	+	.	+
Polygonum convolvulus	.	.	.	+	.	+
Polygonum hydropiper	.	.	.	+	.	.
Polygonum persicaria	.	.	.	+	.	.
Ranunculus sp.		.	.	.	.	+
Reseda lutea	.	.	.	.	.	+
Rumex acetosa	.	.	.	+	.	+
Rumex crispus	.	.	.	+	.	+
Saponaria officinalis	.	.	.	.	.	+
Setaria sp.	+	.	.	.	.	.
Sinapis arvensis	.	.	.	.	.	+
Solanum nigrum	.	.	.	.	.	+
Solanum sp.	+	.	.	.	.	.
Stellaria media		.	.	.	.	+
Thlaspi arvense		.	.	.	.	+
Vicia cracca	.	.	.	+	.	.

Table 7.2: Other plants discovered in neolithic and chalcolithic settlements in the region between the Carpathians and the Dniester, 6th–4th millennia cal BC.

Figure 7.3: Carbonised plant remains from the Starčevo-Criş site of Sacarovca (after Kuzminova et al. 1998): 1 *Triticum dicoccum*, 2 *Prunus cerasifera*, 3 *Prunus insititia*.

insititia and *Prunus domestica* (figure 7.3). Most of the impressions that were identified (a total of 508) were of the genus *Triticum*, but 171 of these could not assigned to species (Kuzminova et al. 1998, p.167). As Zoia Januševič pointed out, in this material *Triticum dicoccum* was also dominant, with 145 impressions (43.0%), and followed by *Triticum monococcum*, with 93 (27.6%) impressions. Similarly, Natalia Kuzminova concludes that *Triticum spelta* and *Triticum compactum* were not grown as pure crops (Kuzminova et al. 1998, pp.168–169).

Although the statistical data based on the identifications of the impressions are less reliable than those based on plant macro-remains, it should be noted that barley was the second most frequently represented crop, with a total of 143 impressions. The impressions of the two varieties of barley, hulled and naked, are almost equal in number (Kuzminova et al. 1998, p.170). Oats are poorly represented, with only 17 impressions, and this seems to confirm Zoia Januševič's statement that it was not grown deliberately, but was a weed that infested the wheat and barley fields (Kuzminova et al. 1998, p.170). A total of 97 impressions of *Panicum miliaceum* have been identified and it seems that this species was of considerable importance in the agriculture of the inhabitants of Sacarovca (Kuzminova et al. 1998, p.167). It is also interesting to note the presence of the seeds of *Setaria* sp., which which commonly grows in fields alongside *Panicum miliaceum*.

The identifications made on the basis of the impressions have been verified by the carbonised remains. Natalia Kuzminova confirmed the presence at Sacarovca of *Triticum dicoccum*, and in addition the existence of several cultivated trees, for example, *Prunus insititia* and *Prunus domestica*. In the carbonised material remains of *Corylus avellana*, *Quercus robur* and *Cornus mas* were also found, which can be added to the list of plants gathered by the inhabitants of Sacarovca (Kuzminova et al. 1998, p.168–171). The species of cultivated fruit trees that were identified are particularly important because they could be evidence of the beginning of the sedentarization process for the Starčevo-Criş tribes. The large range of the plants found at Sacarovca is impressive and there is only one other site of the same date, the Starčevo-Criş settlement at Glăvăneştii Vechi, where plant remains (*Triticum monococcum*) have been identified (Cârciumaru 1996, p.79).

7.4.1 A few observations

Although the information that I have presented comes from a peripheral territory of the great Pre-Sesklo-Karanovo I-Kremikovcy-Starčevo-Körös-Criş cultural complex, it should be noted that the early neolithic populations already possessed the main cultivated plants, namely wheat, barley, pulses and a few species of fruit trees. This observation confirms the hypothesis that the neolithisation of the region was a result of colonization. There is no doubt that acculturation had a significant influence on the new way of life, but this phenomenon took place after the first neolithic colonists settled in the region. In my opinion, the so-called Bug-Dniester culture is the result of the influences exerted by the Starčevo-Criş populations. The existence of an autochthonous neolithic that evolved from of a mesolithic culture in the Bug-Dniester interfluve seems improbable, but it is possible that new research supported by radiocarbon data will clarify this issue.

7.5 Plants cultivated in the middle Neolithic (ca. 5400–5000 cal BC)

The tribes of the Linearbandkeramik culture (LBK) entered the region between the Carpathians and the river Dniester from the north-west, scattering chiefly along the rivers, including the Siret and its tributaries, the middle Prut valley and in the basin of the Răut. They only partially occupied the northern part of the province and for a while cohabited with the Starčevo-Criş tribes, and also with those of the so-called Bug-Dniester culture. The LBK communities settled mainly in the regions with podzol-like (forest) soils and chernozems, preferring the river meadows with alluvial soils. It is also interesting to note that approximately 40% of the LBK sites overlie or are in the vicinity of several older Starčevo-Criş settlements.

Although over 80 LBK settlements have been discovered in the area between the Carpathians and the Dniester, the information about the plants cultivated by the inhabitants is still rather limited. At Ruseştii Noi II and Gura Camenca impressions of *Triticum dicoccum* ears have been discovered and at the LBK settlements at Ruseştii Noi I, level 2, Ruseştii Noi II, Brăneşti I and Gura-Camenca impressions of ears, paleae and caryopses of *Triticum spelta* have been also identified (Larina 1994, 1999). The absence of barley from the list of plants cultivated by the LBK tribes in this region is interesting, but it is probably due to sampling issues.

7.6 Plants cultivated in the Chalcolithic (ca. 5000–3600 cal BC)

In the sub-Carpathian region of Moldavia and in the southeast of Transylvania, in the first quarter of the 5th millennium cal BC, a phenomenon of cultural synthesis between the communities of the Boian south culture and those of the LBK took place that led to the development of the Pre-Cucuteni culture. This dynamic new culture spread eastwards and covered most of the region. The communities of the Stoicani-Aldeni-Bolgrad culture occupied the south of the territory, between the Carpathians and Dniester, for a period of about two centuries, and then were replaced by the tribes of the Cernavoda I culture, which originated in the steppe (figure 7.1).

Over 3,000 Pre-Cucuteni and Cucuteni sites have been identified in the area between the Carpathians and Dniester during 120 years of research. As a result of the large number of settlements and the vast excavations, there has been recovery of an impressive quantity of plant macro-remains, which has generated an extremely rich body of information. In the area between the river Prut and the Dniester the data derived largely from the identification of impressions left by the cultivated plants on the burnt pottery and adobe. In the area between the Carpathians

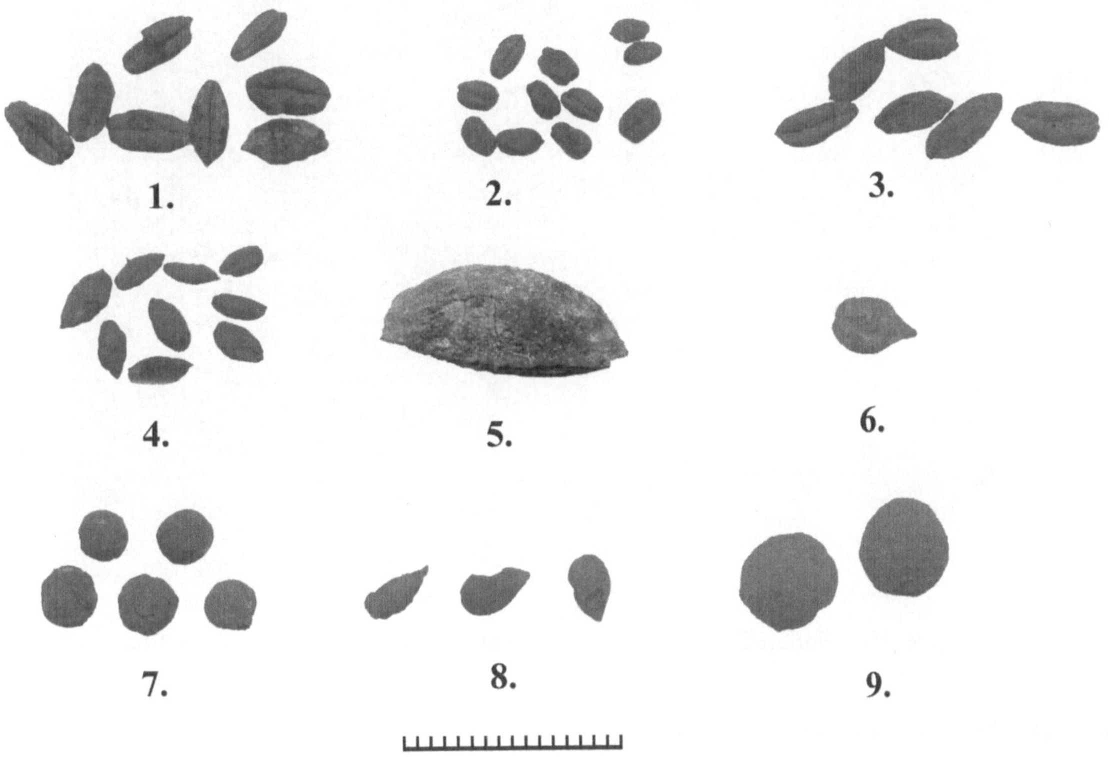

Figure 7.4: Plant remains from the chalcolithic tell of Poduri-Dealul Ghindaru (scale in mm). 1 *Triticum aestivum*, 2 *Triticum aestivum* ssp. *compactum*, 3 *Triticum dicoccum*, 4 *Hordeum vulgare*, 5 *Coriandrum sativum*, 6 *Malus/ Pyrus*, 7 *Cerasus avium*, 8 *Prunus domestica*, 9 *Vitis vinifera*.

and the Prut the study of impressions was used to a lesser extent, and instead the abundant carbonised material was the source of information on the plant economies.

7.6.1 Pre-Cucuteni culture (ca. 5000–4700 cal BC)

For the Pre-Cucuteni culture we possess several tens of thousands of identifications of impressions and an equally large number of carbonised plant remains. The quantification of all discoveries is impossible, therefore I shall describe here only a few exceptional cases. It should be mentioned that the largest cereal deposits were discovered in the tell at Poduri-Dealul Ghindaru, where there were extremely good conditions for the preservation of plant materials (figure 7.4). Thus, in habitation no. 31 no fewer than 10 deposits of carbonised cereals were found, and of these, three were in adobe storage bins. The bins were approximately one-metre-square and with a frame of about 10 to 45 centimetres high (figure 7.5) (Cârciumaru and Monah 1985; Monah and Monah 2002). Of interest in construction 44 at Poduri was the fact that a square adobe structure was found on the clay platform on its western side, which contained five grinders, fixed on clay stands with boards painted in white. The construction was described by archaeologists as a 'mill', however other interpretations were not excluded. To the east of this structure, there were four more that were tronconic in shape (55-cm diameter at the base, preserved to a height of 35 cm), which were thought to be silos (figure 7.6) (Monah 2002). These contained a large amount of carbonised cereals from which 34 kilograms were sampled. It is interesting to note

Figure 7.5: Poduri-Dealul Ghindaru 'cereal bin' from the Pre-Cucuteni level.

that the silos were specialised: two of them had been used to deposit wheat while the other two contained stores of barley (Cârciumaru and Monah 1985).

The Pre-Cucuteni cereal deposits are characterized by a remarkable purity. Generally, mixtures of wheat and barley are rare and when they are found one of the species is usually present in low frequencies. The purity of deposits could also be explained by the existence of separate fields and storage facilities. At Poduri the crops seemed to be particularly 'pure', because the seeds of segetal plants were very rare (Monah et al. 1987). The low frequencies of weeds might be an indication that separate ears were harvested, but also that there was frequent rotation of the tilled fields, such that weeds could not become established. The fact that the Pre-Cucuteni inhabitants frequently had rich harvests could also be deduced from the presence of specific features, for example, storage bins and silos, and large storage vessels, that were used for the deposition of cereals. Evidence for a subsistence economy based on cereal crops comes also from the presence of numerous grinders, as well as areas especially designated for grinding.

Although millet (*Panicum miliaceum*) has been confirmed in the identifications for Pre-Cucuteni culture, this species seems to have been cultivated less frequently than other cereals (Januševič 1976, pp.152–153). Of note is that at Târpeşti, pollen analysis indicates the presence of some fields of *Fagopyrum esculentum* during this period (Marinescu-Bîlcu et al. 1981, p.651).

Carbonised remains and impressions from other cultivated plants were found in the Pre-Cucuteni sites in Bessarabia. For example, from the Pre-Cucuteni III stratum at Ruseştii Noi I there were several fragments of carbonised seeds of *Armeniaca vulgaris*; Zoia Januševič suggests that this species was brought from Anatolia via the Balkan Peninsula (Januševič 1976, pp.183–184). Carbonised kernals of *Prunus cerasifera* were discovered in the same context—another species that was thought to have been introduced to the Carpathian-Dniester region by way of the Balkans (Januševič 1976, pp.184–185). Several kernals of *Prunus insititia* were also found in the same Pre-Cucuteni stratum (Januševič 1976, pp.185–186). I will address the complex issue of whether or not there was cultivation of trees and shrubs during the Chalcolithic in the Carpathian-Dniester region on another occasion.

Figure 7.6: Poduri-Dealul Ghindaru: A. silo in situ, B. reconstruction of the Pre-Cucuteni silo.

7.6.2 Stoicani-Aldeni-Bolograd culture (ca. 4800–4300 cal BC)

The Stoicani-Aldeni-Bolgrad culture is the result of a synthesis between Pre-Cucuteni-Cucuteni and Gumelniţa cultures. It was partially contemporary with the last phase of Pre-Cucuteni culture, the Cucuteni phases A1-3 and the first phases of Gumelniţa culture. The settlements, which have not been studied very intensely, are only found in the Carpathian-Dniester area. The archaeobotanical information is based on the identifications of plant impressions from two sites: Vulcăneşti II and Lopăiţca, where the following cultivated plant species have been identified: *Triticum monococcum*, *Triticum dicoccum*, *Triticum spelta*, *Hordeum vulgare*, *Hordeum vulgare* var. *coeleste*, *Avena* sp. and *Panicum miliaceum* (Janušević and Kuzminova 1989, p.44).

7.6.3 Cucuteni culture (ca. 4700–3600 cal BC)

The Cucuteni culture is the last great chalcolithic civilization of what Marija Gimbutas called 'Old Europe'. At this time the Chalcolithic reached its zenith in Southeastern Europe. Approximately 3,000 Cucuteni settlements were discovered in the Carpathian-Dniester area, which indicates an exceptionally high demographic density. The Cucuteni tribes had a highly developed system of agriculture and cultivated a wide range of crops. The archaeobotanical information we have at present is based mainly on identifications of carbonised plant remains and the impressions on pottery and adobe.

Carbonised plant remains have been discovered in several different contexts, for example, cereal stores, food residues and especially ritual deposits. The largest amounts of carbonised cereals were found in the tell at Poduri-Dealul Ghindaru, a site which also has the greatest range of taxa. The cereals were found in painted vessels and storage vessels, spread on the platforms of the houses and in cult pits. I present only a few cases here, which are of particular interest.

A mass of carbonised cereals that formed a 30 cm thick layer was found in the bottom of a pit, with a diameter of 0.80 m, in the Cucuteni A2 stratum in the tell at Poduri. In the middle of this mass there was a child's cranium, aged between three and four years (Monah, Monah 1996: 52-54). The cereals comprised the caryopses of *Triticum dicoccum* and *Hordeum vulgare*, the latter being the dominant species. In another pit in the tell, dated to Cucuteni phase B, there was a large storage vessel, the remains of two painted vessels, several agricultural tools and a number of plant remains, which included caryopses of *Hordeum vulgare*, *Triticum dicoccum* and *Secale cereale*, as well as seeds of wild fruits (Monah and Monah 1997, pp.59–60).

Another interesting discovery has been made recently at Poduri. A painted amphora, in which there were 248 fruits of *Coriandrum sativum*, was found on the ground floor of a two-storeyed construction. Near the amphora, there was a wooden receptacle (possibly a cask), at the bottom of which was a mixture of the fruits of *Coriandrum sativum* and *Sambucus nigra*. Coriander fruits were also found in other levels of the tell. Worthy of note at Poduri is the discovery of a kernal of *Prunus domestica* and a seed of *Vitis vinifera* in the Cucuteni A2 stratum. *Vitis vinifera* is also present in the Cucuteni B site at Varvarovka XV in Bessarabia (Januševič 1976, pp.186–187).

There are some differences in the cereal crops between the eastern and the western parts of the territory. In the west, in the Cucuteni phase A, *Triticum dicoccum* and *Triticum compactum* seem to have been used more frequently, while only low percentages of *Triticum monococcum* and *Triticum spelta* have been recorded. Presence only is confirmed for *Triticum compactum*, *Triticum durum* and *Triticum dicoccoides*. In the eastern territories, it seems that *Triticum monococcum*, *Triticum dicoccum* and *Triticum spelta* were common, while the presence of *Triticum compactum* is recorded only sporadically.

From the identifications of the plant macro-remains, it appears that barley (*Hordeum vulgare* and *Hordeum vulgare* var. *coeleste*) was as important as wheat. *Secale cereale* and *Avena sativa* were far less common, however, and seem not to have been intentionally cultivated. *Panicum miliaceum* appears to have been used frequently, especially in the eastern part of the territory.

7.7 Summary

In summary, the Carpathian-Dniester territory was neolithized later than the Balkan Peninsula and is likely a result of colonization by tribes of the Starčevo-Criş complex. In the area east of the Dniester, the neolithisation took place mainly by acculturation and, in my opinion, the Bugo-Nistreană culture is a "barbarized" aspect of the Starčevo-Criş culture.

The Starčevo-Criş tribes brought with them cultigens, of which the main crops were cereals (wheat, barley and millet), pulses (pea, *Pisum sativum* and bitter vetch, *Vicia ervilia*), and *Fagopyrum esculentum* (buckwheat). Some fruit trees were also cultivated, for example, *Prunus domestica* and *Armeniaca vulgaris*. By the Chalcolithic, the Pre-Cucuteni and Cucuteni tribes practised a remarkably sophisticated form of agriculture, and at this time other economic plants were also introduced (e.g., *Vitis vinifera* and *Coriandrum sativum*). The cultivation of some trees and shrubs (for example, *Vitis vinifera*) indicates a high degree of sedentism for the Pre-Cucuteni and Cucuteni tribes. However, towards the end of the Atlantic period—that is, by ca. 3000 cal BC—agriculture in the Carpathian-Dniester area entered a period of crisis.

References

Cârciumaru, M. (1996). *Paleoetnobotanica. Studii în preistoria şi protoistoria României.* Iaşi: Glasul Bucovinei-Helios.

Cârciumaru, M. and F. Monah (1985). Raport preliminar privind seminţele carbonizate de la Poduri-Dealul Ghindaru, judeţul Bacău. *Memoria Antiquitatis IX–XI*, 699–708.

Januševič, Z. V. (1976). *Kul'turnye rastenija Iugo-Zapada SSSR. Po paleobotaničeskim issledovanijam.* Kišinev: Ştiinţa.

Januševič, Z. V. (1986). *Kul'turnye Rastenija Severnogo Pričernomor'ja. Paleoetnobotaničeskie Issledovanija.* Kišinev, Moldova: Sţiinţa.

Januševič, Z. V. and N. N. Kuzminova (1989). Vozniknovenie i razvitie zemledel'ja v Severnom Pričernomor'e po paleobotaničeskim dannym. *Botaničeskie issledovanja. Flora i rastitel'nost' 5*, 38–49.

Kremeneckij, K. (1991). *Paleoekologija drevnejšikh zemledel'cev i skotovodov Russkoj Ravniny.* Moskva: Akademija Nauk SSSR. Institut Geografij.

Kuzminova, N., V. Dergačev, and O. Larina (1998). Paleoetnobotaničeskie issledovanija na poselenii Sakarovka I. *Revista Arheologică 2*, 166–182.

Larina, O. (1994). Neoliticul pe teritoriul Republicii Moldova. *Thraco-Dacica XI*(1–2), 41–66.

Larina, O. (1999). Kul'tura linejno-lentočnoj keramiki Pruto-Dnestrovskogo regiona . *Stratum plus 2*, 10–140. Confirm reference.

Marinescu-Bîlcu, S., A. Bolomey, M. Cârciumaru, and M. Muraru (1984). Ecological, economic and behavioural aspects of the Cucuteni A4 community at Draguseni. *Dacia N.S. XXVIII*(1-2), 41–46.

Marinescu-Bîlcu, S., M. Cârciumaru, and A. Muraru (1981). Contribution to the ecology of pre- and proto-historic habitation at Tirpesti. *Dacia N.S. XXV*, 7–33.

Markevič, V. (1974). *Bugo-Dnestrovskaja kul'tura na teritorii Moldavii.* Kišinev: Ştiinţa.

Monah, D. (2002). Découvertes de pains et restes d'aliments céréaliers en Europe de l'est et centrale. *Civilisations 49*(1–2), 77–99.

Monah, F., I. Bara, and D. Monah (1987). Observaţii asupra compoziţiei depozitelor de cereale din sşezarea Precucuteni III de la Poduri—Dealul Ghindaru. *Memoria Antiquitatis XV–XVII*, 249–262.

Monah, F. and D. Monah (1997). Stadiul cercetărilor arheobotanice pentru eneoliticul din Moldova de vest. *Memoria Antiquitatis XXI*, 297–316.

Monah, F. and D. Monah (2002). Les céréales cultivés par les population néo-énéolithiques de la Moldavie. *Civilisations 49*(1–2), 67–76.

Popovici, R. (1932). Beiträge zur Waldgeschichte Nord-Rumäniens. *Buletinul Facultăţii de Ştiinţe Cernăuti VI*, 229–250.

Popovici, R. (1934). Pădurile paleo şi neolitice din nordul României. *Buletinul Facultăţii de Ştiinţe Cernăuti VIII*, 277–295.

Volontir, N. (1989). K istorii rastitel'nost' iuga moldavii v golocene. In *Četvertičnyi period. Paleoetnologija i arkheologija*, pp. 90–97. Kišinev: Ştiinţa.

Chapter 8

Seed and fruit remains associated with neolithic origins in the Carpathian Basin

Ferenc Gyulai
Institute for Agrobotany, Tápiószele, Hungary

8.1 Introduction

The beginning of the Neolithic in Hungary partly coincides with the beginning of the Atlantic phase of the early Holocene, dated to ca. 6500–3000 cal BC (Kalicz 1993). The warm, dry climate of the preceding Boreal phase, with its frequent continental extremes, was disadvantageous to plant cultivation, whereas during the Atlantic phase (also called the Atlantic oak phase in Hungary), which was climatically similar to that of the warm, humid and temperate sub-Mediterranean, conditions were much more favourable.

On the basis of pollen analysis carried out at locations in the Great Hungarian Plain (Komlódi 1966) it has been possible to reconstruct the vegetation of the Atlantic phase (Lacza 1991). The plain is largely covered by loess and, due to the changing course of the River Tisza in the Holocene, the surface is highly dissected. It has always been floristically open from the north (North Hungarian Range), the west (Trans-Danubia) and the south (western and eastern Illyricum and the Ukrainian steppes). With time, many species that were related both climatically and ecologically migrated or were brought into the area.

Under similar conditions, therefore, the techniques and technology of plant cultivation spread rapidly from Southwest Asia towards Central Europe (also via another route towards the Caucasus) and reached this region within 2000 years. The similarities between the archaeological finds from the sites of the Karanovo I culture in Bulgaria and those of the Starčevo culture in Serbia support the assumption that one of the main routes for the spread of knowledge was up into mainland Greece, through Thessalia and towards the Balkans. Possible alternative routes are via the valleys of the Strimon (Strumm) and the Marica-Tunja, which are parallel to the valley of the Vardar (Axios)-Morava. The first farmers to reach the Great Hungarian Plain and Transylvania were the peoples of the Körös-Starčevo and the Criş cultures, who were related to the early neolithic farmers from the Balkans.

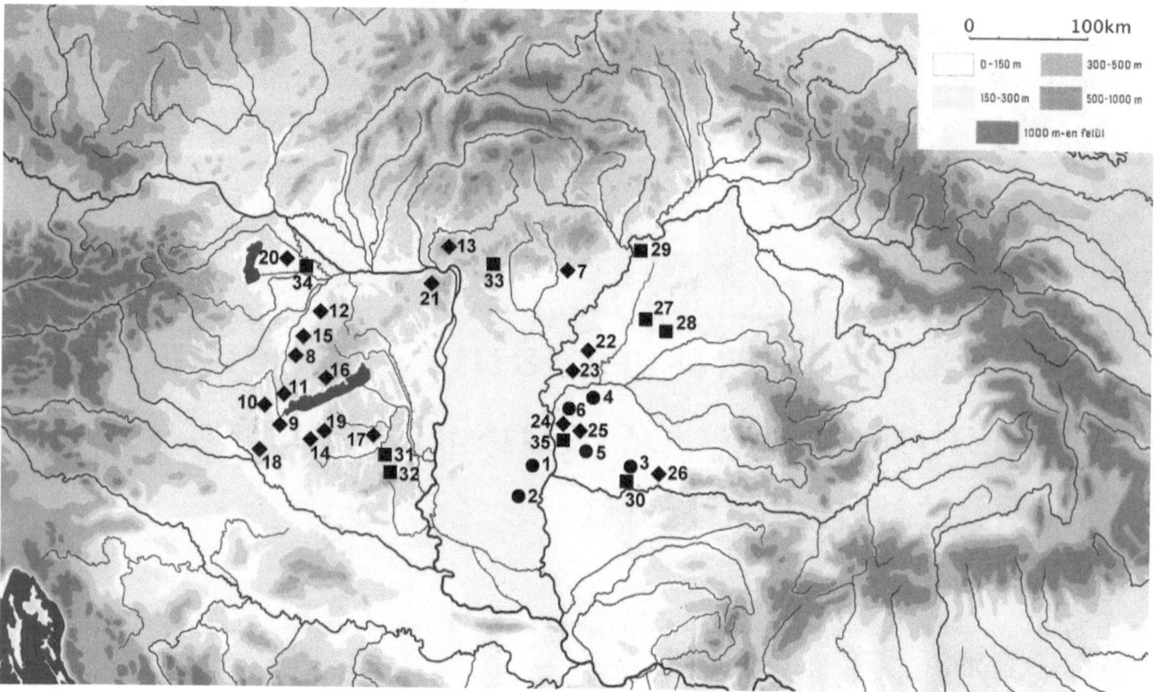

Figure 8.1: Location of Carpathian Basin sites discussed in the text. For site codes, please see table 8.1.

8.2 Neolithic agriculture in the Carpathian Basin

8.2.1 Agriculture in the early Neolithic

The earliest agriculturalists of Central and Southeastern Europe were the people of the Körös-Starčevo culture, dated to the period 6000–5300 cal BC. They were connected with groups in the Balkans and had a subsistence system based on crop and livestock husbandry with a heavy dependence on fishing. They settled in the fertile open lands of the Great Hungarian Plain near the Danube, avoiding the sandy areas that occur in the region (figure 8.1). Soil exhaustion and migratory farming are thought to be reasons why their settlements were short-lived, but there is evidence of repeated occupation at the sites.

The plant remains found on sites of the Körös-Starčevo culture were mostly preserved as imprints of seeds, spikelets, rachises and glumes in fragments of burnt wattle-and-daub from house walls and pottery (Hartyányi et al. 1968; Hartyányi and Nováki 1974; Füzes 1990). To date no cereal grains have been found.

A majority of the silicone rubber casts of the imprints found on the pottery fragments from the sites of Szeged-Gyálarét (O. Trogmayer in 1964) and Röszke-Lúdvár (excavated by O. Trogmayer in 1964/65) were identified as einkorn, and fewer were of emmer. The pottery fragments found at Battonya-Basarága (excavated by J. Szabó 1975) and Gyomaendrőd (excavated by J. Makkay in 1990) did not bear the imprints of any cereals other than einkorn. Interestingly, there were no imprints of barley; however, it probably should be included as one of the plants cultivated at this time (table 8.1).

One of the textile imprints found at Gyomaendrőd was identified by M. Füzes as linen—which is one of the earliest finds of this species in Europe (table 8.3). The pottery fragments of the Körös culture exhibit numerous textile imprints (Makkay 2001).

Sites	*Triticum monococcum*	*Triticum dicoccum*	*Triticum spelta*	*Triticum aestivum*	*Triticum compactum*	*Triticum* sp.	*Hordeum distichum*	*Hordeum tetrastichum*	*Hordeum hexastichum*	*Hordeum nudum*	*Hordeum* sp.	*Panicum miliaceum*	*Setaria italica*	Cerealia	*Lens culinaris*	*Pisum sativum*	*Lathyrus sativus*	*Vicia sativa*	*Vicia faba*	*Linum usitatissimum*
Early Neolithic																				
1 Szeged-Gyálarét	+	+	.	.	.	.	.	.	.	.	.	.	.	.	.	.	.	.	.	.
2 Röszke-Lúdvár	+	+	.	.	.	.	.	.	.	.	.	.	.	.	.	.	.	.	.	.
3 Battonya-Basarága	+	.	.	.	.	.	.	.	.	.	.	.	.	.	.	.	.	.	.	.
4 Gyomaendrőd	+	.	.	.	.	.	.	.	.	.	.	.	.	.	.	.	.	.	.	.
5 Hódmezővásárhely-Gorzsa	.	.	.	.	.	+	.	.	.	.	.	.	.	.	.	.	.	.	.	.
6 Szarvas-Szappanos	.	.	.	.	.	+	.	.	.	.	.	.	.	.	.	.	.	.	.	.

Table 8.1: Plant remains recovered from neolithic sites in Hungary.

Sites	*Triticum monococcum*	*Triticum dicoccum*	*Triticum spelta*	*Triticum aestivum*	*Triticum compactum*	*Triticum* sp.	*Hordeum distichum*	*Hordeum tetrastichum*	*Hordeum hexastichum*	*Hordeum nudum*	*Hordeum* sp.	*Panicum miliaceum*	*Setaria italica*	Cerealia	*Lens culinaris*	*Pisum sativum*	*Lathyrus sativus*	*Vicia sativa*	*Vicia faba*	*Linum usitatissimum*
Middle Neolithic																				
7 Füzesabony-Gubakút	.	+	.	.	.	.	.	.	.	.	+	.	.	.	.	.	.	.	.	.
8 Tapolca-Plébániakert	+	+	.	.	.	.	.	.	.	.	.	.	.	.	.	.	.	.	.	.
9 Fenékpuszta-Vámház	+	+	.	.	.	.	.	.	.	.	.	.	.	.	.	.	.	.	.	.
10 Alsópáhok-Kátyánalji dől	+	+	.	.	.	.	.	.	.	.	.	+	.	.	.	.	.	.	.	.
11 Keszthely-Zsidi út	+	.	.	.	.	.	.	.	.	.	.	.	.	.	.	.	.	.	.	.
12 Pápa-Vaszar	+	.	.	.	.	.	.	.	.	.	.	.	.	.	.	.	.	.	.	.
13 Szob-Kilenec	+	.	.	.	.	.	.	.	.	.	.	.	.	.	.	.	.	.	.	.
14 Szenyér-Mesztegny	+	.	.	.	.	.	.	.	.	.	.	.	.	.	.	.	.	.	.	.
15 Hegyesd-Ágói dől	.	.	.	.	.	.	.	.	.	+	.	.	.	.	.	.	.	.	.	.
16 Zánka-vasútállomás	+	+	.	+	+	.	+	.	.	.	.	+	.	.	.	.	.	.	.	.
17 Pári-Altäcker	+	+	.	.	.	.	.	.	.	.	.	.	.	.	.	.	.	.	.	.
18 Becsehely-Újmajor	+	+	+	+	+	.	.	.	.	.	.	.	.	.	.	.	.	.	.	.
19 Marcali-Lókpuszta	.	+	.	.	.	.	.	+	.	.	.	.	.	.	.	.	.	.	.	.
20 Mosonszentmiklós-Pálmajor	.	.	.	.	.	.	.	.	.	.	+	.	.	.	.	+	.	.	.	.
21 Bicske-Galagonyás	.	+	.	.	.	.	.	.	+	.	.	.	.	.	.	.	.	.	.	.
22 Dévaványa-Réhelyi dől	+	+	.	.	.	.	.	.	.	+	.	.	.	.	+	.	.	.	.	.
23 Szeghalom-Kovácsihalom	.	.	.	.	.	.	.	.	.	.	.	.	.	+	.	.	.	.	.	.
24 Szegvár-Tűzköves	.	+	.	.	.	.	.	.	.	.	.	.	.	.	.	.	.	.	.	.
25 Hódmezővásárhely-Kökénydomb	.	.	.	.	.	+	.	.	.	.	.	.	.	.	.	.	.	.	.	.

Table 8.1, cont.

Sites	*Triticum monococcum*	*Triticum dicoccum*	*Triticum spelta*	*Triticum aestivum*	*Triticum compactum*	*Triticum* sp.	*Hordeum distichum*	*Hordeum tetrastichum*	*Hordeum hexastichum*	*Hordeum nudum*	*Hordeum* sp.	*Panicum miliaceum*	*Setaria italica*	Cerealia	*Lens culinaris*	*Pisum sativum*	*Lathyrus sativus*	*Vicia sativa*	*Vicia faba*	*Linum usitatissimum*
Late Neolithic																				
26 Battonya-Parázstanya	+	+	.	.	.	.	.	.	.	+	.	.	.	.	+	+	+	.	.	.
27 Berettyóújfalu-Szilhalom	+	+	.	.	.	.	.	.	.	+	.	.	.	.	.	.	.	.	.	.
28 Berettyóújfalu-Herpály	+	+	.	.	.	.	.	.	.	.	.	+	.	.	.	+	.	.	.	.
29 Tiszapolgár-Csőszhalom	+	+	+	+	.	.	.	.	+	.	.	+	.	.	+	+	.	+	.	.
30 Battonya-Vertán major	+	+	.	.	.	.	.	.	.	.	.	.	.	.	.	.	.	.	.	.
31 Lengyel	+	.	.	+	+	.	.	.	.	.	+	+	+	.	+	.	+	.	+	.
32 Zengővárkony	.	.	+	.	.	.	+	.	.	.	.	.	.	.	.	.	.	.	.	.
33 Aszód-Papi földek	+	+	.	.	.	.	.	.	.	.	+	.	.	.	.	.	.	.	.	.
34 Börcs-Paphomlok	.	.	.	.	.	.	.	.	+	.	.	+	.	.	.	.	.	.	.	.
35 Hódmezővásárhely-Cukortanya	.	.	.	.	.	.	.	.	.	.	.	.	.	+	.	.	.	.	.	.

Table 8.1, cont.

Taxon	Early Neolithic 6000–5300 cal BC	Middle Neolithic 5300–4700 cal BC	Late Neolithic 4700–4300 cal BC
Hordeum hexastichum	.	+	++
Hordeum distichum	.	.	+
Hordeum nudum	.	+	+++
Hordeum vulgare	+	++	+
Triticum monococcum	+	++	+++
Triticum dicoccum	+	+++	+++
Triticum spelta	.	+	+
Triticum compactum	.	+	+
Triticum aestivum	.	+	++
Triticum aestivum/durum	.	.	++
Panicum miliaceum	.	+	++
Setaria italica	.	+	+
Gruel/other food remains	.	+	++

Table 8.2: Relative abundance of cereal remains recovered from neolithic sites in Hungary (+=sporadic; ++=little; +++=abundant).

Taxon	Early Neolithic 6000–5300 cal BC	Middle Neolithic 5300–4700 cal BC	Late Neolithic 4700–4300 cal BC
Lathyrus sativus	.	+	+++
Lens culinaris	.	+	+++
Linum usitatissimum	+	.	++
Pisum sativum	.	+	+++
Vicia ervilia	.	.	+
Vicia faba	.	.	+
Vicia sativa	.	.	+

Table 8.3: Relative abundance of the remains of leguminous and fibre plants recovered from neolithic sites in Hungary (+=sporadic; ++=little; +++=abundant).

Taxon	Early Neolithic 6000–5300 cal BC	Middle Neolithic 5300–4700 cal BC	Late Neolithic 4700–4300 cal BC
Crataegus monogyna	.	.	+
Cornus mas	+	.	+
Corylus avellana	+++	.	+
Malus sylvestris	.	+	+
Prunus cerasifera	.	.	+
Prunus domestica?	.	.	+
Prunus spinosa	.	.	++
Prunus avium	.	.	+
Quercus pubescens	.	.	+++
Vitis sylvestris	.	+	+

Table 8.4: Relative abundance of the remains of gathered plants recovered from neolithic sites in Hungary (+=sporadic; ++=little; +++=abundant).

In addition to imprints, there are macro-remains to prove that gathered plant foods were also an important resource for the people of the Körös-Starčevo culture (table 8.4). For example, a finger-thick layer of hazelnut shells was found in one of the pits at the site of Méhtelek-Nádas (excavated by N. Kalicz and J. Makkay in 1973).

The cornelian cherry (*Cornus mas*) stones found at the Battonya-Basarága site are no less interesting. Szeged-Gyálarét is the only site of the Körös-Starčevo culture that has imprints other than those of cereals on pottery fragments; imprints of vetchling (*Lathyrus* sp.) and brome grass (*Bromus* sp.) (table 8.5) were found near those of hulled wheat grains, possibly indicating their status as weeds, and as such they would constitute the earliest find of a Hungarian weed community. Unfortunately, identification of these taxa to species was not possible on the basis of the imprints.

Evidence of the settlement and agricultural activities of the Körös culture is available from land snails (Sümegi et al. 1998) and from plant remains found in the lake sediments of the Great Hungarian Plain dating to the same period, which indicate that there was felling of trees and burning (Willis 1997). Bones of rabbits—animals that are found in open grassy areas—are common and this is also connected to the increasing proportion of open areas in the forests (Vörös 1980). The frequent remains of bovines also indicate a considerable interest in the use of domestic cattle (Bökönyi 1959, 1974; Bartosiewicz 1999).

8.2.2 Agriculture in the middle Neolithic

Northern Great Plain

The excavation of the Linearbandkeramik site at Füzesabony-Gubakút on the Great Hungarian Plain in 1995 by L. Domboróczky unearthed archaeobotanical material from one of the earliest archaeological cultures of the northern part of this region (i.e., the middle of 6th millennium cal BC) (Domboróczky 1997). Despite the large quantity of samples, only a very small number of carbonised plant remains have been found. This implies that plant cultivation here was on a modest scale. The remains comprised hulled cereals: barley and emmer (the number of barley grains was four times the number of wheat grains), and a single weed species, spring wild oat (*Avena fatua*). It is thought that this species arrived in the region from Southwest Asia as a weed of cereal crops.

Taxon	Early Neolithic 6000–5300 cal BC	Middle Neolithic 5300–4700 cal BC	Late Neolithic 4700–4300 cal BC
Aethusa cynapium	.	.	+
Agrimonia eupatoria	.	+	.
Agrostemma githago	.	+	+
Amaranthus lividus	.	.	+
Atriplex patula	.	.	+
Avena fatua	.	+	+
Avena nuda	.	.	+
Bromus arvensis	.	+	++
Bromus inermis	.	.	+
Bromus mollis	.	.	+
Bromus secalinus	.	.	+
Bromus tectorum	.	+	.
Bromus sp.	+	.	.
Camelina sativa	.	.	+
Chenopodium album	.	+	++
Chenopodium hybridum	.	.	+
Cichorium intybus	.	.	+
Convolvulus arvensis	.	.	+
Cuscuta europaea	.	.	+
Digitaria ischaemum	.	.	+
Digitaria sanguinalis	.	.	+
Echinochloa crus-gallis	.	.	+
Fallopia convolvulus	.	+	+
Galium mollugo	.	.	+
Galium spurium	.	+	+
Galium verum	.	.	+
Hordeum murinum	.	.	+
Lathyrus sp.	+	.	.
Malva sylvestris	.	.	+
Melampyrum arvense	.	.	+
Polygonum aviculare	.	.	+
Ranunculus repens	.	.	+
Sambucus ebulus	.	.	+
Saponaria officinalis	.	.	+
Scleranthus annuus	.	.	+
Setaria viridis/verticillata	.	.	++
Silene alba	.	.	+
Solanum nigrum	.	.	+
Trifolium arvense	.	.	+
Triticum boeoticum	.	+	.
Vicia angustifolia	.	.	+
Vicia cracca	.	.	+
Vicia sp.	.	+	.
Viola arvensis	.	.	+
Xanthium strumarium	.	.	+

Table 8.5: Relative abundance of the remains of weed and ruderal plants recovered from neolithic sites in Hungary (+=sporadic; ++=little; +++=abundant).

Trans-Danubian

Sites of the Linearbandkeramik culture in the region to the west of the Danube, dated to the middle Neolithic, are located on the fertile loess lands (table 8.1). The material from the so-called Trans-Danubian Group (e.g., the sites of Zánka-Vasútállomás, Tapolca-Plébániakert, Hegyesd-Ágói Dűlő, Pápa-Vaszar, Alsópáhok-Kátyánalja Dűlő, Bazsi, Keszthely-Fenékpuszta Vámház, Keszthely-Zsidi Út, Magyaratád, Szenyér-Mesztegnyő, etc.) is closely related to the cultural region covering Western and Central Europe (Füzes 1990, 1991). The majority of the plant remains again were preserved as imprints in pottery and daub fragments. The number of sites that have produced plant imprints is almost four times as many as those that have also yielded macro-remains.

Our knowledge of neolithic farming has increased significantly as a result of the identification of plant remains preserved as imprints and also carbonised seed and fruit remains found in burnt houses and various waste and storage pits. Based on this evidence it has been possible to define a culture that had considerable understanding of the requirements of plant cultivation. Mostly einkorn and emmer were cultivated, but during this period the number of cereals increased and new cultivated plants appeared, for example, common/bread wheat and club wheat, possibly also spelt wheat and common millet, as well as leguminous plants (table 8.2). The greater number of segetal weeds (e.g., corn cockle, brome grass, bedstraw, vetch, black bindweed and fat hen) is also associated with the increasing significance of plant cultivation. The imprints of rush mats and common reed stalks indicate that, in addition, the people living here endeavoured to utilise the plant species in their natural environment.

In 1990, pottery fragments were collected at the site of Hegyesd-Ágói Dűlő for the Tapolcai Városi Museum. The fractured surface of one of the fragments showed the imprint of a grain of naked barley (Füzes 1991). Füzes (1990) also observed the imprints of rush matting and possibly the stem of the wild vine (*Vitis sylvestris*) on a pot fragment from Magyaratád (table 8.4).

During the reconstruction of the train station at Zánka in 1964, a number of neolithic pits were unearthed in the course of the excavation. An unexpectedly large quantity of carbonised remains of a range of different cereals was found, these comprised intact einkorn and emmer grains and rachis fragments, and possible two-rowed naked and hulled barley grains (cf. *Hordeum vulgare* subsp. *distichum*). The site also has the earliest-known finds of seeds of common millet (*Panicum miliaceum*), common/bread wheat (*Triticum aestivum*) and club wheat (*Triticum aestivum* subsp. *compactum*) in the Carpathian Basin. There were large numbers of imprints in the daub fragments and on the basis of the taxa represented in these it could be concluded that the most important cereals at the time were emmer and einkorn. Füzes also identified spikelets of spelt wheat and wild einkorn (*Triticum boeoticum*).

Even at this early stage weeds were hindering the work of the farmers. The carbonised remains of weeds often associated with the cereal species listed above were also found at the site, including the first occurrence of corn cockle (*Agrostemma githago*) in Hungary (table 8.5) and sterile/barren brome grass (*Bromus* cf. *sterilis*). Füzes (1990) identified the stalks and spikes of common reed (*Phragmites australis*) and the leaves of narrow-leaved cat's-tail (*Typha angustifolia*) in imprints in the daub fragments from the walls and possibly the roofs of the houses (table 8.6).

In 1968, at the Pári-Altäcker Dűlő site excavated by I. Torma, a layer of carbonised cereals (ca. five litres) was found in the bottom of a pit and according to the identifications made by B. P. Hartyányi, the great majority comprised emmer grains, with some einkorn and also small quantities of weed species, e.g., brome grass (*Bromus* sp.), bedstraw (*Galium* sp.) and black bindweed (*Fallopia convolvulus*) (Hartyányi and Nováki 1974).

Material recovered by flotation from a sample obtained at the site at Mosonszentmiklós-Pálmajor by A. Figler, provided evidence that the Linearbandkeramik population also consumed

Taxon	Early Neolithic 6000–5300 cal BC	Middle Neolithic 5300–4700 cal BC	Late Neolithic 4700–4300 cal BC
Alchemilla vulgaris	.	.	+
Astragalus glycyphyllos	.	.	+
Berberis vulgaris	.	.	+
Carex hirta	.	.	+
Carex vulpina	.	.	+
Dianthus sp.	.	.	+
Eleocharis palustris	.	.	+
Glyceria maxima	.	.	+
Lychnis flos-cuculi	.	.	+
Malva sylvestris	.	.	+
Medicago lupulina	.	.	+
Medicago minima	.	.	+
Molinia caerulea	.	.	+
Phleum pratense	.	.	+
Phragmites australis	.	++	+
Pimpinella saxifraga	.	.	+
Pinus sylvestris	.	.	+
Plantago lanceolata	.	.	+
Poa pratensis	.	.	+
Polygonum minus	.	.	+
Polygonum mite	.	.	+
Rumex hydrolapathum	.	.	+
Rumex sanguineus	.	.	+
Sanguisorba officinalis	.	.	+
Schoenoplectus lacustris	.	.	+
Schoenoplectus maritimus	.	.	+
Sparganium erectum	.	.	+
Trifolium pratense	.	.	++
Trifolium repens	.	.	+
Typha angustifolia	.	+	+
Typha latifolia	.	+	+

Table 8.6: Relative abundance of remains of plants representing the natural flora from neolithic sites in Hungary (+=sporadic; ++=little; +++=abundant).

leguminous plants; peas (*Pisum sativum*) (table 8.3) were found together with a small quantity of barley.

With some interruptions, excavations at the LBK settlement in Bicske have been in progress for several decades, from 1930 to 1977, and in the later part of that period under the direction of J. Makkay. The plant imprints on the surface of the pottery fragments of already restored so-called *notenkopf* ornamented pots from the 1976 excavation were processed using a computer image processing system (Gyulai 1996). This enabled identification of several taxa including emmer and six-rowed barley and their weeds (rye brome and fat hen).

Eastern region

We also have a relatively large number of plant remains from the middle neolithic sites in the eastern region of the country. At the site of the so-called Szakálhát-Szilmeg Group, barley and the hulled wheats (emmer, einkorn) were the primary sources of food, together with a new 'kitchen garden' plant, the small seeded lentil.

I. Ecsedi conducted an excavation at Dévaványa-Réhelyi Dűlő in 1970. Among the fragments of large storage vessels in two houses there were large quantities of two-rowed naked barley, einkorn and emmer grains (Hartyányi et al. 1968). Presumably the quantities of grains also provide an indication of their relative significance in the diet at this time. A few small-seeded lentils were also found. In the 10 litres of cereal grains there were hardly any weed seeds and only a few seeds of black bindweed and fat hen were found.

8.2.3 Agriculture in the late Neolithic

The late neolithic sites in Hungary are relatively rich in plant remains. Archaeobotanical finds are mostly in carbonised form, from burnt-out houses and pits (for waste and grain storage).

Cultivation spread further in the late Neolithic (table 8.1) and the plant remains from this period comprise larger quantities of a greater range of species. This is related to the increased number and size of settlements.

The botanical evidence from sites of the Tisza culture suggests a settled population cultivating cereals (einkorn, emmer, two- and six-rowed barley) and leguminous plants (common lentil, grass pea and pea) (table 8.2).

During the excavation of the prehistoric tell at the Battonya-Parázstanya site in 1975, J. G. Szénászky collected a significant quantity of botanical material from the pits and houses. There were remains not only of seeds, but also of flour and meal. The majority of the fifty samples (all from different locations) contained emmer, however einkorn and naked barley were somewhat less common. Of the leguminous plants, grass pea was the most common, with fewer remains of common lentil, and the least frequent was pea. Flax was also cultivated. The presence of cornelian cherry was evidence of gathered wild plant resources. There were only a very small number of weed seeds (e.g., corn cockle, black bindweed and fat hen) among the stored cereals and leguminous plants.

The botanical material from the tell settlements at Tisza-Herpály-Berettyóvölgy (Berettyóújfalu-Szihalom, Berettyóújfalu-Herpály) contained both the hulled wheat species (einkorn and emmer) recorded for the Tisza culture. At Berettyóújfalu-Szilhalom (excavated by M.Sz. Máthé in 1976), according to the observations of B. P. Hartyányi, the emmer grains fall into two categories: the first is an elongated type that is comparable with the emmer grains from the Dévaványa-Réhelyi Dűlő site; the second is broader and is similar to the emmer from Pári-Altäcker Dűlő. A few grains of common/bread wheat were also found. In addition to the einkorn and emmer grains that occurred in roughly equal quantities in the samples from various locations, there was approximately five times that quantity of two-rowed naked barley (*Hordeum vulgare* subsp. *distichum* var. *nudum*). The significance of the field pea (*Pisum sativum* subsp. *arvense*) is also indicated by the fact that the seeds constitute about half the total quantity of finds (Hartyányi and Sz. Máthé 1980) (table 8.3). Complete, carbonised crab apple (*Malus sylvestris*) fruits were also recovered at the site.

In addition to the hulled wheats (einkorn and emmer), common millet was also found at the Berettyóújfalu-Herpály site (excavated by N. Kalicz and P. Raczky in 1978–82). Here the field pea was also the most important leguminous plant. Radiocarbon dates provided a date range for the Berettyóújfalu-Herpály site of 6570–6270 bp (Hertelendi et al. 1997).

During the excavation of one of the houses of Berettyóújfalu-Herpály in 1981, Kalicz and Raczky found a storage context with 150 carbonised seeds, which were predominantly field peas (*Pisum sativum* cf. subsp. *arvense*), and with some bitter vetch (*Vicia ervilia*). This is the earliest known occurrence of bitter vetch in Hungary; it only became widespread later, during the middle Bronze Age (e.g., at the tells of Bölcske-Vörösgyír, Tiszaalpár-Várdomb). The presence of this species, a native of Southwest Asia and the eastern Mediterranean, in the Neolithic indicates possible links with cultures further to the south.

The samples from the tell settlements of Tisza-Herpály-Berettyóvölgy all contained peas and at Berettyóújfalu-Szilhalom peas constituted almost half of all the seed remains. Peas were also found at the late neolithic tell settlement of the Herpály culture at Tiszapolgár-Csőszhalom (excavated in 1995). According to the radiocarbon dates, the occupation at the site was estimated to have lasted from 6700–6370 bp (Hertelendi et al. 1997).

The botanical material obtained by flotation indicates a population that had acquired considerable knowledge about plant cultivation. Their main crop was emmer; einkorn and six-rowed barley (*Hordeum vulgare* subsp. *hexastichum*) played a less prominent role. There were also sporadic finds of the grains of naked wheat, spelt wheat and common millet. It was not possible to decide whether the naked wheat was common/bread wheat or durum wheat due to the similarity of the grains. As expected, the cereals were found in association with leguminous plants, including small seeded lentil, pea and common vetch (*Vicia* cf. *sativa*). Carbonised stones of the cherry or myrabolan plum (*Prunus* cf. *cerasifera* subsp. *myrobalana*) and cornelian cherry (*Cornus mas*) were also found (table 8.4). These were probably gathered for consumption, as were the acorns (*Quercus* sp.), sloes (*Prunus spinosa*) and crab apples (*Malus sylvestris*). The weeds indicate intensive cereal production, for example, black bindweed and fat hen, and in addition some weed species that appeared for the first time in the Hungarian Neolithic: summer vetch (*Vicia angustifolia*), copse-bindweed (*Fallopia dumetorum*), false cleavers (*Galium spurium*), barnyard grass (*Echinochloa crus-galli*) and annual woundwort (*Stachys annua*) (table 8.5). There were also other wild taxa that would have been growing around the site, including danewort (*Sambucus ebulus*), hairy sedge (*Carex hirta*), great water dock (*Rumex hydrolapathum*) and common bulrush (*Schoenoplectus lacustris*) (table 8.6). These species indicate that there were watercourses near the tell settlement and that they were also in the vicinity of the arable land.

The Battonya-Vertán Major site is approximately the same date as the site at Tiszapolgár, although it does not have the same abundance of plant species. In 1982, during the excavation of a large house, J.G. Szénászky found almost 20 g of carbonised plant remains. The majority were emmer grains, with some einkorn, as well as carbonised stones of cornelian cherry and seeds of fat hen, danewort, amaranth (*Amaranthus* sp.) and members of the buckwheat family (Polygonaceae) (Hartyányi 1989).

The plant remains from the sites associated with the Lengyel culture in the Trans-Danubian region indicate that the population still cultivated plants, but less intensively than in the previous period.

At the Lengyel site, which gave the culture its name, M. Wosinsky conducted several excavations between 1885 and 1890. The dating of the finds raised several questions later (see Hartyányi et al. 1968). In 1890, I. Deininger—the founder of Hungarian archaeobotany—collected botanical material at the site. His identifications of the remains from fireplaces, pits and pots were published in 1892. He recorded the following species: many-rowed barley, common/bread wheat, club wheat, einkorn, common millet, foxtail millet (*Setaria italica*), horsebean (*Vicia faba*), grass pea (*Lathyrus sativus*), lentil, spring wild oat (*Avena fatua*), rye brome (*Bromus secalinus*), field brome (*Bromus arvensis*), barnyard grass (*Echinochloa crus-galli*), tufted vetch (*Vicia cracca*), common milkvetch (*Astragalus glycyphyllos*), corn cockle (*Agrostemma githago*), cowherb (*Vaccaria hispanica*), pink (*Dianthus* sp.), ribwort plantain (*Plantago lanceolata*), dodder (*Cuscuta* sp.), European barberry (*Berberis vulgaris*), plum (*Prunus* sp.), cornelian cherry (*Cornus mas*), Scots pine (*Pinus silvestris*), flax (*Linum usitatissimum*), and sedge (*Carex* sp.).

The Lengyel culture site at Zengővárkony was excavated by J. Dombay in the 1940s. The imprints in the daub fragments from the site were studied by E. Gubányi in 1947 and identified as two-rowed barley and spelt wheat (Hartyányi et al. 1968). At a later date Füzes (1990) found additional imprints and he claimed to recognise the mazzard cherry (*Prunus avium* subsp. *sylvestris*) and the imprint of a mat made from narrow-leaved cat's-tail.

Pits belonging to the older phase of the Lengyel culture excavated by N. Kalicz (1960–82) at the Aszód-Papi Földek site yielded pottery fragments on which (Füzes 1990) also found plant imprints. He counted a total of 2,500 imprints of plant taxa on the 14 fragments he examined, but it was possible to assign species identifications for only a small proportion of these. There were also husked grains of barley and some grains of naked oat (*Avena* cf. *nuda*).

M. Füzes collected mazzard cherry stones from the burnt layer of the late neolithic (or possibly the copper age) flint mine opened by L. Vértes at Sümeg-Mogyorósdomb.

The seeds of the wild vine, the fruit of which was collected in the area during the Neolithic, have been found at the Lengyel culture site of Sé (near Szombathely) (Füzes 1990). The nutrient-rich acorns (roasted) were used for several millennia as an emergency food in times of famine, or sometimes as a speciality for people working or gathering in the forest. This is certainly indicated by evidence from the residential site at Moha-Homokbánya (probably Lengyel culture), excavated by A. Kralovánszky in 1961, where a carbonised store of acorns (identified as pubescent oak—*Quercus pubescens*) was found in one of the underground dwellings (Hartyányi et al. 1968).

The climate gradually deteriorated at the end of the Neolithic (Zólyomi 1980). During the 1994 excavation by A. Figler at Börcs-Paphomlok only two species of cultivated cereals were found in the layers belonging to the 3rd phase of the Lengyel culture: six-rowed barley and common millet. On the basis of the absence of bread wheat species from the record at this time, the population of the settlement is believed to have managed livestock rather than cultivated crops.

8.3 Conclusion

The Carpathian Basin played a bridging role in the spread of crop-based agriculture between the Fertile Crescent (Southwest Asia) via the Balkans to Central Europe.

Since the Pleistocene, the Carpathian Basin has been characterised as a mosaic with regard to its climate, soils and plant life. At the macro level, the area is the meeting-point of three different climates: the influence of the continental region fades from east-to-west, while west-to-east the oceanic influence grows weaker and south-to-north the sub-Mediterranean effect diminishes.

The rapid spread of domesticated plants, and the techniques and technologies associated with plant cultivation, can be accounted for by the similarity of environments; for example, arboreal cover on the Great Hungarian Plain during the Atlantic phase was probably similar to the present-day Balkans (and in particular, the Crimean peninsula). It was characterised by the common oak–hornbeam association (*Querco robori–Carpinetum*), which can be regarded as the continuation of the vegetation of the Ukrainian steppes.

Sites dating to the early Neolithic (Körös-Starčevo culture) have been found in the southern regions of the Great Hungarian Plain and Trans-Danubia. It seems that beyond the southern

Balkans, the only place and time where the required ecological conditions for the production of domesticated plants and for animal-keeping existed was in that specific region of the Carpathian Basin. The advantageous conditions that prevailed in the Carpathian Basin because of the mixing of components of western, eastern and southern climates, and which facilitated the initial adaptation of plant species, were not present elsewhere. This ultimately curtailed the further spread of plant cultivation (Kertész and Sümegi 2001).

References

Bartosiewicz, L. (1999). Őskori állattartás és környezet a Kárpát-medencében [Prehistoric animal-keeping and environment in the Carpatian-basin]. In G. Füleki (Ed.), *A táj változásai a Kárpát-medencében a történelmi események hatására*, pp. 195–200. Budapest-Gödöllő: Szent István Egyetem.

Bökönyi, S. (1959). Die frühalluviale Wirbeltierfauna Ungarns. *Acta Archaeologica Hungarica 11*, 39–102.

Bökönyi, S. (1974). *History of Domestic Mammals in Central and Eastern Europe.* Budapest: Akadémiai Kiadó.

Domboróczky, L. (1997). Füzesabony-Gubakút. Újkökori falu a Kr. e. VI. évezredből [A neolithic village from the 6th millennium BC]. In R. Raczky, T. Kovács, and A. Anders (Eds.), *Utak a Múltba: Az M3-as Autópálya Régészeti Leletmentései [Paths into the Past: Resuce Excavations on the M3 Motorway]*, pp. 19–27. Budapest: Magyar Nemzeti Múzeum and ELTE Régészettud. Intézet.

Füzes, M. (1990). A földmívelés kezdeti szakaszának (neolitikum és rézkor) növényleletei Magyarországon (Archaeobotanikai vázlat) [Plant remains from the early phase of plant cultivation—Neolithic and Copper Age in Hungary. An archaeobotanical outline]. *Tapolcai Városi Múz. Közlem. 1*, 139–238.

Füzes, M. (1991). A Dunántúl korai növénytermesztése és növényleletei. A Starčevo kultúra és a "Tapolcai csoport" [Early plant breeding and plant remains in Transdanubien. The Starčevo culture and the "Tapolca-Gruppe"]. *Bibliotheca Musei Tapolcensis 1*, 267–362.

Gyulai, F. (1996). Using image analysis in the evaluation of plant imprints found on sherds from the neolithic site of Bicske. In J. Makkay, E. Starnini, and M. Tulok (Eds.), *Excavations at Bicske-Galagonyás (Part III): The Notenkopf and Sopot-Bicske Cultural Phases*, pp. 258–263. Trieste: Svevo.

Hartyányi, B. (1989). Növényleletek a battonya-parázstanyai neolitikus lakótelepen [Pflanzenfunde der neolitischen Wohnsiedlung bei Battonya—Parázs-Gehöft]. *Magy. Mezőg. Múz. Közlem 1989*, 39–67.

Hartyányi, B. and G. Nováki (1974). Növényi mag- és termésleletek Magyarországon az újkőkortól a XVIII. sz.-ig II. [Samen- und Fruchtfunde in Ungarn von der Jungsteinzeit bis zum XVIII. Jahrhundert II.]. *Magy. Mezőg. Múz. Közlem 1974*, 23–73.

Hartyányi, B., G. Nováki, and Á. Patay (1968). Növényi mag- és termésleletek Magyarországon az újkőkortól a XVIII. sz.-ig I. [Samen- und Fruchtfunde in Ungarn von der Jungsteinzeit bis zum XVIII. Jahrhundert I]. *Magy. Mezőg. Múz. Közlem 1968*, 5–85.

Hertelendi, E., É. Svingor, P. Raczky, and F. Horváth (1997). Radiocarbon chronology of the Neolithic and time span of tell settlements in eastern Hungary based on calibrated radiocarbon dates. In L. Költő and L. Bartosiewicz (Eds.), *Archaeometrical Research in Hungary II*, pp. 61–70. Budapest: Hungarian National Museum.

Kalicz, N. (1993). The early phases of the Neolithic in western Hungary (Transdanubja). *Poročilo Oraziskovanju Paleolita, Neolita i Euneolita v Sloveniji XXI*, 85–135.

Kertész, R. and P. Sümegi (2001). Theories, critiques and a model: why did the expansion of the Körös-Starčevo culture stop in the centre of the Carpathian Basin? In R. Kertész and J. Makkay (Eds.), *From the Mesolithic to the Neolithic. Proceedings of the International Archaeological Conference held in the Damjanich Museum of Szolnok, September 22-27, 1996*, pp. 225–246. Budapest: Archaeolingua.

Komlódi, M. (1966). Adatok az Alföld negyedkori klíma- és vegetációtörténetéhez I [Quaternary climatic changes and vegetational history of the Great Hungarian Plain I]. *Bot. Közlem. 53*, 191–201.

Lacza, J. (1991). *Reconstruction of the History of the Postglacial Flora and Vegetation in the Region between Danube and Tisza and Tiszántúl.* Kiev: N.G. Kholodny Institute of Botany Ac. Sci. Ukr. SSR.

Makkay, J. (2001). *Textile Impressions and Related Finds of the Early Neolithic Körös Culture in Hungary.* Budapest: Published by the Author.

Sümegi, P., E. Hertelendi, E. Magyari, and M. Molnár (1998). Evolution of the environment in the Carpathian Basin during the last 30.000 years BP and its effects on the ancient habitats of the different cultures. In L. Költő and L. Bartosiewicz (Eds.), *Archaeometrical Research in Hungary II*, pp. 183–193. Budapest: Hungarian National Museum.

Vörös, I. (1980). Zoological and palaeoeconomical investigation on the archaeozoological material of the early Neolithic Körös culture. *Folia Archaeologica 31*, 35–62.

Willis, K. (1997). The impact of early agriculture upon the Hungarian landscape. In J. Chapman and P. Dolukhanov (Eds.), *Landscapes in Flux: Central and Eastern Europe in Antiquity*, Colloquia Pontica 3, pp. 193–209. Oxford: Oxbow Books Ltd.

Zólyomi, B. (1980). Landwirtschaftliche Kultur und Wandlung der Vegetation im Holozän am Balaton. *Phytocoenologia 7*, 121–126.

Chapter 9

Neolithic agriculture in Italy: an update of archaeobotanical data with particular emphasis on northern settlements

Mauro Rottoli and Andrea Pessina
Musei Civici Como, Italy

9.1 Introduction

One of the main problems that has limited the general understanding of early neolithic agriculture in Italy has been the restricted dissemination of archaeobotanical data, largely because there are few recent publications in English (for two examples see Biagi and Nisbet 1987; Biagi et al. 1993). This paper addresses this problem by summarizing the latest data from Italian neolithic sites.

Over the last few years new archaeobotanical data have been collected from well-known archaeological contexts. For example, Friuli (northeastern Italy) is particularly interesting; here systematic research is producing important data on the first neolithic communities and on the development of the first farming sites in the area (see Pessina and Muscio 1998, 2000; Ferrari and Visentini 2002). Elsewhere in Italy, there has been infrequent recovery of plant remains. There are, however, exceptional sites such as La Marmotta, Rome (Rottoli 1993, 2002) and a series of new sites in Puglia, southern Italy (Fiorentino et al. 2000; Costantini 2002), which are producing extremely interesting evidence of the emergence and development of agriculture. This increasingly large body of data is also modifying the traditional hypothesis, particularly as advocated for central and northern Italy.

Until the end of the 1980s, it was assumed that the first neolithic cultures from northern Italy (Vhò, Fiorano, Gaban, and Fagnigola) did not practice agriculture and the economy was basically mesolithic in character, with a heavy dependence on hunting and gathering. In particular, the abundance of hazelnut shells in mesolithic and early neolithic sites led researchers to propose that there was continuity in subsistence patterns based on wild plant exploitation. Archaeobotanical reports at the time suggested that cereals were progressively introduced into the early neolithic diet, but it was only during the middle Neolithic that agriculture became a fundamental element of subsistence (Barker et al. 1987). More recent data, described below, raises doubts with this hypothesis and, conversely, neolithisation for most of northern Italy now appears to have been a very rapid phenomenon.

9.2 Aspects of the Italian early Neolithic

By about 7000 bp (i.e., ca. 6000–5800 cal BC), early neolithic communities were established in Italy. The earliest sites with impressed pottery are situated in the south (Basilicata, Puglia and Calabria) and along the Adriatic and Ionic coasts. In these areas there was direct contact with groups that had arrived in Europe after having spread westwards from the Near East. These first farming groups were foreigners that had crossed the sea from the east and landed in the Italian Peninsula bringing with them new plants and animals.

Between 6900 and 6500 bp, peoples using impressed pottery settled along the Ligurian coast, in the northwestern part of the country. Just before 6500 bp, groups belonging to an Adriatic tradition colonized areas of Abruzzo and Marche, and continued to spread northwards into Romagna. However, they did not reach the Po Plain (Improta and Pessina 1998).

It is during this period that new cultures developed in the northern regions; in most cases, their origins are only partially known. The central part of the Po Plain and the northern part of Tuscany witnessed the development of one of the best-known of these, the Fiorano culture (6600–6000 bp). It played an important role in the management of the well-developed network controlling the supply of raw materials during the Neolithic. For example, Fiorano groups seem to have dominated the best flint sources in the Lessini Mountains near Verona (Pessina 1998), which were exploited by many neolithic communities from the North.

At the same time, at Sammardenchia (Udine) in Friuli, northeastern Italy, a new neolithic group is well-documented, which was influenced by both the Fiorano and the Danilo cultures (the latter known along the eastern Adriatic coast, in the northwestern part of the Balkans) (Pessina et al. 1998). From an archaeobotanical point of view, settlements in Friuli (Sammardenchia, Piancada, Valer, Fagnigola and Pavia di Udine) are the best studied neolithic sites from northern Italy (Castelletti and Rottoli 2002). Archaeological research has shown an intense neolithic occupation of this region. Recent surveys have also brought to light several villages on the plains, which were often occupied for many centuries. ^{14}C dates also confirmed that the earliest neolithic occupation of this area took place at about 6700-6500 bp.

9.3 North Italy

9.3.1 Wood charcoal data

Wood charcoal analyses from northern Italian sites indicate that the first neolithic occupation occurred in a heavily forested plain (Carugati et al. 1996; Castelletti and Maspero 1992; Castelletti and Carugati 1995). The initial phases of the Neolithic are characterized by the presence of a mixed oak forest in which caducifolious oaks (*Quercus* sez. ROBUR), ash (*Fraxinus* sp.) and maple (*Acer* sp.) were the main tree species. Other species, for example, elm (*Ulmus* sp.), poplar (*Populus* sp.), and more occasionally alder (*Alnus glutinosa/incana*) and beech (*Fagus sylvatica*) were also represented. On sites in mountainous areas (i.e., at the outer limit of frontal moraine, pre-Alps) such as Isolino di Varese (Castiglioni and Rottoli 2000) and Pizzo di Bodio (Castelletti and Madella 1994), fir (*Abies alba*) is well attested. In addition, lime (*Tilia* sp.) has been identified at other sites such as Lugo di Romagna (Rottoli, unpublished) and Roncade (Pignatelli and Rottoli 1996).

A rapid increase of Pomoideae (e.g., pear, apple, hawthorn and whitebeam), hazel (*Corylus avellana*) and cherry/sloe (*Prunus* spp.), already present in the natural vegetation, marks the beginning of deforestation, woodland exploitation and the selection of certain tree species for fuel and raw material. Later on hornbeam (*Carpinus betulus*) is also represented (Castelletti and Maspero 1992).

9.3.2 Plant remains (seeds and fruits)

Sammardenchia-Cueis (Udine, Friuli-Venezia Giulia, northeastern Italy)

The settlement: The sites of Sammardenchia are located on a terrace of tectonic origin situated on the Friuli plain, near Udine (Fontana 1999). Archaeological research carried out over the last 20 years in the area has shown a recurring pattern of occupation with cyclical settlement in small sites on the same rich and fertile land (Ferrari and Pessina 1996). Additionally, pedological analyses of different levels of the site have revealed that neolithic occupation extended over more than 650 ha.

Approximately 30 ^{14}C dates between 6570 ± 74 bp and 5684 ± 58 bp are available from Sammardenchia (Improta and Pessina 1998; unpublished). During the first phases, the influence of the Fiorano culture is evident, whereas for the latter phases the stimulus of the Danilo culture prevails. The analysis of the raw material suggests a complex series of relationships and contacts with different areas (Delpino et al. 1999; Pessina and D'Amico 1999; Pessina 1998). The flint clearly originated in the Lessini Mountains, some 150 km away from the site, whereas the stones used for polishing (jade and other rocks) were collected in the northwestern part of Italy, at a distance of ca. 400–500 km. The obsidian has been studied and also has an exogenous origin as it comes from the island of Lipari and the Carpathian mountains. In addition, there are also artefacts (e.g, axes and chisels) with clear connections to the Danubian.

The Sammardenchia sites have been only partially conserved; in most cases the upper part of the structures has been destroyed by modern agricultural practices. Several types of ditches and pits used for rubbish dumping have been preserved. Originally the pits probably had many functions and it has been suggested that prior to their utilization as rubbish pits, they were storage silos or were merely used for clay extraction. Recently, a large 10 cm thick blackened area has been excavated ('structure 126'). A type of a well-cistern for water storage, dated to the end of the early Neolithic phase (first centuries of the 6th millennium bp), has been found in the middle of this area (Cermesoni et al. 1999). In the earliest phase at Sammardenchia there was also evidence of an interesting ditch with an opening passage that enclosed one of the sites.

Archaeobotanical analysis: More than 100 samples collected from pits and from the large blackened area have been investigated (Rottoli 1999, Rottoli, unpublished). Charred plant remains were sparsely distributed and, in many cases, poorly preserved. Approximately 19,000 remains have been identified including many cereal species and legumes. Cereals are represented by hulled and possibly also naked barley (*Hordeum vulgare*), emmer (*Triticum dicoccum*), einkorn (*Triticum monococcum*) and free-threshing wheats (*Triticum aestivum/durum*). The 'new' hulled wheat species described by Jones et al. (2000) is also present in the samples.

Barley, emmer and einkorn are the dominant species whereas free-threshing wheats are scarce. The new wheat seems to have been an important crop; however, as the remains are very fragmented, it is difficult to ascertain its real significance. There is only sparse evidence of rye (*Secale cereale*), spelt (*Triticum* cf. *spelta*), and millets (*Panicum/Setaria*) and interpretation of their function is therefore problematic. It is likely that these species were weeds in the cereal fields and had not yet been cultivated as 'true' crops. *Bromus* sp. is particularly frequent and this raises the question of its possible cultivation (see also Bakels, this volume). The other weed species (*Galium aparine*, Polygonaceae, *Rumex* sp., *Silene* cf. *vulgaris*) are represented by just a few seeds.

There is an abundance and diversity of pulses, including pea (*Pisum* sp.), lentil (*Lens* sp.), vetches (*Vicia sativa* agg.), bitter vetch (*Vicia ervilia*) and grass pea (*Lathyrus sativus/cicera*). The most frequent of these are vetches, lentils and peas. The cultivation of broad beans (*Vicia* cf. *faba*) has yet to be confirmed. Other remains include fruits and nuts, for example, hazelnuts (*Corylus avellana*), apples fruits (*Malus domestica*), acorns (*Quercus* sp.), bramble stones (*Rubus fruticosus* agg.), hawthorn seeds (*Crataegus* sp.), plum stones (*Prunus insititia/spinosa*) and grape pips (*Vitis vinifera*). A fragment of walnut shell (*Juglans regia*) whose distribution is focused on the Balkans (Rottoli, unpublished) has also been found. Finally, a single seed of flax (*Linum usitatissimum*) has been identified.

Piancada and other sites in Friuli (northeastern Italy)

A large artificial canal connected to several smaller channels (probably constructed for drainage purposes) and a series of small pits have been found at the site of Piancada, located a few kilometres from the upper Adriatic coast. The canal has yielded numerous faunal remains (not preserved at Sammardenchia), which include sheep/goat, bovids, pigs and a distinctive wild fauna. Also worthy of note is the discovery of a burial from the early Neolithic (Ferrari and Pessina 1996). Further evidence for this period in Friuli can be found at Fagnigola, Valer and Pavia di Udine (Pessina et al. 2004). However, little additional archaeobotanical information has been ascertained from these sites (Carugati 1993, 1994, Rottoli, unpublished, Pessina et al. 2004).

To summarise: cereals are always represented by barley, emmer and einkorn. Of note is that the einkorn is slightly more abundant at Piancada. Free-threshing wheats are either absent or rare. Spelt (*Triticum* cf. *spelta*, spikelet bases) has been tentatively identified at Pavia di Udine and Piancada, whereas the new hulled wheat is only present at Piancada. Samples from Piancada include *Lens* sp., *Pisum* sp. and *Vicia sativa* agg., and at Pavia di Udine only *Pisum* sp. has been retrieved. Hazelnuts are the most commonly occurring wild resource. Other wild species, including *Quercus* sp., *Malus* sp., *Vitis vinifera*, *Prunus spinosa* agg., *Rubus fruticosus* agg., *Crataegus* sp. and *Sambucus* sp. are only represented by a few specimens. Seeds of arable weeds are scarce (e.g. *Vicia* sp., *Fallopia convolvulus*, *Rumex* sp. and members of the Panicoideae, Poaceae, and Caryophyllaceae).

Other sites in northern Italy

In the recent past, very few other early neolithic sites from northern Italy have been investigated (figure 9.1), these include: Pizzo di Bodio (Castelletti and Madella 1994), Isorella (Nisbet 2000), Ostiano-Dugali Alti (Nisbet 1995) and Vhò di Piadena (Castelletti and Maspero 1992). The numbers of plant remains analyzed are often very few and in most cases only caryopses have been identified. This obviously limits what can be concluded and interpreted from the archaeobotanical data. At almost every site, barley, emmer and einkorn are the main crop species, while free-threshing wheats and legumes are poorly represented.

New analyses have recently been undertaken at the site of Lugo di Romagna (Rottoli, unpublished)—a classic site of the Fiorano culture—dated between 6626–5630 bp. Preservation seems to be exceptional for the region; different structures were identified including the remains of a ditch, a large palisade and a well-preserved house that had collapsed after a fire (Degasperi et al. 1998). A sample was retrieved inside the house from under a series of vases that were destroyed during the fire. The sample was composed of clean emmer grains ready for consumption. Chaff remains have not yet been found, indicating that dehusking may have taken place outside the house. Emmer grains from this site are relatively small in size and rounded at one end, with clear marks of the glumes visible on their surfaces.

In total more than 5,000 macro-remains have been identified at Lugo di Romagna of which emmer is the dominant species, followed by barley and free-threshing wheats; einkorn is only sparsely represented. Pulses, represented by peas and lentils, and gathered wild food resources, e.g., acorns and hazelnuts, are also present.

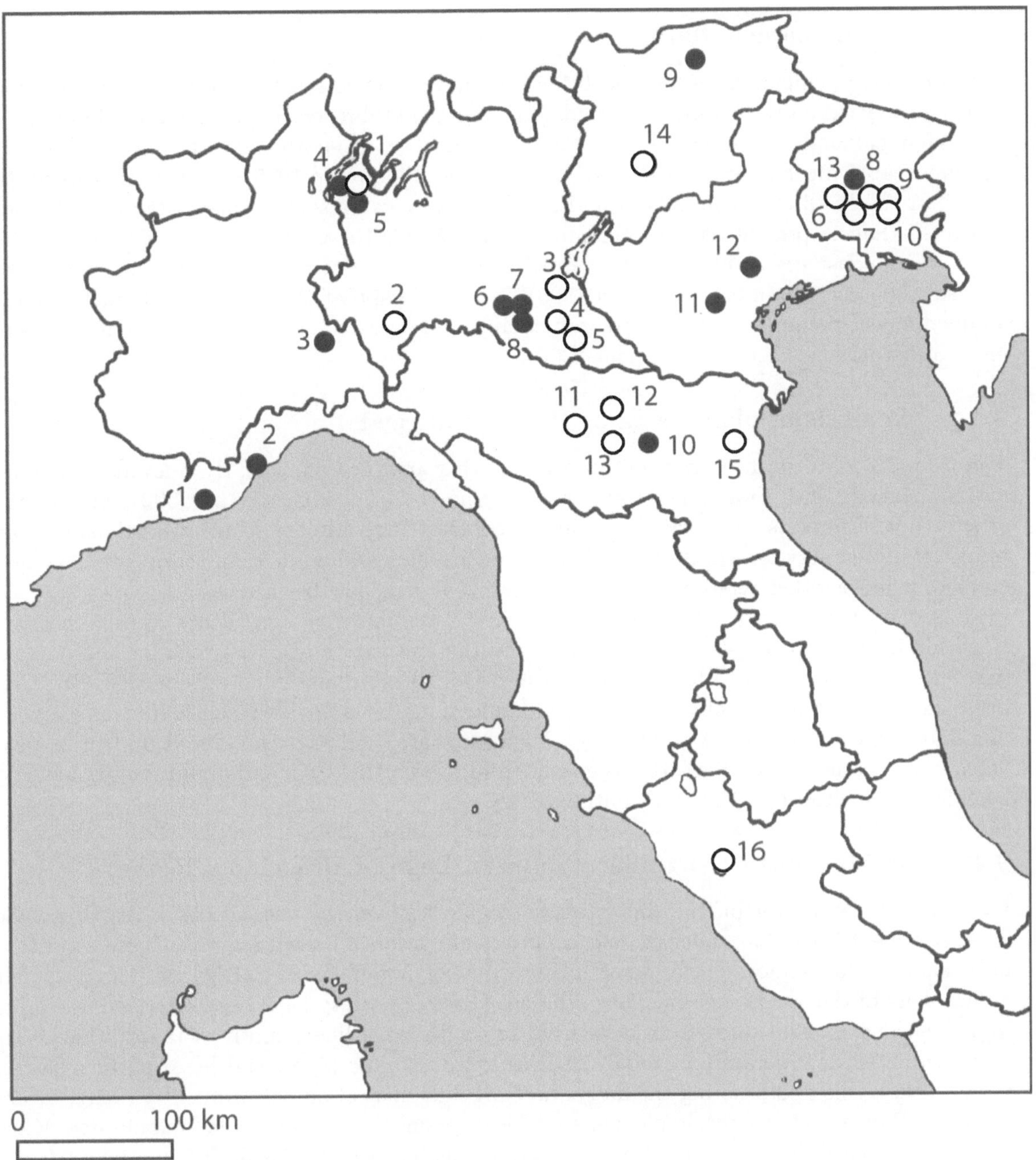

Figure 9.1: Location of northern Italian neolithic sites with carpological remains. White dots=early neolithic sites: 1. Pizzo di Bodio; 2. Cecima; 3. Isorella; 4. Ostiano-Dugali Alti; 5. Vhò di Piadena; 6. Fagnigola; 7. Piancada; 8. Valer; 9. Pavia di Udine; 10. Sammardenchia; 11. Albinea; 12. Savignano; 13. Chiozza di Scandiano; 14. Vela di Trento; 15. Lugo di Romagna; 16. La Marmotta. Black dots=late neolithic sites: 1. Arma dell'Aquila; 2. Arene Candide; 3. Casalnoceto; 4. Isolino di Varese; 5. Lagozza di Besnate; 6. Rivarolo Mantovano; 7. Acquanegra sul Mosio; 8. Casatico di Marcaria; 9. Barbiano; 10. Spilamberto; 11. Fimon-Molino Casarotto; 12. Monte Covolo; 13. Bannia-Palazzine di Sopra; 14. Vela di Trento.

9.4 Central and southern Italy

9.4.1 Wood charcoal data

Data from several hundreds of early neolithic sites in central and southern Italy is abundant, however, information on the environment and subsistence is still incomplete. The limited number of wood charcoal analyses from the Italian Peninsula suggests an enormous variety of environments and landscapes during the Neolithic (see, for example, Fiorentino 1995). On the other hand, the spread of the Mediterranean maquis and evergreen species appears to have been slightly more restricted than in present day. In some areas, for example the Campagna Romana near Rome, there was a considerable increase of arboreal species, and during the late Neolithic the forests reached their maximum extent (Magri 1995). The landscape at this time was characterized by an open forest, dominated in parts by caducifolius and evergreen oaks, hazel and elm, and with areas of herbaceous vegetation (Celant 2000).

9.4.2 Analysis of plant remains (fruits and seeds)

The available data from early neolithic sites in this area are particularly uneven (Costantini and Stancanelli 1994; Fiorentino et al. 2000; Perrino et al. 2000; Costantini 2002). The number of sites investigated is relatively high but, in contrast, the number of identifications of plant remains is often very low. In most cases only cereal grains and a few chaff fragments have been studied, whereas other types of remains are present in very low frequencies.

Based on the existing information, it is fair to assume that agriculture during the early Neolithic in southern Italy was characterized by the cultivation of barley, emmer and einkorn; free-threshing wheats (both tetraploids and hexaploids) are represented infrequently. Spelt has been identified at a few sites but these identifications need to be reconsidered as they are unreliable, and are based on grains only. Pulses are rare, and are represented by lentils, peas, vetches and broad beans (the latter is absent from the north). Flax and opium poppy have not been found on sites in central and southern Italy.

9.4.3 La Marmotta (Anguillara Sabazia, Lake of Bracciano, Rome)

La Marmotta was built on the shores of a volcanic lake on the present-day Lake Bracciano. The settlement is ca. 8 m under the water, after sinking for unknown reasons. These conditions have resulted in extraordinarily good preservation of organic material by waterlogging. The archaeology of the site is very complex; different types of pottery have been retrieved, including impressed and painted wares and incised pottery with large linear motives. In addition to the extremely abundant ceramic material (such as small ceramic pipes, and large pottery storage vessels, etc.), it is worth noting the finds of a boat made out of an oak trunk 10 m long, several models of boats and wooden implements (sickles, spoons, etc.) (Fugazzola Delpino and Mineo 1995; Fugazzola Delpino 1998). Several ^{14}C dates ranging between 6855 ± 65 bp and 6310 ± 75 bp have been obtained.

The analysis of the archaeobotanical samples from La Marmotta has produced an interesting body of data. The analyses of wood charcoal indicate a distinct type of vegetation, described as hot and humid Mediterranean maquis and characterized by the dominant presence of *Laurus nobilis* (including many large trees that were cut and used for the palisades), also with different types of caducifolious oaks, ash, hornbeam and beech (Rottoli and Fugazzola Delpino 2000).

Cereal cultivation at La Marmotta is characterized by the presence of hulled barley (the two-row form, *Hordeum distichum*), emmer, einkorn and archaic free-threshing wheat, probably *Triticum turgidum* or *Triticum durum*. The finds of *Triticum aestivum* have yet to be confirmed. The new hulled wheat species has not been identified. Legumes are well represented by a wide variety of species, including lentil, pea, grass pea and vetches. Finds of bitter vetch are dubious and broad beans are absent from the records. Flax is documented from the site but the number of specimens is low. The presence of opium poppy (*Papaver somniferum*) is particularly noteworthy, the remains are very abundant and comprise both charred and uncharred seeds, charred capsules and stigmatic discs. The seed and fruit morphology exhibit characteristics that are half-way between the wild and cultivated forms and it was not possible, therefore, to distinguish between the two. However, the site is within the primary distribution area of the wild species (*Papaver somniferum* subsp. *setigerum*).

There are large numbers of specimens representing a wide variety of species gathered for their fruits/nuts, for example, *Corylus avellana*, *Malus* sp., *Pyrus* sp., *Quercus* sp., *Ficus carica*, *Prunus spinosa/domestica*, *Fragaria vesca* and *Cornus mas*. Particularly interesting is the high frequency of grape pips (*Vitis vinifera*); this species is not only represented by pips with the characteristics of the wild forms, but also by wood and charcoal. The vine was probably a key element of the peculiar vegetation surrounding the site and grapes would have certainly been collected for food. No evidence has been gathered regarding the practice of viticulture.

Of interest also is the atypical presence of two different types of thistles (*Carthamus lanatus* and *Silybum marianum*). The frequency of both charred and uncharred specimens would seem to indicate their use as food. From *Silybum* it is possible to extract an edible and medicinal oil, however, there is no information about present-day uses of *Carthamus lanatus*. It is likely that this species played a similar role to that of *Carthamus tinctorius* (safflower), a plant commonly used in later periods. The plant assemblage is also composed of more than 50 other species characteristic of the local environment (e.g, wetland, riverine, segetal and forest species, etc.) (Rottoli 1993, 2002).

9.5 An archaeobotanical review of the early Italian Neolithic

It is difficult to summarize the archaeobotanical information from the earliest phases of the Neolithic in Italy (table 9.1); firstly, sites with studies of plant remains are limited, secondly, the sites are located in an enormous variety of environments with their own characteristics, and finally, the enormous complexity of the cultural relationships between Italy and the Mediterranean and the Balkans does not allow for easy synthesis of the data.

However, it is possible to highlight some general trends for the earliest agriculture in Italy:

- The main cereals are barley (in most cases the six-rowed hulled forms dominate although six-rowed naked and the two-rowed hulled forms are also present), emmer and einkorn. Free-threshing wheats are documented but always in low numbers and are more important in central and southern Italy but, as a general rule, evidence indicates that their cultivation is restricted at least during the earliest period. The new hulled wheat species is only present in northern Italy; it appears in the extreme northeastern part of the country and on sites where there are cultural elements characteristic of the eastern part of Central Europe. The presence of this particular wheat species may well prove to be linked to a particular route that connects Central Europe and the Balkans to Italy. However, much more information is needed to substantiate this hypothesis.

- Emmer caryopses exhibit different morphological characteristics according to their place of origin. This possibly suggests the presence of different varieties adapted to different geographic and climatic conditions, but again, more morphological and genetic information is needed to validate or reject this hypothesis (Fiorentino et al. 2000).

- Spelt, rye and millets are only occasionally present, and do not seem to have been cultivated during the Neolithic.

	Site and map code														
	1 Piz	2 Cec	3 Iso	4 Ost	5 Vhò	6 Fag	7 Pia	8 Val	9 Pav	10 Sam	11 Alb	12 Sav	13 Chi	14 LaV	15 Lug
Cereals															
Hordeum vulgare s.l.	+	+	+	+	+	+	+	+	+	+	+	+	+	+	+
Triticum dicoccum	+	.	+	.	.	+	+	+	+	+	.	.	.	.	+
Triticum monococcum	.	.	.	.	+	+	+	+	+	+	.	.	.	.	+
Triticum aestivum/durum	.	.	.	.	+	+	.	.	.	+	.	.	.	.	+
Triticum cf. *spelta*	.	.	.	.	.	.	?	.	?	?	.	.	.	.	
"New" glume wheat	.	.	.	.	.	.	+	.	.	+	.	.	.	.	
Triticum sp.	+	.	+	.	.	+	+	+	+	+	.	+	.	+	
Bromus sp.	.	.	.	.	.	.	.	+	.	+	.	.	.	.	
Other cultivated plants															
Linum usitatissimum	.	.	.	.	.	.	.	.	.	+	.	.	.	.	
Pulses															
Lens culinaris	.	.	.	.	.	.	+	.	.	+	.	.	.	.	
Lathyrus cicera/sativus	.	.	.	.	.	.	.	.	.	+	.	.	.	.	
Pisum sativum	.	.	.	.	.	.	+	.	+	+	.	.	.	.	+
Vicia ervilia	.	.	.	.	.	.	.	.	.	+	.	.	.	.	
Vicia cf. *faba*	.	.	.	.	.	.	.	.	.	+	.	.	.	.	
Vicia sativa agg.	.	.	.	.	.	.	+	.	.	+	.	.	.	.	
Leguminosae	+	.	.	.	.	.	+	.	.	+	.	.	.	.	
Fruits															
Corylus avellana	+	.	.	+	+	+	+	+	+	+	.	.	.	+	+
Cornus mas	.	.	.	.	.	.	.	.	.	+	.	.	.	.	+
Cornus sanguinea	.	.	.	.	.	.	.	.	.	+	.	.	.	+	
Malus sp.	.	.	.	.	.	.	+	.	.	+	.	.	.	.	+
Prunus insititia/spinosa	.	.	.	.	.	.	+	.	.	+	.	.	.	.	+
Prunus sp.	.	.	.	.	.	.	+	.	.	+	.	.	.	.	+
Quercus sp.	.	.	.	.	+	.	+	.	.	+	.	.	.	.	+
Rubus fruticosus agg.	.	.	.	.	.	.	+	.	.	+	.	.	.	.	+
Sambucus ebulus	.	.	.	.	.	+	.	.	.	+	.	.	.	.	+
Sambucus nigra/racemosa	.	.	.	.	.	.	.	.	.	+	.	.	.	.	+
Solanum dulcamara	.	.	.	.	.	.	.	.	.	.	.	.	.	.	+
Vitis vinifera sylvestris	+	.	.	.	.	.	+	.	.	+	.	.	.	.	+

Sam= Sammardenchia; Pia= Piancada; Pav= Pavia di Udine; Fag= Fagnigola; Val= Valer; LaV=La Vela; Piz= Pizzo di Bodio; Iso= Isorella; Ost= Ostiano;Vhò= Vhò di Piadena; Lug= Luogo di Romagna; Sav= Savignano; Cec= Cecima; Alb= Albinea; Chi= Chiozza.

Table 9.1: Carpological remains from early neolithic sites in northern Italy

- The range of pulses is varied and includes lentil, pea, grass pea, bitter vetch, broad bean and vetches (chickpeas have not been recorded). Overall, the number of legumes is always lower than that of cereals.
- There are a few finds of flax from only a small number of sites.
- Poppy has only been recorded at La Marmotta.

9.6 The latest neolithic phases

Compared to the early Neolithic, the number of archaeobotanical studies from later phases (ca. 5500–4500 bp) is even more limited (Ferrari and Visentini 2002). The major constraints again are the small numbers of samples taken and identifications that have been made at each site. Consequently, information is partial and highly biased towards casual finds. In spite of these deficiencies, new investigations in northeastern Italy, for example Bannia-Palazzine di Sopra (Cottini et al. 1997; Cottini and Rottoli nd) and Palù di Livenza (Corti et al. 1998). in the central and southern regions, together with the old bibliographic data, suggest an agricultural system characterized by the following aspects:

- Cereal cultivation follows the pattern initiated during the early Neolithic period, however, there seems to be an increase in free-threshing wheats particularly in central and southern Italy (Costantini 2002). At present, data from the north are insufficient to evaluate the importance of single species. The identification of millets and spelt are still uncertain.
- The same pulse species are present in this period as in the early phases. The cultivation of broad beans in northern Italy has yet to be confirmed.
- New evidence is recorded for finds of poppy and flax. Opium poppy is present in at least three waterlogged sites in the north, whereas flax is well attested and has been found in high numbers, at Palù di Livenza (Corti et al. 1998), Lagozza and Isolino di Varese (Castelletti 1975, 1976, Castelletti 1990). Such an increase in flax cultivation may be related to a development of textile activities, as indicated also by the numerous weight looms.
- There is also an increase in fruit remains, for example, *Malus* sp. at Spilamberto (Castelletti et al. 1998) and *Cornus mas* at Bannia (Cottini et al. 1997; Cottini and Rottoli nd). This particular interest in fruits may perhaps be explained as a result of a progressive varietal selection during this period. An alternative explanation might be the evolution of storage and processing techniques. The increase of *Cornus mas* fruits, already present during the early Neolithic, foresees the extremely heavy consumption of fruits in northern Italy during the Bronze Age, which may be related to the preparation of a fermented drink (Rottoli 2001). This increase in the use of fruit does not seem to have included grapes, of which there are few finds throughout the Neolithic.
- In contrast, there is a decrease in hazelnut consumption. *Corylus avellana* was one of the key elements of the early neolithic vegetation in northern Italy. It might be that the spread of other oil-rich species such as flax or poppy had led to a reduction in the exploitation of this species.

Acknowledgments

The authors thank Leonor Peña Chocarro for translating this paper into English.

References

Barker, G. W. W., P. Biagi, L. Castelletti, M. Cremaschi, and R. Nisbet (1987). Sussistenza, economia ed ambiente nel neolitico dell'Italia settentrionale. In A. Revedin (Ed.), *Atti della XXVI Riunione Scientifica dell'Istituto Italiano di Preistoria e Protostoria, Firenze 7-10 Nov. 1985*, Firenze, pp. 103–118. Istituto Italiano di Preistoria e Protostoria.

Biagi, P., M. Cremaschi, and R. Nisbet (1993). Soil exploitation and early agriculture in northern Italy. *The Holocene 3*(2), 164–168.

Biagi, P. and R. Nisbet (1987). The earliest farming communities in northern Italy. In J. Guilaine, J. Courtin, J.-L. Roudil, and J.-L. Vernet (Eds.), *Premières Communautés Paysannes en Méditerranée Occidentale. Actes du Colloque International du C.N.R.S., Montpellier, 26–29 Avril 1983*, Paris, pp. 447–454. C.N.R.S.

Carugati, M. G. (1993). Il neolitico antico in Friuli attraverso lo studio dei resti vegetali carbonizzati di tre siti: Fagnigola (PN), Valer (PN) e Sammardenchia (UD). *Quaderni Friulani di Archeologia 3*, 15–22.

Carugati, M. G. (1994). Nota sui resti vegetali carbonizzati del sito neolitico di Valer (Azzano Decimo - Pordenone). *Atti della Società di Preistoria e Protosoria del Friuli-Venezia Giulia 1993*(VIII), 115–120.

Carugati, M. G., L. Castelletti, and M. Rottoli (1996). L'agricoltura nel primo neolitico del Friuli. le ricerche a Sammardenchia, Fagnigola e Valer. In *Sammardenchia e i Primi Agricoltori del Friuli*, Udine, pp. 103–112. Credito Cooperativo – Banca di Credito Cooperativo di Basiliano.

Castelletti, L. (1975). Reperti di resti vegetali macroscopici nell'Italia settentrionale. In *Congresso Nazionale di Storia dell'Agricoltura, Milano 7-9 maggio 1971*, Firenze, pp. 93–102. Accademia Economico-Agraria dei Georgofili.

Castelletti, L. (1976). Agricoltura neolitica a sud delle Alpi. *Atti del Centro Studi e Documentazione Italia Romana, 1975-76 7*, 105–115.

Castelletti, L. (1990). Relazione preliminare sui resti macroscopici vegetali dell'Isolino di Varese: Scavi 1977-85. In P. Biagi (Ed.), *The Neolithisation of the Alpine Region*, Monografie di "Natura Bresciana" 13, Brescia, pp. 207–212. Vannini.

Castelletti, L. and M. G. Carugati (1995). I resti vegetali del sito neolitico di Sammardenchia di Pozzuolo del Friuli (Udine). In A. Revedin (Ed.), *Preistoria e Protostoria del Friuli Venezia Giulia e dell'Istria, Atti della XXIX Riunione Scientifica dell'Istituto Italiano di Preistoria e Protostoria, Trieste, 28-30 settembre 1990*, Firenze, pp. 167–174.

Castelletti, L., E. Castiglioni, L. Leoni, and M. Rottoli (1998). Resti botanici dai contesti del neolitico medio-recente, appendice 3. *Bullettino di Paletnologia Italiana (n. s.) 89*(7), 191–200.

Castelletti, L. and M. Madella (1994). Appendix 3: Note sugli scavi di Pizzo di Bodio (Varese), 1985-88. *Preistoria Alpina (1991) 27*, 236–238.

Castelletti, L. and A. Maspero (1992). Analisi di resti vegetali di Campo Ceresole del Vhò di Piadena e di altri siti neolitici padani. *Natura Bresciana: Annali del Museo Civico di Scienze Naturali, Brescia (1990-91) 27*, 289–305.

Castelletti, L. and M. Rottoli (2002). Nuovi dati sull'agricoltura e l'ambiente del Neolitico dell'Italia settentrionale. In *Atti della XXXIII Riunione Scientifica dell'Istituto Italiano di Preistoria e Protostoria, Trento 21-24 ottobre 1997*, pp. 271–277.

Celant, A. (2000). Nuovi dati archeobotanici su ambiente e agricoltura nel Neolitico del Lazio: un esempio dalla Campagna Romana. In A. Pessina and G. Muscio (Eds.), *La Neolitizzazione Tra Oriente e Occidente, Atti del Convegno di Studi, Udine 23-24 Aprile 1999*, pp. 355–364. Comune di Udine: Edizioni del Museo Friulano di Storia Naturale.

Cermesoni, B., A. Ferrari, P. Mazzieri, and A. Pessina (1999). Considerazioni sui materiali ceramici e litici. In A. Ferrari and A. Pessina (Eds.), *Sammardenchia Cueis. Contributi per la Conoscenza di una Comunità del Primo Neolitico*, pp. 231–258. Udine: Edizioni del Museo Friulano di Storia Naturale, Comune di Udine.

Corti, P., N. Martinelli, R. Micheli, E. M. Kokelj, G. Petrucci, A. Riedel, M. Rottoli, P. Visentini, and S. Vitri (1998). Siti umidi tardoneolitici: nuovi dati da Palù di Livenza (Friuli-Venezia Giulia, Italia). In A. A. et al. (Ed.), *Atti del XIII Congresso dell'Unione Internazionale delle Scienze Preistoriche e Protostoriche, Forlì 8-14 Settembre 1996*, pp. 1379–1391. Forlì: A. B. A. C. O. Edizioni.

Costantini, L. (2002). Italia centro-meridionale. In G. Forni and A. Marcone (Eds.), *Storia dell'Agricoltura Italiana, Vol. I, L'età Antica, 1° Preistoria*, pp. 221–234. Firenze: Accademia dei Georgofili, Edizioni Polistampa.

Costantini, L. and M. Stancanelli (1994). La preistoria agricola dell'italia centro-meridionale: il contributo delle indagini archeobotaniche. *Origini 18*, 149–244.

Cottini, M., A. Ferrari, P. Pellegatti, G. Petrucci, M. Rottoli, G. Tasca, and P. Visentini (1997). Bannia, Palazzine di Sopra (Fiume Veneto, Pordenone): scavo 1995. *Atti della Società per la Preistoria e Protostoria della Regione Friuli-Venezia Giulia 10*, 119–150.

Cottini, M. and M. Rottoli (nd). Le analisi archeobotaniche. In P. Visentini (Ed.), *Il sito Neolitico di Bannia Palazzine di Sopra.* Pordenone: Museo delle Scienze, Comune di Pordenone. In Press.

Degasperi, N., A. Ferrari, and G. Steffe (1998). Il sito neolitico di Lugo di Romagna. In A. Pessina and G. Muscio (Eds.), *Settemila Anni fa il Primo Pane. Ambienti e Culture delle Prime Comunità neolitiche. Catalogo della Mostra, Udine, Museo Friulano di Storia Naturale*, pp. 185–191. Udine: Museo Friulano di Storia Naturale.

Delpino, C., A. Ferrari, and P. Mazzieri (1999). Le rocce silicee scheggiate di Sammardenchia Cueis (scavi 1994–1998): provenienza e dispersione. In A. Ferrari and A. Pessina (Eds.), *Sammardenchia Cueis. Contributi per la Conoscenza di una Comunità del Primo Neolitico*, pp. 275–290. Udine: Edizioni del Museo Friulano di Storia Naturale 41.

Ferrari, A. and A. Pessina (Eds.) (1996). *Sammardenchia e i Primi Agricoltori del Friuli.* Tavagnacco, Udine: Banca di Credito Cooperativo di Basiliano.

Ferrari, A. and P. Visentini (Eds.) (2002). *Il Declino del Mondo Neolitico. Ricerche in Italia Centro-Settentrionale fra Aspetti Peninsulari, Occidentali e Nord-Alpini. Atti del Convegno, Pordenone 5-7 Aprile 2001.* Quaderni del Museo Archeologico del Friuli Occidentale 2. Pordenone: Museo delle Scienze, Comune di Pordenone.

Fiorentino, G. (1995). New perspectives in anthracological analysis. Palaeoecological and technological implications of charcoals found in the neolithic flintmine at La Defensola (Vieste, Apulia, Italy). *Quaternaria Nova 5*, 99–128.

Fiorentino, G., I. M. Muntoni, and F. Radina (2000). La neolitizzazione delle murge baresi: ambienti, insediamenti e attività produttive. In A. Pessina and G. Muscio (Eds.), *La Neolitizzazione tra Oriente e Occidente, Atti del Convegno di Studi, Udine 23-24 Aprile 1999*, pp. 381–412. Comune di Udine: Edizioni del Museo Friulano di Storia Naturale.

Fontana, A. (1999). Aspetti geomorfologici dell'area di Sammardenchia. In A. Ferrari and A. Pessina (Eds.), *Sammardenchia Cueis. Contributi per la Conoscenza di una Comunità del Primo Neolitico*, pp. 11–22. Udine: Edizioni del Museo Friulano di Storia Naturale 41.

Fugazzola Delpino, M. A. (1998). La vita quotidiana del neolitico. il sito della Marmotta sul lago di Bracciano. In A. Pessina and G. Muscio (Eds.), *Settemila anni fa il primo pane. Ambienti e culture delle prime comunità neolitiche. Catalogo della Mostra, Udine, Museo Friulano di Storia Naturale*, pp. 192–211. Udine: Museo Friulano di Storia Naturale.

Fugazzola Delpino, M. A. and M. Mineo (1995). La piroga neolitica di Bracciano (La Marmotta 1). *Bullettino di Paletnologia Italiana n. s. II 86*, 197–266.

Improta, S. and A. Pessina (1998). La neolitizzazione dell'Italia settentrionale. il nuovo quadro cronologico. In A. Pessina and G. Muscio (Eds.), *Settemila Anni fa il Primo Pane. Ambienti e Culture delle Prime Comunità Neolitiche. Catalogo della Mostra, Udine, Museo Friulano di Storia Naturale*, pp. 107–115. Udine: Museo Friulano di Storia Naturale.

Jones, G., S. Valamoti, and M. Charles (2000). Early crop diversity: a 'new' glume wheat from northern Greece. *Vegetation History and Archaeobotany 9*, 133–146.

Magri, D. (1995). Some questions on the late-Holocene vegetation of Europe. *The Holocene 5*(3), 354–360.

Nisbet, R. (1995). I resti macrobotanici. In *L'Insediamento Neolitico di Ostiano-Dugali Alti (Cremona) nel suo Contesto Ambientale ed Economico*, Monografie di Natura Bresciana 22, pp. 104–106. Brescia: Museo Civico di Scienze Naturali di Brescia.

Nisbet, R. (2000). Aspetti forestali e agricoltura ad Isorella, in E. Starnini et al. "Nuovi dati sul neolitico antico della Pianura Padana centrale dal sito di Isorella (Brescia)". In A. Pessina and G. Muscio (Eds.), *La Neolitizzazione tra Oriente e Occidente, Atti del Convegno di Studi, Udine 23-24 Aprile 1999*, pp. 240–247. Comune di Udine: Edizioni del Museo Friulano di Storia Naturale.

Perrino, P., K. Hammer, G. Laghetti, B. Margiotta, S. Cifarelli, and G. Fiorentino (2000). Farro in Italia meridionale: dal neolitico ai tempi moderni. In A. Pessina and G. Muscio (Eds.), *La Neolitizzazione Tra Oriente e Occidente, Atti del Convegno di Studi, Udine 23-24 Aprile 1999*, pp. 425–438. Comune di Udine: Edizioni del Museo Friulano di Storia Naturale.

Pessina, A. (1998). Aspetti culturali e problematiche del primo neolitico dell'Italia settentrionale. In A. Pessina and G. Muscio (Eds.), *Settemila Anni fa il Primo Pane. Ambienti e Culture delle Prime Comunità neolitiche. Catalogo della Mostra, Udine, Museo Friulano di Storia Naturale*, pp. 95–105. Udine: Museo Friulano di Storia Naturale.

Pessina, A. and C. D'Amico (1999). L'industria in pietra levigata del sito neolitico di Sammardenchia (Pozzuolo del Friuli, Udine). Aspetti archeologici e petroarcheometrici. In A. Ferrari and A. Pessina (Eds.), *Sammardenchia Cueis. Contributi per la Conoscenza di una Comunità del Primo Neolitico*, Volume 41, pp. 23–92. Udine: Edizioni del Museo Friulano di Storia Naturale.

Pessina, A., A. Ferrari, and A. Fontana (1998). Le prime popolazioni agricole del Friuli. In A. Pessina and G. Muscio (Eds.), *Settemila Anni fa il Primo Pane. Ambienti e Culture delle Prime Comunità Neolitiche. Catalogo della Mostra, Udine, Museo Friulano di Storia Naturale*, pp. 133–145. Udine: Museo Friulano di Storia Naturale.

Pessina, A., G. C. Fiappo, and M. Rottoli (2004). Un sito neolitico a Pavia di Udine, proprietà Paolini. Nuovi dati sull'inizio dell'agricoltura in Friuli. *Gortania, Atti del Museo Friulano di Storia Naturale, Udine 25*, 237–280.

Pessina, A. and G. Muscio (Eds.) (1998). *Settemila Anni fa il Primo Pane. Ambienti e Culture delle Prime Comunità Neolitiche. Catalogo della Mostra, Udine, Museo Friulano di Storia Naturale.* Udine: Museo Friulano di Storia Naturale.

Pessina, A. and G. Muscio (Eds.) (2000). *La Neolitizzazione tra Oriente e Occidente. Atti del Convegno di Studi, Udine 23-24 Aprile 1999.* Udine: Edizioni del Museo Friulano di Storia Naturale, Comune di Udine.

Pignatelli, O. and M. Rottoli (1996). Analisi archeobotaniche. In E. B. Citton (Ed.), *Indagine Interdisciplinare nell'Insediamento Neolitico di Roncade (Treviso), Località Biancade*, pp. 113–119. Quaderni di Archeologia del Veneto.

Rottoli, M. (1993). "La Marmotta", Anguillara Sabazia (RM). Scavi 1989. Analisi paletnobotaniche: prime risultanze. Appendice 1. *Bullettino di Paletnologia Italiana n. s. II 84*, 305–315.

Rottoli, M. (1999). I resti vegetali di Sammardenchia - Cûeis (Udine), insediamento del neolitico antico. In A. Ferrari and A. Pessina (Eds.), *Sammardenchia Cueis. Contributi per la Conoscenza di una Comunità del Primo Neolitico*, pp. 307–326. Udine: Edizioni del Museo Friulano di Storia Naturale 41.

Rottoli, M. (2001). Analisi archeobotaniche: i macroresti vegetali. In P. Frontini (Ed.), *Castellaro del Vhò. Campagne di Scavo 1996-1999*, pp. 175–195. Milano: Comune di Milano, Settore Cultura Musei e Mostre, Raccolte Archeologiche e Numismatiche.

Rottoli, M. (2002). Zafferanone selvatico (*Carthamus lanatus*) e cardo della madonna (*Silybum marianum*), piante raccolte o coltivate nel neolitico antico a "La Marmotta"? *Bullettino di Paletnologia Italiana (2000-2001) 91–92*, 47–61.

Rottoli, M. and M. A. Fugazzola Delpino (2000). Lauro-Carpinetum: A particular forest formation in the Roman countryside; archaeological data and actual diffusion. In J. L. Vernet and M. Chabal (Eds.), *Nouvelles Approches Méthodologiques, Histoire de la Végétation et des Usages du Bois Depuis la Préhistoire, Second Colloque International d'Anthracologie, Paris 13-16 Septempber 2000 (Résumés)*, Paris, pp. 45. C.N.R.S.

Chapter 10

Crop evolution: new evidence from the Neolithic of west Mediterranean Europe

Ramon Buxó
Museu d'Arqueologia de Catalunya, Girona, Spain

10.1 Introduction

The process of neolithisation in the west Mediterranean—whereby a farming economy progressively replaced one of hunting and gathering—was complex and was due in part to the geographical diversity of the territory. The inadequacy of information about the process in this part of the Iberian Peninsula has no doubt exaggerated our present view of its complexity.

The first problem to focus on is how to define the reasons for the initial impetus behind the process. Most archaeologists are in agreement that the mesolithic substratum of the west Mediterranean coasts is characterised by the presence of a geometric stone industry (Fortea 1973; Juan-Cubanilles 1990). To date, this facies is unknown in Catalonia (northeast Spain) and the suggestion that Neolithisation had its roots in earlier, technologically more complex groups such as those that inhabited the cave of Filador in Margalef del Montsant (García-Argüelles et al. 1990), or the open-air site of Font del Ros in Berga (Pallarés et al. 1997), is not supported by appropriate equivalent evidence in the surrounding territories.

If we try to assess the problem in the perspective of indigenist and migrationist theories, both are in agreement that the origin of the Neolithic has a Mediterranean dimension that has its roots in the Pre-Pottery Neolithic B (PPNB) of the Near East. The neolithic culture on the Adriatic and Sicilian coasts at 7000 cal BC is identified by Impressed Ware pottery. This culture spread westwards and agriculture (i.e., crops and livestock) reached the coasts of the Iberian Peninsula by the beginning of 6000 cal BC; in this region the culture was characterised by Cardial Ware pottery.

It is possible to identify different points along the coast where small neolithic communites settled. Neolithisation was not uniform in these areas but it can be considered to represent a limited colonisation process that involved in a direct or indirect way acculturation of indigenous hunter-gatherers (Martí et al. 1997). One of these foci for the penetration of the Cardial Ware culture seems to have been the central Catalonian coast, as deduced from the distribution of sites that have produced this type of pottery (figure 10.1).

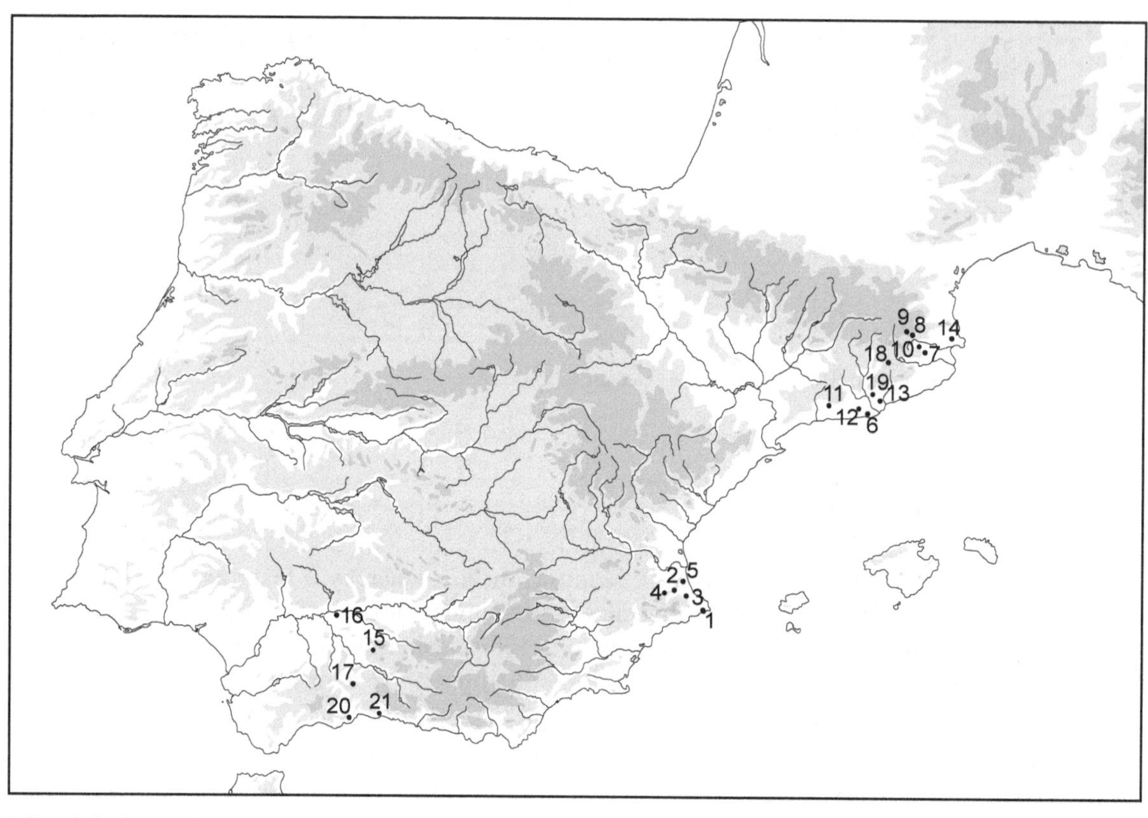

1. Cova de les Cendres (Moraira, Alicante)
2. Cova de l'Or (Beniarrés, Alicante)
3. Cova de Recambra (Alicante)
4. Cova de la Sarsa (Bocairent, Valencia)
5. Cova del Llop (Gandía, Valencia)
6. Cova de Can Sadurní (Begues, Barcelona)
7. La Draga (Banyoles, Girona)
8. Plansallosa (Tortellà, Girona)
9. Cova 120 (Sales de Llierca, Girona)
10. Cova de Pau (Serinyà, Girona)
11. Les Guixeres de Vilobí (Vilobí, Barcelona)
12. Can Tintorer (Gavà, Barcelona)
13. Montmeló (Montmeló, Barcelona)
14. Ca N'Isach (Palau Saverdera, Girona)
15. Cueva de los Murciélagos (Zuheros, Córdoba)
16. Cueva de los Mármoles (Córdoba)
17. Cueva del Toro (Antequera, Málaga)
18. Cova del Toll (Moià, Barcelona)
19. Bóbila Madurell (Sabadell, Barcelona)
20. Cueva del Bajoncillo (Málaga)
21. Cueva de Nerja (Nerja, Málaga)

Figure 10.1: Location of the early and middle neolithic sites of the Mediterranean Iberian Peninsula with archaeobotanical evidence.

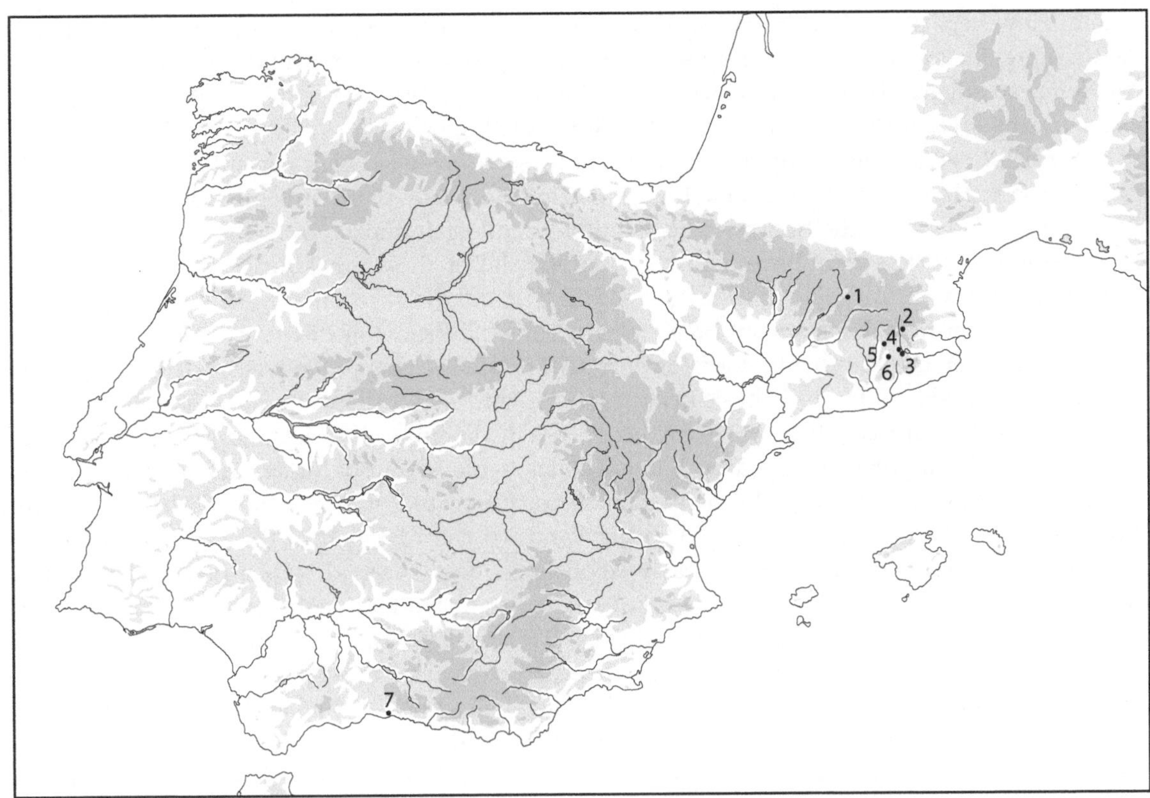

1. Balma Margineda (Andorra)
2. Sota Palou (Campdevànol, Ripollès)
3. Cingle Vermell (Vilanova de Sau)
4. Roc del Migdia (Vilanova de Sau, Osona)
5. Font del Ros (Berga, Berguedà)
6. Gai (Moià, Bages)
7. Cueva de Nerja (Nerja, Málaga)

Figure 10.2: Location of the epipalaeolithic/mesolithic sites with archaeobotanical evidence referred to in the text.

10.2 First evolutionary phase

The range of wild plant foods exploited in the Mesolithic was rich and diverse, for example, blackthorn/sloe (*Prunus spinosa*) was identified in the Gai (Moià, Bages), the Font del Ros (Berga, Berguedà) and in the Balma Margineda (Andorra); remains of hazelnut endocarp (*Corylus avellana*) were found in the Cingle Vermell (Vilanova de Sau, Osona), Sota Palou (Campdevànol, Ripollès), Roc del Migdia (Vilanova de Sau, Osona), Font del Ros and the Balma Margineda; acorns (*Quercus* sp.) were discovered at the Balma Margineda, the Cingle Vermell and at the Roc del Migdia; walnuts (*Juglans regia*) and pine kernels (*Pinus pinea*) were identified at the Cingle Vermell, the cave of Nerja (Bernabeu et al. 1993) and in the Balma Margineda; and remains of wild apple (*Malus sylvestris*) and wild pear (*Pyrus sylvestris*) were recovered from the Font del Ros (Marinval 1995; Buxó 1997; Holden et al. 1995; Terradas 1995) (figure 10.2).

Cereals, including different varieties of wheat and barley, and legumes (e.g., peas and beans), as well as livestock such as cattle, goats and pigs, together with the knowledge of crop and animal husbandry, were introduced with the first neolithic culture. Small groups of colonists would initially have been established in the littoral regions, the central prelittoral valleys and on the plains. These groups had a strong cultural unity and the presence on sites of artefacts such as marble bracelets demonstrated that there were almost certainly trade networks between them. However, the archaeobotanical record indicates how species of domestic plants increased without replacing the wild species used in the Mesolithic. Thus it would seem that there is no evidence to suggest there was in situ domestication of plants that were locally available in the wild in this area of the Iberian Peninsula (Buxó 2002a).

Neolithic settlements were initially focused in locations with high agricultural potential that were close to rivers and areas of marshland. During this early phase it is possible that the groups used caves and rocky shelters prior to gaining complete influence over the territory, however, they would doubtless soon have realised the potential of these areas for hunting and collecting forest resources.

The pioneering colonists would have carried out incursions of discovery and advance in the surrounding territories. On the basis of current knowledge this process seems to have occurred both rapidly and on a frequent basis in comparison with the occupation of the littoral and prelittoral plains. In contrast, the neolithisation process in the upper interior and pre-Pyrenees region was diachronic and irregular because it involved the replacement of hunter-gatherer communities by isolated groups of neolithic farmers. This is demonstrated, for example, at the site of Balma Margineda in Andorra (carbon dated, from wood, to the first half of 6th millennium cal BC) where the spectrum of plants in the early Neolithic is characterised by the presence of domestic crops together with the remains of gathered resources; although in comparison to the Mesolithic there is a reduction in the number of wild taxa (Marinval 1995). The domestic plants comprised cereals (a compact type of free-threshing wheat, naked barley and emmer) and peas.

In Catalonia six sites have been recorded from the early Neolithic and four from the middle Neolithic. Most studies have been focused on sites located in the eastern Pyrenees and in the Mediterranean coastal areas, with the exception being those sites in the interior and in the Catalonian Pyrenees, and very few in the south and west of Catalonia. The sites are distributed as follows:

Early Neolithic (second half of the 6th millennium cal BC to the second half of 5th millennium cal BC—figure 10.3): La Draga (Banyoles, Pla de l'Estany), Plansallosa (Tortellà, la Garrotxa), Bauma del Serrat del Pont (Tortellà, la Garrotxa), Cova 120 (Sales de Llierca, la Garrotxa), Can Sadurní (Begues, Baix Llobregat), Guixeres de Vilobí (Vilobí, Alt Penedès). La Draga, Plansallosa and Guixeres de Vilobí, are open air sites and Bauma del Serrat del Pont, Cova 120 and Can Sadurní are caves.

Middle Neolithic (first half of 4th millennium cal BC—figure 10.4): Can Tintorer (Gavà, Baix Llobregat), Cova del Toll (Moià, Bages), Cova de Pau (Serinyà, Pla de l'Estany), Ca N'Isach (Palau Saverdera, Alt Empordà). Can Tintorer, Cova del Toll and Cova de Pau are caves, and Ca N'Isach is an open air site.

Cereal species comprise 32% (based on chaff and grains) of the total of number of taxa identified in the early Neolithic; pulses make up 21% of the total, and the gathered wild species represent 47%. Six cereal taxa have been identified (figure 10.5): free-threshing wheat (*Triticum durum*), a compact type of free-threshing wheat (*Triticum aestivum/durum compactum* type), hulled barley (*Hordeum vulgare*) and naked barley (*Hordeum vulgare* var. *nudum*), and two species of hulled wheats: emmer (*Triticum dicoccum*) and einkorn (*Triticum monococcum*). The pulses in the early Neolithic sites are represented by four species: grass pea (*Lathyrus cicera/sativus*), lentil (*Lens culinaris*), pea (*Pisum sativum*) and broad bean (*Vicia faba*).

Comparison of the percentage presence of the cereals (figure 10.5) indicates that free-threshing wheat occurs slightly more frequently than the two types of barley (the value for both is 16.2%); however, the differences between the taxa in terms of the numbers of identified items are not as pronounced. Emmer wheat is less common than these three cereals and so it is not possible to confirm whether or not it played an important role in early neolithic agriculture. The low

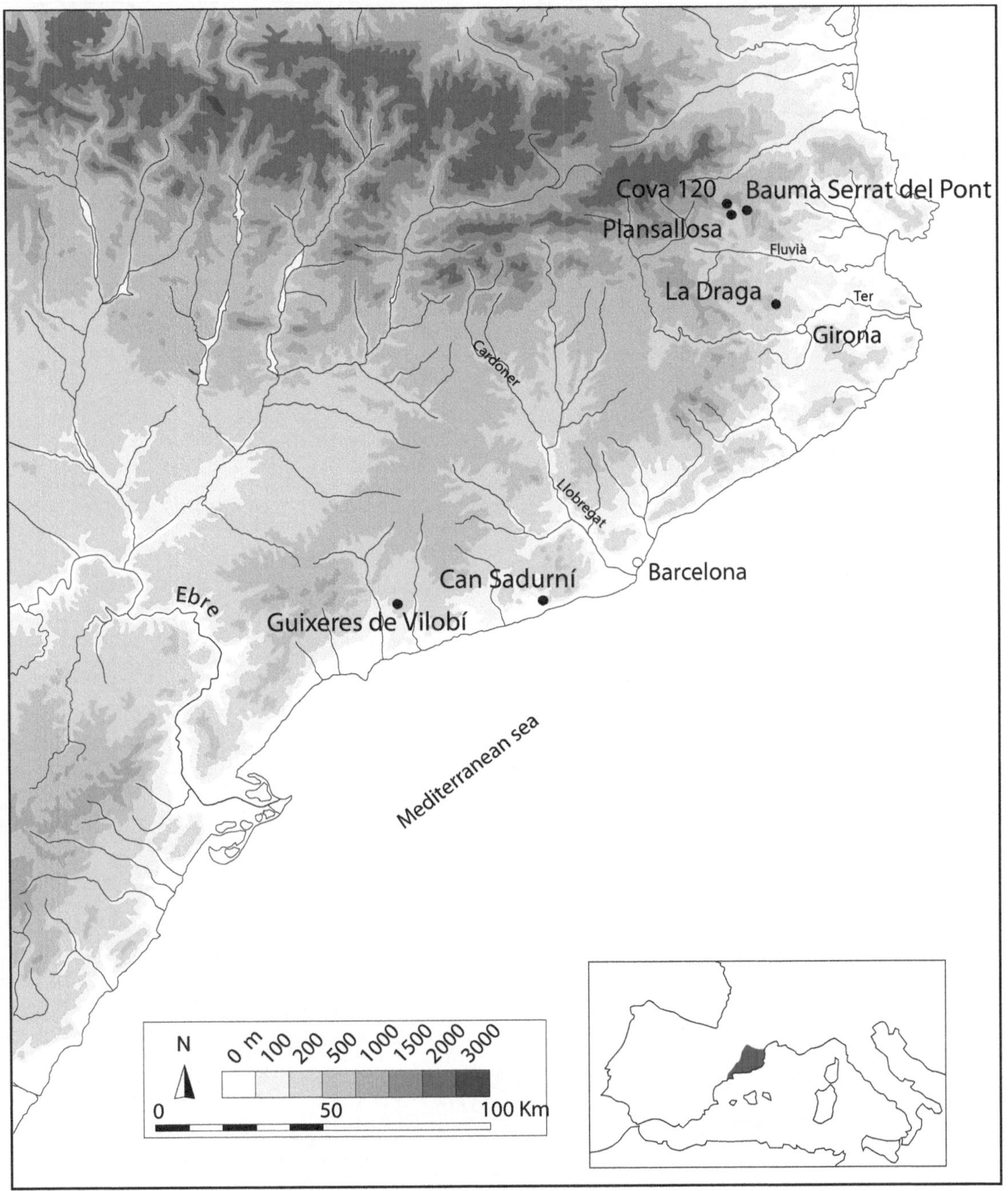

Figure 10.3: Location of the early neolithic sites from northeast Spain with archaeobotanical evidence referred to in the text.

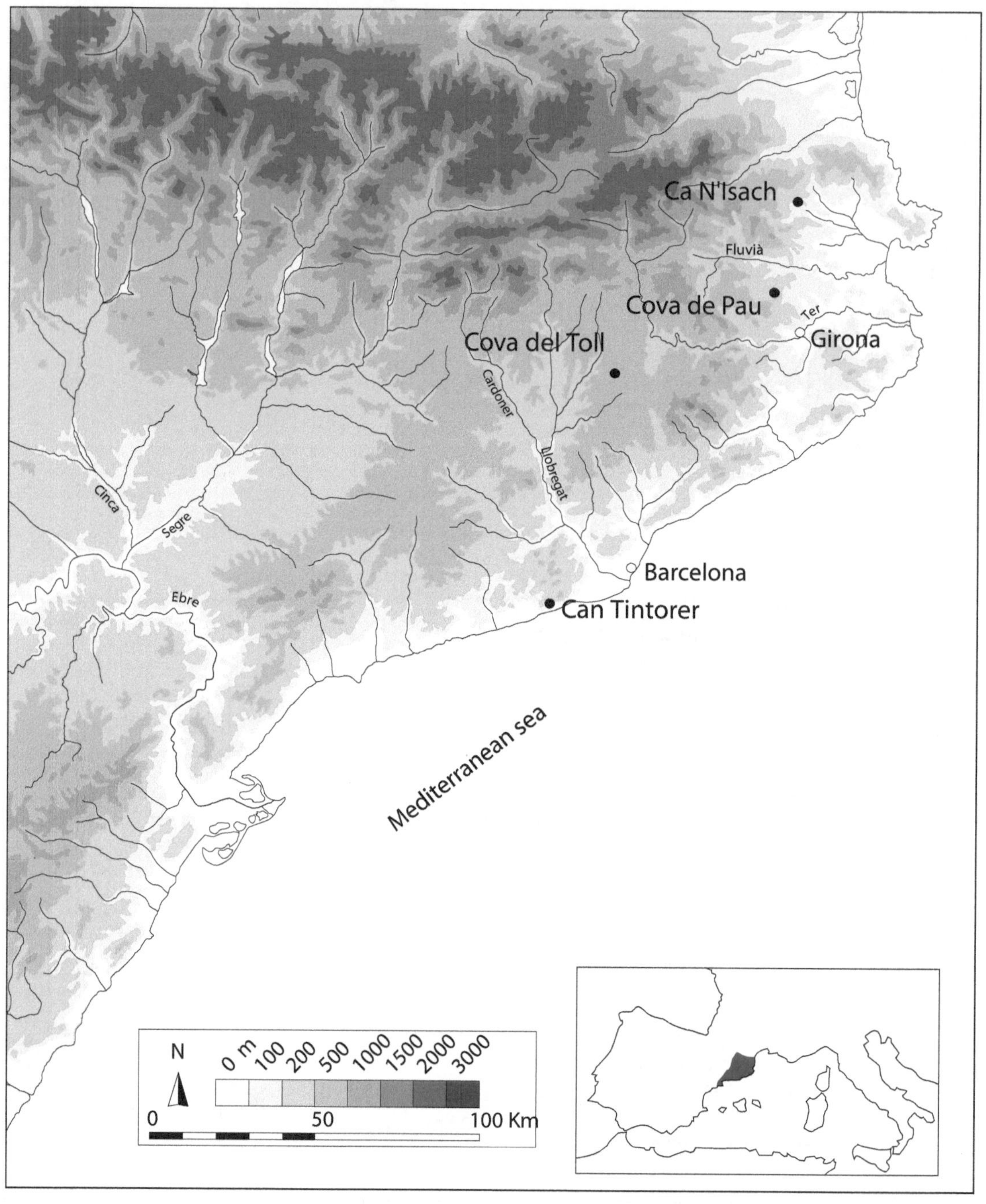

Figure 10.4: Location of the middle Neolithic sites from northeast Spain with archaeobotanical evidence referred to in the text.

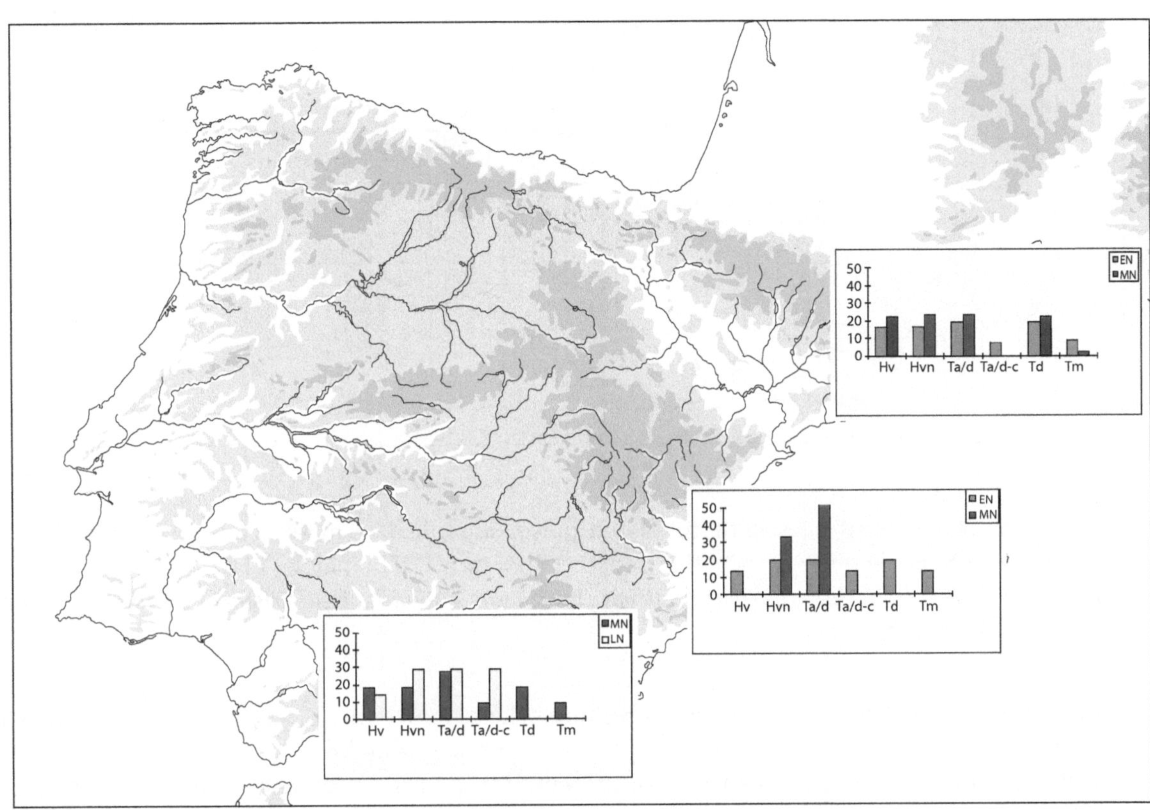

Figure 10.5: Percentage presence of cereals identified from six sites in three regions of Mediterranean Spain in different neolithic periods (EN=early, MN=middle, LN=late). Cereals: Hv= *Hordeum vulgare*, Hvn= *Hordeum vulgare* var. *nudum*, Ta/d= *Triticum aestivum/ durum*, Ta/d-c= *Triticum aestivum/ durum* var. *compactum*, Td= *Triticum dicoccum*, Tm= *Triticum monococcum*.

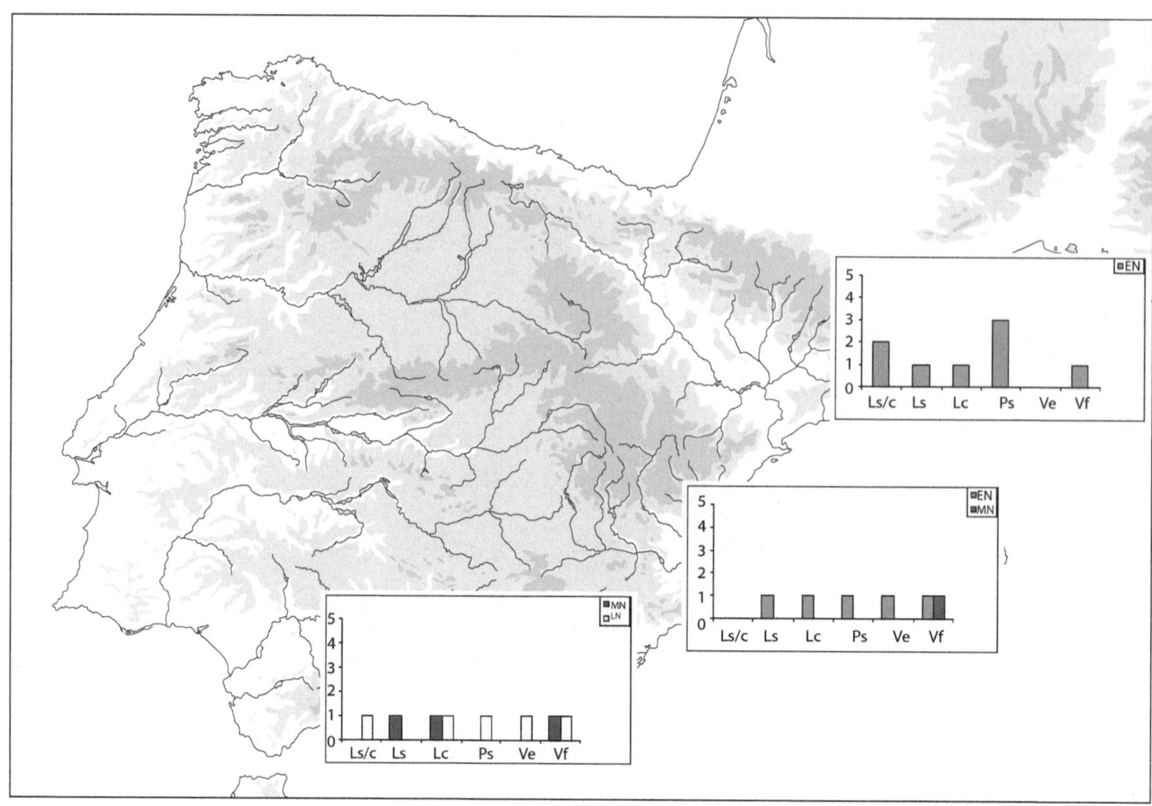

Figure 10.6: Numbers of sites with cultivated pulses identified from six sites in three regions of Mediterranean Spain in different neolithic periods (EN=early, MN=middle, LN=late). Pulses: Ls/c= *Lathyrus sativus/cicera*, Ls= *Lathyrus sativus*, Lc= *Lens culinaris*, Ps= *Pisum sativum*, Ve= *Vicia ervilia*, Vf= *Vicia faba*.

frequencies recorded for compact free-threshing wheat and einkorn confirmed their status as secondary crops .

Of the four cultivated pulses that are present in the early Neolithic (figure 10.6), pea and lentil are the most common species (the percentage presence of both is 8.1%), broad bean and grass pea occur less frequently and are in only 2.7% of all samples. It is possible, therefore, to confirm that pulses were being cultivated in the Iberian Peninsula as early as the cereals, albeit on a smaller scale. On this basis it is suggested that their role in early neolithic agriculture was secondary to that of the cereals.

The first evidence of domesticated plant remains recorded in Catalonia is from the site called La Draga (5300–5150 cal BC) (Bosch et al. 2000). The agricultural system at La Draga appears to have been both complex and varied (Bosch et al. 2000). The inhabitants of the settlement cultivated cereals that were common in the early Neolithic of the western Mediterranean: free-threshing wheat, a compact type of free-threshing wheat, emmer wheat, and both naked and hulled barley; very large quantities of free-threshing wheat grains and chaff were found at the site (more than 500,000 grains in 200 samples, the majority of which were free-threshing), but there were only small amounts of other cereal species (figures 10.7 and 10.8). The cereals must have been cultivated in the great flat expanses of the nearby Pla d'Usall and Martís. They were processed at the settlement and stored in large ceramic jars (Buxó et al. 2000, p.129).

The archaeological evidence suggests that many of the cereal grains were roasted in the fireplaces; triangular wooden structures, which were found next to the fireplaces, could have

Figure 10.7: Carbonised kernels of *Triticum aestivum/durum* from La Draga (terrestrial area).

Figure 10.8: Rachis fragment of *Triticum durum* from La Draga.

acted as small platforms for supporting the jars that contained the grain during the roasting process.

The presence at the site of both of broad beans and peas indicates that pulses were cultivated, possibly on the marshy land on the shores of the lake. Gathered fruits and berries are well documented in the samples at La Draga; for example, there is an abundance of hazelnuts, acorns, pine nuts, blackberries, sloes, wild grapes (*Vitis vinifera* var. *sylvestris*) and wild apples or pears (Buxó et al. 2000) (table 10.1). These wild resources would doubtless have been collected in the nearby woods.

The archaebotanical analyses carried out on samples from two levels in the cave of Can Sadurní confirmed the presence of cultivated taxa together with gathered wild products (Blasco et al. 1999). In the level corresponding to the early Neolithic, that is, the late Cardial/Epicardial (dated between 5280 and 4705 cal BC), there were a number of domesticated species of which the most frequently represented were the cereals: free-threshing wheat, spelt, hulled and naked barley, and einkorn. Lentisc (*Pistacia lentiscus*) was also identified in this level.

The next level in the cave comprises two associated post-cardial phases (dated between 4885 and 3875 cal BC). Cereal remains (e.g., free-threshing wheat, emmer, einkorn, and hulled and naked barley) as well as pulses (e.g., pea, lentil, grass pea and vetch (*Vicia sativa*)) have been recorded in both phases. The remains of wild plants included lentisc, wild grape, fig (*Ficus carica*) and acorn.

There appear not to be many differences between the various levels dated to the early Neolithic at the cave of Can Sadurní in terms of the plants represented. Free-threshing wheat is the most common species in both levels, together with hulled wheats (especially emmer) and barley (especially naked barley). We have identified acorns as well as fleshy fruits (e.g., wild grape) together with the stones of the strawberry tree (*Arbutus unedo*) and lentisc.

Cultivated plant remains have also been identified at Cova 120 (Agustí et al. 1987) and Plansallosa (Bosch et al. 1998). Naked and hulled barley, and free-threshing wheat were identified and the last-named taxon was the most common. The frequency of occurrence of the hulled wheats (i.e., emmer) at the site was much lower than that at the cave of Can Sadurní.

In the layer dated to the early Neolithic at the site of Balma del Serrat near to La Garrotxa, however, there are only records of the remains of gathered wild plant resources (Buxó 2002b, p.83).

The earliest record of cultivated plants in the eastern zone of the Iberian Peninsula is dated to the latter half of the 6th millennium cal BC, corresponding with early neolithic settlement in that region (figure 10.9). These settlements are located in caves close to the coast but it is assumed that they were in coexistence with open-air sites. The groups practised agriculture and cultivated several different cereal and pulse species (Buxó 1997). Free-threshing wheat, naked and hulled barley were the most important cereals and the hulled wheats (emmer and einkorn) were only slightly less common; the pulses included grass pea, lentil, pea and broad bean. The remains of gathered wild plants occurred with a similar frequency to that of the earlier period and acorns were the most common taxon (figure 10.5 and 10.6).

Archaeobotanical analyses undertaken at the well-known neolithic site of the Cova de l'Or (Beniarrés, Comtat) demonstrated the presence of five cereals species: free-threshing wheat, emmer, einkorn, and hulled and naked barley, but also the absence of cultivated pulses (Hopf 1966). Another site from this period is the Cova de la Sarsa (Bocairent, Vall d'Albaida) where emmer wheat and free-threshing wheat have been identified (López 1980). In the early levels at a third cave site, the Cova de Cendres (Moraira-Teulada, Marina Baixa, 7540±140 bp or 6600–6050 cal BC), there were the remains of cereals, predominantly free-threshing wheat and emmer but with some naked and hulled barley, as well as a small amount of einkorn. Pulses were represented by the remains of grass pea, lentil, pea and broad bean. There was also evidence of the remains of plants gathered from the wild, for example, acorns and wild olives (*Olea europaea*

Taxon	Terrestrial area	Underwater area
Cereals		
Hordeum vulgare nudum	+	.
Hordeum vulgare	+	.
Triticum dicoccum	+	+
Triticum aestivum t. *compactum*	+	.
Triticum durum	+	+
Pulses		
Pisum sativum	+	.
Vicia faba	+	.
Wild plants (fruits)		
Cornus sanguinea	.	+
Corylus avellana	.	+
Pinus pinea	+	.
Prunus spinosa	.	+
Quercus sp.	.	+
Rubus fruticosus	+	+
Vitis vinifera sylvestris	+	+
Other wild plants		
Acer sp.	.	+
Alisma plantago-aquatica	+	.
Alnus glutinosa	.	+
cf.*Apium repens* type	+	.
cf.*Carex canescens* type	+	.
Chenopodium album	+	.
Cladium mariscus	+	.
Cupressaceae	+	.
Cyperaceae/Polygonaceae	+	.
Cyperus cf. *longus*	+	.
Fraxinus cf. *excelsior*	.	+
Hypericum sp.	+	.
Mentha arvensis/aquatica	+	.
Papaver rhoeas/dubium	+	.
Physalis alkekengi	+	.
cf. *Platanus* sp.	.	+
Polygonum cf. *persicaria*	+	.
Ranunculus sceleratus type	+	.
Ranunculus sp.	+	.
cf. *Salix* sp.	.	+
Stellaria sp.	+	.
Taraxacum officinale	+	.
Tilia platyphyllos	.	+
Verbena officinalis	+	.

Table 10.1: List of taxa from terrestrial and underwater samples from La Draga.

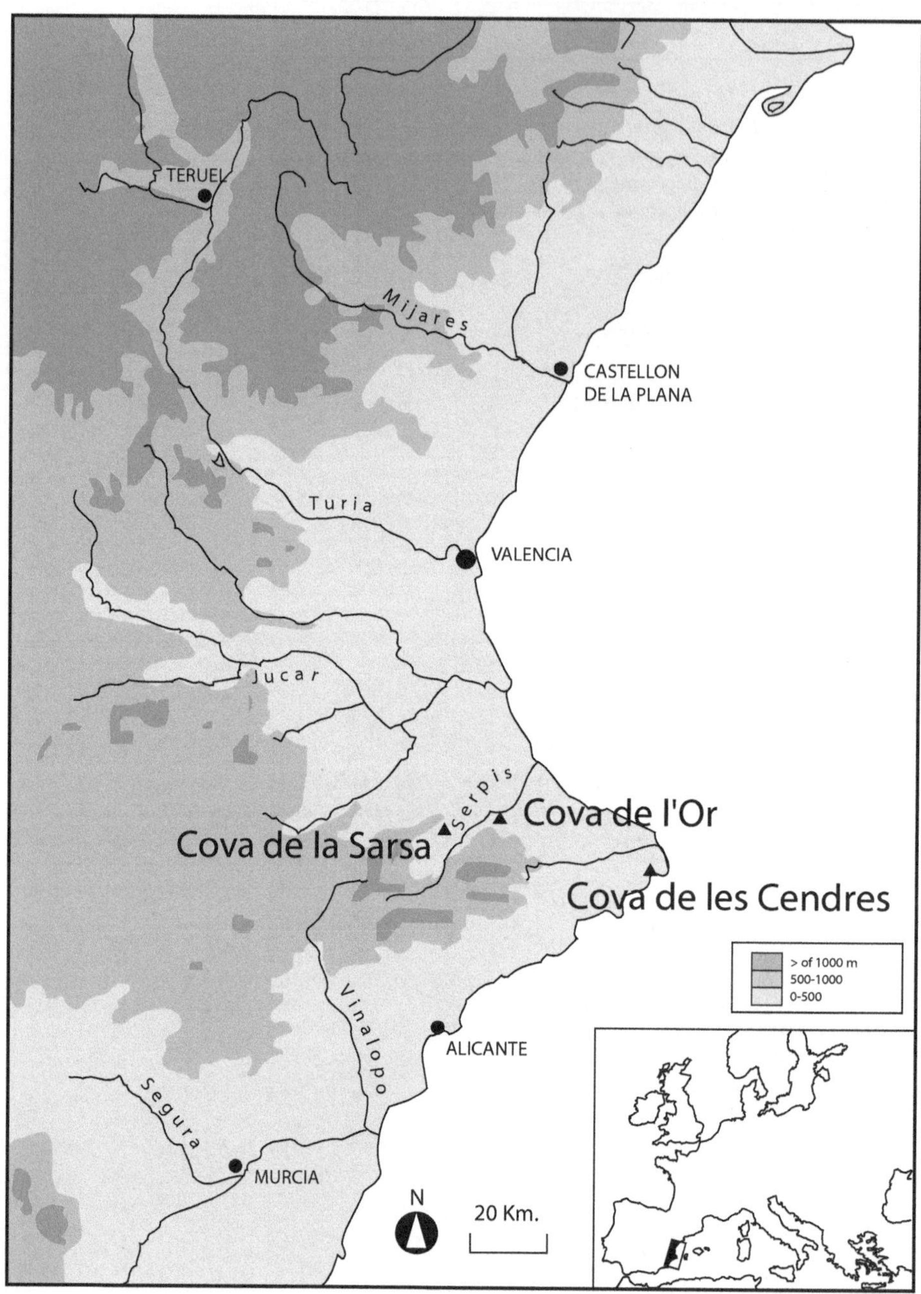

Figure 10.9: Location of the Neolithic sites from the east of Spain with archaeobotanical data referred to in the text.

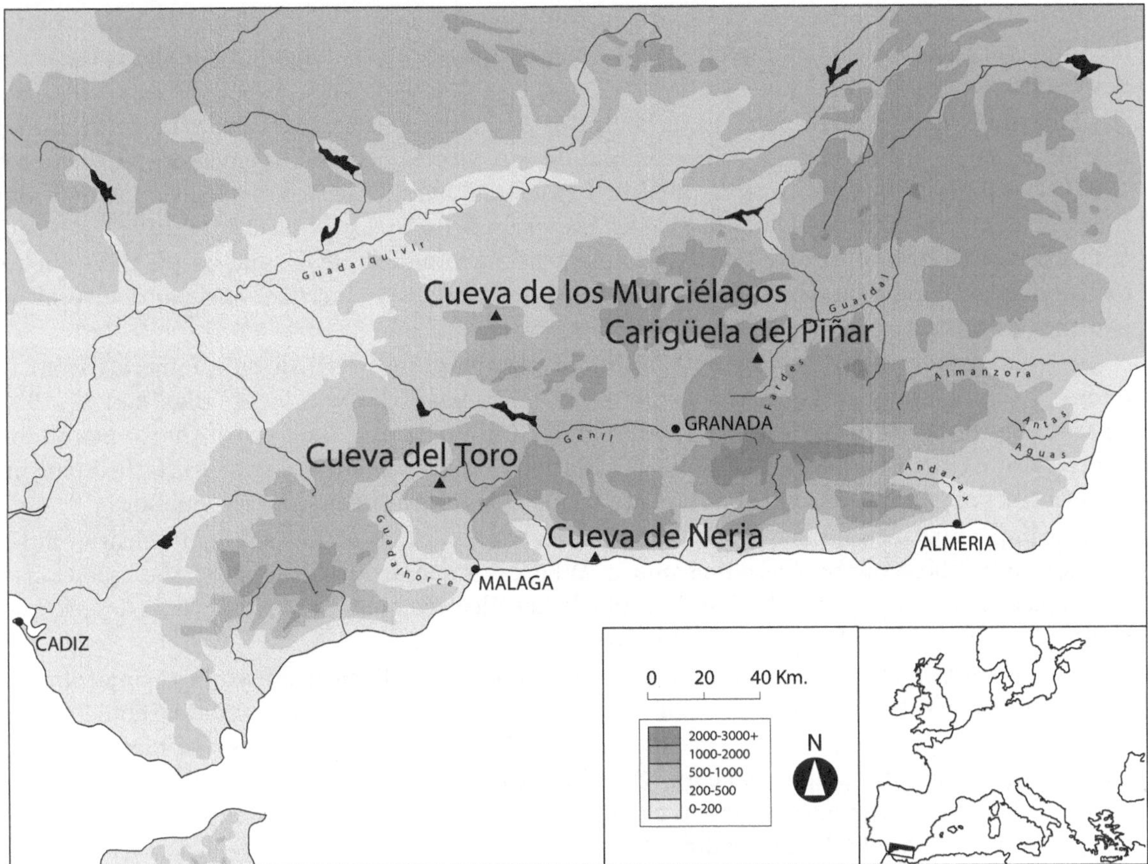

Figure 10.10: Location of the neolithic sites from the southeast of Spain with archaeobotanical data referred to in the text.

var. *oleaster*) (Buxó 1997). Results from some recent excavations in Valencian (at Mas d'Is and Falguera) have produced samples of naked and hulled barley, free-threshing wheat and hulled wheats (emmer and einkorn) (G. Pérez, pers. com.).

Cardial Ware pottery has also been recorded in the south of the Iberian Peninsula but from a later phase than in the eastern zone (figure 10.10). Early archaeobotanical results from this region are represented by few samples ranging in the second half of the 5th millennium cal BC (hulled and naked barley, free-threshing wheat, einkorn and few pulses were found). The most representative's sites are Cueva de Nerja, Carigüela del Piñar, Cueva de los Murciélagos and Los Castillejos (Buxó 1997; Peña-Chocarro 1999, N. Rovira pers. com.).

10.3 Second evolutionary phase

By the end of the first half of the 5th millennium cal BC neolithisation of the region seems to have been complete and the best agricultural lands of the coasts, interiors valleys and plains had been occupied.

It is possible to distinguish three groups in Catalonia in the first phase of the 5th millennium: the first (Montboló's group) is located in the north and south of the central east Pyrenees, between the rivers Tet and Ter; the second (Molinot's group) is located in the central Catalonian littoral and prelittoral areas, which was the nucleus for the Cardial Ware culture; and the third

(Amposta's group) is located in the south surrounding the mouth of the river Ebro. Information on these three groups is incomplete and as a consequence it is difficult to make comparisons between them.

The three groups correspond to agricultural populations who lived in settlements near to the plains or in low-lying areas, for example, at the site of Barranc d'en Fabra (Bosch et al. 1992); the later phase of Guixeres de Viloví (Mestres 1998); the occupation deposits of Sant Pau del Camp (Granados et al. 1993); and the early layer at Ca N'Isach (Tarrús et al. 1995). It has been noted that they also often occupied the surrounding caves, and use of these shelters for keeping ovicaprid herds would have enabled the groups to increase their agricultural and livestock potential.

At this time variscite (hydrated aluminium phosphate, a greenish-blue mineral) from the mines of Can Tintorer in Gavà was exploited to make beads for necklaces (Blasco et al. 1992). These types of beads have been found throughout Catalonia and they prove that minerals/artifacts were exchanged between all the populations in the region regardless of their cultural affinities. These necklaces made of variscite beads would have been durable symbols.

It is important to point out that in this phase certain burial practices marked a clear evolution in relation to those of the preceding phase. Whereas some groups continued the traditions of their epicardial ancestors and buried individuals in collective funerary caves, others adopted new burial practices with different types of necropoli.

The quantities of plant remains preserved in middle neolithic contexts are small and the quality of preservation is poor, therefore any interpretations based on the data should be considered with due caution. Cereal species comprise 57% of the total number of identified plant taxa and gathered wild plants make up 43% of the total. To date, no pulses have been identified, however, because of the inadequacy of the archaeobotanical record any comparisons made with the previous period may be misleading.

The cereal taxa included free-threshing wheat, two types of barley (naked and hulled), emmer and einkorn. However, it should be noted that the above-mentioned four sites from the middle Neolithic (Can Tintorer, Cova del Toll, Cova de Pau and Ca N'Isach) have in common the absence of einkorn and cultivated pulses. The frequency of occurrence of four of the five cereal species (free-threshing wheat, naked and hulled barley and emmer) is very similar, all are present in 20 to 25% of the samples from this period. The increase in frequencies from the early Neolithic could indicate that stable climatic conditions at this time enabled greater agricultural productivity (figure 10.5).

There were similar quantities of the three cereal species recorded at Can Tintorer, as well as some grains of emmer wheat (Buxó et al. 1991; Villalba et al. 1986). Two types of barley (naked and hulled) and emmer wheat were identified in the samples from Cova del Toll, but free-threshing wheat was absent.

The archaeobotanical data from Ca N'Isach are less interesting because only one seed of free-thresing wheat was identified, however, pollen analyses indicated the presence of Cerealia pollen, which suggested that there were cultivated fields in the vicinity of the site (Tarrús et al. 1995, p.430). In the early Neolithic the fruits/nuts of nine wild plant taxa were identified: strawberry tree (*Arbutus unedo*), dogwood (*Cornus sanguinea*), hazel (*Corylus avellana*), fig (*Ficus* sp.), lentisc (*Pistacia lentiscus*), pine (*Pinus* sp.), blackthorn/sloe (*Prunus spinosa*), acorn (*Quercus* sp.) and wild grape (*Vitis vinifera* var. *sylvestris*). In the middle Neolithic only three wild species were recorded: wild olive (*Olea europaea* var. *oleaster*), acorn (*Quercus* sp.) and wild grape (*Vitis vinifera* var. *sylvestris*).

At Cueva de Los Murciélagos (Zuheros, Córdoba) in southern Spain, free-thresing wheat was particularly common together with naked barley and emmer (Hopf 1974) (figure 10.5). The remains of a compact type of free-threshing wheat, hulled-barley and acorns were identified in a later study of another set of samples from the same site (López 1980). Peña-Chocarro (1999) also carried out analyses on samples from an early neolithic level (dating from the middle of the 5th to the beginning of the 4th millennium cal BC) and she listed free-threshing wheat, hulled barley and the remains of wild olive stones. In a later level (the first half of the 4th millennium cal BC), the by-products from threshing hulled wheat (i.e., spikelets of emmer) were identified, as well as weeds of cultivated fields, and the remains of acorns and fruits.

In the Cueva de Nerja (Nerja, Málaga) a silo containing the remains of naked barley and free-threshing wheat was recorded (Hopf and Pellicer 1970). Wild olive stones and one cotyledon of acorn were also identified from the site.

Finally, it is important to include the Cueva del Toro (Antequera, Málaga) in this review as the stratigraphic sequence covers occupation from the Neolithic to Roman and Arabian times, including copper and bronze age levels. The most common cultivated plants in the middle Neolithic (between 4500/4300 and 3700/3600 cal BC) and late Neolithic (between 3700/3600 and 3300/3200 cal BC) were naked barley and free-threshing wheat, and it has been confirmed that in the layers associated with the beginning of the Bronze Age naked barley disappears probably at the expense of the hulled form, which increases. There are very few remains of hulled wheats, and emmer was found in just one sample. Broad bean and lentil are the most commonly occurring pulse species and there are less frequent records of grass pea, bitter vetch (*Vicia ervilia*) and pea.

10.4 Conclusions

From the earliest stages of the adoption of the agriculture in Iberia the main crops were the cereals (i.e., wheat and barley), the most important of which were free-threshing wheat, and naked and hulled barley. The hulled wheats (i.e., emmer and einkorn) were represented only slightly less frequently at this time. The pulses included grass pea, lentil, pea and broad bean. The remains of gathered wild plants appear in a similar frequency to that in the Epipalaeolithic/Mesolithic, and acorns were the most common taxon. It seems, therefore, that the gathering of wild plant resources continued to be important after agriculture had been adopted, possibly to supplement the products of the harvests and to provide a buffer against low crop yields in any given year.

References

Agustí, B., G. Alcalde, F. Burjachs, R. Buxó, N. Juan-Muns, J. Oller, M. T. Ros, J. M. Rueda, and A. T. Toledo (1987). *Dinàmica de la Utilització de la Cova 120 per l'Home en els Darrers 6000 Anys.* Serie Monogràfica 7. Girona: Centre d'Investigacions Arqueològiques.

Bernabeu, J., J. E. Aura, and E. Badal (1993). *Al Oeste del Edén. Las Primeras Sociedades Agrícolas en la Europa Mediterránea.* Ha. Universal 4, Prehistoria. Madrid: Ed. Sintesis.

Blasco, A., M. Edo, and M. J. Villalba (1992). La cal·laïta: l'ús dels minerals verds durant el neolític a Catalunya a partir de la difractometria de raigs X. In *9è Colloqui Internacional d'Arqueologia de Puigcerdà, Estat de la Investigació Sobre el Neolític a Catalunya*, pp. 206–208. Andorra: Institut d'Estudis Ceretans.

Blasco, A., M. Edo, M. J. Villalba, R. Buxó, J. Juan, and M. Saña (1999). Del cardial al postcardial en la cueva de Can Sadurní (Begues, Barcelona). Primeros datos sobre su secuencia estratigráfica, paleoeconómica y ambiental. In *Actes del II Congrés del Neolític a la Península Ibèrica*, Saguntum, Extra-2, pp. 59–68. València: Universitat de València.

Bosch, A., R. Buxó, A. Palomo, M. Buch, J. Mateu, E. Tabernero, and J. Casadevall (1998). *El Poblat Neolític de Plansallosa. L'Explotació del Territori dels Primers Agricultors-Ramaders de l'Alta Garrotxa.* Publicacions Eventuals d'Arqueologia de la Garrotxa 5. Olot: Museu Comarcal de la Garrotxa.

Bosch, A., J. Chinchilla, and J. Tarrús (Eds.) (2000). *El Poblat Lacustre Neolitic de la Draga. Excavaciones de 1990 a 1998.* Monografies del CASC 2. Girona: Museu d'Arqueologia de Catalunya–Centre d'Arqueologia Subaquàtica de Catalunya.

Bosch, J., A. Forcadell, and M. del mar Villalbí (1992). Les estructures d'hàbitat a l'assentament del Barranc de Fabra (Montsià). In *9è Colloqui Internacional d'Arqueologia de Puigcerdà, Estat de la Investigació Sobre el Neolític a Catalunya*, pp. 121–122. Andorra: Institut d'Estudis Ceretans.

Buxó, R. (1997). *Arqueología de las Plantas. La Explotación Económica de las Semillas y los Frutos en el Marco Mediterráneo de la Península Ibérica.* Barcelona: Ed. Critica.

Buxó, R. (2002a). De la recollecció a l'agricultura. Una evolució decisiva en les societats prehistòriques. *L'Avenç 274*, 53–59.

Buxó, R. (2002b). Gestió i Explotació dels Recursos Vegetals: Llavors i fruits. In G. Alcalde, M. Molist, and M. Saña (Eds.), *Procés d'Ocupació de la Bauma del Serrat del Pont (La Garrotas) Entre el 5480 i 2900 cal AC*, Publicacions Eventuals d'Arqueologia de la Garrotxa 7, pp. 83. Olot: Museu Comarcal de la Garrotxa.

Buxó, R., M. Català, and M. J. Villalba (1991). Llavors i fruits en un conjunt funerari situat en la galeria d'accés a la Mina 28 de Can Tintorer (Gavà). *Cypsela 9*, 65–72.

Buxó, R., N. Rovira i Buendia, and C. Saüch (2000). Les Restes Vegetals de Llavors i Fruits. In A. Bosch i Lloret, C. Sanchez, and J. Tarrus i Galter (Eds.), *El Poblat Lacustre Neolįtic de la Draga. Excavaciones de 1990 a 1998*, Volume 2 of *Monografies del CASC*, pp. 129–140. Girona: Museu d'Arqueologia de Catalunya.

Fortea, J. (1973). *Los Complejos Microlaminares y Geométricos del Epipaleolítico Mediterráneo Español.* Salamanca: Universidad de Salmanca.

García-Argüelles, P., M. Bergada, and R. Doce (1990). El estrato 4 del Filador (Priorato, Tarragona): un ejemplo de la transición Epipaleolítico-Neolítico en el Sur de Cataluña. *Saguntum (PLAV) 23*, 61–76.

Granados, J. O., F. Puig, and A. Farré (1993). La intervenció arqueològica a Sant Pau del Camp: un nou jaciment prehistòric al Pla de Barcelona. In *Tribuna d'Arqueologia 1991-1992*, pp. 27–32. Barcelona: Departamento de Cultura de la Generalitat de Catalunya.

Holden, T. G., J. G. Hather, and J. P. N. Watson (1995). Mesolithic plant exploitation at the Roc del Migdia, Catalonia. *Journal of Archaeological Science 22*, 769–778.

Hopf, M. (1966). *Triticum monococcum* L. y *Triticum dicoccum* Schubl, en el neolítico antiguo español. *Archivo de Prehistoria Levantina 11*, 53–80.

Hopf, M. (1974). Neolithische Pflanzenreste aus der Höhle Los Murciélagos bei Züheros, Córdoba. *Madrider Mitteilungen 15*, 9–27.

Hopf, M. and M. Pellicer (1970). Neolithische Getreidefunde in der Höhle von Nerja (Málaga). *Madrider Mitteilungen 11*, 18–34.

Juan-Cubanilles, J. (1990). Substrat Epipaléolithique et Néolithisation en Espagne: apport des industris lithiques à l'identification des traditions culturelles. In D. Cahen and M. Otte (Eds.), *Rubané & Cardial: Actes du Colloque de Liège, 1988*, pp. 417–436. Liège: Etudes et Recherches Archéologiques de l'Université de Liège.

López, P. (1980). Estudio de semillas prehistóricas en algunos yacimientos españoles. *Trabajos de Prehistoria 37*, 419–432.

Marinval, P. (1995). Recol·lecció i agricultura de l'epipaleolític al neolític antic: anàlisi carpològica de la Balma de la Marginada. In J. Guilaine and M. Martzluff (Eds.), *Les Excavacions a la Balma de la Margineda (1979-1991)*, Andorra la Vella, Edicions del Govern d'Andorra, vol. III, pp. 65–82. Andorra: Govern d'Andorra.

Martí, B., J. Juan, J. Bernabeu, P. Fumanal, M. Dupré, M. Hernández, E. Badal, E. Grau, and E. Vento (1997). *El Neolític Valencià. Els Primers Agricultors i Ramaders.* València: SIP, Diputació de València.

Mestres, J. (1998). El Neolítico Antiguo en Cataluña. In V. Baldellou, I. Mestre, B. Martí, and J. Juan Cabanilles (Eds.), *El Neolítico Antiguo (Los Primeros Agricultores y Ganaderos en Aragón, Cataluña y Valencia)*, pp. 21–25. Huesca: Diputación de Huesca.

Pallarés, M., A. Bordas, and R. Mora (1997). El proceso de neolitización en Ios Pirineos Orientales. Un modelo de continuidad entre los cazadores-recolectores y los primeros grupos agropastoriles. *Trabajos de Prehistoria 54*, 121–141.

Peña-Chocarro, L. (1999). *Prehistoric Agriculture in Southern Spain during the Neolithic and the Bronze Age: The Application of Ethnographic Models.* BAR International Series S818. Oxford: Archaeopress.

Tarrús, J., J. Chinchilla, O. Mercadal, and S. Aliaga (1995). Fases estructurals i cronològiques a l'hàbitat neolític de Ca N'Isach (Palau-Saverdera, Alt Empordà). I Congrés del Neolític a la Península Ibèrica. Gavà-Bellaterra 1995. *Rubricatum 1*, 429–438.

Terradas, X. (1995). *Las Estrategias de Gestión de los Recursos Líticos del Prepirineo Catalán en el IX Milenio BP: El Asentamiento Prehistórico de la Font del Ros (Berga, Barcelona).* Treballs d'Arqueologia 3. Bellaterra: Universitat Autònoma de Barcelona.

Villalba, M. J., L. Bañolas, J. Arenas, and M. Alonso (1986). *Les Mines de Can Tintorer (Gavà): Excavacions 1978-1980.* Excavacions Arqueològiques de Catalunya 6. Barcelona: Departament de Cultura, Generalitat de Catalunya.

Chapter 11

Early agriculture in central and southern Spain

Leonor Peña-Chocarro
Musei Civici Como, Italy

To Gordon Hillman, with all my gratitude, for his enthusiasm, wisdom and encouragement.

11.1 Introduction

Research into the origins of agriculture for most of the Iberian peninsula is still at its early stages. It is true, however, that for the Mediterranean coast, where most of the investigations have been concentrated, there is considerable knowledge about the agriculture of the first farming communities. For example, Hopf's study of one of the major neolithic sites from the eastern coast, Cova de l'Or (Hopf 1966), was a major milestone in the investigation of prehistoric farming in the Iberian Peninsula. However, on-site systematic recovery techniques did not start until the 1980s when flotation began to be used at archaeological sites. These investigations have started to produce a new body of data (see Buxó, this volume, and Zapata et al. nd for the latest reviews), which will certainly improve our knowledge of this initial phase of food production. Conversely, for the rest of the peninsula, archaeobotanical data are still very sparse (Buxó 1997; Hopf 1987, 1991; Peña-Chocarro 1999; Zapata 2002; Zapata et al. nd; Zapata, this volume), and our understanding of the adoption of agriculture is advancing very slowly.

The purpose of this paper is to examine the available archaeobotanical data from two large areas: the central (Meseta) and the most southern (Andalucía) parts of the Iberian Peninsula.

11.2 Central Spain: the Meseta

The Meseta is the large plateau that occupies the central part of the Iberian Peninsula, covering more than one-third of its total area. This large territory, situated at 600 m elevation, is divided up by the Sistema Central mountain chain into the two so-called north and south mesetas. With an east-west orientation, the Meseta is traversed by two of the main rivers of the Iberian Peninsula: the Duero in the north and the Tagus in the south. The present region of Castilla-León occupies the northern part, whereas the regions of Madrid, Castilla-La Mancha and Extremadura extend over the southern part.

Research into the Neolithic in this region has taken longer to develop than in the eastern and southern coastal areas of Spain. In fact, up to the last decade, neolithic data were very scarce, and they were mostly represented by finds from surface collections. This limited material, together with the lack of stratigraphic evidence and absolute dates, led researchers to suggest that the neolithic stimulus that reached the Meseta from peripheral areas of Iberia arrived in this region after a great delay. For example, it was, assumed that the first occupation of this territory during the Holocene was that of the 'megalithic people' during the 4th millennium BC (Delibes 1977).

It is only recently that considerable effort has been put into the study of the neolithic communities in these areas, and several surveys have demonstrated the importance of this period in the northern Meseta (Estremera 1999, 2003; Kunst and Rojo 1999, 2000; Rojo and Estremera 2000; Rojo and Kunst 1999; Rojo et al. nd). By the beginning of the 1990s a total of 53 sites had been located in the region (Iglesias et al. 1996) and at the end of that decade, 54 new sites had been identified in the Ambrona Valley (Soria) alone. For the southern Meseta, new investigations (Bueno et al. 1999, 2000, 2002; Cerrillo 1999; Cerrillo et al. 2002, nd; Díaz del Río 2001; Díaz del Río and Consuegra 1999; González-Cordero and Cerrillo 2001; Jiménez Guijarro 1998, 1999; Jiménez Sanz et al. 1997) are also contributing to a better understanding of this period.

11.2.1 Theories about the spread of the Neolithic

Two main models have been proposed in an attempt to explain the process of neolithisation in this area:

1. In the first model, the spread of the Neolithic is explained as a result of the colonization of a previously depopulated region. This is particularly defended (Delibes and Manzano 2000; Estremera 2003; Kunst and Rojo 1999; Rojo and Estremera 2000) for the north Meseta along the Duero Valley, where intensive survey has failed to detect the presence of pre-neolithic communities. Fully developed neolithic groups would have arrived in the interior of the Iberian Peninsula from peripheral areas such as Andalucía, the Mediterranean coast or Portugal, bringing new cultural elements, resources and techniques. Two of the best examples of the neolithic occupation of this region, La Vaquera Cave (Segovia) and sites in the Ambrona Valley (Soria), show clear parallels with areas of Andalucía (southern Spain) in the first case, and to the Ebro Valley in the latter.

2. The second model sees the Neolithic as a result of a diffusionist process but places emphasis upon the role of an indigenous mesolithic hunter-gatherer population (Jiménez Guijarro 1998, 1999; Jiménez Sanz et al. 1997). Interaction between different groups (farming and non-farming) would have led to the exchange of both materials and ideas, and the adoption of innovations by hunter-gatherers. The economic transformations that characterize the Neolithic would have then developed within groups of hunter-gatherers through the cultural diffusion of exogenous elements. The homogeneity of the material culture, with little difference at a regional scale, is understood as the outcome of a process of indirect acculturation.

Whichever model is correct, it seems clear that the Neolithic is the result of some kind of diffusionist process, which allowed new technologies, modes of subsistence, and cultural elements to arrive in the interior of Iberia. However, as Estremera (2003) has recently pointed out, our knowledge of how the first farming communities established in the Meseta is as yet limited. After two decades of research, information comes from just a few excavated sites and material from surface collection. Furthermore, archaeobotanical data are restricted to a few sites where systematic recovery of plant remains has been carried out.

Also at issue is the lack of research into hunter-gatherer prehistory. Even if the number of mesolithic sites in the interior of Spain has recently increased, our knowledge is still far from complete. We know very little about the interactions between the last mesolithic foragers and the processes by which new elements were adopted within the cultural traditions of the indigenous

Site	^{14}C bp	cal BC (2σ)	Material	Taxon
La Vaquera (Segovia) Phase I	6120±160	5460–4690	charcoal	*H. vulgare*
				H. vulgare var. *nudum*
				T. aestivum/durum
				T. cf. *monococcum*
La Vaquera (Segovia) Phase II	5800±30	4770–4550	charcoal	*T. dicoccum*
	6440±50	5480–5320	acorn	*Lens* sp.
	6080±70	5220–4790	acorn	*V. sativa*
La Lámpara (Soria)	6871±33	5840–5670	bone	*T.* cf. *monococcum*
	6144±46	5250–4940	bone	*T.* cf. *dicoccum*
				P. somniferum/setigerum
La Revilla del Campo (Soria)	6809±37	5770–5630	charcoal	*T.* cf. *monococcum*
	6202±31	5280–5050	bone	*T. monococcum/dicoccum*
				L. usitatissimum
Los Barruecos (Cáceres)	6060±50	5204–4801	charcoal	Cereal pollen
	6080±40	5204–4847		*H. vulgare* phytolith

Table 11.1: Plant remains and associated ^{14}C dates from early neolithic sites in the Meseta. For La Vaquera see Estremera (2003) and López et al. (2003); La Lámpara Rojo et al. (nd); Stika (1999, nd); La Revilla del Campo Rojo et al. (nd); Stika (1999, nd); Los Barruecos Cerrillo et al. (nd).

population. It is likely that in such a large area several different processes were involved, thus increasing the complexity of the transformation.

11.2.2 Archaeobotanical evidence: the north Meseta

Paradoxically, even if domesticated plants were a protagonist of the economic changes that characterized the transition from hunter-gatherers to farmers, very little is known about the range of plants that were cultivated by the first neolithic farmers of the Meseta. Evidence for plant subsistence is still too limited, and only three sites have been investigated: La Vaquera Cave (Segovia), and two sites within the Ambrona Valley in Soria (La Lámpara and La Revilla del Campo) (figure 11.1). Table 11.1 shows the cultivated plant remains recovered from the three sites together with ^{14}C dates. No other evidence for the use of plant remains has been published.

La Vaquera Cave (Torreiglesias, Segovia) has recently been excavated by M.S. Estremera (1999; 2003). The cave is located in a transitional area between the Sistema Central chain and the Duero Valley, at c. 960 m elevation. This strategic position between two different ecosystems would certainly have been attractive to the group that established there; the valleys of the rivers Pirón and Viejo with abundant water and pasture lands that were available even in summer, and the dry lands of the Meseta where farming was possible. The habitation space or Sala A occupies an area of c. 60 m and has almost 6 metres of accumulated sediment. The stratigraphic sequence includes levels from the Neolithic, Chalcolithic and early Bronze Age. Three different phases can be distinguished within the Neolithic (Estremera 1999, 2003; Rojo and Estremera 2000): (a) phase I dated between 5300–4700 cal BC; (b) phase II between 4600–3600 cal BC; and, (c) phase III between 3500–3000 cal BC.

The assemblage of plant remains studied by A.M. Arnanz (López et al. 2003) includes several cultivated plants. In phase I cereals are represented by hulled and naked barley (*Hordeum vulgare* and *Hordeum vulgare* var. *nudum*), hulled wheats dominated by emmer (*Triticum dicoccum*), a few grains of einkorn (*Triticum monococcum*) and free-threshing wheats (*Triticum aestivum/durum* including a compact form). Although emmer is quite abundant in the first part of phase I, free-threshing wheats are the dominant cereals throughout the sequence. Arnanz has

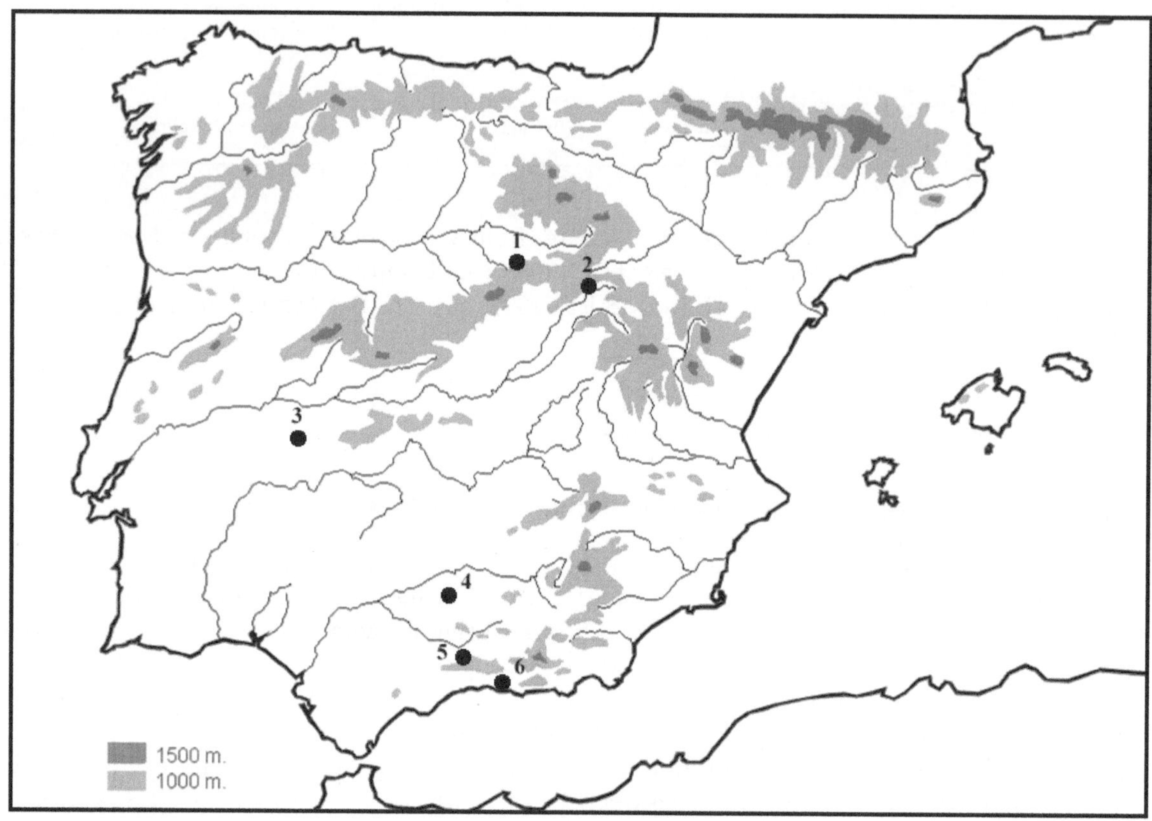

Figure 11.1: Location of the sites with early agriculture mentioned in the text. 1. Cueva de La Vaquera; 2. La Lámpara and La Revilla del Campo; 3. Los Barruecos; 4. Cueva de Los Murciélagos; 5. Cueva del Toro; 6. Cueva de Nerja.

also identified several legumes: *Lens* sp., *Vicia sativa* and some possible members of the genus *Vicia*. The remaining species are mostly weeds from arable fields, with some wild fruits such as wild grapes (*Vitis vinifera* ssp. *sylvestris*) and acorns (*Quercus* sp.).

Although numbers are not particularly high, this assemblage suggests that between the end of the 6th millennium cal BC and the beginning of the 5th millennium cal BC, the inhabitants of La Vaquera Cave were certainly farmers. This is also supported by the lithic industry, including 24% of the blades from phases I and II that had cereal gloss (Estremera 2003). The wide array of cultivated plants, comprising both cereals and legumes, suggests a well-developed agricultural system. It is interesting to note that cereal pollen appears only from phase II, and is absent in the first phase of occupation of the cave. On the other hand, chaff remains are almost missing. This lack of both pollen and chaff has led Estremera to suggest that the cave was probably occupied during the summer months, when people with cattle would have actually moved to the pastures carrying grain reserves with them. However, I would argue that the absence of chaff might be the result of crop processing that was carried out away from the cave, or a consequence of taphonomic or preservation factors.

Early neolithic plant remains have also been recovered from two different sites of the Ambrona Complex: La Lámpara and La Revilla del Campo located in Soria province. These sites belong to a series of 33 identified along the Valley of Ambrona, currently investigated by the University of Valladolid and the German Archaeological Institute (Kunst and Rojo 1999, 2000; Rojo and Estremera 2000; Rojo and Kunst 1999; Rojo et al. nd). Forty ^{14}C dates have been obtained from both locations and these demonstrate that the area was inhabited during the first half of the 6th millennium cal BC (although if only dates from short-lived materials are taken into account, initial occupation is delayed until the last centuries of the 6th millennium cal BC). According to Rojo et al. (nd), these early dates support the idea that neolithisation of the interior of the Iberian Peninsula was older than currently thought. Perhaps, most significantly, they suggest an extremely rapid diffusion of agriculture throughout the Peninsula.

The application of a systematic program of flotation has facilitated the recovery of archaeobotanical samples that are being studied by H.P. Stika. The analysis of the plant remains from the two sites is still in progress but it seems that hulled wheats (both einkorn and emmer) played a key role in agriculture. Barley is also present together with two other cultivated plants: poppy (*Papaver somniferum/setigerum*) and flax (*Linum usitatissimum*) (Stika 1999, nd; Zapata et al. nd). According to Stika (pers. comm.), hulled wheats might be over-represented as their chaff was used for tempering. Because the remains are badly preserved, he also suggests that free-threshing wheats may be present amongst the categories '*Triticum* sp.' and 'Cerealia indet.'.

11.2.3 Archaeobotanical evidence: the south Meseta

As in the case of the north plateau, evidence of early agriculture in the southern Meseta is almost absent; there are few early neolithic sites and systematic sampling and recovery has been rarely applied. In Castilla-La Mancha, the oldest evidence of agriculture (there being none from the early Neolithic) comes from two sites in Toledo province: Azután, and El Castillejo, where palynological analyses have been carried out. In both cases, cereal pollen together with clear signs of deforestation appears during the late 5th millennium and beginnings of the 4th cal BC (Bueno et al. 2002).

In the province of Madrid, data on the neolithic occupation are again very limited. Díaz del Río and Consuegra (1999) and Díaz del Río (2001) argue that this gap contrasts markedly with the high number of sites identified for the Chalcolithic. In fact, they suggest that such an important number of chalcolithic sites are likely to be preceded by a earlier neolithic development. The only neolithic site from this province where systematic recovery of plant remains has been carried out is La Deseada (Rivas-Vaciamadrid), and several archaeobotanical samples were studied by the author but none of them contained plant remains (Díaz del Río and Consuegra 1999). Some other new sites from this province are currently being studied by the author.

For Extremadura, recent work (Cerrillo 1999; Cerrillo et al. 2002, nd; González-Cordero and Cerrillo 2001) points towards a neolithic occupation of the territory by the end of the 6th millennium cal BC. Two sites have been investigated: the open-air site of Los Barruecos (Malpartida de Cáceres) and El Conejar Cave (Cáceres) (figure 11.1). From Los Barruecos, two AMS dates have been obtained (6080±40 bp and 6060±50 bp, Cerrillo et al. nd). Several archaeobotanical samples from the site have been studied but none of them yielded plant remains (Cerrillo et al. nd). The palynological analysis by J.A. López has revealed the presence of Cerealia pollen in the earliest layers. In addition to which J. Juan (Cerrillo et al. nd) has also identified *Hordeum vulgare* phytoliths in a quern (table 11.1). A small assemblage of cereals containing *Triticum dicoccum* and *Triticum aestivum/durum* was recovered from the nearby El Conejar Cave (Cerrillo 1999). Even though in the absence of absolute dating it is impossible to ascertain whether the plant material from this site is neolithic, most of the archaeological remains seem to belong to the early Neolithic period (these have not been included in table 11.2).

11.3 Southern Spain: Andalucía

Andalucía is a large region that occupies the southern part of Spain, and includes the present provinces of Huelva, Cádiz, Sevilla, Córdoba, Jaén, Málaga, Granada and Almería. It is separated from the central plateau by the great natural barrier of the Sierra Morena mountain chain and the Guadalquivir river cuts through the territory from east to west.

Neolithic research in this area has advanced greatly over the last few years. New sites have been investigated and the amount of information is increasing considerably. However, as Gavilán (1997) has noted there are still controversial aspects regarding the origins, cultural development and subsistence patterns.

Different theories have been proposed to explain the transition to food production in this area. The dual model put forward by Bernabeu 1996 for the eastern coast is, however, the most widely accepted for this region. According to his theory, the diversity and variability of the material culture observed in the archaeological record is the result of two different traditions: a neolithic of exogenous origin, and a local mesolithic, the intertwining of which would have led to a direct acculturation of the mesolithic groups. The Cardial ware, characteristic of the western Mediterranean first neolithic, is residual for most parts of this territory. Instead, the red slipped ware ('almagra pottery') appears as the key element of the neolithic horizon from the region.

This is not, however, the only accepted model. Based on the very early dates of two sites in western Andalucía, Dehesilla (Cádiz) and Cueva Chica de Santiago (Sevilla), whose chronologies go back as far as the late 7th millennium and the beginning of the 6th millennium cal B.C, some scholars (Acosta 1986, 1995; Acosta and Pellicer 1990) have defended an alternative position, emphasizing the autochthonous character of the Andalousian Neolithic. In addition, Ramos Muñoz et al. (1995) point out the clear continuity in the lithic industries from as early as the late Solutrean period and, as such, stresses the importance of the mesolithic substratum. However, since most of the material discussed comes from surface collection, this argument has been criticized (Gavilán 1997).

The significant number of archaeological sites from the Neolithic period in Andalucía contrast significantly with the few absolute dates available. In addition, one of the problems faced by archaeologists is the lack of data on subsistence and particularly on agriculture. Compared to the eastern Mediterranean coast of Iberia, archaeobotanical studies from Andalucía are very scant; three sites concentrated in a small area in the centre of the region (provinces of Córdoba and Málaga) are the only ones where plant remains have been systematically recovered. The remaining part of Andalucía remains uninvestigated in this way. However, despite these de-

Site		^{14}C bp	cal BC (2σ)	Material	Taxon
Cueva de Los Murciélagos (Córdoba)		6430±130	5620–5070	charcoal	*T. aestivum/durum*
		6295±45	5370–5080	charcoal	*T. dicoccum*
		6190±130	5460–4800	cereal	*H. vulgare nudum*
		6170±130	5460–4780	cereal	*H. vulgare*
		6025±45	5040–4780	cereal	*P. sommniferum/setigerum*
		5980±130	5220–4540	cereal	*P. sommniferum/setigerum*
Cueva del Toro (Málaga)	Layer IV	6400±280	5840–4710		*T. aestivum/durum*
	Layer IIIb	5450±120	4500–4000		*T. dicoccum*
	Layer IIIa	5380±45	4330–4050		*H. vugare nudum*
		5320±230	4600–3650		*H. vugare*
		5250±60	4230–3960		*V. faba minor*
		5205±40	4220–3950		*L. culinaris*
		5200±60	4220–3810		*L. sativus*
					P. sativum
					V. ervilia
Cueva de Nerja (Málaga)		5065±140	4230–3540	cereal	*T. aestivum/durum*
		6420±60	5480–5310	charcoal	*H. vugare nudum*
					Vicia/Lathyrus

Table 11.2: Plant remains and associated ^{14}C dates from early neolithic sites in Andalucía. For Cueva de Los Murciélagos see Hopf and Muñoz (1974); López (1980); Peña-Chocarro (1999); Cueva del Toro (Buxó 1997); Cueva de Nerja (Jordà Pardo et al. 1990).

ficiencies in research, data suggest that the first domesticated plants and animals started to appear in the archaeological record by ca. 5500 cal BC.

11.3.1 Archaeobotanical data: Cueva de Los Murciélagos (Córdoba, Spain), Cueva de El Toro (Málaga, Spain) and Cueva de Nerja (Málaga)

The earliest evidence of agriculture comes from the middle of the 6th millennium cal BC (ca. 5500 cal BC). Three cave sites have provided important archaeobotanical data: Cueva de Los Murciélagos (Córdoba), Cueva del Toro (Málaga) and Cueva de Nerja (Málaga) (figure 11.1). A fully developed agriculture is attested, including both the cultivation of cereals and legumes. Cereals are represented by free-threshing wheats, emmer and hulled and naked barley, and the legumes comprise lentils, broad beans, grass pea, peas and bitter vetch. Table 11.2 shows the species identified from the three caves with the available ^{14}C dates.

Cueva de los Murciélagos is a large cave, situated at 980 m elevation within the mountain range of Sierras Subbéticas in Córdoba province. The landscape is characterized by a karstic relief with abundant steep cliffs. The stratigraphy shows a long sequence from the Medieval period to the Palaeolithic, with levels assigned to the Mousterian, Upper Palaeolithic, Neolithic, Chalcolithic, Bronze Age, Roman and Medieval. During the excavation of a small chamber ('Cueva Chica') in the 1960s, there was recovery of plant remains from a large concentration identified as a silo (Quadra-Salcedo and Vicent 1964; Vicent and Muñoz 1973). The assemblage, composed of emmer (*Triticum dicoccum*), free-threshing wheats (*Triticum durum/aestivum*) and naked barley (*Hordeum vulgare nudum*), was studied by M. Hopf (Hopf 1974; Hopf and Muñoz 1974), and at a later date a small sample was examined by P. López (1980). ^{14}C dates were obtained from both charcoal and seeds showing that domesticated plants were already cultivated by ca. 5500 cal BC. During the 1990s a new area of the cave was excavated (Gavilán 1991; Gavilán and Vera 1992; Gavilán et al. 1994, 1996), and flotation was carried out; the new assemblage (Peña-Chocarro 1999), dated again to ca. 5500 cal BC, contained free-threshing wheats, barley (*Hordeum vulgare*), and poppy (*Papaver somniferum*) seeds.

Cueva del Toro is located within the Torcal Massif (Málaga). The stratigraphic sequence extends from the Neolithic to Medieval times (Martín Socas et al. 1993, 1999). Plant remains were systematically recovered using flotation and were studied by Buxó 1997. The earliest neolithic stratum (layer IV), dated between 5840–4710 cal BC, yielded free-threshing wheats, naked barley, grass pea (*Lathyrus sativus*), lentil (*Lens culinaris*) and broad beans (*Vicia faba*).

Cueva de Nerja is situated on the eastern coast of Málaga province. The excavations by Pellicer (1963; 1987; Pellicer and Acosta 1986) revealed stratigraphy from the Palaeolithic to the Chalcolithic. Plant remains were studied from a large concentration, which was dated to the late Neolithic (Hopf 1970), and free-threshing wheats and hulled barley were the main components. Plant remains have recently been examined by G. Pérez Jordà (Jordà Pardo et al. 1990) from a new excavation and this study has revealed the presence of naked barley and possibly two specimens on *Vicia/Lathyrus* in a context dated to 5480–5310 cal BC.

Later, during the 5th millennium cal BC, both hulled wheats (grain and chaff) and free-threshing wheats are present at Murciélagos. Rachis fragments of bread wheat (*Triticum aestivum*) and barley have also been identified. As in the previous millennium, the spectrum of cultivated plants from Cueva del Toro remains the same, whereas from Cueva de Nerja no archaeobotanical data are available from this period.

Although they have not been included in table 11.2 because of lack of absolute dates, I would like to note the assemblage recovered from Cueva de los Mármoles (Córdoba), situated a few kilometres away from the Cueva de los Murciélagos, where the preliminary analysis by E. Martos (unpublished) showed the presence of *Triticum dicoccum*, *Triticum aestivum* and *Hordeum vulgare* from middle neolithic levels.

A great range of wild fruits has been identified in addition to the cultivated plants. Species from the earliest levels include acorn (*Quercus* sp.), wild olive (*Olea europaea* ssp. *oleaster*) and caper (*Capparis* sp.). From this evidence it seems likely that acorns played an important role in the diet of the neolithic groups. Charcoal analysis from Los Murciélagos (Rodríguez Ariza 1996) indicates an intense use of strawberry tree (*Arbutus unedo*) and oak (*Quercus ilex/coccifera*) wood during the Neolithic. Again, in Cueva del Toro, acorns are documented from the 5th millennium onwards (Buxó 1997).

Agrarian practices

Data on agrarian practices is limited because of the kind of remains that have been recovered; our archaeobotanical samples are composed largely of cereal grains with very few remains of chaff and arable weeds. In the case of Cueva de Los Murciélagos, one source of evidence comes from the combination of lithic studies and experimental work (Ibáñez and González Urquijo 1995; González et al. 1994; Ibáñez et al. 2000). Micro-wear analysis of a group of blades identified as sickle elements has demonstrated the presence of clear gloss traces. In addition, on some of the specimens the vestiges of the adhesive used to insert the blades into the hafts have been preserved, which has allowed reconstruction of the way the piece was obliquely inserted into a wooden shaft. This mode of insertion is also confirmed by the wear pattern. Some experimental harvesting using flint sickles was also undertaken, the main objective being to reproduce within an optimal framework the conditions of prehistoric hulled wheat harvesting. Comparison between both sets of flints (experimental and archaeological) revealed that cereals in Cueva de los Murciélagos were probably harvested low on the straw, using sickles. This is supported by the strong striation present on the archaeological samples, which resembles most closely that present on the experimental flints.

11.4 *Papaver somniferum* in the Iberian Neolithic: archaeobotanical evidence

Poppy seeds have been recovered from both study areas. In Andalucía, seeds associated with a hearth have been identified in the early neolithic levels (ca. 5500 cal BC) from Cueva de Los Murciélagos (Peña-Chocarro 1999) and remains are also present in later phases. In addition, poppy seeds have been retrieved from Cueva del Toro (Buxó 1997) from the late 5th millennium cal BC. Other early finds include the *Papaver somniferum* capsules described by Neuweiler 1935 from the Cueva de los Murciélagos (Albuñol, Granada - a different cave from the one described here), which correspond to wild poppy. In the interior of the Peninsula, Stika has identified poppy seeds from a context dated to the 6th millennium cal B.C (Zapata et al. nd). These early finds provide an interesting subject for research because, as yet, the area where cultivation of *Papaver somniferum* began is unknown. In the archaeobotanical record for the Near East there are no remains of poppy dating to the Neolithic. While recovery techniques may be a factor determining whether or not poppy is recovered in archaeological contexts, it seems unlikely that this applies to the Near East where systematic flotation has long been practised.

It has been suggested (Schultze-Motel 1979; Bakels 1982, 1992; Zohary and Hopf 1993) that the domestication of *Papaver somniferum* probably occurred somewhere in the western Mediterranean. This hypothesis is supported by the present-day distribution of the wild *Papapver somniferum* ssp. *setigerum*, which is concentrated in the west and central Mediterranean area and where the earliest finds have been recorded. In addition to the examples mentioned from Spain, Rottoli (1993) refers to both seeds and capsules found in early neolithic contexts at the site of La Marmotta in Italy. Neolithic *Papaver somniferum* seeds have also been recovered from Bandkeramik levels in many sites of the Netherlands, Germany, northern France and Switzerland (Bakels 1982, 1992; Knörzer 1967, 1995; Heim and Jadin 1998, amongst others).

On the evidence obtained from the few sites cited above it is difficult to ascertain whether our poppy seeds belong to a wild or a cultivated form. The wild opium poppy comprises both diploid and tetraploid chromosome races (Zohary and Hopf 1993), and distinction between wild and cultivated seeds is difficult. The work of Hammer and Fritsch (1977) and Fritsch (1979) emphasized the extreme variability of the subspecies *somniferum*, which covers the variability found in both diploid and tetraploid races of the subspecies *setigerum*.

Although more data are needed for a full understanding of the domestication process of *Papaver somniferum*, the available information indicates that the earliest contexts where poppy seeds and/or capsules are found are in the Italian and Iberian Peninsulas. This certainly supports the idea of its local domestication but only further research will help to elucidate this fascinating subject.

11.5 Concluding remarks

According to the available data regarding early agriculture, there are several observations I would like to make:

1. Different hypotheses have been proposed to explain the adoption of agriculture in these regions, ranging from highly autochthonous models where acculturated hunter-gatherers would have played a major role, to models of cultural diffusion accompanied by demic movements. Although the debate has now persisted for a long time, there are still great uncertainties. For example, one of the problems is the lack of knowledge of the mesolithic substratum; and as many authors have already stated (Acosta 1987; Asquerino 1987; Chapman 1990; Fortea and Martí Oliver 1984) more intensive research on the Mesolithic

is needed for a full understanding of the different processes that took place during the transition to food production in both areas.

2. The available data suggest that a fully developed agriculture based on the cultivation of a wide range of cereals and legumes was present between 5500-5000 cal BC. While in Andalucía the spectrum of cultivated plants is very similar to that of the eastern coast with a great diversity of species, in the Meseta (at least in the northern part) the situation is very distinct between the two areas investigated. In La Vaquera free-threshing wheats predominate over the hulled wheats, and conversely, in the Ambrona Valley (sites of La Lámpara and La Revilla del Campo), agriculture seems to have concentrated on the cultivation of hulled wheats. There are very few finds of barley from these sites.

3. Central to this discussion is the question of what we actually know about the sort of agriculture that took place at these sites. Information on crop processing or agrarian practices is limited since chaff remains and weed seeds are poorly represented in the archaeobotanical record. For Cueva de los Murciélagos, micro-wear analysis and experimental work have allowed to suggest the use of sickles for harvesting low on the straw.

4. The *Papaver somniferum* seeds from the Meseta and Andalucía are amongst the earliest remains of this species in Europe.

5. Despite all the difficulties, the archaeobotanical dataset and the increasing number of absolute dates suggest that the diffusion of agriculture to the interior of the Iberian Peninsula was an extremely rapid process. The early dates (6th millennium cal BC) obtained from both the Meseta and Andalucía are similar to those from the Mediterranean coast. However, the question of the source of agriculture and the role of indigenous peoples in the neolithisation of these regions is still open to major debate. There are many gaps in our research and the archaeobotanical record is incomplete. Although some systematic sampling and recovery has been initiated, flotation has not yet become common practice. So, a major challenge marks the future of forthcoming excavations: the regular retrieval of plant remains from dated contexts. Only after intensive research will we be in a position to discuss the sources of regional variability which seems to characterize the first neolithic communities of these regions.

Acknowledgements

The author has a post-doctoral contract at the CSIC in Madrid, within the Programa I3P funded by the European Social Fund.

Many people have contributed with data and discussion to this paper. I would like to thank the following: Ana Arnanz, Enrique Cerrillo, Pedro Díaz del Río, Beatriz Gavilán, Hans Peter Stika and Juan Carlos Vera for providing with information and comments on this paper. I am extremely grateful to Lydia Zapata for many inspiring discussions on prehistoric agriculture and for making comments on this manuscript. Valter Morelli has helped with technical support.

References

Acosta, P. (1986). El neolíthico en Andalucía occidental. Estado actual. In *Actas del Congreso Homenaje a Luis Siret*, pp. 136–151. Sevilla: Consejería de Cultura de la Junta de Andalucía.

Acosta, P. (1987). El neolítico antiguo en el suroeste español. La Cueva de la Dehesilla (Cádiz). In J. Guilaine, J. Courtin, J.-L. Roudil, and J.-L. Vernet (Eds.), *Premières Communautés*

Paysannes en Méditerranée Occidentale. Actes du Colloque International du C.N.R.S., Montpellier, 26–29 Avril 1983, pp. 653–659. Paris: C.N.R.S.

Acosta, P. (1995). Las culturas del Neolítico y Calcolítico en Andalucía Occidental. *Espacio, Tiempo y Forma. Serie 1, Prehistoria y Arqueología 8*, 33–80.

Acosta, P. and M. Pellicer (1990). *La Cueva de La Dehesilla (Jérez de la Frontera). Las primeras civilizaciones productoras en Andalucía Occidental.* Jerez de la Frontera: CSIC.

Asquerino, M. D. (1987). El neolítico en Andalucía: estado actual de su conocimiento. *Trabajos de Prehistoria 44*, 63–85.

Bakels, C. (1982). Der Mohn, die Linearbandkerammik und das westliche Mittelmeergebiet. *Archäologisches Korrespondenzblatt 12*, 11–13.

Bakels, C. (1992). Fruits and seeds from the Linearbandkerammik settlement at Meindling, Germany, with special reference to *Papaver somniferum*. *Analecta Praehistorica Leidensia 25*, 55–68.

Bernabeu, J. (1996). Indigenismo y migracionismo. Aspectos de la neolitización en la fachada oriental de la Península Ibérica. *Trabajos de Prehistoria 53*(2), 37–54.

Bueno, P., R. D. Balbín, and R. Barroso (2000). Una necrópolis Ciempozuelos con cuevas artificiales al interior de la Península. *Estudos Pre-Historicos 8*, 49–80.

Bueno, P., R. D. Balbín, R. Barroso, J. M. Rojas, R. Villa, R. Félix, and S. Rovira (1999). Neolítico y Calcolítico en Huecas (Toledo). El Túmulo de Castillejo. Campaña. *Trabajos de Prehistoria 56*(2), 141–160.

Bueno, P., R. Barroso, R. De Balbín, M. Campo, F. Etxeberría, A. González, L. Herrasti, J. Juan, P. López-García, A. López-García, J. C. Matamala, and B. Sánchez (2002). Áreas habitacionales y funerarias en el Neolítico de la cuenca interior del tajo: la provincia de Toledo. *Trabajos de Prehistoria 59*, 65–79.

Buxó, R. (1997). *Arqueología de las Plantas. La Explotación Económica de las Semillas y los Frutos en el Marco Mediterráneo de la Península Ibérica.* Barcelona: Ed. Critica.

Cerrillo, E. (1999). La Cueva de El Conejar (Cáceres): avance al estudio de las primeras sociedades productoras en la Península cacereña. *Zephyrus 52*, 107–128.

Cerrillo, E., A. Prada, A. G. Cordero, and F. J. Heras (2002). La secuencia cultural de las primeras sociedades productoras en Extremadura: una datación absoluta del yacimiento de Los Barruecos (Malpartida de Cáceres, Cáceres). *Trabajos de Prehistoria 59*, 101–111.

Cerrillo, E., A. Prada, A. Gallardo, A. G. Cordero, F. J. Heras, and M. E. S. Barba (nd). Los Barruecos y las primeras comunidades agrícolas del Tajo interior. Campañas de excavación 2001 y 2002. In P. Arias, R. Ontañón, and C. García-Moncó (Eds.), *Actas del III Congreso del Neolítico de la Península Ibérica.* Santander: Servicio de Publicaciones de la Universidad de Cantabria. In Press.

Cerrillo, E., A. Prada, A. Gallardo, A. G. Cordero, A. Morales, J. A. L. Sáez, P. L. García, A. Arnanz, J. Pastor, J. Juran, and J. C. Matamala (nd). Bases económicas y ambientales para el estudio de las comunidades neolíticas del centro-oeste peninsular: perspectivas desde el yacimiento de Los Barruecos. In P. Arias, R. Ontañón, and C. García-Moncó (Eds.), *Actas del III Congreso del Neolítico de la Península Ibérica.* Santander: Servicio de Publicaciones de la Universidad de Cantabria. In Press.

Chapman, R. (1990). *Emerging Complexity. The Later Prehistory of South-East Spain, Iberia and the West Mediterranean.* New Studies in Archaeology. Cambridge: Cambridge University Press.

Delibes, G. (1977). El poblamiento eneolítico en la Meseta Norte. *Sautuola 2*, 141–151.

Delibes, G. and J. F. Manzano (2000). La trayectoria cultural de la Prehistoria Reciente (6400–2500 BP) en la submeseta norte Española: principales hitos de un proyecto. In V. Oliveira Jorge (Ed.), *Actas do 3.° Congresso de Arqueologia Peninsular*, Volume IV: Pré-História Recente da Península Ibérica, pp. 95–122. Porto: Adecap.

Díaz del Río, P. (2001). *La formación del paisaje agrario: Madrid en el III y II milenios BC.* Arqueología, Paleontología y Etnografía, 9. Madrid: Comunidad de Madrid.

Díaz del Río, P. and S. Consuegra (1999). Primeras evidencias de estructuras de habitación y almacenaje neolíticas en el entorno de la Campiña madrileña: el yacimiento de "La Deseada" (Rivas-Vaciamadrid, Madrid). II Congrés del Neolític a la Península Ibèrica. *Saguntum-PLAV Extra 2*, 251–257. Actes del II Congrés del Neolític a la Península Ibèrica, Universitat de València, 7-9 d'Abril de 1999.

Estremera, M. S. (1999). Sobre la trayectoria del Neolítico Interior: precisiones a la secuencia de la Cueva de La Vaquera (Torreiglesias, Segovia). II Congrés del Neolític a la Península Ibèrica. II Congrés del Neolític a la Península Ibèrica. *Saguntum-PLAV Extra 2*, 245–250. Actes del II Congrés del Neolític a la Península Ibèrica, Universitat de València, 7-9 d'Abril de 1999.

Estremera, M. S. (Ed.) (2003). *Primeros Agricultores y Ganaderos en la Meseta Norte: el Neolítico de la Cueva de La Vaquera (Torreiglesias, Segovia)*, Volume 11. Memorias. Arqueología en Castilla y León. Zamora: Junta de Castilla y León.

Fortea, J. and B. Martí Oliver (1984). Consideraciones sobre los inicios del Neolítico en el Mediterráneo español. *Zephyrus 37/38*, 167–199.

Fritsch, R. (1979). Zur Samenmorphologie des Kultumohns (*Papaver somniferum* L.). *Kulturpflanze 27*, 217–227.

Gavilán, B. (1991). Avance preliminar sobre la excavación arqueológica de urgencia de la Cueva de los Murciélagos de Zuheros (Córdoba). *Antiquitas 2*, 23–30.

Gavilán, B. (1997). Reflexiones sobre el Neolítico andaluz. *Spal 6*, 23–33.

Gavilán, B. and J. C. Vera (1992). Breve avance sobre los resultados obtenidos en la excavación arqueológica de urgencia en la Cueva de los Murciélagos de Zuheros (Córdoba). *Antiquitas 3*, 23–30.

Gavilán, B., J. C. Vera, L. Peña-Chocarro, J. J. Cepillo, M. R. Delgado, and C. Marfil (1994). Preliminares sobre la tercera campaña de excavación arqueológica de urgencia en la Cueva de los Murciélagos de Zuheros (Córdoba). *Antiquitas 5*, 5–12.

Gavilán, B., J. C. Vera, L. Peña-Chocarro, and M. M. I. Cornellà (1996). El V y IV milenios en Andalucía Central: La Cueva de Los Murciélagos de Zuheros (Córdoba). *Rubricatum 1*, 323–327.

González, J. E., J. J. Ibáñez, L. Peña-Chocarro, B. Gavilán, and J. C. Vera (1994). Harvesting tasks in the neolithic levels of 'Los Murcielagos' Cave. An archeobotanical and functional approach. *Helinium 34*, 321–344.

González-Cordero, A. and E. Cerrillo (2001). El proceso de neolitización en la comarca estremeña de La Vera. *Madrider Mitteilungen 42*, 1–31.

Hammer, K. and R. Fritsch (1977). Zur Frage nach der Ursprungsart des Kulturmohns (*Papaver somniferum* L.). . *Kulturpflanze 25*, 113–124.

Heim, J. and I. Jadin (1998). Sur les traces de l'orge et du pavot. L'agriculture danubienne de Hesbaye sous influence, entre Rhin et Bassin parisien? *Anthropologie et Préhistoire 109*, 187–205.

Hopf, M. (1966). *Triticum monococcum* L. y *Triticum dicoccum* Schubl, en el neolítico antiguo español. *Archivo de Prehistoria Levantina 11*, 53–80.

Hopf, M. (1970). Neolitische Getreidefundein der Höhle von Nerja (Prov. Málaga). *Madrider Mitteilungen 11*, 9–27.

Hopf, M. (1974). Breve informe sobre el cereal neolítico de la Cueva de Zuheros. *Trabajos de Prehistoria 31*, 295–296.

Hopf, M. (1987). Les débuts de l'agriculture et la difussion des plantes cultivées dans la Péninsule Ibérique. In J. Guilaine, J. Courtin, J.-L. Roudil, and J.-L. Vernet (Eds.), *Premières Communautés Paysannes en Méditerranée Occidentale. Actes du Colloque International du C.N.R.S., Montpellier, 26–29 Avril 1983*, pp. 267–274. Paris: C.N.R.S.

Hopf, M. (1991). South and Southwest Europe. In W. van Zeist, K. Wasylikowa, and K.-E. Behre (Eds.), *Progress in Old World Palaeoethnobotany: A Retrospective View on the Occassion of 20 Years of the International Work Group for Palaeoethnobotany*, pp. 241–277. Rotterdam: A. A. Balkema.

Hopf, M. and A. M. Muñoz (1974). Neolithische Pflanzenreste aus der Höhle Los Murciélagos bei Zuheros, Prov. Córdoba. *Madrider Mitteilungen 15*, 9–27.

Ibáñez, J. J. and J. E. González Urquijo (1995). El uso de los útiles de sílex de los niveles neolíticos de la cueva de Los Murciélagos (Zuheros, Córdoba). Primeros resultados. *Rubricatum 1*, 169–176.

Ibáñez, J. J., J. E. González Urquijo, L. Peña-Chocarro, B. Gavilán, and J. C. Vera (2000). El aprovechamiento de los recursos vegetales en los niveles neolíticos del yacimiento de Los Murciélagos (Zuheros, Córdoba). Estudio arqueobotánico y de la función del utillaje. *Complutum 11*, 171–190.

Iglesias, J. C., M. A. Rojo, and V. Álvarez (1996). Estado de la cuestión sobre el Neolítico en la Submeseta norte. I Congrés del Neolític a la Península Ibèrica, Gavá-Bellaterra. *Rubricatum 1*, 721–734.

Jiménez Guijarro, J. (1998). La neolitización de la Cuenca Alta del Tajo. *Complutum 9*, 27–47.

Jiménez Guijarro, J. (1999). El proceso de neolitización del interior peninsular. II Congrés del Neolític a la Península Ibèrica. *Saguntum-PLAV Extra 2*, 493–501. Actes del II Congrés del Neolític a la Península Ibèrica, Universitat de València, 7-9 d'Abril de 1999.

Jiménez Sanz, P. J., J. J. Alcolea, M. A. García, and J. M. Jiménez Guijarro (1997). Nuevos datos sobre el Neolítico meseteño: la provincia de Guadalajara. In R. de Balbín Berhmann and P. Bueno Ramírez (Eds.), *II Congreso de Arqueología Peninsular (1996), Zamora, 1997*, Volume 2, pp. 33–47. Zamora: Fundación Rei Alfonso Henriques.

Jordà Pardo, J. F., J. E. Aura, and F. Jordà Cerdà (1990). El límite Pleistoceno-Holoceno en el yacimiento de la Cueva de Nerja (Málaga). *Geogaceta 8*, 102–104.

Knörzer, K.-H. (1967). Subfossile Pflanzenreste von bandkeramischen Fundstellen im Rheinland. *Archaeo-Physika 2*, 3–29.

Knörzer, K.-H. (1995). Pflanzenfunde aus dem bandkeramischen Brunnen von Küchhoven bei Erkelenz Vorbericht. In H. Kroll and R. Pasternak (Eds.), *Res Archaeobotanicae: 9th Symposium IWGP*, pp. 81–86. Kiel: Oetker-Voges Verlag.

Kunst, M. and M. A. Rojo (1999). El Valle de Ambrona: un ejemplo de la primera colonización Neolítica de las tierras del interior peninsular. *Saguntum Extra 2*, 503–512. Actes del II Congrés del Neolític a la Península Ibèrica, Universitat de València, 7-9 d'Abril de 1999.

Kunst, M. and M. A. Rojo (2000). Ambrona 1998. Die neolithische Fundkarte und 14 C-Datierungen. *Madrider Mitteilungen 41*, 1–31.

López, P. (1980). Estudio de semillas prehistóricas en algunos yacimientos españoles. *Trabajos de Prehistoria 37*, 419–432.

López, P., A. M. Arnanz, R. Macías, P. Uzquiano, and P. Gil (2003). Arqueobotánica de la Cueva de La Vaquera. In M. S. Estremera (Ed.), *Primeros Agricultores y Ganaderos en la Meseta Norte: el Neolítico de la Cueva de La Vaquera (Torreiglesias, Segovia)*, Volume 11. Memorias. Arqueología en Castilla y León, pp. 247–255. Zamora: Junta de Castilla y León.

Martín Socas, D., R. Buxó, M. D. Camalich, and A. Goñi (1999). Estrategias subsistenciales en Andalucía Oriental durante el Neolítico. II Congreso de Neolític a la Peninsula Ibèrica. *Saguntum-PLAV Extra 2*, 25–30. Actes del II Congrés del Neolític a la Península Ibèrica, Universitat de València, 7-9 d'Abril de 1999.

Martín Socas, D., M. D. Camalich, P. González, and A. Mederos (1993). El Neolítico en la comarca de Antequera (Málaga). In *Investigaciones Arqueológicas en Andalucía (1985-1992). Proyectos*, pp. 273–284. Huelva: Consejería de Cultura de la Junta de Andalucía.

Neuweiler, E. (1935). Nachtrage urgeschichtlicher Pflanzen. *Vierteljahrsschrift der Naturforschenden den Gesellschaft in Zürich 80*, 98–112.

Pellicer, M. (1963). *Estratigrafía prehistórica de la Cueva de Nerja*, Volume 16 of *Excavaciones Arqueológicas Españolas.* Madrid: Ministerio de Educación y Ciencia.

Pellicer, M. (1987). El neolítico de la Cueva de Nerja (Málaga). In J. Guilaine, J. Courtin, J.-L. Roudil, and J.-L. Vernet (Eds.), *Premières Communautés Paysannes en Méditerranée Occidentale. Actes du Colloque International du C.N.R.S., Montpellier, 26–29 Avril 1983*, pp. 639–643. Paris: C.N.R.S.

Pellicer, M. and P. Acosta (1986). Neolítico y Calcolítico de la Cueva de Nerja. In J. J. Pardo (Ed.), *La Prehistoria de la Cueva de Nerja (Málaga). Trabajos sobre la Cueva de Nerja*, Volume 1, pp. 341–450. Málaga: Patronato de la Cueva de Nerja.

Peña-Chocarro, L. (1999). *Prehistoric Agriculture in Southern Spain during the Neolithic and the Bronze Age: The Application of Ethnographic Models.* BAR International Series S818. Oxford: Archaeopress.

Quadra-Salcedo, A. and A. M. Vicent (1964). Informe de las excavaciones en la Cueva de los Murciélagos de Zuheros (Córdoba). Primera campaña, noviembre 1962. *Noticiero Arqueológico Hispano VI*, 68–72.

Ramos Muñoz, J., V. Castañeda Fernández, M. Pérez Rodríguez, M. Lazarich González, C. Martínez Peces, M. Montañés Caballero, J. M. Lozano Moya, and D. Calderón Estrada (1995). Los Charcones. Un poblado agrícola del III y II milenios AC. *Almoraima 13*, 33–50.

Rodríguez Ariza, M. O. (1996). Análisis antracológicos de yacimientos neolíticos de Andalucía. *Rubricatum 1*, 73–83.

Rojo, M. A. and M. S. Estremera (2000). El Valle de Ambrona y la Cueva de la Vaquera: testimonios de la primera ocupación neolítica en la cuenca del Duero. In V. Oliveira Jorge (Ed.), *Actas do 3.° Congresso de Arqueologia Peninsular*, pp. 81–95. Porto: Adecap.

Rojo, M. A. and M. Kunst (1999). Zur Neolithisierung des Inneren der Iberischen Halbinsel. *Madrider Mitteilungen 40*, 1–52.

Rojo, M. A., M. Kunst, I. García, R. Garrido, and G. Morán (nd). La neolitización de la Meseta Norte a la luz del C-14: análisis de 40 dataciones absolutas de dos yacimientos domésticos del valle de Ambrona, Soria, España. In P. Arias, R. Ontañón, and C. García-Moncó (Eds.), *Actas del III Congreso del Neolítico de la Península Ibérica.* Santander: Servicio de Publicaciones de la Universidad de Cantabria. In Press.

Rottoli, M. (1993). "La Marmotta", Anguillara Sabazia (RM). Scavi 1989. Analisi paletnobotaniche: prime risultanze. Appendice 1. *Bullettino di Paletnologia Italiana n. s. II 84*, 305–315.

Schultze-Motel, J. (1979). Die urgeschichtlichen Reste des Schlafmohns (*Papaver somniferum* L.) und die Entstehung der Art. *Kulturpflanze 27*, 207–213.

Stika, H.-P. (1999). Erste archäobotanische Ergebnisse zu den neolithischen Ausgrabungen 1997 in Ambrona, Prov. Soria. *Madrider Mitteilungen 40*, 61–65.

Stika, H.-P. (n.d.). Early Neoltihic Agriculture in Ambrona, Prov. of Soria (Spain). In Prep.

Vicent, A. M. and A. M. Muñoz (1973). *Segunda campaña de excavaciones en la Cueva de los Murciélagos. Zuheros (Córdoba) 1969*, Volume 77 of *Excavaciones Arqueológicas de España.* Massachusetts: Ministerio de Educación y Ciencia.

Zapata, L. (2002). Charcoal analyses from Basque archaeological sites: new data to understand the presence of *Quercus ilex* in a damp environment. In S. Thiébault (Ed.), *Charcoal Analysis: Methodological Approaches, Palaeoecological Results and Wood Uses. Proceedings of the Second International Meeting of Anthracology, Paris, September 2000*, BAR International Series 1063, pp. 121–126. Oxford: Archaeopress.

Zapata, L., L. Peña-Chocarro, G. Pérez-Jordá, and H. P. Stika (nd). Difusión de la agricultura en la Península Ibérica. In P. Arias, R. Ontañón, and C. García-Moncó (Eds.), *Actas del III Congreso del Neolítico de la Península Ibérica.* Santander: Servicio de Publicaciones de la Universidad de Cantabria. In Press.

Zohary, D. and M. Hopf (1993). *Domestication of Plants in the Old World* (2nd ed.). Oxford: Claredon Press.

Chapter 12

First farmers along the coast of the Bay of Biscay

Lydia Zapata Peña
Universidad del Pas Vasco/Euskal Herriko Unibertsitatea, Vitoria-Gasteiz, Spain

Gordon Hillmanentzat,
bihotz-bihotzez,
irakatsitako eta emandako
guztiarengatik eskertuz.

12.1 The historical process: first farming in the Iberian peninsula

Models for the origin of the early Neolithic in the Iberian peninsula are similar to those that have been put forward for the adoption of the Neolithic across Europe. In the Valencia and Aragón regions of Iberia, the Mesolithic-Neolithic transition has been interpreted in two ways. One of these is what has been called a dual-model hypothesis whereby two communities coexisted: one fully neolithic, allochthonous, represented by sites with Cardial ceramics and no Epipalaeolithic industrial tradition; and one of indigenous groups who maintained hunter-gatherer ways of life, but over time adopted some neolithic traits such as ceramics (e.g., Bernabeu 1996, 2002; Fortea et al. 1987; Juan-Cubanilles 1992; Martí 1992, 1998; Martí et al. 1987; Utrilla 2002, among others). The latter hypothesis has been contested by researchers who prefer to interpret the archaeological evidence as representing contemporaneous *facies* of the same population carrying out different activities in different sites (Barandiarán and Cava 1992, 2000, 2001). A similar picture has been suggested for the centre of Portugal—a neolithic maritime colonisation of areas uninhabited by mesolithic people (Zilhao 2001). Other regions like the northern Meseta and northern Portugal witnessed the colonization of a previously depopulated area (Carvalho 2002; Delibes and Manzano 2000; Kunst and Rojo 1999), although alternative interpretations have also been put forward.

In this paper I shall focus on archaeobotantical evidence for farming in the Atlantic region of northern Iberia—the coastal fringe along the Bay of Biscay in the Basque-Cantabrian area (figure 12.1). Modern Basques are genetically one of the best-studied populations in Europe. A non-Indoeuropean language is still spoken in the region and, historically, the area has not been subject to major in-migration. Basques stand as a genetic outlier for several loci besides the Rh-negative allele, and some genetic studies have interpreted that Basques represent a pre-neolithic population that has survived in isolation (Calafell and Bertranpetit 1993, 1994).

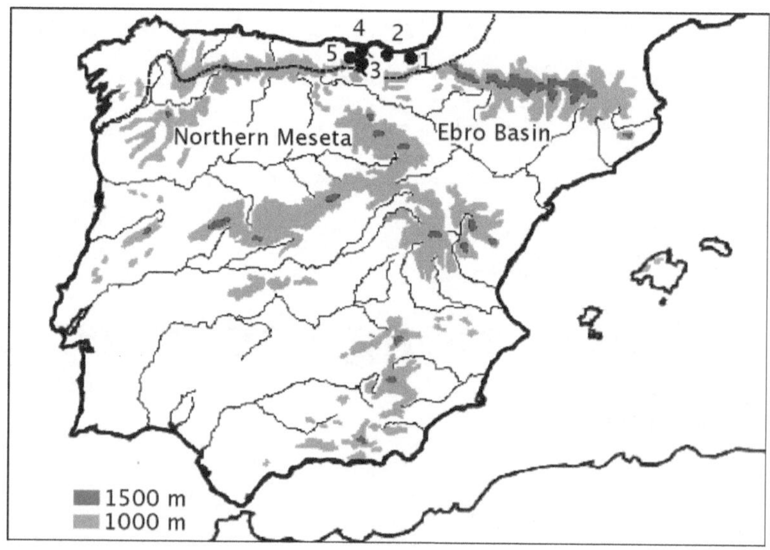

Figure 12.1: Location of the sites discussed in the text: 1. Herriko Barra; 2. Kobaederra; 3. Arenaza; 4. Pico Ramos; 5. El Mirón.

This would support the picture of the adoption of neolithic ways of life and farming through cultural diffusion (Comas et al. 1996). However, available genetic data are not straightforward since mitochondrial studies do not identify differences between the Basques and the bulk of other Europeans (Barbujani 2002). Their outlier status may thus be partly the result of reduced neolithic penetration as well as considerable genetic drift due to isolation and small population size (Richards 2003, p.153).

Archaeological evidence, on the other hand, shows that a model of cultural isolation is not tenable for this region. The adoption of farming, according to all authors, would have been an indigenous process (Alday nd; Arias 1999; Barandiarán and Cava 2001; Zilhao 2000) although naturally developing from outside stimuli, in a process of trait adoption-diffusion. If we consider the entire Basque territory, including the Upper Ebro Valley, continuity can be appreciated from the Mesolithic to the Neolithic in the settlements and in other aspects such as lithic industry (Alday 1999, 2000, nd; Barandiarán and Cava 2001; López Quintana nd). The number of early neolithic sites is not very high compared to mesolithic ones (Barandiarán and Cava 2001, p.525), and there is little evidence of population increase.

12.2 Geographical setting

In the north of the Iberian peninsula there is a narrow coastal fringe that belongs to the Euro-Siberian phytogeographical region (figure 12.1). It is an area with an oceanic climate that is mild and very wet (ca. 1000–1700 mm/year) with no summer drought. The landscape is dominated by a mosaic of hills, mountains and small valleys. To the south, at the limit with the Mediterranean watershed, there is a more abrupt relief with mountains at heights of over 1000 m. The Pyrenees have their lowest altitudes in the east, so in prehistory the Basque region, far from being isolated, was one of the best corridors for people and plants to enter and leave Iberia.

Potential vegetation is currently composed of different types of forest; those characterised by *Quercus robur* are the most extensive, while *Fagus sylvatica* predominates in the montane areas. Under certain conditions, other communities are also present, including forests of evergreen oak (*Quercus* spp.), *Quercus pyrenaica, Betula* and *Alnus* (Aseginolaza et al. 1989; Loidi et al. 1997).

Pollen and charcoal data also indicate that there were extensive *Corylus* and mixed deciduous *Quercus* forests during the middle Holocene when farming was adopted (Iriarte 1994a; Peñalba 1994; Zapata 2002b).

12.3 Megaliths, and some assumptions about the neolithic

Northern coastal territories in Iberia have traditionally been considered culturally marginal and isolated during the Neolithic period. It has been assumed that: (i) there was continuity between the Mesolithic and the Neolithic periods, with a long period of hunting-gathering even when neolithic and chalcolithic material culture was available, and; (ii) there was a difficulty in adopting cereal agriculture due to the geographic conditions of the Atlantic valleys, which were better suited to pastoralism. As a consequence, even when agriculture was adopted, agrarian practices would not have been a major contribution to prehistoric human subsistence. It is only recently that these assumptions have begun to be refuted on the basis of historical and ethnographic information (Peña-Chocarro 1999; Peña-Chocarro and Zapata 2003), and archaeobotanical data (Iriarte et al. 2005; Zapata 2002b).

Research on the Neolithic has mainly been focused on megaliths, the most visible structures of the period. In recent decades, excavations have been carried out not only on single monuments but there have also been long-term projects on megalithic necropoli and those areas where there are concentrations of these monuments (see, for example, de Blas 1997; Díez Castillo 1995; Fábregas and Suárez 1999; Mujika and Armendáriz 1991; Yarritu and Gorrotxategi 1995). Their significance in relation to neolithic practices in Cantabrian Iberia has long been debated. In the past, certain authors have associated megaliths with the beginning of a production economy; on the contrary, some assumed that they represented the expansion of farming societies, while others regarded them simply as sacred or meeting areas with a symbolic and social meaning, rather than having any economic significance (for example, among others, Alday 1999, 2000; Alday et al. 1996; Andrés 1990, 1997; Arias 1997; de Blas 1997; González Morales 1992, 1996; Jarman et al. 1982; Yarritu and Gorrotxategi 1995). Thus far these questions have been discussed with very little reference to bioarchaeological data.

Megaliths in the Atlantic territory along the Bay of Biscay were first built from ca. 4200–4000 cal BC. Their construction continued for some time after this, and they were in use at least until the 3rd millennium cal BC (Alday and Mujika 1999; Arias 1999; Arias et al. 1999). Recent bioarchaeological information has shown that domestic plants and animals were present from at least ca. 5200–4700 cal BC and thus predate their initial construction. Here we shall focus on the obscure millennium (ca. 5200–5000 to 4200–4000 cal BC) prior to the building of the first megaliths, reviewing evidence from sites for which archaeobotanical data are available. We shall discuss two types of contemporaneous sites: (i) those with an emphasis on the procurement of wild resources, such as the sites of Herriko Barra and Pico Ramos, and; (ii) those with abundant domestic elements, for example, Arenaza, Kobaederra and El Mirón (figure 12.1).

12.4 Sites that focus on wild resources

Herriko Barra is a coastal open-air site (300 m from the beach) located in the urban area of Zarautz (Gipuzkoa). J.A. Mujika conducted a rescue excavation, the results of which have only been partially published (Alday and Mujika 1999; Iriarte et al. 2005). Three ^{14}C dates on bone are available for Level C, where archaeological material was present: 5810±170, 5960±95 and 6010±90 bp.

It is a site with no domestic animals or ceramics. The lithic industry is based on notches and denticulates (28%), side scrapers (17%), end scrapers (14%), truncations (9%) borers (5%) and geometric microliths with simple bifacial retouch (12%). This form of retouch has been considered a good chronological indicator for the Neolithic in Iberia and France (Barandiarán and Cava 2001; Marchand 2000; Utrilla 2002), although some authors reasonably contest the significance of lithic material on its own to define cultural entities (González Morales 1996). Red deer (*Cervus elaphus*) accounts for more than 90% of the mammalian fauna; other animals present are aurochs (*Bos primigenius*), boar (*Sus scrofa*) and roe deer (*Capreolus capreolus*). Female red deer seem to be have been hunted preferentially and the presence of some newborn individuals shows that hunting took place at least at the end of the spring or beginning of the summer. Many faunal remains show anthropogenic manipulation (e.g., evidence of fire, breaking of bones for marrow extraction, incisions and slaughter marks) (Mariezkurrena and Altuna 1995). In addition, the bones of five species of birds have been identified with evidence of cut marks, including the great auk (*Pinguinus impennis*). This is a bird that could not fly and would have been easily captured by humans (Elorza and Sánchez Marco 1993). Marine molluscs were also retrieved although they are very badly preserved. Thus, Herriko Barra seems to be a site where there was specialisation in hunting and gathering activities. However, cereal pollen has recently been identified and the presence of domestic crops in the vicinity has been assumed (Iriarte et al. 2005).

Pico Ramos (excavated by the author) is a very small cave located above the Barbadun river mouth, to the west of Bilbao. There are two archaeological contexts: a chalcolithic collective burial ground (Zapata 1995) and a shell midden, which has been dated with wood charcoal to 5860±65 bp (4910–4540 cal BC), and with oysters to 6040±90 bp (4710–4330 cal BC). It has no domestic animals or ceramics. The lithic industry from the site is scarce and the most diagnostic artefacts are geometric microliths with simple bifacial retouch, which according to microwear analyses, were used as projectile points. Some other lithic implements have butchery marks while others, unretouched and with very light wear, have only been used for a short period of time (Ibáñez and Zapata 2001). Level 4 from Pico Ramos is defined by the abundance of marine molluscs. The majority of these comprise three species of limpets (57%). Winkles are less frequent (23%), as are mussels, oysters and clams (6–7% each) and there are a few examples of crab and sea urchin. The abundance of gastropods (80%) indicates that rocky areas were the most frequently exploited biotype. The presence of bivalves (e.g., mussels and oysters) and clams is evidence that both fixed and mobile substrates were exploited. In some cases (10–25%), there were traces of burning, probably related to the way the marine resources were eaten (Moreno 2003). The clearest evidence of the seasonality of the site comes from an unpublished study on oysters carried out by N. Milner, in which she shows that they were gathered throughout the spring and early summer. Other faunal remains from the site include red deer (*Cervus elaphus*) with human cut marks and carnivore bite marks, which indicates the use of this resource by animals once abandoned by humans. The presence of great auk (*Pinguinus impennis*) suggests, as at Herriko Barra, that there was human activity in the cave sometime between May and August when this bird came inland to nest.

Analyses of pollen and macro-remains have been carried out at the site. M. Iriarte carried out the pollen analysis but only the chalcolithic samples have thus far been published (Iriarte 1994b). All samples suggest an open landscape with values of arboreal pollen lower than 20%. Arboreal taxa represented in Level 4 are *Corylus avellana*, *Quercus* subg. *Quercus*, and *Alnus*. Cereal pollen has only been identified in two of the samples from the chalcolithic burial context (ca. 4250 bp). Macro-remains were retrieved from all the sediment excavated in Level 4 (a total volume of 1693.5 litres) using flotation. Wood charcoal was practically the only archaeobotanical material present in the samples. Four hundred and forty three fragments have been analysed from the shell midden and 419 have been identified (table 12.1). The main taxon (92.8 %) is *Quercus* subg. *Quercus*, which anatomically includes all deciduous and semi-deciduous *Quercus* that currently grow in the region. A few fragments of evergreen oaks are also present, as well as hazel, strawberry tree (*Arbutus unedo*) and Rosaceae (e.g., blackthorn) and Pomoideae.

Taxon	n=
Quercus ilex–Q. coccifera	5
Quercus subg. *Quercus*	389
Corylus avellana	2
Arbutus unedo	8
Pomoideae	9
Prunus spinosa	6
Indeterminate	24
Total	**443**

Table 12.1: Numbers of fragments of wood charcoal by taxa, from Level 4 of the cave shell-midden of Pico Ramos

Other macro-remains include hazelnut shells and one barley (*Hordeum* sp.) grain, which has been AMS radiocarbon dated to 5370±40 bp (4330–4050 cal BC). In spite of the presence of this caryopsis, which provided a later date than the wood charcoal and oyster derived dates, I think that the evidence from Level 4 from Pico Ramos indicates an occupation focussed on the procurement and primary processing and consumption of wild resources. There are many other cave shell middens and contexts in the Atlantic valleys of the Basque Country and Cantabria that could most likely be ascribed to a category of sites specialising in the acquisition of wild resources. However, most are poorly-known levels with little bio-archaeological information and no archaeobotanical data. They have radiocarbon dates that cover the 4th and 5th millennium cal BC and only wild resources have been identified, in particular, sea molluscs and red deer. Some of them have ceramics but others do not. The lithic industry includes microliths. Some examples in Cantabria and the Basque Country coast are: Arenillas (Arias 1996; Arias and Ontañón 1996), La Trecha (González Morales 1995, 1996), La Chora (Straus et al. 2002), El Cubío Redondo (Ruiz Cobo and Smith 2001), Tarrerón (Apellániz 1971), Los Gitanos (Ontañón 2000), level AmK-s from Kobeaga II (Quintana 2000), Marizulo (Alday and Mujika 1999; Altuna 1980; Cava 1978) and Mouligna (Chauchat 1974).

12.5 Sites with abundant domestic remains

A limiting factor when discussing this group of sites is that the data are either based on old excavations (e.g., Arenaza) or are from sites where work is still in progress (e.g., Kobaederra and El Mirón).

Arenaza is a large cave located in the same fluvial valley as Pico Ramos. It has a long sequence but here only the following levels will be discussed:

- Level II: Mesolithic with geometric microliths and sea molluscs. Faunal remains are composed exclusively of wild animals, and red deer dominates the assemblage. This level has not been dated but it is later than 9600±180 bp, which is the radiocarbon date for the underlying stratum (Apellániz and Altuna 1975a; Arias and Altuna 1999).

- Sub-level IC2: Context with impressed and possibly cardial pottery (Apellániz and Altuna 1975b, p.195). Domestic animals constitute 79% of the macro-mammals and ovicaprids dominate (as 73.4% of domestic fauna), followed by pig (28.6%) and bovines (23.1%) (Altuna 1980). Three samples of domestic bovine (*Bos taurus*) were recently radiocarbon dated to: 5755±65 bp, 6040±75 bp and 10860±120 bp (this last date was rejected because of incompatibility with the context from which it comes (Arias and Altuna 1999).

- Sub-level IC1: Sub-level with human remains and pottery with impressions. The lithic material includes geometric microliths, and domestic fauna comprise 97.8% of the remains of macro-mammals (Altuna 1980). Two radiocarbon dates are available (4965±195 bp and 4730±110 bp) that place this context in the late Neolithic period of the region.

Kobaederra is another large cave located in the mouth of the river Gernika (Arias 1991; Arribas and Berganza 1984; Barandiarán 1967, 1979; Loriana 1943; Marcos 1982; Nolte 1962). Recently a new excavation has been carried out in order to define better the chronological sequence and retrieve bio-archaeological information. The excavation was conducted in a false lateral gallery, and three neolithic levels with dates between 4700 and 4000 cal BC were defined (Zapata et al. 1997). Pollen had not preserved at the site but wood charcoal remains, quantified by fragment number, shows that *Quercus* subg. *Quercus* was the main firewood used at the site (table 12.2). Over time the percentage of deciduous oaks decreases in favour of shrubby taxa, possibly indicating an opening up of the forest as a result of human activity (Zapata 2002a).

All excavated sediment was processed by flotation and the samples are currently under analysis. On the basis of the available data it is obvious that carpological remains are extremely scarce; they consist of acorns, fragments of *Corylus* pericarp, Rosaceae pomes and a few grains of barley (*Hordeum vulgare*) and emmer (*Triticum dicoccum*) (Zapata 2002b). One grain of *Hordeum* was ^{14}C dated to 5375±90 bp (table 12.3). Faunal remains are being studied but preliminary results indicate a very high percentage of domestic animals (J. Altuna and K. Mariezkurrena, work in progress).

El Mirón is a large cave located at ca. 250 m above sea level in one of the coastal ranges of the Cantabrian Cordillera, at the eastern edge of the province of Cantabria. Since 1996 excavation has been concentrated in different areas of the large entrance hall (10 m wide × 30 m deep × 13 m high). The cultural sequence includes contexts pertaining to the Mousterian, Upper Palaeolithic, Azilian, Mesolithic, Neolithic, Chalcolithic and Bronze Age periods, and with traces of Medieval occupation (Morales and Straus 2000; Straus and González Morales 2003).

The earliest neolithic contexts have radiocarbon dates of ca. 5790–5500 bp. The site has provided a good set of archaeozoological data, and domestic animals significantly dominate in all the neolithic contexts. Seasonality evidence derived from the ovicaprids shows that the cave was occupied throughout the year (Altuna and Mariezkurrena 2003).

Various types of archaeobotanical analyses have been undertaken at El Mirón, including investigations of charcoal, pollen and seeds (studied by L. Zapata, M.J Iriarte and L. Peña-Chocarro, respectively). Wood charcoal analysis indicates the use of deciduous oak as the main fuel during the Neolithic, with significant representation of *Corylus*, *Fraxinus* and Rosaceae. In spite of systematic recovery techniques and flotation of practically all the excavated sediment, there are very few carpological remains. A few grains of three wheat species have been identified, namely einkorn (*Triticum monococcum*), emmer wheat (*Triticum dicoccum*) and a naked wheat (*Triticum aestivum/durum*) (Peña-Chocarro et al. 2005).

In spite of all the sampling issues associated with radiocarbon dating, it appears that the contexts from the sites we have discussed overlap to a considerable degree (figure 12.2). This is quite obvious in the case of the bones of wild animals dated in Herriko Barra and the domestic bovines from Arenaza.

12.6 First agriculture

On the basis of the evidence discussed above, it is possible to propose at least two hypotheses for the period ca. 5200–4600 cal BC: (i) human groups with different subsistence patterns coexisted

Taxon	Level I 4405±55 bp	Level II 5200±110 bp 5460±60 bp	Level III 5820±240 bp	Level IV 5630±100 bp
Laurus nobilis	4	6	2	.
Fagus sylvatica	1	.	.	.
Quercus ilex-Q. coccifera	1	.	1	.
Quercus subg. *Quercus*	24	86	118	311
Quercus	2	.	.	.
Corylus avellana	6	3	3	7
Arbutus unedo	9	11	5	17
cf. *Arbutus unedo*	.	1	.	1
Rosaceae Pomoideae	3	.	1	.
Prunus spinosa	1	2	1	.
Prunus cf. *avium*	3	1	.	.
Prunus	.	.	1	.
cf. *Prunus*	1	.	.	.
Rhamnus alaternus–Phillyrea	9	1	.	.
cf. *Rhamnus alaternus–Phillyrea*	.	1	.	.
Fraxinus excelsior	5	7	5	8
Indeterminate	6	7	4	16
Total	**75**	**126**	**141**	**360**

Table 12.2: Numbers of fragments of wood charcoal by taxa from Levels at Kobaederra (with associated dates bp).

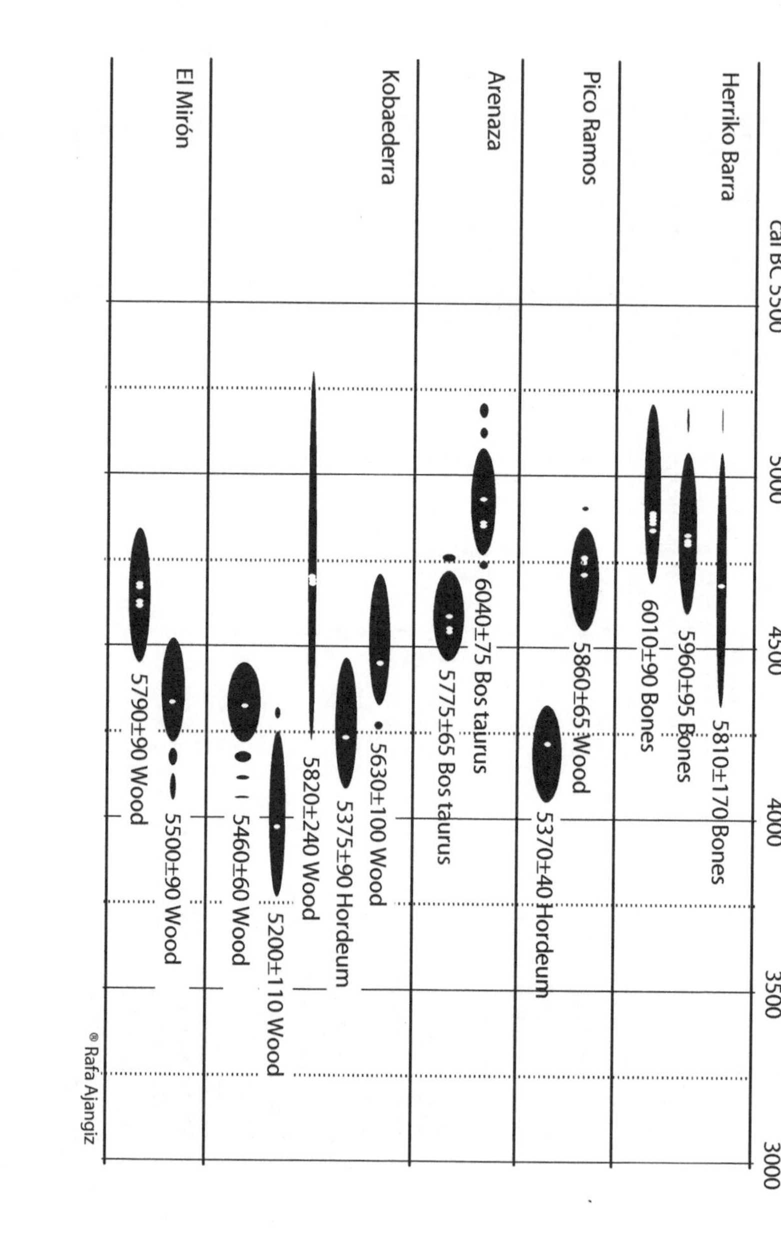

Figure 12.2: Radiocarbon dates for sites discussed in text calibrated according to Calib 4.3 (Stuiver and Reimer 2000).

and some groups 'resisted' the adoption of farming; or (ii) the same groups performed different activities at different sites. The data tend to support the second of these two hypotheses:

1. There is evidence to suggest that there was specialisation of activities carried out at each type of site. At Pico Ramos and Herriko Barra the focus was on a limited range of activities, namely hunting red deer and exploiting coastal resources, e.g., molluscs and birds. In contrast, at large caves with long sequences (Arenaza, Kobaederra and El Mirón) there is evidence of a greater diversity of resources and activities. Archaeozoological datasets show a predominance of domestic fauna (ca. 80–97% ungulates), although wild resources such as sea molluscs are also abundant. Cereals are present but they are extremely scarce.

2. Seasonality indicators for the procurement of wild products (e.g., red deer in Herriko Barra, oysters in Pico Ramos and great auk on both sites) suggest that hunting/gathering took place at the same time of the year, that is, at the end of the spring and beginning of the summer. If the people occupying these sites had begun to practice farming, this would have been the most critical time of the year: just before the harvesting of cereals, at which time they might have had to rely on wild resources (see a discussion for other areas in Milner 2002a and Milner 2002b). This hypothesis will need to be substantiated with data about the management of domestic animals; at El Mirón, for example, ovicaprids were killed all year-round, so in this case there might not have been a seasonal shortage.

3. In other European coastal regions a 'resistance' to farming has been documented amongst late hunter-gatherers who occupied different territories to those of incoming famers. For example, late mesolithic Scandinavian contexts have evidence of imported neolithic items and no signs of domestication for over a thousand years ca. 4600–3900/3500 cal BC (Price 2000), and in the loess areas of the Netherlands agriculture is documented from 5300 cal BC but in the adjacent Rhine-Meuse delta there is no evidence until after 4300 cal BC (Bakels 2000; Brinkkemper et al. 1999). In contrast to this, however, sites on the Basque-Cantabrian coast are juxtaposed in the same physical space and we can see no similar patterning that would indicate territorial divisions.

12.6.1 Chronology and foci of first agriculture in Cantabrian Iberia

According to the available data, crops spread throughout the Iberian peninsula very quickly (Zapata et al. nd, 2004). In spite of the lack of carpological data for large areas (i.e., the western half of the peninsula), there are no significant delays between most parts of the region (Pyrenees, Catalonia, Valencia region, Andalucía and the northern Meseta). Radiocarbon dates available for distant regions overlap in range, and thus it seems that the resolution of the evidence is not sufficient to isolate more precisely the routes and rhythms of this process.

If we consider the earliest radiocarbon dates carried out on cereals (prior to 4000 cal BC), we can see that there is a time lag between Cantabrian Iberia and the rest of the country; in the other regions the earliest dates are from at least ca. 5500–5200 cal BC, while in the north the range is from ca. 5200–4700 cal BC for cereal pollen in Herriko Barra, or from 4300–4000 cal BC according to radiocarbon dates on cereal grains (table 12.3). Arias (1999), in accordance with Price (1996, 2000) amongst others, has suggested that Atlantic European coastal regions had a long availability phase in which hunter-gatherers 'resisted' the adoption of agriculture in preference for the exploitation of wild resources. However, it is quite likely that in this case the delay is merely due to the inadequate knowledge and lack of sampling of contexts from the 5th millennium cal BC.

There may have been several different routes via which domestic plants reached the coast of the Bay of Biscay, although traditionally the Ebro basin has been considered the only one. In the Upper Ebro Valley there is evidence of very early neolithic contexts (Alday 2000) and,

Site	Date bp	cal BC (2 σ)	Material dated	Associated remains	Reference
Herriko Barra (Gipuzkoa)	5810±170 5960±95 6010±90	5190–4330 5200–4560 5210–4690	bone	cereal pollen	Iriarte et al. 2005
Lumentxa (Bizkaia)	5180±70	4220–3800	wood	*Hordeum vulgare*	Zapata 2002a
Kobaederra (Bizkaia)	5630±100 5375±90	4720–4260 4360–3990	wood *Hordeum* grain	*Hordeum vulgare* *Triticum dicoccum*	Zapata 2002a
Pico Ramos (Bizkaia)	5370±40	4330–4050	*Hordeum* grain	*Hordeum vulgare*	Zapata, in progress
El Mirón (Cantabria)	5500±90 5790±90	4520–4050 4840–4410	*T. dicoccum* grain wood	*Triticum aestivum/durum* *Triticum dicoccum* *Triticum monococcum*	Peña-Chocarro et al. 2005

Table 12.3: Neolithic sites with cereals in northern Atlantic Iberia.

from the nearby northern Meseta, archaeobotanical data suggest dates of ca. 5500 cal BC for the first agricultural settlements (López et al. 2003; Stika 1999; Zapata et al. nd). However, it could also be argued that the route was via southern France and the Pyrenees, where there are common industrial elements that show systematic trans-Pyrenean relationships (Barandiarán and Cava 2000; Cava 1990, 1994, pp.86–88); (Utrilla 2002). Research carried out in the last two decades refutes the model of late neolithisation of western France (Roussot-Larroque 1990). In Andorra and the southern slopes of the Pyrenees there is very early adoption of farming practices; cereal grains have been identified at Balma Margineda in an early neolithic context dated to 6850±160 and 6670±120 bp (Marinval 1995). There are also other pre-Pyrenean sites with indirect evidence of early farming (Rodanés and Ramón 1995; Utrilla 2002).

We think it is likely, therefore, instead of there being a unilinear movement of people and/or plants via the Ebro valley, that the introduction of domesticates to the coast of the Bay of Biscay was more complex, with southern France, the Pyrenees and even the Meseta playing key roles.

12.6.2 Crops and agrarian practices: an Atlantic specificity?

The original focus for Atlantic agriculture in Iberia is necessarily Mediterranean—from southern France and the Pyrenees, the Ebro basin or the northern Meseta—and we know that early neolithic agriculture in the western Mediterranean was extremely varied with a large variety of cereals and pulses being cultivated (Buxó 1997). However, the information we have about crops for northern Iberia is extremely limited, for example, pollen at Herriko Barra and a few cereal grains at the cave sites. Barley and emmer wheat seem to be the most frequent taxa, although free-threshing wheat has recently been identified in El Mirón (L. Peña-Chocarro, work in progress). We have previously suggested that Atlantic agriculture in Iberia might have peculiar features due to the ecological setting where it developed (that is, mountainous regions with extremely high rainfall). Given the greater climatic and edaphic tolerance of hulled wheats and barley it is possible that these crops were preferred (Zapata and Peña-Chocarro 2005). In fact this is the only region in Iberia where *Triticum dicoccum* is still grown under traditional cultivation regimes because farmers claim that nothing else responds so well to the mountainous humid conditions (Peña-Chocarro 1999; Peña-Chocarro and Zapata 1998). Obviously, this hypothesis will have to be tested in the future with more archaeobotanical information.

12.6.3 Agrarian practices: harvesting

Information on agrarian practices is practically nonexistent because archaeobotanical finds in our samples are limited to cereal grains with only a few weeds or chaff items. In contrast to Mediterranean Spain where lithic sickle elements are common (Bosch et al. 1999; González Urquijo et al. 1994), sickle elements from archaeological contexts in northern Spain are very rare and only become relatively abundant from the Chalcolithic period (Cava 1986, 1990). My colleagues and I have undertaken experimental and ethnographic research in Spain and Morocco in order to tackle this problem and we have suggested alternative methods for harvesting during the Neolithic in this region (Ibáñez et al. 2001). For example:

- Uprooting the whole plant: This is usually done when the plant is short and difficult to cut with a sickle or when the full length of straw needs to be used, for example, in handcrafts or for thatching. This method is used by the Bedouins in Jordan (Simms and Russell 1997), by farmers in Lanzarote in the Canary Islands and in the Rif (Morocco) with einkorn and naked wheat (Peña-Chocarro et al. 2000).

- Plucking the ears off by hand: This is the most time consuming technique and is used today in Asturias, northern Spain, in fields of *Triticum dicoccum* and *Triticum spelta*, because the semi-fragile rachises of the ears are easily broken.

- Plucking the ears with *mesorias* (two sticks joined with a string): This method is still in use in Asturias and it has been described in mountainous areas of Nepal, Caucasus and Spain where hulled wheat is grown (Peña-Chocarro 1999; Toffin 1983); breaking off the ears with this implement is three times slower than with sickles.

There are several factors that determine whether or not farmers use one of these alternative techniques, even if he or she is familiar with the sickle:

- The cultivation of hulled wheat: This is a prerequisite for the use of mesorias, as the semi-fragile rachis allows the ear to be easily torn from the stalk.
- Ecological conditions: In wet areas, cereals mature slowly and harvesting can take up to three weeks. Moreover, if the fields are spaced out at different heights, cereals will mature at different rates and more time consuming techniques can be used.
- Processing: When using mesorias, or plucking by hand, the ears are separated from the plant while still in the field, so the straw remains and can be cut, burnt or eaten by the animals.
- Size of the fields: Some harvesting techniques are significantly faster (i.e., by sickle) than others (i.e., by uprooting or plucking the ears by hand) and farmers tend to use the more time consuming methods only when dealing with small fields. Alternatively, the use of the sickles could indicate intensification (Simms and Russell 1997) and larger fields (Ibáñez et al. 2001).
- The use of the straw: Uprooting the plant can be done in order to use the full length of the straw for thatching or for a variety of handcrafts.
- The skill of the farmer: Old or young children, who are not skilled at using sickles or mesorias, sometimes prefer to pluck the ear off by hand.

In sum, the lack of lithic sickles on neolithic archaeological sites in Atlantic Iberia in northern Spain, plus the presence of hulled wheats and the humid and mountainous conditions, suggest that alternative harvesting techniques were being used during this period (Ibáñez et al. 2001). This also suggests that the fields being cultivated were not large.

12.7 Conclusions

The main conclusions of this work are that the indigenous populations along the coast of the Bay of Biscay adopted farming. According to pollen data, the first crops were present from at least ca. 5200–4700 cal BC and macro-remains (caryopses of wheat and barley) have only been identified in contexts dated from ca. 4700 cal BC. Thus, there is a time lag between the first agriculture in the region compared with that in the other Iberian regions (Catalonia, Valencia, northern Meseta and Andalucía), where radiocarbon dates carried out on cereals are as early as ca. 5600–5100 cal BC. This gap may, however, only be due to the lack of archaeological information for the 4th millennium cal BC in the Basque-Cantabrian region. I do not think it is a question of 'resistance' to the adoption of farming, as has been documented in other European coastal territories. In fact, the high values of domestic fauna in Arenaza, Kobaederra and El Mirón from 5100–4500 cal BC suggest that there may have been an earlier transitional phase.

Although the first crops of this region are unquestionably from the western Mediterranean—via southern France and the Pyrenees, the Ebro Valley or the northern Meseta—I cannot discount the possibility of the development of a particular Atlantic agrarian system, manifest in features such as the selection of crops more adapted to humid conditions (barley and hulled wheat) and specific harvesting techniques.

There might have been a long period during the 5th millennium cal BC when wild resources (red deer hunting, molluscs and plant gathering) played an important role in human subsistence. Although seasonality of farmed resources may have been a key factor forcing people to rely more on wild products during the spring, we should not consider subsistence strategies as the only driving force in the maintenance of hunting and gathering practices during the Neolithic. We should not ignore the social, symbolic and cultural role hunting may have played in any farming society.

In spite of intensive sampling strategies at the sites of Kobaederra, Pico Ramos and El Mirón, there are very few remains of crops in contexts from the 5th millennium. However, domestic fauna are abundant on large sites, e.g., Arenaza, Kobaederra and El Mirón. In addition to more general taphonomic problems that affect plant preservation, at the cave sites where we are working there may have been a focus on pastoral resources and this may be biasing the general picture for the region during the Neolithic period. It is very likely that we are lacking evidence from the open-air sites where agrarian practices would have been concentrated.

Many of these observations, particularly those reliant on archaeobotanical investigations, are based on very little data. I hope that the increase in archaeobotanical research in southern European regions will allow us to address these questions and to better understand neolithic farming in the near future.

Acknowledgements

The author is in receipt of a postdoctoral research grant from the Basque Government (Ref. BFI01.12) and her work is part of the Research Group of the University of the Basque Country UPV/EHU 9/UPV00155.13014570/2002. My thanks to the Institute of Archaeology (UCL) for funding my archaeobotanical training through a Cultural Exchange Fellowship supervised by Professor D. Harris and Dr. J. Hather. I will never forget Gordon Hillman's teaching and enthusiasm regarding early farming. I also thank J. Altuna, K. Mariezkurrena and N. Milner for providing unpublished data and L. Straus and M. González Morales for information from El Mirón. The work at Kobaederra was funded by the Basque Government (PU97/7). Thanks to L. Peña-Chocarro for sharing and discussing her archaeobotanical and ethnographic data. Ethnographic research in Morocco was carried out within the project *Las primeras comunidades campesinas de la Región Cantábrica. El aporte de la etnoarqueología en Marruecos*, funded by Fundación Marcelino Botín. R. Ajangiz and J. Meadows have helped in representing the calibrated dates.

References

Alday, A. (1999). Dudas, manipulaciones y certezas para el mesoneolítico vasco. *Zephyrus 52*, 129–174.

Alday, A. (2000). El neolítico en el País Vasco: pensando la marginalidad. In V. Oliveira Jorge (Ed.), *Actas do 3.° Congresso de Arqueologia Peninsular*, Volume III: Neolitização e Megalitismo da Península Ibérica, pp. 97–113. Porto: Adecap.

Alday, A. (n.d.). Temas del neolítico vasco (i): territorios y economía. In *Actas del III Congresso del Neolítico en la Península Ibérica*. Santander: Universidad de Cantabria. In Press.

Alday, A., A. Cava, and J. Mujika (1996). El IV milenio en el País Vasco: transformaciones culturales. *Rubricatum 1*(2), 745–755.

Alday, A. and J. Mujika (1999). Nuevos datos de cronología absoluta concerniente al Holoceno medio en el área vasca. In *XXIV Congreso Nacional de Arqueología*, Volume 2, pp. 95–106. Murcia: Instituto de Patrimonio Histórico.

Altuna, J. (1980). Historia de la domesticación animal en el País Vasco, desde sus orígenes hasta la romanización. *Munibe (Antropologia-Arkeologia) 32*(1-2), 7–88.

Altuna, J. and K. Mariezkurrena (2003). Restos de macromamíferos del yacimiento prehistórico del Mirón (Cantabria). Niveles con cerámica y Mesolítico. Unpublished report.

Andrés, M. T. (1990). El fenómeno dolménico en el País Vasco. *Munibe (Antropologia-Arkeologia) 42*, 141–152.

Andrés, M. T. (1997). Fases de implantación y uso dolménico en la cuenca alta y media del Ebro. In A. Rodríguez Casal (Ed.), *O Neolítico Atlántico e as Orixes do Megalitism*, pp. 431–444. Santiago de Compostela: Universidade de Santiago de Compostela.

Apellániz, J. (1971). El mesolítico de la cueva de Tarrerón y su datación por el C14. *Munibe 23*(1), 91–104.

Apellániz, J. and J. Altuna (1975a). Excavaciones en la cueva de Arenaza I (San Pedro de Galdames, Vizcaya). primera campaña, 1972. neolítico y mesolítico final. noticiario arqueológico Hispánico. *Noticiario Arqueológico Hispánico, Prehistoria 4*, 123–156.

Apellániz, J. and J. Altuna (1975b). Memoria de la iii campaña de excavaciones arqueológicas en la cueva de arenaza i (san pedro de galdames, vizcaya). *Noticiario Arqueológico Hispánico, Prehistoria 4*, 185–197.

Arias, P. (1991). Las industrias neolíticas de Kobaederra (Ereño, Bizkaia). *Munibe (Antropologia-Arkeologia) 43*, 87–103.

Arias, P. (1996). Los concheros con cerámica de la costa cantábrica y la neolitización del norte de la Península Ibérica. In A. Moure (Ed.), *"El Hombre Fósil" 80 Años Después*, pp. 391–415. Santander: Universidad de Cantabria.

Arias, P. (1997). ¿Nacimiento o consolidación? El papel del fenómeno megalítico en los procesos de neolitización de la región cantábrica. In A. Rodríguez Casal (Ed.), *O Neolítico Atlántico e as Orixes do Megalitism*, pp. 371–389. Santiago de Compostela: Universidade de Santiago de Compostela.

Arias, P. (1999). The origins of the neolithic along the Atlantic coast of continental Europe: A survey. *Journal of World Prehistory 13*(4), 403–464.

Arias, P. and J. Altuna (1999). Nuevas dataciones absolutas para el Neolítico de la cueva de Arenaza (Bizkaia). *Munibe (Antropologia-Arkeologia) 51*, 161–171.

Arias, P., J. Altuna, Á. Armendariz, J. González Urquijo, J. Ibáñez, R. Ontañón, and L. Zapata (1999). Nuevas aportaciones al conocimiento de las primeras sociedades productoras de la región Cantábrica. *Saguntum Extra 2*, 549–557. Actes del II Congrés del Neolític a la Península Ibèrica, Universitat de València, 7-9 d'Abril de 1999.

Arias, P. and R. Ontañón (1996). El Neolítico en Cantabria. ensayo de caracterización industrial. *Rubricatum 1*(2), 735–744.

Arribas, J. and E. Berganza (1984). Algunos útiles pulimentados del País Vasco. *Munibe (Antropologia-Arkeologia) 36*, 59–66.

Aseginolaza, C., D. Gómez, X. Lizaur, G. Montserrat, G. Morante, M. Salaverria, and P. Uribe-Echevarria (1989). *Vegetación de la Comunidad Autónoma del País Vasco.* Vitoria-Gasteiz: Gobierno Vasco.

Bakels, C. (2000). The neolithization of the Netherlands: two ways, one result. In A. Fairbairn (Ed.), *Plants in Neolithic Britain and Beyond*, pp. 101–106. Oxford: Oxbow Books.

Barandiarán, I. (1967). *El Paleomesolítico del Pirineo Occidental.* Zaragoza: Universidad de Zaragoza.

Barandiarán, I. and A. Cava (1992). Caracteres industriales del Epipaleolítico y Neolítico en Aragón: su referencia los yacimientos levantinos. In P. Utrilla (Ed.), *Aragón/Litoral Mediterráneo: Intercambios Culturales durante la Prehistoria*, pp. 181–196. Zaragoza: Institución Fernando El Católico.

Barandiarán, I. and A. Cava (2000). A propósito de unas fechas del Bajo Aragón: reflexiones sobre el Mesolítico y el Neolítico en la cuenca del Ebro. *SPAL 9*, 293–326.

Barandiarán, I. and A. Cava (Eds.) (2001). *Cazadores-Recolectores en el Pirineo Navarro. El Sitio de Aizpea Entre 8000 y 6000 Años Antes de Ahora.* Anejos de Veleia. Series maior 10. Vitoria-Gasteiz: Universidad del País Vasco.

Barandiarán, J. M. (1979). *El Hombre Prehistórico en el País Vasco.* San Sebastián: Ediciones Vascas.

Barbujani, G. (2002). Geographic patterns of nuclear and mitochondrial DNA variation in Europe. In *XV Congreso Estudios Vascos*, Volume 1, pp. 161–169. Donostia-San Sebastián: Eusko Ikaskuntza-Sociedad de Estudios Vascos.

Bernabeu, J. (1996). Indigenismo y migracionismo. Aspectos de la neolitización en la fachada oriental de la Península Ibérica. *Trabajos de Prehistoria 53*(2), 37–54.

Bernabeu, J. (2002). The social and symbolic context of neolithization. *Saguntum Extra 5*, 209–233.

Bosch, Á., J. Chinchilla, and J. Tarrús (1999). La Draga un poblado del Neolítico antiguo en el lago de Banyoles (Girona, Catalunya). ii congrés del Neolític a la Península Ibèrica. *Saguntum Extra 2*, 315–321.

Brinkkemper, O., W. Hogestijn, H. Peeters, D. Visser, and C. Whitton (1999). The early Neolithic site at Hoge Vaart, Almere, the Netherlands, with particular reference to non-diffusion of crop plants, and the significance of site function and sample location. *Vegetation History and Archaeobotany 8*, 79–86.

Buxó, R. (1997). *Arqueología de las Plantas. La Explotación Económica de las Semillas y los Frutos en el Marco Mediterráneo de la Península Ibérica.* Barcelona: Ed. Critica.

Calafell, F. and J. Bertranpetit (1993). The genetic history of the Iberian Peninsula: a simulation. *Current Anthropology 34*, 735–745.

Calafell, F. and J. Bertranpetit (1994). Principal component analysis of gene frequencies and the origin of Basques. *American Journal of Physical Anthropology 93*, 201–215.

Carvalho, A. (2002). Current perspectives on the transition from the Mesolithic to the Neolithic in Portugal. *Saguntum Extra 5*, 235–250.

Cava, A. (1978). El depósito arqueológico de la cueva de Marizulo (Guipúzcoa). *Munibe 4*, 55–172.

Cava, A. (1986). La industria lítica de la prehistoria reciente en la cuenca del Ebro. *Boletín del Museo de Zaragoza 5*, 5–72.

Cava, A. (1990). El Neolítico en el País Vasco. *Munibe (Antropologia-Arkeologia) 42*, 97–106.

Cava, A. (1994). El Mesolítico en la cuenca del Ebro. un estado de la cuestión. *Zephyrus XLVII*, 65–91.

Chauchat, C. (1974). Datations C14 concernant le site de Mouligna, Bidart (Pyrénées-Atlantiques). *Bulletin de la Société Préhistorique Française 71*, 140.

Comas, D., F. Calafell, and J. Bertranpetit (1996). La detección del impacto genético de la expansión del Neolítico: estado de la cuestión. *Rubricatum 1*, 557–562.

de Blas, M. (1997). Megalitos en la región cantábrica: una visión de conjunto. In A. Rodríguez Casal (Ed.), *O Neolítico Atlántico e as Orixes do Megalitism*, pp. 311–334. Santiago de Compostela: Universidade de Santiago de Compostela.

Delibes, G. and J. F. Manzano (2000). La trayectoria cultural de la Prehistoria Reciente (6400–2500 BP) en la submeseta norte Española: principales hitos de un proyecto. In V. Oliveira Jorge (Ed.), *Actas do 3.° Congresso de Arqueologia Peninsular*, Volume IV: Pré-História Recente da Península Ibérica, pp. 95–122. Porto: Adecap.

Díez Castillo, A. (1995). El asentamiento de la Peña Oviedo (Camaleño, Cantabria): la colonización de las áreas montañosas de la cornisa cantábrica. Cuadernos de Sección. *Prehistoria-Arqueología 6*, 105–120.

Elorza, M. and A. Sánchez Marco (1993). Postglacial fossil great auk and associated avian fauna from the Biscay Bay. *Munibe (Antropologia-Arkeologia) 45*, 179–185.

Fábregas, R. and J. Suárez (1999). El proceso de neolitización en Galicia. II congrés del Neolític a la Península Ibèrica. *Saguntum Extra 2*, 541–548. Actes del II Congrés del Neolític a la Península Ibèrica, Universitat de València, 7-9 d'Abril de 1999.

Fortea, J., B. Martí, M. Fumanal, M. Dupré, and M. Pérez Ripoll (1987). Epipaleolítico y Neolitización en la zona oriental de la Península Ibérica. In J. Guilaine, J. Courtin, J. Roudil, and J. Vernet (Eds.), *Premières Communautés Paysannes en Méditerranée Occidentale*, pp. 581–591. Paris: C.N.R.S.

González Morales, M. (1992). Mesolíticos y megalíticos: la evidencia arqueológica de los cambios en las formas productivas en el paso al megalítico en la Costa Cantábrica. In A. Moure (Ed.), *Elefantes, Ciervos y Ovicaprinos. Economía y aprovechamiento del Medio en la Prehistoria de España y Portugal*, pp. 185–202. Santander: Universidad de Cantabria.

González Morales, M. (1995). La transición al Holoceno en la región Cantábrica: el contraste con el modelo del Mediterráneo Español. In V. Villaverde (Ed.), *Los Últimos Cazadores. Transformaciones Culturales y Económicas Durante el Tardiglaciar y el inicio del Holoceno en el Ámbito Mediterráneo*, pp. 63–78. Alicante: Diputación de Alicante.

González Morales, M. (1996). La transición al Neolítico en la Costa Cantábrica: la evidencia arqueológica. *Rubricatum 13*(2), 879–885.

González Urquijo, J., J. Ibáñez, L. Peña, B. Gavilán, and J. Vera (1994). Cereal harvesting during the Neolithic of the Murciélagos site in Zuheros (Córdoba, Spain). *Helinium XXXIV*(2), 322–341.

Ibáñez, J., J. González Urquijo, L. Peña-Chocarro, L. Zapata, and V. Beugnier (2001). Harvesting without sickles. Neolithic examples from humid mountain areas. In S. Beyries and P. Petrequin (Eds.), *Ethno-Archaeology and its Transfers*, Number 983 in BAR International Series, pp. 23–36. Oxford: Archaeopress.

Ibáñez, J. and L. Zapata (2001). La función de los útiles de sílex del yacimiento de Pico Ramos (Muskiz, Bizkaia). *Isturitz 11*, 245–257.

Iriarte, M. (1994a). *El paisaje vegetal de la prehistoria reciente en el Alto Valle del Ebro y sus estribaciones Atlánticas. Datos polínicos. Antropización del paisaje y primeros estadios de la economía de producción.* Ph. D. thesis, Vitoria-Gasteiz: Departamento de Geografía, Prehistoria y Arqueología. University of the Basque Country.

Iriarte, M. (1994b). Estudio palinológico del nivel sepulcral del yacimiento arqueológico de Pico Ramos (Muskiz, Bizkaia). *Cuadernos de Sección de la Sociedad de Estudios Vascos/Eusko Ikaskuntza. Prehistoria-Arqueología 5*, 161–179.

Iriarte, M., J. Mujika, and A. Tarriño (2005). Herriko Barra (Zarautz-Gipuzkoa): Caractérisation industrielle et économique des premiers groupes de producteurs sur le littoral Basque. In G. Marchand and A. Tresset (Eds.), *Unité et diversité des processus de néolithisation sur la façade atlantique de l'Europe (7e-4e millénaires avant J.-C.)*, Monographie de la Société Préhistorique Française, pp. 127–136. Paris: Société Préhistorique Française.

Jarman, M. R., G. N. Bailey, and H. N. Jarman (1982). *Early European Agriculture: Its Foundations and Developments.* Cambridge: Cambridge University Press.

Juan-Cubanilles, J. (1992). La neolitización de la vertiente Mediterránea peninsular. Modelos y problemas. In P. Utrilla (Ed.), *Aragón/Litoral Mediterráneo: Intercambios Culturales Durante la Prehistoria*, pp. 255–268. Zaragoza: Institución Fernando.

Kunst, M. and M. A. Rojo (1999). El Valle de Ambrona: un ejemplo de la primera colonización Neolítica de las tierras del interior peninsular. *Saguntum Extra 2*, 503–512. Actes del II Congrés del Neolític a la Península Ibèrica, Universitat de València, 7-9 d'Abril de 1999.

Loidi, J., I. Biurrun, and M. Herrera (1997). La vegetación del centro-septentrional de España. *Itinera Geobotanica 5*, 161–168.

López, P., A. M. Arnanz, R. Macías, P. Uzquiano, and P. Gil (2003). Arqueobotánica de la Cueva de La Vaquera. In M. S. Estremera (Ed.), *Primeros Agricultores y Ganaderos en la Meseta Norte: el Neolítico de la Cueva de La Vaquera (Torreiglesias, Segovia)*, Volume 11. Memorias. Arqueología en Castilla y León, pp. 247–255. Zamora: Junta de Castilla y León.

López Quintana, J. C. (nd). Organización del territorio durante la transición al Neolítico en el Cantábrico oriental: los ejemplos de Urdaibai y Gorbeia. In P. Arias, R. Ontañón, and C. García-Moncó (Eds.), *Actas del III Congreso del Neolítico en la Península Ibérica.* Universidad de Cantabria: Santander. In Press.

Loriana, M. (1943). Las industrias paleolíticas de Berroberría. *Archivo Español de Arqueología 16*, 194–206.

Marchand, G. (2000). La néolithisation de l'ouest de la France: aires culturelles et transferts techniques dans l'industrie lithique. *Bulletin de la Société Préhistorique Française 97*(3), 377–403.

Marcos, J. (1982). *Carta arqueológica de Vizcaya, Primera parte. Yacimientos en cueva*, Volume 8 of *Cuadernos de Arqueología de Deusto*. Bilbao: Universidad de Deusto.

Mariezkurrena, K. and J. Altuna (1995). Fauna de mamíferos del yacimiento costero de herriko barra (zarautz, país vasco). *Munibe (Antropologia-Arkeologia) 47*, 23–32.

Marinval, P. (1995). Recol·lecció i agricultura de l'epipaleolític al neolític antic: anàlisi carpològica de la Balma de la Marginada. In J. Guilaine and M. Martzluff (Eds.), *Les Excavacions a la Balma de la Margineda (1979-1991)*, Andorra la Vella, Edicions del Govern d'Andorra, vol. III, pp. 65–82. Andorra: Govern d'Andorra.

Martí, B. (Ed.) (1992). *Economía y Medio Ambiente en el Neolítico del País Valenciano*. Santander: Universidad de Cantabria.

Martí, B. (1998). El Neolítico. In I. Barandiarán, B. Martí, M. del Rincón, and J. L. Maya (Eds.), *Prehistoria de la Península Ibérica*, pp. 121–191. Barcelona: Ariel.

Martí, B., J. Fortea, J. Bernabeu, M. Pérez Ripoll, J. D. Acuna, F. Robles, and M. Gallart (1987). El neolítico antiguo en la zona oriental de la Península Ibérica. In J. Guilaine, J. Courtin, J.-L. Roudil, and J.-L. Vernet (Eds.), *Premières Communautés Paysannes en Méditerranée Occidentale. Actes du Colloque International du C.N.R.S., Montpellier, 26–29 Avril 1983*, pp. 581–591. Paris: C.N.R.S.

Milner, N. (2002a). *Incremental growth of the European oyster,* Ostrea edulis. *Seasonality Information from Danish Kitchenmiddens*. BAR International Series 1057. Oxford: Archaeopress.

Milner, N. (2002b). Oysters, cockles and kitchenmiddens: changing practices at the Mesolithic/Neolithic Transition. In P. Miracle and N. Milner (Eds.), *Consuming Passions and Patterns of Consumption*, pp. 89–96. Cambridge: McDonald Institute for Archaeological Research.

Morales, M. R. G. and L. Straus (2000). La Cueva del Mirón (Ramales de La Victoria, Cantabria): excavaciones 1996-1999. *Trabajos de Prehistoria 57*(1), 121–133.

Moreno, R. (2003). El conchero de Pico Ramos. Unpublished report.

Mujika, J. A. and A. Armendáriz (1991). Excavaciones en la estación megalítica de Murumendi (Beasain, Gipuzkoa). *Munibe (Antropologia-Arkeologia) 43*, 105–165.

Nolte, E. (1962). Materiales procedentes de la cueva de Gaizkoba (Cortézubi, Vizcaya). *Anuario de Eusko Folklore 19*, 237–240.

Ontañón, R. (2000). Investigaciones arqueológicas en Montealegre (Sámano, Castro Urdiales). In R. Ontañón (Ed.), *Actuaciones Arqueológicas en Cantabria*, pp. 279–282. Gobierno de Cantabria: Santander.

Peña-Chocarro, L. (1999). *Prehistoric Agriculture in Southern Spain during the Neolithic and the Bronze Age: The Application of Ethnographic Models*. BAR International Series S818. Oxford: Archaeopress.

Peña-Chocarro, L. and L. Zapata (1998). Hulled wheats in Spain: history of minor cereals. In A. A. Jaradat (Ed.), *Triticeae III*, pp. 45–52. New Hampshire, USA: Science Publishers Inc.

Peña-Chocarro, L. and L. Zapata (2003). El cultivo del trigo en el siglo XX en la Euskal Herria atlántica: apuntes etnoarqueológicos. *Zainak 22*, 217–230.

Peña-Chocarro, L., L. Zapata, G. J. E., and J. J. Ibáñez Estévez (2000). Agricultura, alimentación y uso del combustible: aplicación de modelos etnográficos en arqueobotánica. *Saguntum Extra 3*, 403–420.

Peña-Chocarro, L., L. Zapata, M. J. Iriarte, M. González Morales, and L. G. Straus (2005). The oldest agriculture in northern atlantic Spain: new evidence from El Mirón Cave (Ramales de la Victoria, Cantabria). *Journal of Archaeological Science 32*, 579–587.

Peñalba, M. C. (1994). The history of the holocene vegetation in northern Spain from pollen analysis. *Journal of Ecology 8*(4), 815–832.

Price, T. (1996). The first farmers of southern Scandinavia. In D. R. Harris (Ed.), *The Origins and Spread of Agriculture and Pastoralism in Eurasia*, pp. 346–362. London: UCL Press.

Price, T. (2000). The introduction of farming in northern Europe. In T. Price (Ed.), *Europe's First Farmers*, pp. 260–300. Cambridge: Cambridge University Press.

Quintana, J. C. L. (2000). El yacimiento prehistórico de la cueva de Kobeaga II (Ispaster, Bizkaia): cazadores-recolectores en el País Vasco atlántico durante el VIII y VII milenio b.p. *Illunzar 4*, 83–162.

Richards, M. (2003). The neolithic invasion of Europe. *Annual Review of Anthropology 32*, 135–162.

Rodanés, J. M. and N. Ramón (1995). El Neolítico Antiguo en Aragón: hábitat y territorio. *Zephyrus XLVIII*, 101–128.

Roussot-Larroque, J. (1990). Rubané et cardial: le poids de l'ouest. In D. Cahen and M. Otte (Eds.), *Rubané et Cardial*, ERAUL 39, pp. 315–360. Liège: Université de Liège.

Ruiz Cobo, J. and P. Smith (2001). El yacimiento del Cubío Redondo (Matienzo, Ruesga): una estación mesolítca de montaña en Cantabria. *Munibe (Antropologia-Arkeologia) 53*, 31–55.

Simms, S. and K. Russell (1997). Bedouin hand harvesting of wheat and barley: implications for early cultivation in Southwestern Asia. *Current Anthropology 38*(4), 696–702.

Stika, H.-P. (1999). Erste archäobotanische Ergebnisse zu den neolithischen Ausgrabungen 1997 in Ambrona, Prov. Soria. *Madrider Mitteilungen 40*, 61–65.

Straus, L. and M. González Morales (2003). El Mirón Cave and the 14C chronology of Cantabrian Spain. *Radiocarbon 45*(1), 41–58.

Straus, L., M. González Morales, M. Fano, and M. García-Gelabert (2002). Last glacial human settlement in eastern Cantabria (northern Spain). *Journal of Archaeological Science 29*, 1403–1414.

Stuiver, M. and P. Reimer (2000). CALIB Radiocarbon Calibration. `http://radiocarbon.pa.qub.ac.uk/calib/`.

Toffin, G. (1983). Moisson aux baguettes au Népal central. *Objets et Mondes 23*(3-4), 173–176.

Utrilla, P. (2002). Epipaleolíticos y neolíticos del Valle del Ebro. *Saguntum Extra 52*, 179–208.

Yarritu, M. J. and X. Gorrotxategi (1995). El megalitismo en el Cantábrico Oriental. Investigaciones arqueológicas en las necrópolis megalíticas de Karrantza (Bizkaia), 1979-1994. La necrópolis de Ordunte (Valle de Mena, Burgos), 1991-1994. Cuadernos de Sección. *Prehistoria-Arqueología 6*, 155–198.

Zapata, L. (1995). La excavación del depósito sepulcral calcolítico de la cueva Pico Ramos (Muskiz, Bizkaia). La industria ósea y los elementos de adorno. *Munibe (Antropologia-Arkeologia) 47*, 35–90.

Zapata, L. (2002a). Charcoal analyses from Basque archaeological sites: new data to understand the presence of *Quercus ilex* in a damp environment. In S. Thiébault (Ed.), *Charcoal Analysis: Methodological Approaches, Palaeoecological Results and Wood Uses. Proceedings of the Second International Meeting of Anthracology, Paris, September 2000*, BAR International Series 1063, pp. 121–126. Oxford: Archaeopress.

Zapata, L. (2002b). *Origen de la Agricultura en el País Vasco y Transformaciones en el Paisaje: Análisis de Restos Vegetales Arqueológicos. Kobie. Anejo 4.* Bilbao: Diputación Foral de Bizkaia.

Zapata, L., J. Ibáñez, and J. González Urquijo (1997). El yacimiento de la cueva de Kobaederra (Oma, Kortezubi, Bizkaia). Resultados preliminares de las campañas de excavación 1995-97. *Munibe (Antropologia-Arkeologia) 49*, 51–63.

Zapata, L. and L. Peña-Chocarro (2005). L'agriculture néolithique de la façade Atlantique Européenne. In G. Marchand and A. Tresset (Eds.), *Unité et diversité des processus de néolithisation sur la façade atlantique de l'Europe (7e-4e millénaires avant J.-C.)*, Monographie de la Société Préhistorique Française, pp. 189–199. Paris: Société Préhistorique Française.

Zapata, L., L. Peña-Chocarro, G. Pérez-Jordá, and H.-P. Stika (2004). Early neolithic agriculture in the Iberian peninsula. *Journal of World Prehistory 18*(4), 283–325.

Zapata, L., L. Peña-Chocarro, G. Pérez-Jordá, and H. P. Stika (nd). Difusión de la agricultura en la Península Ibérica. In P. Arias, R. Ontañón, and C. García-Moncó (Eds.), *Actas del III Congreso del Neolítico de la Península Ibérica.* Santander: Servicio de Publicaciones de la Universidad de Cantabria. In Press.

Zilhao, J. (2000). From the Mesolithic to the Neolithic in the Iberian Peninsula. In T. Price (Ed.), *Europe's First Farmers*, pp. 144–182. Cambridge: Cambridge University Press.

Zilhao, J. (2001). Radiocarbon evidence for maritime pioneer colonization at the origins of farming in west Mediterranean Europe. *Proceedings of the National Academy of Science 98*(24), 14180–14185.

Chapter 13

Early agriculture and subsistence in Austria: a review of neolithic plant records

Marianne Kohler-Schneider
Universität für Bodenkultur, Vienna, Austria

13.1 Introduction and methodological remarks

This paper is a review of archaeobotanical records from the Neolithic in Austria. The focus is clearly on eastern Austria, where most neolithic sites are located, where the diversity of neolithic cultures is greatest and where most archaeological and some archaeobotanical work has been done (figure 13.1). Research in the western, alpine parts of the country has recently been stimulated by the discovery of the 'ice-man'. This obviously represents an extraordinary find, and the numerous publications of Bortenschlager, Oeggl and colleagues should be referred to for details of the palaeobotanical studies carried out on the body and its surroundings (Bortenschlager et al. 1992; Bortenschlager 2000; Dickson et al. 1996, 2000; Oeggl and Schoch 1995, 2000; Oeggl 2000; Oeggl et al. 2000; Pfeifer and Oeggl 2000). Apart from the ice-man, little archaeobotanical work has been done on neolithic sites in the Austrian alps (however, see Walde 1999), but this does not necessarily imply that this region was unimportant in the early development of agriculture. It must be stressed that the recent findings of Erny-Rodmann et al. (1997) concerning possible 'agricultural' activities in the late Mesolithic of Switzerland have their Austrian counterpart. From two peat bogs in northern Tyrol (Katzenloch near Seefeld and Kirchbichl near Wörgl) Wahlmüller (1985) has presented surprisingly early evidence of human impact on the vegetation; most remarkable are the pollen records of Cerealia and *Centaurea cyanus* (the latter only in Kirchbichl) dating from 7410–7100 bp and 8100 bp respectively. Much remains to be discovered in this region, most importantly, the settlements of these early 'forager-farmers'. There is perhaps greater likelihood of finding the sites in the alpine regions, where the cultural landscape has undergone fewer transformations than in the lowlands.

In eastern Austria there is a great discrepancy between the intensity of archaeological research on neolithic sites and the amount of archaeobotanical work that has been undertaken. This is mainly due to the fact that after a promising start in the 1930s, when E. Hofmann and H. Werneck were local pioneers of archaeobotany, there was a hiatus of more than 50 years before research was resumed on a larger scale again. As neolithic sites are not the most prolific in terms of preserved plant material, they do not figure prominently in recent work. For climatic and pedological reasons, most excavations in eastern Austria are on dry soils, where there is a

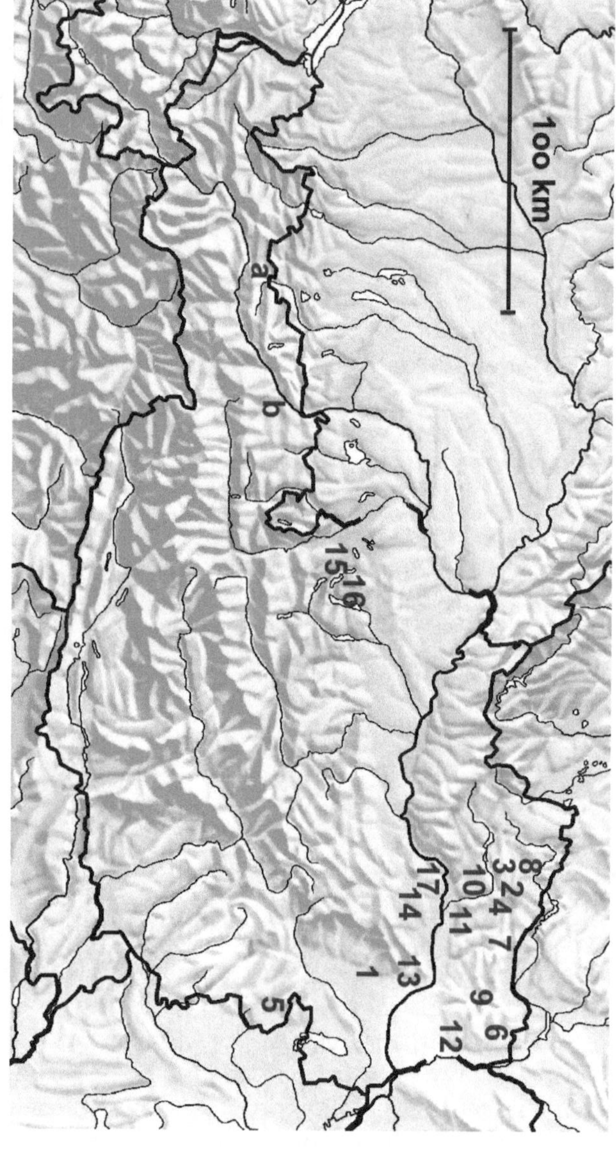

Figure 13.1: Neolithic sites in Austria where archaeobotanical work has been undertaken. Early Neolithic: 1 Brunn/Wolfholz, 2 Frauenhofen, 3 Strögen, 4 Rosenburg, 5 Neckenmarkt, 6 Herrnbaumgarten, 7 Pulkau, 8 Poigen, 9 Schletz/Asparn. Middle Neolithic: 2 Frauenhofen, 10 Kamegg, 11 Strass. Late Neolithic: 12 Grub an der March, 13 Wien/Gumpendorfer Straße, 14 Ossarn, 15 Mondsee (See, Scharfling, Mooswinkl), 16 Attersee (Miesling II, Weyregg), 17 Kleiner Anzingerberg. Off-site pollen records with early human impact: a Katzenloch, b Kirchbichl.

reduced likelihood of good preservation of botanical macro-remains on the oldest sites. Neolithic plant material from waterlogged contexts has only been recovered from the lakeshore settlements of the Salzkammergut region.

The following review of published and unpublished material is ordered chronologically, according to the sequence of the different regional cultures. The results of archaeobotanical investigations are presented after a brief summary of the distribution and dating of each culture. If not otherwise stated, all radiocarbon dates are calibrated. Older publications (pre-1950) are only quoted in a few instances and when they constitute the only source of information or are of particular relevance to the discussions.

13.2 Results

13.2.1 Early Neolithic

Early neolithic settlements are entirely confined to the lowland and hill landscapes of eastern Austria (i.e., Vienna Basin, Weinviertel loess hills, eastern Waldviertel, northern and central Burgenland, for distribution maps see Lenneis 1995). The oldest known settlement (Brunn/Wolfholz, Lower Austria) dates from the middle of the 6th millennium cal BC (Stadler 1995). Early neolithic sites are classified as belonging either to the Vornotenkopfkeramik, representing the early Linearbandkeramik, or to the Notenkopfkeramik, representing a somewhat later stage.

Earlier Linearbandkeramik (*Vornotenkopfkeramik*)

There are more than 50 known sites in Austria with dates ranging from 5480 to 4960 cal BC (Lenneis 1995; Lenneis et al. 1996). Archaeobotanical work has been carried out on five sites: Frauenhofen/Neue Breiten (Bakels 1986), Strögen, Neckenmarkt, Rosenburg (Kreuz 1990; for Rosenburg also Brinkkemper nd) and Brunn/Wolfholz (Kohler-Schneider unpub.). At the latter named site, the analysis of several hundred soil samples (a total volume of more than 40,000 litres) has yielded practically no plant remains except wood charcoal. On the other sites the preservation of charred plant material was slightly better and the following taxa were identified: cultivated plants were represented by emmer (*Triticum dicoccum*), einkorn (*Triticum monococcum*), pea (*Pisum sativum*) and lentil (*Lens culinaris*); while barley (*Hordeum sativum*), poppy (*Papaver somniferum*) and linseed (*Linum usitatissimum*) were lacking from the records. The absence of both linseed and barley can be accounted for; the former because of its poor preservation in charred form, and the latter because of its possible status at this time as a rare weed in cereal fields (Kreuz, this volume). The absence of poppy is considered to be a characteristic feature of early Linearbandkeramik sites (Jacomet and Kreuz 1999). Finds of wild plants include the usual array of weeds and ruderals found on sites of this period (*Chenopodium album, Fallopia convolvulus, Veronica arvensis,* and members of the Solanaceae family) and a number of fruit-bearing shrubs (*Cornus sanguinea, Corylus avellana, Prunus spinosa,* cf. *Rosa* sp., and *Sambucus* sp.). The most remarkable wild plant find is charred hornbeam-nuts (*Carpinus betulus*) from Rosenburg (Kreuz 1990; Brinkkemper nd). This confirms the findings of Peschke (1972), who stated that the hornbeam, which is well documented for its late return from the glacial forest refugia, had reached northeastern Austria at quite an early date.

Later Linearbandkeramik (*Notenkopfkeramik*)

More than 280 sites are known from Austria with dates ranging from 5280 to 4960 cal BC (Lenneis et al. 1996). During this phase the regions that were settled remained essentially the same as before (with some minor extensions into the pre-alpine areas of Upper Austria), however, the amount of cultivated land within these regions increased substantially. Despite the much higher number of known sites, archaeobotanical investigations have been undertaken in only five locations. In four of these: Herrnbaumgarten (Hopf in Felgenhauer 1965), Poigen (Hopf 1977), Pulkau (Hopf 1980) and Frauenhofen/Milchtaschen (Bakels 1984), the preserved plant material was very limited, consisting mostly of imprints in daub and ceramics, and a few charred plant remains. The imprints show that wheat chaff from both einkorn and emmer was used with fine culm fragments of wild grasses for tempering daub and ceramics. An additional and very remarkable find of two charred halves of wild apple (*Malus sylvestris*) including pips has been made in Poigen (Hopf 1977). The apple might have been cut in two for drying, a procedure that is well documented from Swiss lakeshore settlements (Jacomet et al. 1989).

A representative number of plant specimens could only be obtained from Schletz/Asparn (Schneider 1994; Kohler-Schneider nd), where more than 30,000 litres of soil have been floated. Schletz/ Asparn is an outstanding archaeological site in the Weinviertel loess-hills, dating predominantly from the later Linearbandkeramik (Windl 1997). Here, botanical macro-remains were recovered from long houses, ovens and various pits, and from special structures such as the oval ring ditch enclosing the settlement. Of particular interest were the contents of a well (over 7.5-m deep) that was constructed with a wooden frame. AMS ^{14}C dating of two cereal grain samples from its basal layers gave dates of 5264–4945 cal BC and 5266–4983 cal BC, thus confirming the assessments made on the basis of the ceramics. So far, only preliminary archaeobotanical results are available for Schletz/Asparn (Schneider 1994; Kohler-Schneider nd); according to these, the full neolithic spectrum of cultivated plants was present at the site. Einkorn and emmer were the dominant cereals, einkorn being mostly represented by the two-grained form. This corresponds to the findings of Kreuz and Boenke (2002), who concluded that two-grained einkorn gained importance in Central Europe only from the late Linearbandkeramik onwards. Barley was also present but in lower numbers overall. Naked wheat (*Triticum aestivum/durum/turgidum*, some *compactum* types) was found in a few samples. Of the pulses, pea and lentil were recorded, and the oil plants included linseed (*Linum usitatissimum*) and possibly gold of pleasure (cf. *Camelina sativa*), as well as poppy (*Papaver somniferum* ssp. *setigerum*) (figure 13.2). The latter was recovered among potsherds dating from the latest phase of the Notenkopfkeramik culture and represents by far the oldest record of poppy from Austria. The provisional list of wild plants includes mostly weeds, e.g., *Chenopodium album, Chenopodium hybridum, Fallopia convolvulus, Polygonum aviculare, Bromus secalinus,* wild Paniceae, and also a few possibly gathered plants, e.g., *Cornus mas*, *Sambucus nigra*, *Sambucus ebulus*, *Solanum dulcamara* and *Physalis alkekengi.* Some taxa, at present only assigned to family, await further identification (Poaceae, Cyperaceae, Rubiaceae, Brassicaceae, Lamiaceae, Fabaceae, and Caryophyllaceae).

13.2.2 Middle Neolithic

The so called *Stichbandkeramik* marks the transition from the early to middle neolithic. Sites of this culture play only a marginal role on Austrian territory and are closely associated with the Lengyel culture, to which the majority of Austrian middle neolithic sites belong. The Lengyel culture is also known as *Bemaltkeramik* and these sites were distributed from western Hungary across southwestern Slovakia to eastern Austria and Moravia (Neugebauer-Maresch 1995). The Austrian sites belong to the distinctive Moravian-East Austrian Group (*Mährisch-Ostösterreichische Gruppe*, abbreviated to MOG). Impressive circular ring-ditches constitute a peculiar feature of many MOG sites. They have often been compared to the 'henge' monuments of Western Europe although their function is still unclear (Trnka 1991).

Stichbandkeramik

The 20 known Austrian sites date from 4910–4600 cal BC (Stadler 1995; Lenneis 1995) and are mostly located in Lower Austria, north of the Danube (with outliers in Salzburg and Upper

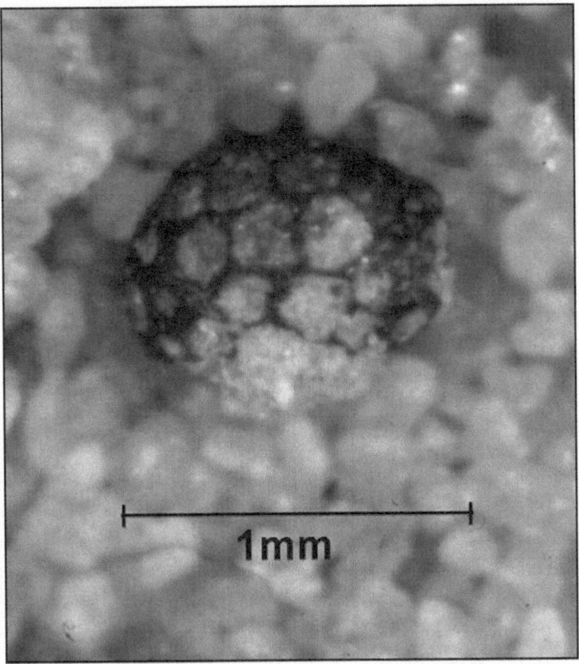

Figure 13.2: Charred seed of poppy (*Papaver somniferum* ssp. *setigerum*) from the Notenkopfkeramik phase of the Linearbandkeramik settlement Schletz/Asparn.

Austria). Only the site of Frauenhofen/Neue Breiten has been the focus of archaeobotanical study (Bakels 1986). Here, einkorn and emmer chaff was identified in imprints in daub and ceramics, and from charred specimens. According to Bakels, the paucity of finds is due to the fact that the archaeologists took only a few small soil samples.

Lengyel, MOG

More than 300 sites are known in Austria, dating from 4900–4300 cal BC (Stadler 1995). So far, only two sites have been investigated archaeobotanically: Kamegg (Link 2004; Kohler-Schneider and Link nd) and Strass (Kohler-Schneider unpub.). At Kamegg, which belongs to the earlier phases of the Lengyel culture (MOG Ia and Ib), 7,000 litres of soil were sampled from 49 pits and the double ring ditch-construction, and yielded approximately 4,000 identifiable plant remains. Of the cultivated plants einkorn and emmer occurred in relatively high numbers, at high frequencies and in roughly equal proportions (both grain and chaff remains); they were obviously the main cereals of the settlement. By contrast, there were only traces of barley (*Hordeum sativum*) and naked wheat (*Triticum aestivum/durum/turgidum*) remains. While this was to be expected for naked wheat, the virtual absence of barley was surprising. It seems that barley was not regularly cultivated in Kamegg but might have occurred as a 'weed' among the glume wheats. This is also almost certainly the case for common and Italian millet (*Panicum miliaceum* and *Setaria italica*), which have been found at Kamegg. The millet records are so far the oldest from Austria and are evidence of the early spread of these species into Central Europe. The cultivation of pea, lentil and linseed was considered likely, although they were preserved only in small numbers. While the spectrum of cultivated plants from Kamegg might be considered as being qualitatively complete, it gives a rather imbalanced impression in terms of the quantities of different taxa, so that it appears the settlement had to rely to a large extent on its einkorn and emmer crop. It is striking that the poor record of cultivated plants is matched by a comparatively rich array of edible wild plant species, including *Cornus mas, Cornus sanguinea, Crataegus monogyna, Fragaria* sp., cf. *Pyrus pyraster, Physalis alkekengi, Sambucus nigra, Sambucus ebulus* and *Corylus avellana.* As archaeozoological analyses have independently provided a similar result, that is, an unusually diverse assemblage of bones of wild mammals and birds, it seems likely that the inhabitants of Kamegg had to depend on wild plants and animals to overcome temporary difficulties with their crop yields (Link 2004; Schmitzberger nd). Other wild plants from Kamegg include mainly weeds and ruderals: *Centaurea cyanus, Thlaspi arvense, Setaria viridis/verticillata, Echinochloa crus-galli, Lapsana communis, Urtica dioica, Chenopodium album, Chenopodium hybridum, Bromus secalinus, Fallopia convolvulus, Galium aparine* and *Verbascum* sp..

The site of Strass belongs to a later phase of the Lengyel culture (MOG IIa). Unpublished archaeobotanical records from four pits include numerous charred chaff remains and grains of einkorn and emmer, as well as single specimens of pea. One pit contained numerous linseeds and a cluster of charred *Stipa* awns (Kohler-Schneider unpub.).

13.2.3 Late Neolithic

The late Neolithic in Austria lasts about 2,000 years, from 4300–2300 cal BC (Ruttkay 1995). During this long time span there is a succession of several cultures and some of them are no longer confined to the eastern parts of the country, thus leading to a stronger regional differentiation. In eastern Austria we find the Epilengyel and the Baalberg cultures during the early late Neolithic and the Baden culture during the later phase. In western Austria, the Mondsee group is contemporary with the Baden culture. In eastern Austria, with the onset of the final Neolithic, the Baden culture is succeeded by the Jevišovice culture (Mödling/Zöbing), which is roughly contemporary with the Cham culture in the west. The widely distributed Schnurkeramik and Glockenbecher cultures that follow belong to the transition between the Neolithic and Bronze Age. Archaeobotanical investigations have been carried out on sites belonging to the Mondsee group, the Baden and Jevišovice cultures, thus covering mostly the later parts of the late Neolithic.

Baden

Sites of the Baden culture in Austria date from 3600–2900 cal BC (Ruttkay 1995). There is an earlier phase (ca. 3500 cal BC), known as the Boleráz group, with sites located mainly in Lower Austria south of the Danube, in northern Burgenland and in the valley of the river March. Archaeobotanical records come from two sites, Grub an der March (Kohler-Schneider nd) and Wien/Gumpendorferstrasse (Hofmann 1930). In the multiperiod site of Grub, three pits ascribed to the Boleráz group had remains of einkorn, emmer, barley, pea/lentil and, most remarkably, some well preserved chaff remains of spelt (*Triticum spelta*). Additionally, there were a few scattered finds of rye (*Secale cereale*) and grains of common millet (*Panicum miliaceum*). Worthy of note are the awns of *Stipa* sp. and a stone of *Prunus spinosa*, besides the more common finds of *Sambucus ebulus*, *Chenopodium* sp., *Polygonum* sp. and *Galium* sp.. The plant material from Wien/Gumpendorferstrasse consisted of the imprints in daub of barley chaff and some charred barley grains (Hofmann 1930).

The Ossarn group represents a later phase of the Baden culture, with sites located in Lower Austria south of the Danube and in northern and central Burgenland. There is only one old archaeobotanical record, which is mentioned here because of its reputation among Austrian archaeologists, the famous 'Grain Offering of Ossarn'. There is somewhat of a mystery surrounding this 'offering', because the excavator J. Bayer repeatedly wrote about a deposit of '30 kg of wheat', while the archaeobotanist Werneck (who made the first identifications) soberly stated that he only received 25-30 g of grains, which came from a pot containing 30 kg of soil

(Werneck 1949). Werneck identified the grains as einkorn and emmer but a recent reexamination of the material has shown that it consists almost exclusively of one-grained einkorn (with a few two-grained specimens) and two barley caryopses. Among the cereals there were single seeds of *Centaurea cyanus*, cf. *Agrostemma githago*, *Fallopia convolvulus* and wild Poaceae (Kohler-Schneider and Caneppele unpubl.). The grains from Ossarn have been dated to 3350–3100 cal BC (Stadler 1995).

Mondsee

The neolithic lakeshore settlements in the Salzkammergut region of Upper Austria and Salzburg are ascribed to a distinctive local group, the Mondsee group, with sites dating from 3700 to 3300 cal BC (Ruttkay 1995). Lakeshore settlements are known from the Mondsee (three sites: See, Scharfling and Mooswinkl) and Attersee (18 sites: e.g., Seewalchen, Weyregg, Misling and Kammer; Offenberger 1981). Archaeobotanical investigations have been undertaken at the sites of See (Hofmann 1924; Schmidt 1986; Pawlik 1993), Scharfling, Mooswinkl, Misling II and Weyregg (Kral 1976). During the 19th and early 20th centuries plant material from the Mondsee and Attersee lakeshore settlements was freely distributed to a number of private and public collections, and subsequently Werneck (1949) compiled all available information on this widely disseminated material.

The lakeshore finds are extremely rich both in number and diversity. In the waterlogged conditions not only have charred diaspores been preserved but also several other types of botanical macro-remains, as well as various organic artefacts (e.g., ropes, nets, wooden tools, textiles). The list of recorded plant species is thus quite long; Schmidt (1986) mentions 110 taxa from the site of See alone. Therefore only the cultivated plants and some of the more interesting wild species will be described here (the reader is referred to the above mentioned publications for a full account). It must be stressed that quantitative analyses have so far not played a major role in the work on the Austrian lakeshore material.

Records of cultivated plants include emmer and barley (both very numerous), einkorn, linseed, poppy, pea and *Brassica* sp.(whether cultivated or wild is uncertain). A recent reinspection of the charred plant material by S. Jacomet (IPNA, Basel University) and the author has failed to confirm the records of hexaploid naked wheat (*Triticum vulgare*, compactum-type), that were reported by Hoffman (1927), Werneck (1949) and Schachl in Schmidt (1986). Instead, a wheat type was found, that probably represents tetraploid naked wheat and closely resembling specimens (Type B) from the lakeshore settlement Hornstaad Hörnle IA at Bodensee (Maier 2001, table 5). However, the definitive identity of this slightly aberrant wheat type remains to be established.

Of the wild plants, hazelnuts (*Corylus avellana*) were recorded in large quantities; as the excavator remarked, 'one could fill baskets with it' (Much 1872, 1874, 1876). Records of crabapple (*Malus sylvestris*) are almost as numerous, either as entire specimens or as halves, both showing signs of having been dried. In some instances apple slices were obviously arranged on a thread after the core had been removed (Hofmann 1924). The records of yew (*Taxus baccata*), dogwood (*Cornus sanguinea*), acorns (*Quercus robur*), beechnuts (*Fagus sylvatica*), rosehips (*Rosa canina*), raspberries (*Rubus idaeus*), blackberries (*Rubus fruticosus*), strawberries (*Fragaria vesca*) and *Vaccinium* sp. show that a wide range of wild fruits was collected. Edible roots are represented by *Daucus carota* and *Pastinaca sativa*. Several of the recorded herbs may have been used for medicinal purposes, e.g., *Agrimonia* sp., *Pulmonaria* sp. and *Symphytum* sp. (Schmidt 1986).

Jevišovice (*Mödling-Zöbing*)

More than 40 sites of this culture are known from Austria, with dates ranging from 3100 to 2800 cal BC (Krenn-Leeb 2002). Their distribution is rather localised and is focussed on the Dunkelsteiner Wald, the valley of the river Kamp and the Vienna Basin (all in Lower Austria). Jevišovice settlements beyond the Austrian border are found in southwestern and central Moravia. So far only one Austrian site, the settlement of Kleiner Anzingerberg, has been investigated archaeobotanically (Caneppele 2003; Caneppele and Kohler-Schneider 2003; Kohler-Schneider 2003). Here, the flotation of 2,500 litres of soil from the undisturbed floor layer of a burnt house yielded approximately 12,000 charred macro-remains. AMS ^{14}C dating gives an age of 2930–2880 cal BC. The bulk of the plant material was made up of cereals, comprising mostly barley and einkorn. Emmer and, quite unexpectedly, common millet (*Panicum miliaceum*) were also present in considerable quantities. The millet was found almost exclusively among the remnants of the roof construction, where it may have been stored in bundles before the house was destroyed by fire. Judging from the amount of grains, common millet must have had the status of a crop plant; the find is remarkable because prior to this it was assumed that millet cultivation did not reach Central Europe before the Bronze Age (with the exception of Poland: Wasylikowa et al. 1991). The record from Kleiner Anzingerberg thus provides evidence of its westward spread at a remarkably early date. Italian millet (*Setaria italica*) has also been found, but only in modest quantities. It is likely that this species wasn't cultivated yet at Kleiner Anzigerberg, but it may have been a weed in common millet fields. Other finds of cultivated plants include poppy, linseed, pea and lentil. There was also a surprisingly large number of wild plant species typical of semi-natural dry grassland communities (e.g., *Asperula cynanchica, Teucrium chamaedrys*, cf. S*tipa* sp. *Plantago media*). It seems that dry (and perhaps grazed) grasslands featured prominently in the landscape surrounding the settlement. As in most neolithic settlements, the presence of fruit bearing plants indicates their importance as supplementary food resources, for example, *Prunus spinosa*, *Cornus mas*, *Rubus idaeus*, *Rubus caesius*, *Physalis alkekengi*, *Sambucus nigra*, *Sambucus ebulus*, *Viburnum opulus* and *Fragaria/Potentilla*.

Acknowledgements

I am grateful to Elisabeth Ruttkay, Eva Lenneis, Alexandra Krenn-Leeb, Walpurga Antl-Weiser, Hans Reschreiter, Klaus Oeggl, Helmut Windl and Peter Stadler who have provided valuable archaeological background information and kindly gave me access to unpublished data. Peter Stadler has also to be acknowledged for calibrating some ^{14}C data. Johannes Sorz kindly produced the distribution map.

References

Bakels, C. (1984). Abdrücke von Pflanzenresten aus der Siedlung von Frauenhofen, Ried Milchtaschen. *Archaeologia Austriaca 68*, 47.

Bakels, C. (1986). Pflanzenreste aus der Siedlung Frauenhofen, Neue Breiten. *Archaeologia Austriaca 70*, 176.

Bortenschlager, S. (2000). The iceman's environment. In S. Bortenschlager and K. Oeggl (Eds.), *The Man in the Ice. Volume 4: The Iceman and his Natural Environment: Palaeobotanical Results*, pp. 11–24. New York: Springer Wien.

Bortenschlager, S., W. Kofler, K. Oeggl, and W. Schoch (1992). Erste Ergebnisse der Auswertung der vegetabilischen Reste vom Hauslabjochfund. In *Der Mann im Eis, Band 1*, pp. 307–312. Innsbruck: Veröffentlichungen der Universität Innsbruck 187.

Brinkkemper, O. (n.d.). Die pflanzlichen Großreste von Rosenburg, einer Siedlung der ältesten Bandkeramik in Niederösterreich. In Prep.

Caneppele, A. (2003). Verkohlte Pflanzenreste aus einer jungsteinzeitlichen Hütte der Jevišovice-Kultur vom Kleinen Anzingerberg (Dunkelsteiner Wald, Niederösterreich). Unpublished diploma work, Institute for Botany, University of Natural Resources and Applied Life Sciences (BOKU), Vienna.

Caneppele, A. and M. Kohler-Schneider (2003). Landwirtschaft im Endneolithikum. Archäobotanische Untersuchungen am Kleinen Anzingerberg, Niederösterreich. *Archäologie Österreichs 14*(2), 53–58.

Dickson, J. H., S. Bortenschlager, K. Oeggl, R. Porley, and A. McMullen (1996). Mosses and the Tyrolean iceman's provenance. *Proceedings of the Royal Society of London 263*, 567–571.

Dickson, J. H., K. Oeggl, T. G. Holden, L. L. Handley, T. C. O'Connell, and T. Preston (2000). The omnivorous Tyrolean Iceman: colon contents (meat, cereals, pollen, moss and whipworm) and stable isotope analyses. *Philosophical Transactions of the Royal Society of London Biological Science 355*(1404), 1843–1849.

Erny-Rodmann, C., E. Gross-Klee, J. N. Haas, S. Jacomet, and H. Zoller (1997). Früher 'human impact' und Ackerbau im Übergangsbereich Spätmesolithikum-Frühneolithikum im schweizerischen Mittelland. *Jahrbuch der Schweizerischen Gesellschaft für Ur- und Frühgeschichte 80*, 27–56.

Felgenhauer, F. (Ed.) (1965). *Ein 'Tonaltar' der Notenkopfkeramik aus Herrnbaumgarten, p. B. Mistelbach, Niederösterreich.* Archaeologia Austriaca 38.

Hoffman, E. (1927). Die pflanzlichen Reste aus der Station See. In L. Franz and J. Weninger (Eds.), *Die Funde aus den prähistorischen Pfahlbauten im Mondsee*, pp. 87–97. Materialien zur Urgeschichte Österreichs 3.

Hofmann, E. (1924). Pflanzenreste der Mondseer Pfahlbauten. *Sitzungsberichte der Akademie der Wissenschaften in Wien (Mathematisch-Naturwissenschaftliche Klasse, Abteilung I) 133*, 379–409.

Hofmann, E. (1930). Pflanzenreste aus einer neolithischen Wohngrube, Wien 6. , Gumpendorfer Straße. *Wiener Prähistorische Zeitschrift 17*, 119.

Hopf, M. (1977). Pflanzenreste aus der linearbandkeramischen Siedlung Poigen, Ger.-Bez. Horn, Niederösterreich. In E. Lenneis (Ed.), *Siedlungsfunde aus Poigen und Frauenhofen bei Horn*, pp. 97–99. Prähistorische Forschungen 8.

Hopf, M. (1980). Getreideabdrücke in Hüttenlehm von Pulkau. *Archaeologia Austriaca 64*, 108.

Jacomet, S., C. Brombacher, and M. Dick (1989). *Archäobotanik am Zürichsee. Ackerbau, Sammelwirtschaft und Umwelt von neolithischen und bronzezeitlichen Seeufersiedlungen im Raum Zürich.* Zürcher Denkmalpflege, Monographien 7. Zürich: Verlag Orell Füssli.

Jacomet, S. and A. Kreuz (1999). *Archäobotanik. Aufgaben, Methoden und Ergebnisse vegetations—und agrargeschichtlicher Forschung.* UTB für Wissenschaft 8158. Stuttgart: Verlag Eugen Ulmer. With contributions by M. Rösch.

Kohler-Schneider, M. (2003). Klima und Vegetation während des Endneolithikums im Raum Dunkelsteiner Wald – Östliches Alpenvorland. *Archäologie Österreichs 14*(2), 49–52.

Kohler-Schneider, M. (n.d.). Neolithic agriculture in the central Weinviertel loess hills — archaeobotanical investigation of the Linearbandkerammik settlement Schletz/Asparn, Lower Austria. In Prep.

Kohler-Schneider, M. and B. Link (n.d.). Agriculture and subsistence in the middle neolithic settlement of Kamegg, Lower Austria. In Prep.

Kral, F. (1976). Erste Ergebnisse palynologischer und karpologischer Untersuchungen von Proben aus den Pfahlbausiedlungen im Mondsee und Attersee. In H. Mitscha-Märheim, H. Friesinger, and H. Kerchler (Eds.), *Festschrift für Richard Pittioni zum siebzigsten Geburtstag*, Archaeologia Austriaca, Beiheft 13, pp. 277–278. Horn: Franz Deuticke Wien, Ferdinand Berger und Söhne.

Krenn-Leeb, A. (2002). Der Kleine Anzingerberg—ein spannendes Forschungsprojekt. In *Wölbling einst und jetzt*, pp. 61–102. Wölbling: Marktgemeinde Wölbling.

Kreuz, A. (1990). *Die ersten Bauern Mitteleuropas: Eine archäobotanische Untersuchung zu Umwelt und Landwirtschaft der Ältesten Bandkeramik.* Analecta Praehistorica Leidensia 23. Leiden: Leiden University Press.

Kreuz, A. and N. Boenke (2002). The presence of two-grained einkorn at the time of the Bandkeramik culture. *Vegetation History and Archaeobotany 11*, 233–240.

Lenneis, E. (1995). Altneolithikum: Die Bandkeramik. In J. W. Neugebauer (Ed.), *Jungsteinzeit im Osten Österreichs*, Volume 17 of *Forschungsberichte zur Ur- und Frühgeschichte*, pp. 11–56. St. Pölten, Wien: Niederösterreichisches Pressehaus.

Lenneis, E., P. Stadler, and H. Windl (1996). Neue 14C-Daten zum Frühneolithikum in Österreich. *Préhistoire Européenne 8*, 97–116.

Link, B. (2004). Archäobotanische Untersuchung der mittelneolithischen Kreisgrabenanlage Kamegg, Niederösterreich. Unpublished diploma work, Institute for Botany, University of Natural Resources and Applied Life Sciences (BOKU), Vienna.

Maier, U. (2001). Archäobotanische Untersuchungen in der neolithischen Ufersiedlung Hornstaad-Hörnle IA am Bodensee. In U. Maier and R. Vogt (Eds.), *Siedlungsarchäologie im Alpenvorland VI. Botanische und pedologische Untersuchungen zur Ufersiedlung Hornstaad-Hörnle IA*, Forschungen und Berichte zur Vor- und Frühgeschichte in Baden-Württemberg 74, pp. 9–384. Stuttgart: Konrad Theiss Verlag.

Much, M. (1872). Bericht über die Auffindung eines Pfahlbaues im Mondsee. *Mitteilungen der Anthropologischen Gesellschaft Wien 2*, 203.

Much, M. (1874). Bericht über Pfahlbauforschungen in den oberösterreichischen Seen. *Mitteilungen der Anthropologischen Gesellschaft Wien 4*, 293.

Much, M. (1876). Bericht. *Mitteilungen der Anthropologischen Gesellschaft Wien 6*, 161.

Neugebauer-Maresch, C. (1995). Mittelneolithikum: Die Bemaltkeramik. In J. W. Neugebauer (Ed.), *Jungsteinzeit im Osten Österreichs*, pp. 57–107. St. Pölten, Wien: Verlag Niederösterreichisches Pressehaus.

Oeggl, K. (2000). The diet of the iceman. In S. Bortenschlager and K. Oeggl (Eds.), *The Man in the Ice. Volume 4: The Iceman and his Natural Environment: Palaeobotanical Results*, pp. 89–116. New York: Springer Wien.

Oeggl, K., J. H. Dickson, and S. Bortenschlager (2000). Epilogue: The search for explanations and future developments. In S. Bortenschlager and K. Oeggl (Eds.), *The Man in the Ice. Volume 4: The Iceman and his Natural Environment: Palaeobotanical results*, pp. 163–166. New York: Springer Wien.

Oeggl, K. and W. Schoch (1995). Neolithic plant remains discovered together with a mummified corps ("Homo tyrolensis") in the Tyrolean Alps. In H. Kroll and R. Pasternak (Eds.), *Res Archaeobotanicae: 9th Symposium IWGP*, pp. 229–238. Oetker-Voges Verlag.

Oeggl, K. and W. Schoch (2000). Dendrological analyses of artefacts and other remains. In S. Bortenschlager and K. Oeggl (Eds.), *The Man in the Ice. Volume 4: The Iceman and his Natural Environment: Palaeobotanical Results*, pp. 29–62. New York: Springer Wien.

Offenberger, J. (1981). *Die 'Pfahlbauten' der Salzkammergutseen. Das Mondseeland Geshichte und Kultur.* Linz: Ausstellungskatalog des Heimatmuseums Mondsee.

Pawlik, B. (1993). Die botanische Untersuchung der jungneolithischen Feuchtbodensiedlung Station See am Mondsee, Oberösterreich. Unpublished FWF-project report for the Natural History Museum, Vienna (project leader E. Ruttkay).

Peschke, P. (1972). Pollenanalytische Untersuchugen im Waldviertel Niederösterreichs. *Flora 161*, 256–284.

Pfeifer, K. and K. Oeggl (2000). Analysis of the bast used by the iceman as binding material. In S. Bortenschlager and K. Oeggl (Eds.), *The Man in the Ice. Volume 4: The Iceman and his Natural Environment: Palaeobotanical results*, pp. 69–76. New York: Springer Wien.

Ruttkay, E. (1995). Spätneolithikum. In J. W. Neugebauer (Ed.), *Jungsteinzeit im Osten Österreichs*, Forschungsberichte zur Ur- und Frühgeschichte 17, pp. 108–209. St. Pölten and Wien: Verlag Niederösterreichisches Pressehaus.

Schmidt, R. (1986). Palynologie, Stratigraphie und Großreste von Profilen der neolithischen Station See am Mondsee, Oberösterreich. *Archaeologia Austriaca 70*, 227–235.

Schmitzberger, M. (n.d.). Die Tierknochen aus der mittelneolithischen Kreisgrabenanlage und Siedlung von Kamegg, Niederösterreich. In Prep.

Schneider, M. (1994). Verkohlte Pflanzenreste aus einem neolithischen Brunnen in Schletz, Niederösterreich. *Archäologie Österreichs 5*, 18–22.

Stadler, P. (1995). Ein Beitrag zur Absolutchronologie des Neolithikums in Ostösterreich aufgrund der 14C-Daten. In J. W. Neugebauer (Ed.), *Jungsteinzeit im Osten Österreichs*, Volume 17 of *Forschungsberichte zur Ur- und Frühgeschichte*, pp. 210–224. St. Pölten and Wien: Verlag Niederösterreichisches Pressehaus.

Trnka, G. (1991). *Studien zu mittelneolithischen Kreisgrabenanlagen.* Mitteilungen der Prähistorischen Kommission der österreichischen Akademie der Wissenschaften 26. Wien: Verlag der Österreichischen Akademie der Wissenschaften.

Wahlmüller, N. (1985). Der vorgeschichtliche Mensch in Tirol: Neue Aspekte aufgrund der Pollenanalyse. *Veröffentlichungen des Museum Ferdinandeum 65*, 105–120.

Walde, C. (1999). Palynologische Untersuchungen zur Vegetations- und Siedlungsentwicklung im Raum Kramsach-Brixlegg (Tirol, Österreich). *Berichte des naturwissenschaftlich-medizinischen Vereins Innsbruck 86*, 61–79.

Wasylikowa, K., M. Cârciumaru, E. Hajnalová, B. P. Hartyányi, G. Pashkevich, and Z. V. Yanushevich (1991). East-Central Europe. In W. van Zeist, K. Wasylikowa, and K.-E. Behre (Eds.), *Progress in Old World Palaeoethnobotany: A Retrospective View on the Occasion of 20 Years of the International Work Group for Palaeoethnobotany*, pp. 207–239. Rotterdam: A. A. Balkema.

Werneck, H. (1949). *Ur- und frühgeschichtliche Kultur- und Nutzpflanzen in den Ostalpen und am Rande des Böhmerwaldes.* Wels: Herausgegeben vom Amt der Oberösterreichischen Landesregierung in Linz, Kommissionsverlag: Oberösterreichischer Landesverlag.

Windl, H. (1997). Ein Fundplatz überregionaler Bedeutung aus dem Nordosten Niederösterreichs. *Archäologie Österreichs 8*, 34–39.

Chapter 14

Neolithic plant economies in the northern Alpine Foreland from 5500–3500 cal BC

Stefanie Jacomet
Basel University, Switzerland

14.1 Introduction

The region under discussion lies in the northern border area of the Alps, stretching from the western Alps over the French Jura (situated in the French Dep. of Franche-Comté) to eastern (Lower) Bavaria (figure 14.1). The few sites with quantifiable archaeobotanical investigations from the Austrian pre-alpine areas are treated in Kohler-Schneider (this volume).

The region consists largely of rather flat to hilly areas between the alpine mountain range and the Jura (figure 14.2). Most of these areas were glaciated during the last ice age and the geological substratum is therefore predominantly glacial till. However, there are also some areas with loess, for example, in Lower Bavaria and alongside the Hoch-Rhine (to the west of Lake Constance). On the whole, these were regions suitable for agriculture.

Before the appearance of 'true' neolithic agricultural societies at approximately 5500 cal BC, which are evident in aspects of the material culture such as ceramics and polished stone axes, the area was settled by different groups of late mesolithic populations (for an overview see Mazurié de Keroualin 2003, p.48) who had wide-ranging trade networks (for example, Erny-Rodmann et al. 1997; Gronenborn 2003). The earliest 'true' neolithic influences reached the area from two directions, which can be traced back to the two different main 'neolithisation' routes known today (Lüning 2000, p.6). The eastern route is through the Balkans to the Carpathian basin where the Linearbandkeramik culture (LBK) developed. The LBK and succeeding groups are the so-called 'Danubian cultures'. The western route originated in the western part of the Mediterranean basin. Westerly-influenced neolithic cultures have their roots in the Impressa/Cardial-cultures, which developed mainly during the 6th millennium cal BC in the western Mediterranean (for an overview see, for example, Guilaine 2003; Buxó; Peña-Chocarro; Zapata Peña, this volume). There were probably also contacts with the south over passes in the Alps. These influences from the east on the one hand ('Danubian'), and from the (south)west on the other ('Mediterranean'), are an important characteristic of the region being discussed for the whole Neolithic period (figure 14.3).

Several different theories exist about the adaptation of the neolithic 'way of life' in the area concerned, mainly based on weak botanical evidence (a compilation of our current under-

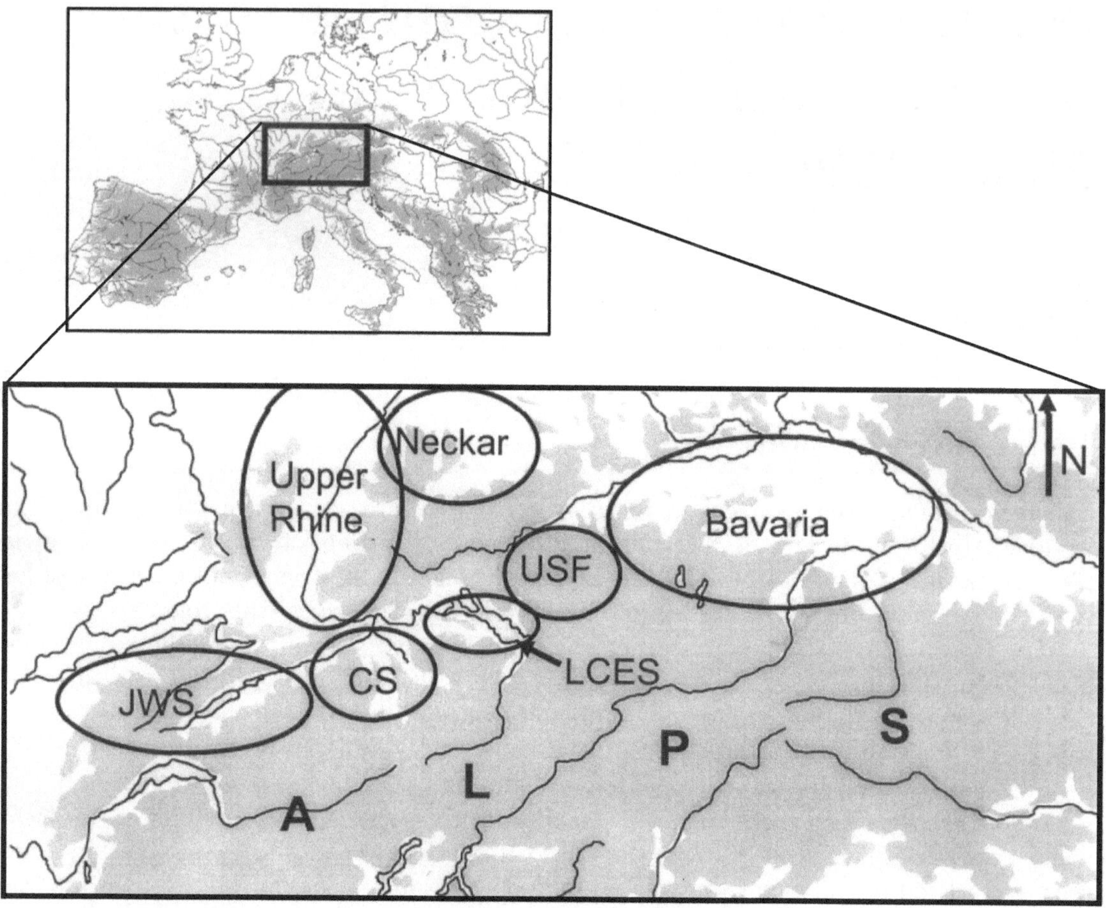

Figure 14.1: Map of the area with location of the regions. JWS=Jura and western Switzerland; CS=Central Switzerland; LCES=Lake Constance and Eastern Switzerland; USF=Upper Swabia, including Federsee. After Hafner and Suter (1997).

Figure 14.2: Reconstruction of the lake shore settlement of Hornstaad Hörnle IA, Lake Constance, around 3910 cal BC, with surrounding landscape. From Dieckmann et al. (1997). Reprinted with permission.

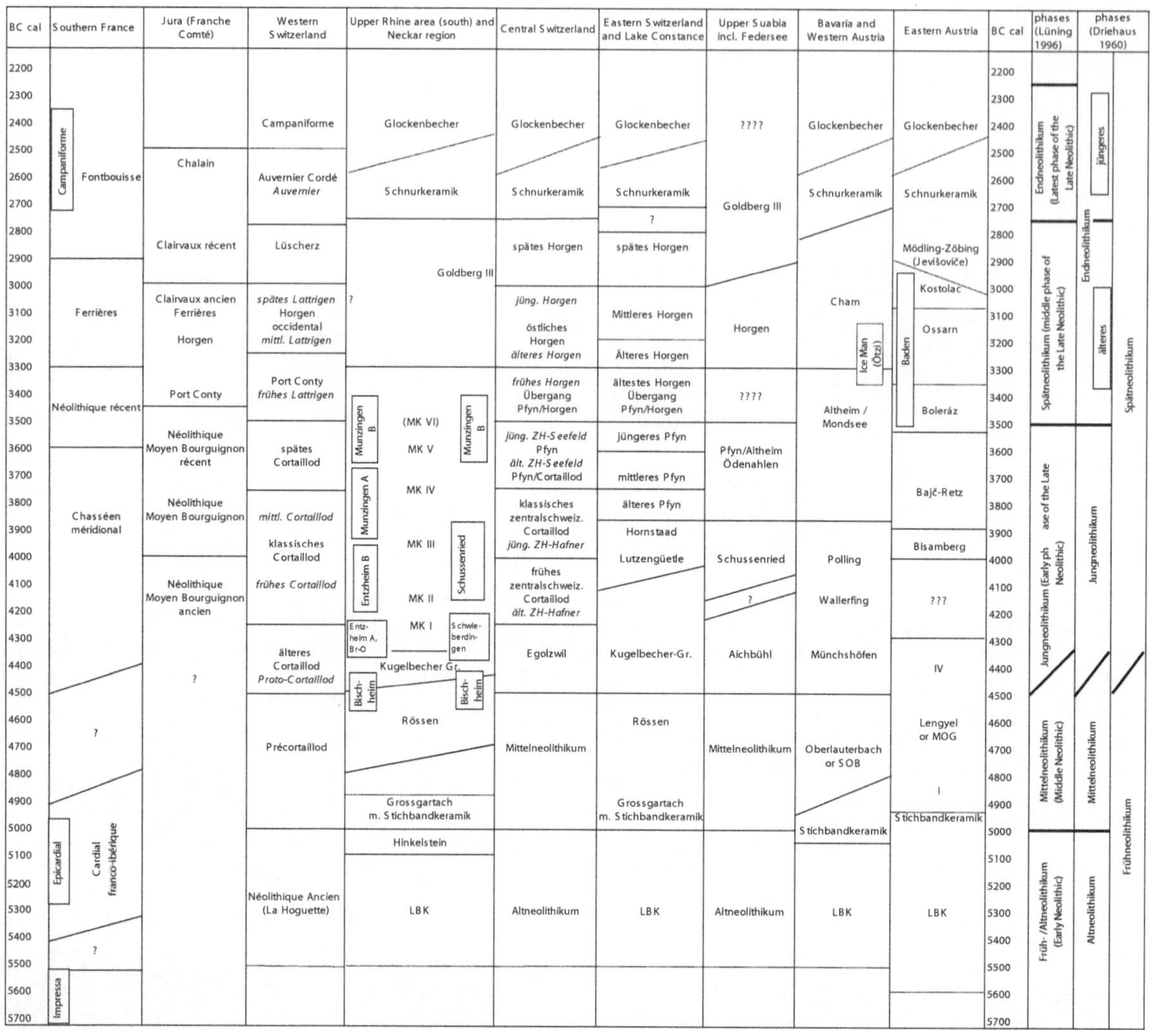

Figure 14.3: Chronological table with the cultural groups mentioned in the text. For the geographical location of the regions, see figure 14.1. MK=Michelsberger Kultur; LBK=Linearbandkeramik; SOB = Südostbayerisches Mittelneolithikum; Br-O=Brubach Oberbergen. After: Strahm (1994, 1997); Hafner and Suter (1997, 2003); Stöckli et al. (1995); Schlichtherle (1988, 1990a, b, 1991), Schlichtherle (pers. comm.); Jeunesse (pers. comm.); Sangmeister (1993); Lüning (1996); Stadler (1995); Driehaus (1960); Lenneis (1995); Ruttkay (1995); Matuschik (pers. comm.); Dieckmann (pers. comm.); van Willigen (pers. comm.). Italics denote new denominations of the cultural groups by Hafner and Suter (1997, 2003) for western and central Switzerland (http://www.jungsteinsite.de).

Figure 14.4: Excavation of a lakeshore settlement (Arbon Bleiche 3, Lake Constance, Switzerland); wooden posts are visible. From: Leuzinger 2000: 124. Photo: AATG, D. Steiner. Used with permission.

standing may be found in Erny-Rodmann et al. 1997); more recently, however, these have been substantiated by newly compiled pollen data (for example, Beckmann 2003). Whether or not late mesolithic populations cultivated cereals will remain unclear until better on-site evidence comes to light.

In the second half of the 5th millennium cal BC there are new developments in the area; the first lakeshore settlements ('lake dwellings' or 'Pfahlbauten') appear in the vicinity of the Alps (figure 14.4; for an overview see, for example, Schlichtherle 1997a). Most of the results presented in this paper come from these types of settlements, which have excellent waterlogged preservation of the plant materials that also include carbonised remains.

Several locally restricted cultural groups are traceable in the archaeological material in the area concerned (figure 14.3; for the archaeological background see citations in the map-legend). For example, in the late 5th and the first half of the 4th millennium cal BC, the westerly-based cultural groups are the Cortaillod, Egolzwil and also the Michelsberg cultures. They remain strongly influenced by the Néolithique Moyen Bourguignon (NMB) and the Chasséen in (south)-western Europe. The Münchshöfen and the Altheim cultures in Bavaria, and also the Aichbühl culture in Upper Swabia are based on (south)eastern roots and are influenced by the Danubian complexes of the Lengyel and Epilengyel cultural groups ('Lengyelisation'). The Schussenried and Pfyn cultures (including the Hornstaad Group) in the Neckar region, Upper Swabia, the Lake Constance region and east-central Switzerland are also mainly Danubian based but, in addition, show some influence from the west.

It is important to note that during the last few years new names have been defined for some of these cultures (these are given in italics in figure 14.3), however, these are not yet commonly cited in the literature (Hafner and Suter 1997; Suter 2002; Hafner and Suter 2003, p.8).

The influences mentioned can be detected in the ceramic assemblages, the sources of the flint raw material, the funerary practices, the imports of seashells and in the first appearance of copper at about 4000 cal BC. The latter comes from the (south)east, reaching the region discussed here via the Danube (see, for example, Strahm 1994; Gross-Klee and Hochuli 2002). New strong easterly influences are detectable at about the middle of the 4th millennium cal BC; for example, those connected with the appearance of wooden wheels (Köninger et al. 2002). Thus, throughout the Neolithic (and of course also earlier) there is good evidence for wide-ranging trade networks.

In summary, the archaeological situation in the study area is rather complicated. All the cultural groups mentioned show influences partly from the west and partly from the east, although the relative contribution from each differs depending on the region in question. Knowledge of these archaeological details is a fundamental element of this paper, the main goals of which are to provide an analytical synthesis and to examine how the cultivated plant spectrum reflects relative eastern and western cultural influences. In the following overview, I consider almost exclusively site-based data derived from the identification of plant macro-remains (i.e., seeds and chaff).

14.2 Materials and methods

14.2.1 Sites with archaeobotanical investigations

Archaeobotanical data in the study region are derived from over 200 neolithic settlements that cover the whole Neolithic period. I have compiled a table of data for 73 settlements with dates in the range 5500–3500 cal BC; reasonably well-quantified data exist for these (either published data or unpublished data made available through personal communications). The basic data are given in tables 14.1 and 14.2. The data recorded from old museum collections have not been included because of the problems with dating and quantification, etc. (see, for example, Blankenhorn and Hopf 1982; Hopf and Blankenhorn 1986; Hofmann 1986). For example, one cannot exclude the possibility that layers were mixed during 'excavations' in the 19th and early 20th centuries. However, I have mentioned these data when they are of particular interest. I have also exluded all references to data published prior to ca. 1970 (for example, Bertsch and Bertsch 1947).

The geographical distribution of the settlements is shown in figures 14.1, 14.5 and 14.6. I define settlement here as meaning not only a site but also a settlement phase, dated as precisely as possible (i.e., several phases of the same site count as several settlements). Dating in the region of the lake dwellings is mainly based on dendrochronology, which enables the creation of a precise chronological framework.

14.2.2 Methods and methodological problems

It is clear from the tables that in the area considered, not only are the preservation conditions of the plant remains different, but also the sampling and processing methods. Fully quantified data are not available from all sites, and representative investigations exist for only a few sites (by 'representative' I mean sites from which I have reliable information on the intra-site patterns; however, this is only very rarely the case as seen in category 1 on tables 14.1 and 14.2). Fortunately, there are some settlements that have been thoroughly investigated, but in each case at least one of the requirements I've stipulated to fulfil a definition of 'representatively investigated' has not been satisfied. In a majority, for example, the evaluation of the archaeological materials and contexts is lacking and so precise dating is not possible (see categories 2, 3 and 4 on tables 14.1 and 14.2; Jacomet 2005). This is due to the fact that in over 90% of the cases the materials analysed from the area considered come from rescue excavations and are not routinely subjected

Settlement	Region	Culture	Beginning cal BC
Aiterhofen (D) Kr. Straubing-Bogen	BA	Bandkeramik LBK	ca. 5500
Altdorf (D) Kr. Landshut	BA	Bandkeramik LBK	ca. 5500
Berg ober Landshut (D) St. Kr. Landshut	BA	Bandkeramik LBK	ca. 5500
Oberpiebing (D) Kr. Straubing-Bogen	BA	Bandkeramik LBK	ca. 5500
Hienheim (D) Kr. Kelheim	BA	Bandkeramik LBK	ca. 5500
Mintraching (D) Kr. Regensburg	BA	Bandkeramik, älteste (LBK I)	ca. 5500
Eggingen (D) Stkr. Ulm	USF	Bandkeramik LBK, mittl.-jüngere	ca. 5200
Bietigheim-Bissingen (D) Kr. Ludwigsburg	N	Bandkeramik LBK	ca. 5500
Ditzingen (D) Kr. Ludwigsburg	N	Bandkeramik LBK	ca. 5500
Heilbronn (D) Klingenberg	N	Bandkeramik LBK	ca. 5500
Ludwigsburg (D) Ossweil, Kr. Ludwilgsburg	N	Bandkeramik LBK	ca. 5500
Marbach (D) Kr. Ludwigsburg	N	Bandkeramik LBK	ca. 5500
Rottenburg (D) Kr. Tübingen Lindele	N	Bandkeramik, älteste (LBK I)	ca. 5500
Herrenberg (D) Affstätt, Kr. Böblingen	N	Bandkeramik LBK	ca. 5500
Weiler z. Stein (D) Gem. Leutenbach, Rems-Murr-Kr.	N	Bandkeramik LBK	ca. 5500
Singen (D) Scharmenseewadel	LCES	Bandkeramik LBK	ca. 5500
Hilzingen (D) Kr. Konstanz	LCES	Bandkeramik LBK ältere (Hegau Stilphase 1)	ca. 5300
Hilzingen (D) Kr. Konstanz	LCES	Bandkeramik LBK ältere-mittl. (Hegau Stilphasen 2-5)	ca. 5250
Hilzingen (D) Kr. Konstanz	LCES	Bandkeramik LBK mittl.(Hegau Stilphase 6)	ca. 5200
Hilzingen (D) Kr. Konstanz	LCES	Bandkeramik LBK mittl.-jüng. (Hegau Stilphasen 7-9)	ca. 5100
Künzing (D) Unternberg, Kr. Deggendorf	BA	Oberlauterbach	ca. 4600
Gaimersheim (D) Kr. Eichstätt	BA	Oberlauterbach	ca. 4600
Schernau (D) Kr. Kitzingen	BA	Bischheimer-Gr., Spätrössen	ca. 4400
Endersbach (D) Rems-Murr-Kreis	N	Grossgartach	ca. 5000
Ilsfeld (D) Kr. Heilbronn	N	Grossgartach/Rössen	ca. 5000
Ditzingen (D) Kr. Ludwigsburg	N	Rössen	ca. 4700
Gonvillars (F) Layer XI, Haute-Saone,	JWS	Mittelneolithikum, früh	ca. 4900

Table 14.1: Summary of early neolithic LBK (before 5000 cal BC) and middle neolithic (5000–4300 cal BC) sites where there has been archaeobotanical investigation. There are two other investigations of LBK settlements not included in the table (Rösch 1992, 1998). Preservation: M = mineral soil site; W = waterlogged. Representativeness: see explanation in the text. See figure 14.1 for region codes (abbreviation for Neckar = N; for Bavaria = BA). Other codes: n.i. = no information available; sample type, s = (plus/minus) systematic sampling, j = judgment samples (eg, stores).

Settlement	Preser- vation	Represent- ativeness	Context	Min. mesh size (mm)	Sample type	no. of samples	total vol. (*l*)	no. of contexts	xno. cereal remains	References
Aiterhofen	M	3	pits	n.i.	s	7	31	7	1636	1
Altdorf	M	4	pits	n.i.	s	5	20	4	95	1
Berg ober Landshut	M	4	pits	n.i.	s	2	7	2	24	1
Oberpiebing	M	5	pits	n.i.	s	3	6	3	25	1
Hienheim	M	2	pits	0.5	s/j?	37	165	many	1510	2
Mintraching	M	1	pits	0.5	s	106	1805	many	141	3
Eggingen	M	1	pits	0.5	0	385	1802	157	1619	4
Bietigheim-Bissingen	M	1	pits	0.3	s/j	74	31	39	4286	5
Ditzingen	M	1	pits	0.3	s	52	100	52	3248	6
Heilbronn	M	3	pits	0.5	s	16	77	6	255	7, 8
Ludwigsburg	M	5	pits	0.3	s	1	2	1	267	9
Marbach	M	5	pits	0.3	s	1	3	1	202	9
Rottenburg	M	5	pits	0.5	s	1	55	1	175	7
Herrenberg	M	5	pits	0.5	s?	1	15	1	344	7
Weiler z. Stein	M	5	pits	0.3	s	3	12	1	225	9
Singen	M	5	pits	0.5	s	2	n.i.	2	190	10
Hilzingen (LBK)	M	5	pits	0.5	s?	4	91	2	102	7
Hilzingen (LBK ältere)	M	3	pits	0.5	s	17	233	7	149	7
Hilzingen (LBK ältere-mittl.)	M	4	pits	0.5	s	6	95	2	217	7
Hilzingen (LBK mittl.-jüng.)	M	3	pits	0.5	s	24	423	8	4717	7
Künzing	M	1	n.i.	0.5	s	54	216	many	74	11
Gaimersheim	M	4	pits?+div.	0.3	2	6	24	6	785	12, 13
Schernau	M	5	pits	n.i.	s	3	n.i.	3	896	14
Endersbach	M	5	pits	0.3	j?	1	5	1	1206	15, 9
Ilsfeld	M	5	pits	0.3	j?	2	3	2	3570	15, 6
Ditzingen	M	5	pits	0.3	s	4	10	3	60	6
Gonvillars	M	5	layer	n.i.	j	n.i.	n.i.	1	100	16, 17

Reference codes: 1. Bakels 1984, 2. Bakels 1978, 3. Kreuz 1990, 4. Gregg 1989, 5. Piening 1989, 6. Piening 1998, 7. Stika 1991, 8. Stika 1996, 9. Piening 1982, 10. Sooss unpub. after Stika 1991, 11. Küster 1991, 12. Küster 1992, 13. Küster 1995, 14. Hopf 1981, 15. Piening 1979, 16. Villaret-von Rochow 1974, 17. Villaret-von Rochow 1970.

Table 14.1, cont.

Settlement	Region	Culture	Beginning cal BC
Gaimersheim (D) Kr. Eichstätt	BA	Münchshöfen	ca. 4100
Vilsbiburg (D) Kr. Landshut	BA	Münchshöfen	ca. 4100
Vilsbiburg (D) Kr. Landshut, Bauplatz König und Lerchenstrasse	BA	Münchshöfen	ca. 4100
Ergolding (D) Fischergasse, Kr. Landshut	BA	Altheim	3685
Ehrenstein (D) Phasen I-III, Kr. Ulm,	USF	Schussenried	3955
Alleshausen (D) Hartöschle Kr. Biberach	USF	Schussenried	3920
Reute (D) Schorrenried Kr. Ravensburg	USF	Pfyn/Altheim	3738
Oedenahlen (D) Riedwiesen, Kr. Biberach	USF	Pfyn/Altheim	3700
Aldingen (D) Gem. Remseck, Kr. Ludwigsburg Halden I	N	Schwieberdinger Gr.	ca. 4350
Hochdorf (D) Gem. Eberdingen, Kr. Ludwigsburg	N	Schussenried	4334
Aldingen (D) Gem. Remseck, Kr. Ludwigsburg Halden II	N	Schussenried	ca. 4000
Freiberg (D) Geisingen, Kr. Ludwigsburg	N	Schussenried	ca. 4000
Grossachsenheim (D) Holderbüschle Kr. Ludwigsburg	N	Schussenried	ca. 4000
Schlösslesfeld (D) Kr. Ludwigsburg	N	Schussenried	3750
Heilbronn (D) Klingenberg	N	Michelsberger MKV	ca. 3600
Hornstaad (D) Hörnle IA Kr. Konstanz	LCES	Pfyn (Hornstaader Gr.)	3917
Thayngen (CH) Weier, Sch. 16-19, Profil III	LCES	Pfyn	3822
Wallhausen (D) Ziegelhütte, Kr. Konstanz	LCES	Pfyn	ca. 3750
Sipplingen (D) Osthafen, Pfyner Schichten, Bodenseekreis	LCES	Pfyn	ca. 3750
Gachnang (CH) Niederwil	LCES	Pfyn	3660
Hornstaad (D) Hörnle IB Kr. Konstanz	LCES	Pfyn	3586
Zürich (CH) Kleiner Hafner, Schichten 5A+B	CS	Egolzwil	4384
Egolzwil (CH) 3, Wauwiler Moos	CS	Egolzwil	4282
Cham (CH) Eslen	CS	Egolzwil/frühes zs. Cortaillod bzw. Zürich Hafner	4225
Zürich (CH) Kleiner Hafner Schichten 4A-C/D	CS	Cortaillod, frühes zs. bzw. Zürich-Hafner	
Zürich (CH) Kleiner Hafner Schichten 4E/F	CS	Cortaillod, klass. zs. bzw. Zürich-Hafner	3968
Zürich (CH) Mozartstrasse Schicht 6	CS	Cortaillod, klass. zs. bzw. Zürich-Hafner	3908
Zürich (CH) Mozartstrasse Schicht 5 u,o	CS	Cortaillod, klass. zs. bzw. Zürich-Hafner	3864
Zürich (CH) Kan San Schicht 9	CS	Pfyn/Cortaillod-Üb. bzw. Zürich Hafner	3827
Seeberg (CH) Burgäschisee-Süd	CS	Cortaillod, klass. zs. bzw. Zürich-Hafner	3760
Zürich (CH) AKAD/Pressehaus, Schicht J	CS	Pfyn. bzw. Zürich-Seefeld	3728
Zürich (CH) KanSan Schicht 7	CS	Pfyn. bzw. Zürich-Seefeld	3719
Zürich (CH) Mozartstrasse Schicht 4B+4A	CS	Pfyn. bzw. Zürich-Seefeld	?3714
Risch (CH) Oberrisch, Aabach	CS	Pfyn. bzw. Zürich-Seefeld	?3710
Cham (CH) St. Andreas	CS	Pfyn (-Cortaillod?)	?3700
Zürich (CH) Mozartstrasse Schicht 4 u,m,o	CS	Pfyn. bzw. Zürich-Seefeld	3668
Zürich (CH) KanSan Schicht 5	CS	Pfyn. bzw. Zürich-Seefeld	3616
Twann (CH) US, E1-2	JWS	Cortaillod classique	3838
Concise (CH) Sous Colachoz (EMS)*	JWS	Cortaillod moyen, Ens. 2	3709
Twann (CH) MS (E3-4)	JWS	Cortaillod tardif	3702
Port (CH) Stüdeli, US	JWS	Cortaillod, spätes	3686
Twann (CH) MS (E5 - 5a)	JWS	Cortaillod tardif	3643
Twann (CH) OS (6-7)	JWS	Cortaillod tardif	3596
Port (CH) Stüdeli, OS	JWS	Cortaillod, spätes	3572
Concise (CH) Sous Colachoz COT2 (Häuf.-Klass.)	JWS	Cortaillod tardif	3567
Clairvaux (F) V Motte Aux Magnins	JWS	Néolithique Moyen Bourguignon récent	3659

Table 14.2: Summary of early *Jungneolithikum* (early part of the late Neolithic, 4300–3500 cal BC) sites where there has been archaeobotanical investigation. For coding, please see table 14.1.

Settlement	Preser-vation	Represent-ativeness	Context	Min. mesh size (mm)	Sample type	no. of samples	total vol (*l*)	no. of contexts	no. cereal remains	References
Gaimersheim (D) Kr. Eichstätt	M	3	pits ? + div.	0.3	s	10	40	6	583	1, 2
Vilsbiburg (D) Kr. Landshut	M	5	pits, oven	n.i.	s?	3	16	3	63	3
Vilsbiburg (D) Kr. Landshut, Bauplatz König und Lerchen-strasse	M	3	n.i.	n.i.	s	18	69	n.i.	44	2
Ergolding (D) Fischergasse, Kr. Landshut	M(W)	2	cultural layer	n.i.	s	38	103	many	3886	4
Ehrenstein (D) Phasen I-III, Kr. Ulm,	W	4	cultural layer	n.i.	s/j	67	>35	many	157373	5
Alleshausen (D) Hartöschle Kr. Biberach	W	5	cultural layer	n.i.	s?	>5	n.i.	many	+	6
Reute (D) Schorrenried Kr. Ravensburg	W	4	cultural layer	0.0	s/j	75	>8	many	1911	7
Oedenahlen (D) Riedwiesen, Kr. Biberach	W	3	cultural layer	0.3	s/j	176	>33	many	9526	8
Aldingen (D) Gem. Remseck, Kr. Ludwigsburg Halden I	M	3	pits	0.3	j	24	27	5	16913	9
Hochdorf (D) Gem. Eberdingen, Kr. Ludwigsburg	M	1	pits	0.3	s/j	142	934	56	57480	10
Aldingen (D) Gem. Remseck, Kr. Ludwigsburg Halden II	M	5	pit	0.3	j?	1	27	1	1398	11
Freiberg (D) Geisingen, Kr. Ludwigsburg	M	5	pits	0.3	j?	4	10	>5	2013	12
Grossachsenheim (D) Holderbüschle Kr. Ludwigsburg	M	5	pits	0.3	s	3	5	3	1754	13
Schlösslesfeld (D) Kr. Ludwigsburg	M	3	pits	n.i.	s	10(?)	50	1	89	14
Heilbronn (D) Klingenberg	M	1	pits, ditch	0.3	s/j	107	>500	many	11789	15
Hornstaad (D) Hörnle IA Kr. Konstanz	W	1	cultural layer	0.3	s/j	382	99	>300	123315	16
Thayngen (CH) Weier, Sch. 16-19, Profil III	W	5	cultural layer	0.4	s	9	n.i.	2	2706	17, 18
Wallhausen (D) Ziegelhütte, Kr. Konstanz	W	5	cultural layer	n.i.	s?	3	n.i.	1	1547	19
Sipplingen (D) Osthafen, Pfyner Schichten, Bodenseekreis	W	2	cultural layer	0.5?	s/j?	44	16	44	40955	20
Gachnang (CH) Niederwil	W	2	cultural layer	0.2	s/j	60	n.i.	many	77418	21
Hornstaad (D) Hörnle IB Kr. Konstanz	W	5	cultural layer	0.3	s	10	3	n.i.	115	22
Zürich (CH) Kleiner Hafner, Schichten 5A+B	W	5	cultural layer	0.0	s	16	6	5	536	23, 24
Egolzwil (CH) 3, Wauwiler Moos	W	2	cultural layer	0.3	s	103	82	>70	170	25
Cham (CH) Eslen	W	2	cultural layer	.	s	16	20	16	33	26
Zürich (CH) Kleiner Hafner Schichten 4A-C/D	W	3	cultural layer	0.3	s/j	34	5	22	12339	23, 24
Zürich (CH) Kleiner Hafner Schichten 4E/F	W	2	cultural layer	0.3	s/j	45	8	35	29589	23, 24
Zürich (CH) Mozartstrasse Schicht 6	W	5	cultural layer	0.3	s/j	39	14	14	6541	23, 24
Zürich (CH) Mozartstrasse Schicht 5 u,o	W	3	cultural layer	0.3	s/j	43	12	31	306	23, 24
Zürich (CH) Kan San Schicht 9	W	2	cultural layer	0.0	s/j	34	16	25	1445	24
Seeberg (CH) Burgäschisee-Süd	W	4	cultural layer	<0.5	s/j	>130	n.i.	many	392	27
Zürich (CH) AKAD/Pressehaus, Schicht J	W	2	cultural layer	0.3	s	128	210	74	11567	28, 23, 24
Zürich (CH) KanSan Schicht 7	W	5	cultural layer	0.0	s/j	9	5	8	22107	24
Zürich (CH) Mozartstrasse Schicht 4B+4A	W	5	cultural layer	0.0	s	11	8	4	200	24
Risch (CH) Oberrisch, Aabach	M(W)	1	cultural layer	0.5	s/(j)	54	668	54	12208	29
Cham (CH) St. Andreas	M(W)	5	cultural layer	0.5	j	4	ca1	4	936	30
Zürich (CH) Mozartstrasse Schicht 4 u,m,o	W	3	cultural layer	0.0	s/j	68	35	24	2791	24
Zürich (CH) KanSan Schicht 5	W	5	cultural layer	0.0	s/j	13	5	10	67	24
Twann (CH) US, E1-2	W	5	cultural layer	0.3	s/j	25	ca3	5	1358	31
Concise (CH) Sous Colachoz (EMS)*	W	2	cultural layer	0.5	s	18	ca 18	5	++	32, 33
Twann (CH) MS (E3-4)	W	5	cultural layer	0.3	s	21	ca 2.5	5	10	31
Port (CH) Stüdeli, US	W	4	cultural layer	0.5	j	21	4	21	15557	24
Twann (CH) MS (E5 - 5a)	W	5	cultural layer	0.3	s/j	21	ca 3	18	3949	31
Twann (CH) OS (6-7)	W	5	cultural layer	0.3	s/j	16	ca 2	9	565	31
Port (CH) Stüdeli, OS	W	4	cultural layer	0.5	j	26	3	26	24711	24
Concise (CH) Sous Colachoz COT2 (Häuf.-Klass.)	W	5	cultural layer	0.5?	n.i.	4	n.i.	n.i.	++	33
Clairvaux (F) V Motte Aux Magnins	W	3	cultural layer	0.2	s	29	ca 4	23	682	34

Reference codes: 1. Küster 1992, 2. Küster 1995, 3. Bakels 1984, 4. Küster 1989, 5. Hopf 1968, 6. Maier 2004, 7. Hafner 1998, 8. Maier 1995, 9. Piening 1986a, 10. Küster 1985, 11. Piening 1992, 12. Piening 1988, 13. Piening 1986b, 14. Hopf 1977, 15. Stika 1996, 16. Maier 2001, 17. Jörgensen 1975, 18. Fredskild 1997, 19. Rösch 1990, 20. Riehl, unpub., 21. van Zeist and Boekschoten-van Helsdingen 1991, 22. Maier, unpub., 23. Jacomet et al. 1989, 24. Brombacher and Jacomet 1997, 25. Bollinger 1994, 26. Martinoli and Jacomet 2002, 27. Villaret-von Rochow 1967, 28. Jacomet 1981, 29. Jacomet, in prep., 30. Jacomet 1986, 31. Amman et al. 1981, 32. Märkle 2000, 33. Karg and Märkle 2002, 33. Lundström-Baudais 1989.

Table 14.2, cont.

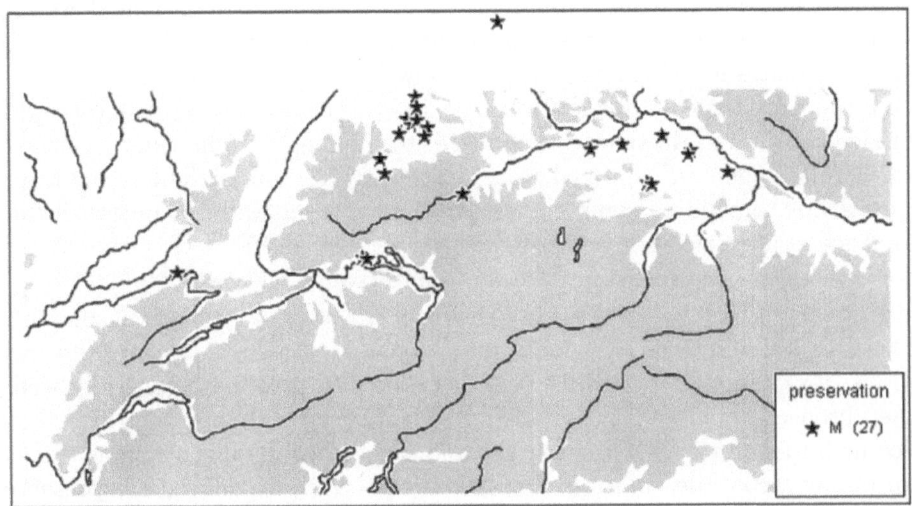

Figure 14.5: Location of early neolithic (LBK) and middle neolithic sites in the region considered. All are situated on dry mineral soils (M). Map: R. Ebersbach, IPAS, Basel University. For geographical context see figure 14.1. For details of sites, see table 14.1.

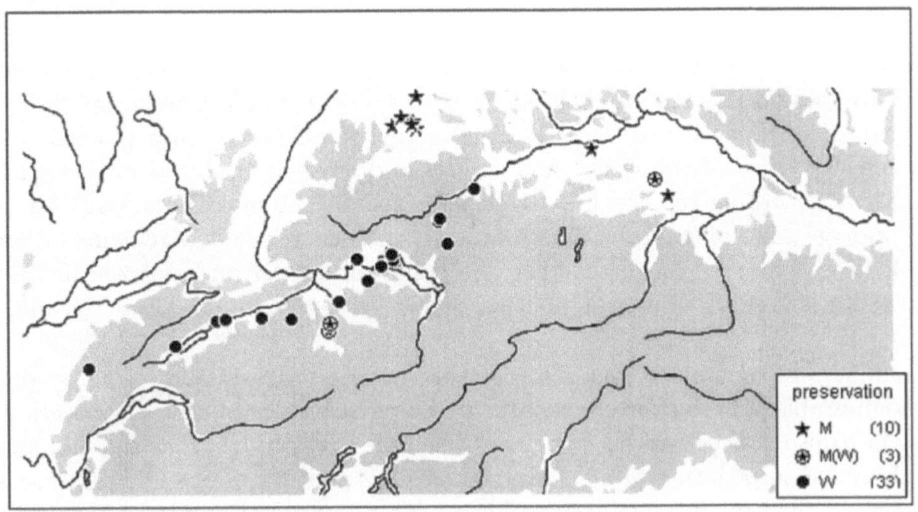

Figure 14.6: Map with sites of the earlier phases of the late Neolithic (4300–3500 cal BC). These are mainly lakeshore settlements with waterlogged (W) preservation. Map: R. Ebersbach, IPAS, Basel University. For geographical context, see figure 14.1. For details of sites, see table 14.2.

to scientific evaluation (and this happens only years later in certain instances). In the whole area, with very few exceptions, no systematic survey-fieldwork has taken place.

Nevertheless, I view this to be a minor problem and with due caution, therefore, the settlements in categories 2–4 can also be taken into consideration. Many of the sites considered belong to category 5 (i.e., sites for which material from few [< five] contexts was analysed) and in these cases, even if there are large numbers of remains, the results are not representative. Nevertheless, I have included the data in the database and refer to them in the following text, the reason being that when there are some investigated sites in a region from a particular time-frame, even one sample can provide some hints about the importance of a cultivated plant (with the caveat that the results from a category 5 site are potentially aberrant and that they should not be deemed suitable for direct comparisons with more representatively collected sites).

Another factor that influences the composition of the dataset is preservation. Only carbonised plant material is available from sites located on mineral soils. As is known from methodological investigations (for example, Willerding 1991), cereals and pulses are generally well represented even in the absence of preservation by waterlogging because there is a greater likelihood that they will come into contact with fire and become carbonised. The contrary is the case with flax and opium poppy seeds; they burn easily because they contain oil and are preserved only very rarely in a recognisable form (see, for example, Jacomet et al. 1989, table 32). More significantly, their remains are very fragile and are likely to become damaged during processing procedures.

The same is the case when one looks at the remains of the gathered plants, which also played an important role in the economy in the Neolithic (see, for example, Jacomet et al. 1989, 193 ff.). There are many millions of seed remains preserved under waterlogged conditions, whereas under mineral soil conditions we regularly find only small quantities of hazelnut shells.

When preservation is by waterlogging it is also necessary to take into consideration the damage caused during the sieving procedure. A lot of the cereal remains in the lake dwellings are represented by uncarbonised chaff, which is very easily destroyed during sieving (Hosch and Zibulski 2003). However, carbonised remains are also found everywhere and these are not as badly affected during the recovery process. Therefore, we consider that the cereal data are reliable.

Because the main goal of this paper is to gain insights into the cereal-growing economy, I have tried to standardize the data based on scores (for example, similar to that of Robinson 2003, p.148). Sites that yielded fewer then 50 cereal remains were not plotted. For all others I have taken the sums of the cereal finds and these have been scored on a logarithmic scale from 1000 to 1 according to their importance at each site (1000 = dominant, 1st place, 100 = major component, 2nd place, 10 = many remains, 3rd place, 1 = few remains or presence only, 4th place). In the figures I have combined the values of emmer and einkorn; in the tables they are listed separately (the results of the quantification are represented on the maps in figures 14.7–14.11).

The numbers of remains of the different species of cultivated and gathered plants are not directly comparable. The best way of making comparisons (at least in the waterlogged layers), therefore, is to use concentrations per litre of sediment analysed. This has not been possible, however, because such data are only rarely published and also, depending on the sieving procedure, some of the remains are likely to have been damaged (Hosch and Zibulski 2003). As a consequence I have relied principally on presence/absence data for the remaining taxa.

When 'naked wheat' is mentioned in the text I am referring to tetraploid or hexaploid free-threshing wheat (*Triticum durum/turgidum* or *aestivum*). Distinguishing between the ploidy levels is not possible when grains alone are preserved; this is only possible on the basis of rachis remains (see, for example, Jacomet and Schlichtherle 1984; Jacomet et al. 1989, p.328; Maier 1996). Identifications of free-threshing wheat made before ca. 1985 (and in some instances also later) are often described as *Triticum aestivum* or *Triticum aestivo-compactum* (for example,

Hopf 1968; Jörgensen 1975, and many others); these have to be considered merely as 'working-terms'. Many of the finds have subsequently turned out to be the tetraploid species.

The different forms of barley were not listed separately, although we have added a note if naked forms were observed. The barley remains in the publications are called *Hordeum vulgare*, and in most cases the six-rowed form is implied. We found no reference in the literature to the presence of two-rowed barley for the time period we are dealing with.

The nomenclature of plant names follows Zohary and Hopf (2000) for the crop plants, and Binz and Heitz (1990) for the wild plants.

14.3 Results

14.3.1 The early Neolithic

Quantifiable data from 20 settlements that date to the second half of the 6th millennium cal BC were compiled in the table (table 14.1, figure 14.5). All 20 settlements belong to the LBK culture. Plant material from different phases was investigated but only rarely is more precise dating available from the archaeobotanical publications. In the region considered here, the LBK settlements are mostly situated in the loess areas of Lower Bavaria and in the Neckar region. In recent years some settlements were also excavated in the Hegau, in southwestern Germany (north of the Hochrhine) (figure 14.5). All sites are situated on dry mineral soils.

Eighteen of these settlements yielded more than 50 cereal remains (table 14.1). Einkorn (*Triticum monococcum*) and emmer (*Triticum dicoccum*) are by far the dominant cereals (table 14.3), which is in agreement with many other LBK sites (see other contributions in this volume), and of particular interest is the presence of two-grained einkorn at Hilzingen (Stika 1991, fig. 8).

Other cereals are rare. Naked wheat (*Triticum aestivum/durum/turgidum*) was found only twice, in a middle LBK-context in the Hochrhine area (Hilzingen: Stika 1991) and in the Neckar region (Bietigheim-Bissingen, phase unknown: Piening 1989, pp.128–129 and fig. 8). It was not possible to specify the ploidy level because only a few grains were discovered. Barley (*Hordeum vulgare*) is present in only nine sites and mostly in very small amounts (i.e., fewer than 10 specimens). In many cases it has been identified as the naked form, with the exception of Ludwigsburg-Ossweil, where large amounts of hulled barley were found (Piening 1982). This latter exception is based only on one sample from one pit, and is not necessarily representative of the site as a whole. The one find of millet (*Panicum miliaceum*) in the oldest LBK in Mintraching, and the two grains of rye (*Secale cereale*) in Marbach (Piening 1982, fig. 2) are worthy of mention, and these most probably represented weeds. Overall, the cereal spectrum is very uniform during the early Neolithic in the whole area; it is questionable whether other cereals were grown besides einkorn and emmer.

In addition to einkorn and emmer, pulses obviously played an important role as food. Pea (*Pisum sativum*) is found in 12 of the settlements; at Hienheim and Hilzingen (younger LBK) there were also large amounts of stored peas (table 14.3). Lentil (*Lens culinaris*) is much more rare; it is present in only six settlements in Bavaria and the Neckar region, one find comes from the oldest LBK phase (Mintraching), and with one exception (Ulm-Eggingen), the samples of lentils comprise less than 10 specimens. The exact importance of lentil, therefore, remains unclear. Even less clear is the significance of five seeds of bitter vetch (*Vicia ervilia*) found at Marbach (Piening 1982: Fig. 3, p. 147).

Flax (*Linum usitatissimum*) seeds were present in seven of the 20 sites, and at three the finds comprised 10 or more specimens (table 14.3). It is likely that the seeds are under-represented in the mineral soils on which these sites are located, and if this was in fact the case it could be deduced that flax-growing was important during the early Neolithic in the whole of the area (see also Stika 1991, pp.91–92). The question of how important flax was will perhaps remain

Settlement	Culture	Preservation	Representativeness	Cereals							
				Einkorn	Emmer	Einkorn/emmer	Spelt	Naked wheat	Barley	Rye	Broomcorn millet
Aiterhofen (D) Kr. Straubing-Bogen	Bandkeramik LBK	M	3	7	91	1538	.	.	.	.	.
Altdorf (D) Kr. Landshut	Bandkeramik LBK	M	4	.	1	94	.	.	.	.	.
Berg ober Landshut (D) St.Kr. Landshut	Bandkeramik LBK	M	4	.	.	24	.	.	.	.	.
Oberpiebing (D) Kr. Straubing-Bogen	Bandkeramik LBK	M	5	1	.	24	.	.	.	.	.
Hienheim (D) Kr. Kelheim	Bandkeramik LBK	M	2	280	158	1072	.	.	.	.	.
Mintraching (D) Kr. Regensburg	Bandkeramik, älteste (LBK I)	M	1	1	.	135	.	.	5	.	1
Eggingen (D) Stkr. Ulm	Bandkeramik LBK, mittl.-jüngere	M	1	58	135	1418	.	.	8	.	.
Bietigheim-Bissingen (D) Kr. Ludwigsburg	Bandkeramik LBK	M	1	2540	1134	603	.	9	9	.	.
Ditzingen (D) Kr. Ludwigsburg	Bandkeramik LBK	M	1	958	1713	573	.	.	4	.	.
Heilbronn (D) Klingenberg	Bandkeramik LBK	M	3	78	97	79	.	.	1	.	.
Ludwigsburg (D) Ossweil, Kr. Ludwilgsburg	Bandkeramik LBK	M	5	2	18	4	.	.	243	.	.
Marbach (D) Kr. Ludwigsburg	Bandkeramik LBK	M	5	57	41	102	.	.	.	2	.
Rottenburg (D) Kr. Tübingen "Lindele"	Bandkeramik, älteste (LBK I)	M	5	32	41	102	.	.	.	.	.
Herrenberg (D) Affstätt, Kr. Böblingen	Bandkeramik LBK	M	5	125	99	117	.	.	3	.	.
Weiler z. Stein (D) Gem. Leutenbach, Rems-Murr-Kr.	Bandkeramik LBK	M	5	66	131	28	.	.	.	.	.
Singen (D) Scharmenseewadel	Bandkeramik LBK	M	5	78	52	60	.	.	.	.	.
Hilzingen (D) Kr. Konstanz	Bandkeramik LBK ältere (Hegau Stilphase 1)	M	5	5	4	93	.	.	.	.	.
Hilzingen (D) Kr. Konstanz	Bandkeramik LBK ältere-mittl. (Hegau Stilphasen 2-5)	M	3	17	15	117	.	.	.	.	.
Hilzingen (D) Kr. Konstanz	Bandkeramik LBK mittl.(Hegau Stilphase 6)	M	4	41	23	151	.	1	1	.	.
Hilzingen (D) Kr. Konstanz	Bandkeramik LBK mittl.-jüng. (Hegau Stilphasen 7-9)	M	3	867	1631	2218	.	.	1	.	.
Künzing (D), Kr. Deggendorf	Oberlauterbach	M	1	35	26	13	.	.	.	.	.
Gaimersheim (D) Kr. Eichstätt	Oberlauterbach	M	4	416	5	364	.	.	.	.	.
Schernau (D) Kr. Kitzingen	Bischheimer-Gr.,Spätrössen	M	5	397	155	.	43cf	120	224	.	.
Endersbach (D) Rems-Murr-Kreis	Grossgartach	M	5	952	79	.	.	.	175	.	.
Ilsfeld (D) Kr. Heilbronn	Grossgartach/Rössen	M	5	1383	301	194	.	927	765	.	.
Ditzingen (D) Kr. Ludwigsburg	Rössen	M	5	26	2	1	.	5	26	.	.
Gonvillars (F) layer XI, Haute-Saone,	Mittelneolithikum, früh	M	5	.	50	.	1cf	25	25	.	.

Table 14.3: Cultivated and gathered plants of the early Neolithic LBK (before 5000 cal BC) and the middle Neolithic (5000–4300 cal BC). Only carbonised remains preserved.

Settlement	Culture	Preservation	Representativeness	Other cultivated plants				Gathered plants			
				Pea	Lentil	Flax	Opium poppy	Hazelnut	Wild apple	Sloe	Elder
Aiterhofen (D) Kr. Straubing-Bogen	Bandkeramik LBK	M	3	4	.	32	.	1	.	.	.
Altdorf (D) Kr. Landshut	Bandkeramik LBK	M	4	.	.	.	.	.	.	.	.
Berg ober Landshut (D) St.Kr. Landshut	Bandkeramik LBK	M	4	.	.	.	.	.	.	.	.
Oberpiebing (D) Kr. Straubing-Bogen	Bandkeramik LBK	M	5	.	.	.	.	.	.	.	.
Hienheim (D) Kr. Kelheim	Bandkeramik LBK	M	2	3237	6	3	.	1	.	.	.
Mintraching (D) Kr. Regensburg	Bandkeramik, älteste (LBK I)	M	1	.	5	.	.	11	.	1	.
Eggingen (D) Stkr. Ulm	Bandkeramik LBK, mittl.-jüngere	M	1	4cf	56	25	2	.	.	.	.
Bietigheim-Bissingen (D) Kr. Ludwigsburg	Bandkeramik LBK	M	1	2	.	.	.	.	.	.	.
Ditzingen (D) Kr. Ludwigsburg	Bandkeramik LBK	M	1	1	.	1	.	.	.	.	.
Heilbronn (D) Klingenberg	Bandkeramik LBK	M	3	130	9	.	.	.	.	.	3
Ludwigsburg (D) Ossweil, Kr. Ludwilgsburg	Bandkeramik LBK	M	5	.	1	.	.	.	.	.	.
Marbach (D) Kr. Ludwigsburg	Bandkeramik LBK	M	5	.	.	.	.	.	.	.	.
Rottenburg (D) Kr. Tübingen "Lindele"	Bandkeramik, älteste (LBK I)	M	5	.	.	.	.	.	.	.	.
Herrenberg (D) Affstätt, Kr. Böblingen	Bandkeramik LBK	M	5	.	.	.	.	.	.	.	.
Weiler z. Stein (D) Gem. Leutenbach, Rems-Murr-Kr.	Bandkeramik LBK	M	5	43	.	.	.	.	.	.	
Singen (D) Scharmenseewadel	Bandkeramik LBK	M	5	4	.	.	.	.	.	.	.
Hilzingen (D) Kr. Konstanz	Bandkeramik LBK ältere (Hegau Stilphase 1)	M	5	78	.	.	.	.	.	.	.
Hilzingen (D) Kr. Konstanz	Bandkeramik LBK ältere-mittl. (Hegau Stilphasen 2-5)	M	3	9	.	8	.	2	.	.	.
Hilzingen (D) Kr. Konstanz	Bandkeramik LBK mittl.(Hegau Stilphase 6)	M	4	3	.	7	.	1	.	.	.
Hilzingen (D) Kr. Konstanz	Bandkeramik LBK mittl.-jüng. (Hegau Stilphasen 7-9)	M	3	3460	.	10	.	12	1	3	.
Künzing (D) Unternberg, Kr. Deggendorf	Oberlauterbach	M	1	.	2	1	.	.	1	.	1
Gaimersheim (D) Kr. Eichstätt	Oberlauterbach	M	4	.	.	.	.	.	.	.	.
Schernau (D) Kr. Kitzingen	Bischheimer-Gr.,Spätrössen	M	5	.	.	1	2	.	.	.	
Endersbach (D) Rems-Murr-Kreis	Grossgartach	M	5	.	.	.	.	.	.	.	.
Ilsfeld (D) Kr. Heilbronn	Grossgartach/Rössen	M	5	.	.	.	.	.	.	.	.
Ditzingen (D) Kr. Ludwigsburg	Rössen	M	5	.	.	.	.	.	.	.	.
Gonvillars (F) layer XI, Haute-Saone,	Mittelneolithikum, früh	M	5	.	.	.	.	.	.	.	

Table 14.3, cont.

unresolved until there is an investigation of a waterlogged early neolithic site from the area under discussion. It is interesting to note that in the LBK well of Kückhoven in the Rhineland (Knörzer 1998, p.232) flax is present in considerable amounts (a total of almost 2,000 remains).

Opium poppy (*Papaver setigerum/somniferum*) was found very rarely, it is present only in Ulm-Eggingen in a middle-LBK context and here there were just two seeds (table 14.3; Gregg 1989, p.373). This is the eastern-most find of poppy in the early Neolithic in our region. However, it is impossible to assess accurately its role because of the unfavourable preservation conditions. In the above-mentioned well of Kückhoven over 6,000 seeds were found (Knörzer 1998, p.232).

The same is valid for the group of gathered plants (table 14.3); the only species that are preserved reasonably well are those that had some chances of becoming carbonised or of being preserved once carbonised. So only hazelnut shells (*Corylus avellana*) occurred regularly. Wild apple (*Malus sylvestris*) was found at one site and sloe (*Prunus spinosa*) was present at two sites. The caryopses of brome grass (*Bromus* spp.), which are almost always found in large amounts, were probably also a food resource; it is possible, however, that they were in fact only weeds. Worth mentioning are the finds of fruits of mistletoe (*Viscum album*) at two sites, in Hilzingen and Ditzingen (Stika 1991, pp.96–98; Piening 1998). From later waterlogged settlements it is recorded that mistletoe was often collected in very large amounts and used as fodder for livestock (Hosch and Jacomet 2004) or for medicinal purposes (Maier 2001, pp.135–138). Interesting finds of ruderal plants, which were probably also used for medicinal purposes, include *Hyoscyamus niger* and *Nepeta cataria* (Piening 1982, 1998). As we know, for instance from most recent analyses of waterlogged remains from wells dated to the LBK (Knörzer 1998), it is likely that gathering of plants also played an important role in the economy during the early Neolithic.

14.3.2 The middle Neolithic

Quantifiable archaeobotanical investigations exist for only seven settlements of the middle Neolithic in the area under consideration (ca. 5000-4300 cal BC.; table 14.1, figure 14.5), although for five of these the analyses are not considered to be representative. Three of the settlements are situated in Bavaria and three are in the Neckar region, so in the same regions as the LBK sites, and only one, the Grotte de Gonvillars (Dép. Haute-Saone), is located in the western-most part of the region. This cave site was excavated in the 1960s (for new insights see Jeunesse and Pétrequin 1997). There are some ongoing investigations of settlements located in the west (including Switzerland), but only very preliminary results are available from these. All the sites included in the table are situated on dry mineral soils, however, one of the newly investigated settlements yielded waterlogged plant material of the Grossgartacher Kultur (Singen-Offwiese: Dieckmann et al. 1998).

Einkorn and emmer continue to be the most important cereals in the middle neolithic settlements (table 14.3). In four of the sites, however, naked wheat is found in much higher numbers than in all the early neolithic sites of the area, and this is also the case at Singen-Offwiese. In addition, naked wheat is known from two sites where rachis remains were found (Ditzingen: Piening 1998, fig. 10; and Singen-Offwiese: Dieckmann et al. 1998) and it is clearly a hexaploid form (*Triticum aestivum*). Hexaploid naked wheat was also found in large amounts in other middle neolithic settlements in Europe (for an overview see Maier 1998).

In the Grotte de Gonvillars settlement there is also evidence of spelt wheat, where grains and also some glume bases were identified as *Triticum* cf. *spelta* (Villaret-von Rochow 1974). These identifications should be re-checked to confirm the presence of the species.

Barley seems to have had greater importance in the middle Neolithic (for example, on the five settlements dated to this period) than in the early Neolithic (table 14.3). Naked barely is the most commonly identified form. In summary, there seem to be some basic changes in the importance of certain crops in the cereal spectra of this period compared with the early Neolithic.

Other cultivated (table 14.3) and gathered plants are very rare. Lentil and flax are present in the only thoroughly investigated site (Künzing-Unternberg in Bavaria, Oberlauterbacher Gruppe); poppy was found in Schernau but pea is absent from the records. Even if there are no on-site data yet available for the Swiss Plateau, agriculture must have also played an important role there during the middle Neolithic. In Zürich in lake sediments below the first lakeshore settlements there is up to 5% cereal pollen (Erny-Rodmann 1995). Finds of single sherds out of context—'Streufunde'—indicate the presence of the Grossgartacher Kultur (shortly after 5000 cal BC; figure 14.3). However, to date there are no traces of the sites that correspond to these pollen data. In summary, the whole area was settled and cultivated in the middle Neolithic, but our knowledge of this period is very sparse.

14.3.3 Late Neolithic 1: 4300–3500 cal BC

From the early part of the late Neolithic (4300-3500 cal BC) there are many more (46) settlements from which plant remains have been investigated (table 14.2, figure 14.6). Ten settlements are situated on dry mineral soils (lower Bavaria, Neckar region) and the remaining 36 are located on lakeshores or in peat bogs. They are spread from the French Jura over the whole Swiss Plateau as far as Bavaria (figure 14.1), and belong to different cultural groups (see figure 14.3). In recent years some sites on mineral soils were excavated that were also in the region of the lake dwellings, but archaeobotanical data from these are not yet available.

Bavaria

Only four settlements of this time period have been investigated in Bavaria; three belong to the Münchshöfener Kultur, which dates from before 4000 cal BC, and one to the later Altheimer Kultur (figure 14.3, table 14.2).

Of the sites of the Münchshöfener Kultur, only one (Gaimersheim) has been investigated reasonably thoroughly. Einkorn and emmer are the most important cereals at this site, naked wheat is absent and barley is present in very low amounts (table 14.4). The other two settlements of this cultural group either yielded below 50 cereal remains or produced evidence resulting from the analysis of material from only very few contexts. Nevertheless, the picture does not change and emmer and einkorn are the most abundant wheat species; barley is absent, however, as is naked wheat.

There are some old investigations of another 10 sites of the Münchshöfener Kultur, which were compiled by Hopf and Blankenhorn (1986; see also above in 'Materials and methods'). The results relating to the wheat species are identical to the more modern evaluations, but for barley they are different and in these older analyses it is frequently recorded as being present. To summarize, on sites of the Münchshöfener Kultur, einkorn and emmer seem to be the most important cereals and naked wheat is absent; barley is present only in very low amounts, so its importance is not clear.

The only settlement of the Altheimer Kultur with archaeobotanical data is Ergolding-Fischergasse (after ca. 3800 cal BC, figure 14.3, table 14.2 and table 14.4), this site has been analysed quite thoroughly. It was formerly a waterlogged peat bog site, but the sediments dried out as a result of land improvement, so mostly carbonised remains were preserved. There are no differences in the occurrence of the wheat and barley species in comparison with the earlier sites. Worth mentioning are some glume bases identified by the author as spelt (*Triticum spelta*), however, in the illustration they look rather emmer-like (Küster 1989, p.24). Of interest also are several grains of millet (*Panicum miliaceum*), which are the first finds of this species since the oldest LBK.

No pulses or opium poppy seeds were found in any of the four sites mentioned. Flax is present in small amounts only in two settlements of the Münchshöfener Kultur (table 14.4). Gathered

Settlement	Culture	Preservation	Representativeness	Cereals					
				Einkorn	Stores of einkorn	Emmer	Stores of emmer	Einkorn/emmer	Stores of einkorn/emmer
Gaimersheim (D) Kr. Eichstätt	Münchshöfen	M	3	11	.	33	.	522	.
Vilsbiburg (D) Kr. Landshut	Münchshöfen	M	5	.	.	.	.	63	.
Vilsbiburg (D) Kr. Landshut, Bauplatz König und Lerchenstrasse	Münchshöfen	M	3	7	.	28	.	9	.
Ergolding (D) Fischergasse, Kr. Landshut	Altheim	M(W)	2	1746	.	314	.	1815	5
Ehrenstein (D) Phasen I-III, Kr. Ulm,	Schussenried	W	4	33537	x	104035	11	+++	7
Alleshausen (D) Hartöschle Kr. Biberach	Schussenried	W	5	1059	.	400	.	689	.
Reute (D) Schorrenried Kr. Ravensburg	Pfyn/Altheim	W	4	32	.	7	.	.	.
Oedenahlen (D) Riedwiesen, Kr. Biberach	Pfyn/Altheim	W	3	1356	.	1482	.	2098	.
Aldingen (D) Gem. Remseck, Kr. Ludwigsburg "Halden I"	Schwieberdinger Gr.	M	3	6043	x	755	.	300	.
Hochdorf (D) Gem. Eberdingen, Kr. Ludwigsburg	Schussenried	M	1	31105	x	1244	.	14178	x
Aldingen (D) Gem. Remseck, Kr. Ludwigsburg "Halden II"	Schussenried	M	5	265	.	72	.	49	.
Freiberg (D) Geisingen, Kr. Ludwigsburg	Schussenried	M	5	1379	.	21	.	2	.
Grossachsenheim (D) Holderbüschle Kr. Ludwigsburg	Schussenried	M	5	1188	.	216	.	313	.
Schlösslesfeld (D) Kr. Ludwigsburg	Schussenried	M	3	24	.	34	.	.	.
Heilbronn (D) Klingenberg	Michelsberger MKV	M	1	913	.	981	.	1348	.
Hornstaad (D) Hörnle IA Kr. Konstanz	Pfyn (Hornstaader Gr.)	W	1	11253	x	1363	.	5404	.
Thayngen (CH) Weier, Sch. 16-19, Profil III	Pfyn	W	5	.	.	18	.	11	.
Wallhausen (D) Ziegelhütte, Kr. Konstanz	Pfyn	W	5	8	.	24	.	.	.
Sipplingen (D) Osthafen, Pfyner Schichten, Bodenseekreis	Pfyn	W	2	1389	.	353	.	1461	.
Gachnang (CH) Niederwil	Pfyn	W	2	.	.	54	.	.	.
Hornstaad (D) Hörnle IB Kr. Konstanz	Pfyn	W	5	.	.	9	.	.	.
Zürich (CH) Kleiner Hafner, Schichten 5A+B	Egolzwil	W	5	174	.	67	.	.	.
Egolzwil (CH) 3, Wauwiler Moos	Egolzwil	W	2	3	.	1	.	1	.
Cham (CH) Eslen	Egolzwil/frühes zs. Cortaillod bzw. Zürich Hafner	W	2	5	.	3	.	3	.
Zürich (CH) Kleiner Hafner Schichten 4A-C/D	Cortaillod, frühes zs. bzw. Zürich-Hafner	W	3	441	.	10	.	.	.
Zürich (CH) Kleiner Hafner Schichten 4E/F	Cortaillod, klass. zs. bzw. Zürich-Hafner	W	2	302	.	315	1	.	.
Zürich (CH) Mozartstrasse Schicht 6	Cortaillod, klass. zs. bzw. Zürich-Hafner	W	5	9	.	94	.	.	.
Zürich (CH) Mozartstrasse Schicht 5 u,o	Cortaillod, klass. zs. bzw. Zürich-Hafner	W	3	17	.	103	.	.	.
Zürich (CH) Kan San Schicht 9	Pfyn/Cortaillod-Üb. bzw. Zürich Hafner	W	2	134	.	312	.	.	.
Seeberg (CH) Burgäschisee-Süd	Cortaillod, klass. zs. bzw. Zürich-Hafner	W	4	7	.	9	.	.	.
Zürich (CH) AKAD/Pressehaus, Schicht J	Pfyn. bzw. Zürich-Seefeld	W	2	89	.	448	.	.	.
Zürich (CH) KanSan Schicht 7	Pfyn. bzw. Zürich-Seefeld	W	5	.	.	19	.	.	.
Zürich (CH) Mozartstrasse Schicht 4B+4A	Pfyn. bzw. Zürich-Seefeld	W	5	.	.	6	.	.	.
Risch (CH) Oberrisch, Aabach	Pfyn. bzw. Zürich-Seefeld	M(W)	1	.	.	14	.	.	.
Cham (CH) St. Andreas	Pfyn (-Cortaillod?)	M(W)	5	.	.	14	.	.	.
Zürich (CH) Mozartstrasse Schicht 4 u,m,o	Pfyn. bzw. Zürich-Seefeld	W	3	.	.	24	.	.	.
Zürich (CH) KanSan Schicht 5	Pfyn. bzw. Zürich-Seefeld	W	5	.	.	2	.	.	.
Twann (CH) US, E1-2	Cortaillod classique	W	5	11	.	4	.	14	.
Concise (CH) Sous Colachoz (EMS)*	Cortaillod moyen, Ens. 2	W	2	4	.	1	.	.	.
Twann (CH) MS (E3-4)	Cortaillod tardif	W	5	.	.	.	.	.	.
Port (CH) Stüdeli, US	Cortaillod, spätes	W	4	391	.	125	.	57	.
Twann (CH) MS (E5 - 5a)	Cortaillod tardif	W	5	5	.	.	.	.	.
Twann (CH) OS (6-7)	Cortaillod tardif	W	5	6	.	57	.	46	1?
Port (CH) Stüdeli, OS	Cortaillod, spätes	W	4	2850	x	13558	x	217	.
Concise (CH) Sous Colachoz COT2*	Cortaillod tardif	W	5	4	.	.	.	.	.
Clairvaux (F) V Motte Aux Magnins	Néolithique Moyen Bourguignon récent	W	3	.	.	655	.	.	.

Table 14.4: Cultivated plants of the early phases of the late Neolithic (4300–3500 cal BC). Carbonised and waterlogged plant remains preserved.

Settlement	Culture	Preservation	Representativeness	Cereals, cont.						
				Spelt	Naked wheat	Stores of naked wheat	Barley	Stores of barley	Broomcorn millet	Italian millet
Gaimersheim (D) Kr. Eichstätt	Münchshöfen	M	3	.	.	.	3	.	.	.
Vilsbiburg (D) Kr. Landshut	Münchshöfen	M	5	.	.	.	.	.	.	.
Vilsbiburg (D) Kr. Landshut, Bauplatz König und Lerchenstrasse	Münchshöfen	M	3	.	.	.	.	.	.	.
Ergolding (D) Fischergasse, Kr. Landshut	Altheim	M(W)	2	5	.	.	1	.	5	.
Ehrenstein (D) Phasen I-III, Kr. Ulm,	Schussenried	W	4	2227cf	9212	x	8362	1	.	.
Alleshausen (D) Hartöschle Kr. Biberach	Schussenried	W	5	.	136	.	91	.	.	.
Reute (D) Schorrenried Kr. Ravensburg	Pfyn/Altheim	W	4	10cf	1102	?	760	?	.	.
Oedenahlen (D) Riedwiesen, Kr. Biberach	Pfyn/Altheim	W	3	.	3727	x	863	.	.	.
Aldingen (D) Gem. Remseck, Kr. Ludwigsburg "Halden I"	Schwieberdinger Gr.	M	3	.	8771	x	1044	.	.	.
Hochdorf (D) Gem. Eberdingen, Kr. Ludwigsburg	Schussenried	M	1	.	87	.	10865	x	.	1
Aldingen (D) Gem. Remseck, Kr. Ludwigsburg "Halden II"	Schussenried	M	5	55cf	956	.	1	.	.	.
Freiberg (D) Geisingen, Kr. Ludwigsburg	Schussenried	M	5	.	.	.	611	.	.	.
Grossachsenheim (D) Holderbüschle Kr. Ludwigsburg	Schussenried	M	5	.	5	.	32	.	.	.
Schlösslesfeld (D) Kr. Ludwigsburg	Schussenried	M	3	.	2	.	29	.	.	.
Heilbronn (D) Klingenberg	Michelsberger MKV	M	1	.	2949	.	5598	x	1	.
Hornstaad (D) Hörnle IA Kr. Konstanz	Pfyn (Hornstaader Gr.)	W	1	.	78687	x	26608	x	.	.
Thayngen (CH) Weier, Sch. 16-19, Profil III	Pfyn	W	5	.	2571	x	106	.	.	.
Wallhausen (D) Ziegelhütte, Kr. Konstanz	Pfyn	W	5	.	1304	?	211	.	.	.
Sipplingen (D) Osthafen, Pfyner Schichten, Bodenseekreis	Pfyn	W	2	.	35815	x	1937	.	.	.
Gachnang (CH) Niederwil	Pfyn	W	2	.	61578	6	15786	3	.	.
Hornstaad (D) Hörnle IB Kr. Konstanz	Pfyn	W	5	.	41	.	65	.	.	.
Zürich (CH) Kleiner Hafner, Schichten 5A+B	Egolzwil	W	5	.	205	.	90	.	.	.
Egolzwil (CH) 3, Wauwiler Moos	Egolzwil	W	2	.	89	.	76	.	.	.
Cham (CH) Eslen	Egolzwil/frühes zs. Cortaillod bzw. Zürich Hafner	W	2	.	5	.	17	.	.	.
Zürich (CH) Kleiner Hafner Schichten 4A-C/D	Cortaillod, frühes zs. bzw. Zürich-Hafner	W	3	.	5477	2	6411	9	.	.
Zürich (CH) Kleiner Hafner Schichten 4E/F	Cortaillod, klass. zs. bzw. Zürich-Hafner	W	2	.	28534	20	438	1	.	.
Zürich (CH) Mozartstrasse Schicht 6	Cortaillod, klass. zs. bzw. Zürich-Hafner	W	5	.	6378	1	60	.	.	.
Zürich (CH) Mozartstrasse Schicht 5 u,o	Cortaillod, klass. zs. bzw. Zürich-Hafner	W	3	.	142	.	44	.	.	.
Zürich (CH) Kan San Schicht 9	Pfyn/Cortaillod-Üb. bzw. Zürich Hafner	W	2	.	669	.	330	.	.	.
Seeberg (CH) Burgäschisee-Süd	Cortaillod, klass. zs. bzw. Zürich-Hafner	W	4	.	193	.	183	.	.	.
Zürich (CH) AKAD/Pressehaus, Schicht J	Pfyn. bzw. Zürich-Seefeld	W	2	54cf	9565	x	1411	x	.	.
Zürich (CH) KanSan Schicht 7	Pfyn. bzw. Zürich-Seefeld	W	5	.	21591	2	497	.	.	.
Zürich (CH) Mozartstrasse Schicht 4B+4A	Pfyn. bzw. Zürich-Seefeld	W	5	.	132	.	62	.	.	.
Risch (CH) Oberrisch, Aabach	Pfyn. bzw. Zürich-Seefeld	M(W)	1	.	6682	x	5512	x	.	.
Cham (CH) St. Andreas	Pfyn (-Cortaillod?)	M(W)	5	.	710	x	212	x	.	.
Zürich (CH) Mozartstrasse Schicht 4 u,m,o	Pfyn. bzw. Zürich-Seefeld	W	3	.	2198	.	569	.	.	.
Zürich (CH) KanSan Schicht 5	Pfyn. bzw. Zürich-Seefeld	W	5	.	31	.	34	.	.	.
Twann (CH) US, E1-2	Cortaillod classique	W	5	.	1315	3	14	.	.	.
Concise (CH) Sous Colachoz (EMS)*	Cortaillod moyen, Ens. 2	W	2	.	3	.	2	.	.	.
Twann (CH) MS (E3-4)	Cortaillod tardif	W	5	.	3	.	7	.	.	.
Port (CH) Stüdeli, US	Cortaillod, spätes	W	4	.	2649	x	12335	x	.	.
Twann (CH) MS (E5 - 5a)	Cortaillod tardif	W	5	.	2947	6	997	1	.	.
Twann (CH) OS (6-7)	Cortaillod tardif	W	5	13	253	1	190	.	.	.
Port (CH) Stüdeli, OS	Cortaillod, spätes	W	4	.	2030	.	6056	x	.	.
Concise (CH) Sous Colachoz COT2*	Cortaillod tardif	W	5	.	3	.	2	.	.	.
Clairvaux (F) V Motte Aux Magnins	Néolithique Moyen Bourguignon récent	W	3	.	22	.	5	.	.	.

Table 14.4, cont.

Settlement	Culture	Preservation	Representativeness	Other cultivated plants						
				Pea	Lentil	Flax	Flax stores	Opium poppy	Celery	Dill
Gaimersheim (D) Kr. Eichstätt	Münchshöfen	M	3	.	.	.	.	.	.	.
Vilsbiburg (D) Kr. Landshut	Münchshöfen	M	5	.	.	1	.	.	.	.
Vilsbiburg (D) Kr. Landshut, Bauplatz König und Lerchenstrasse	Münchshöfen	M	3	.	.	14	.	.	.	.
Ergolding (D) Fischergasse, Kr. Landshut	Altheim	M(W)	2	.	.	.	.	.	.	.
Ehrenstein (D) Phasen I-III, Kr. Ulm,	Schussenried	W	4	.	.	.	.	.	.	.
Alleshausen (D) Hartöschle Kr. Biberach	Schussenried	W	5	17	.	> 368	.	4	.	1
Reute (D) Schorrenried Kr. Ravensburg	Pfyn/Altheim	W	4	+	.	+	.	.	.	.
Oedenahlen (D) Riedwiesen, Kr. Biberach	Pfyn/Altheim	W	3	.	.	2846	12	202	.	.
Aldingen (D) Gem. Remseck, Kr. Ludwigsburg "Halden I"	Schwieberdinger Gr.	M	3	.	.	.	.	1	.	.
Hochdorf (D) Gem. Eberdingen, Kr. Ludwigsburg	Schussenried	M	1	32	.	1	.	121	.	.
Aldingen (D) Gem. Remseck, Kr. Ludwigsburg "Halden II"	Schussenried	M	5	.	.	.	.	.	.	.
Freiberg (D) Geisingen, Kr. Ludwigsburg	Schussenried	M	5	1	.	.	.	.	.	.
Grossachsenheim (D) Holderbüschle Kr. Ludwigsburg	Schussenried	M	5	3	.	.	.	.	.	.
Schlösslesfeld (D) Kr. Ludwigsburg	Schussenried	M	3	.	.	.	.	.	.	.
Heilbronn (D) Klingenberg	Michelsberger MKV	M	1	3169	24	2	.	5	.	.
Hornstaad (D) Hörnle IA Kr. Konstanz	Pfyn (Hornstaader Gr.)	W	1	100	.	49	x	363	1	4
Thayngen (CH) Weier, Sch. 16-19, Profil III	Pfyn	W	5	.	.	181	.	904	.	.
Wallhausen (D) Ziegelhütte, Kr. Konstanz	Pfyn	W	5	.	.	1096	.	699	.	.
Sipplingen (D) Osthafen, Pfyner Schichten, Bodenseekreis	Pfyn	W	2	.	.	4753	.	201	.	.
Gachnang (CH) Niederwil	Pfyn	W	2	2	.	41456	6	29818	15	.
Hornstaad (D) Hörnle IB Kr. Konstanz	Pfyn	W	5	.	.	4845	.	1219	.	.
Zürich (CH) Kleiner Hafner, Schichten 5A+B	Egolzwil	W	5	89	.	27	.	9182	.	.
Egolzwil (CH) 3, Wauwiler Moos	Egolzwil	W	2	29	1cf	338	.	4663	.	.
Cham (CH) Eslen	Egolzwil/frühes zs. Cortaillod bzw. Zürich Hafner	W	2	6	.	10	.	3938	.	.
Zürich (CH) Kleiner Hafner Schichten 4A-C/D	Cortaillod, frühes zs. bzw. Zürich-Hafner	W	3	1	.	31	.	1536	1cf	.
Zürich (CH) Kleiner Hafner Schichten 4E/F	Cortaillod, klass. zs. bzw. Zürich-Hafner	W	2	1	.	1111	.	2057	.	4
Zürich (CH) Mozartstrasse Schicht 6	Cortaillod, klass. zs. bzw. Zürich-Hafner	W	5	1	.	111	.	3095	1cf	1
Zürich (CH) Mozartstrasse Schicht 5 u,o	Cortaillod, klass. zs. bzw. Zürich-Hafner	W	3	3	.	162	.	4142	.	.
Zürich (CH) Kan San Schicht 9	Pfyn/Cortaillod-Üb. bzw. Zürich Hafner	W	2	1	.	2939	.	6053	.	4
Seeberg (CH) Burgäschisee-Süd	Cortaillod, klass. zs. bzw. Zürich-Hafner	W	4	>10	.	>21	.	>3000	.	.
Zürich (CH) AKAD/Pressehaus, Schicht J	Pfyn. bzw. Zürich-Seefeld	W	2	4	.	30822	x	130752	8	.
Zürich (CH) KanSan Schicht 7	Pfyn. bzw. Zürich-Seefeld	W	5	.	.	885	.	25245	.	.
Zürich (CH) Mozartstrasse Schicht 4B+4A	Pfyn. bzw. Zürich-Seefeld	W	5	.	.	74	.	162	.	.
Risch (CH) Oberrisch, Aabach	Pfyn. bzw. Zürich-Seefeld	M(W)	1	.	.	6044	.	60010	.	.
Cham (CH) St. Andreas	Pfyn (-Cortaillod?)	M(W)	5	3	.	.	.	++	.	.
Zürich (CH) Mozartstrasse Schicht 4 u,m,o	Pfyn. bzw. Zürich-Seefeld	W	3	12	.	18158	.	14131	1	2
Zürich (CH) KanSan Schicht 5	Pfyn. bzw. Zürich-Seefeld	W	5	.	.	2586	.	8799	.	.
Twann (CH) US, E1-2	Cortaillod classique	W	5	.	.	1860	.	5245	.	.
Concise (CH) Sous Colachoz (EMS)*	Cortaillod moyen, Ens. 2	W	2	1	.	469	.	2107	.	1
Twann (CH) MS (E3-4)	Cortaillod tardif	W	5	.	.	772	.	5070	.	.
Port (CH) Stüdeli, US	Cortaillod, spätes	W	4	10	.	114	.	430	.	3
Twann (CH) MS (E5 - 5a)	Cortaillod tardif	W	5	1062	.	327	.	2148	.	.
Twann (CH) OS (6-7)	Cortaillod tardif	W	5	.	.	1971	.	2108	.	.
Port (CH) Stüdeli, OS	Cortaillod, spätes	W	4	62	.	1279	2	583	.	.
Concise (CH) Sous Colachoz COT2*	Cortaillod tardif	W	5	—no information available—						
Clairvaux (F) V Motte Aux Magnins	Néolithique Moyen Bourguignon récent	W	3	10	.	225	.	7203	.	.

Table 14.4, cont.

plants are also rare (table 14.5), but this could be due to the preservation conditions or to the fact that very few samples were analysed. In Ergolding-Fischergasse there are some remains of hazelnut, wild apple, *Rubus* sp. and *Sambucus* sp. (table 14.5). The state of research is such that it does not allow us to make interpretations about the role of the gathered plants.

Neckar region

Five of the seven sites in the Neckar region belong to the Schussenrieder Kultur, one to the somewhat earlier Schwieberdinger Gruppe, which is dated to ca. 4000 cal BC. (figure 14.3), and one (Heilbronn-Klingenberg) belongs to a late phase of the Michelsberger Kultur, dated to ca. 3600 cal BC (MKV, figure 14.3, table 14.2, figure 14.6).

Of the Schussenrieder and Schwieberdinger sites only three have been investigated well, although all yielded relatively large amounts of cereal remains (table 14.4). Einkorn or emmer are dominant in two of these three settlements (Hochdorf and Schlösslesfeld). Einkorn is also the predominant wheat species in two of the less thoroughly investigated sites, followed by emmer. Naked wheat is present in small amounts in these settlements but from the publications it is not clear which type it is (there is no picture of the rachis remains in the Hochdorf publication). Barley is more abundant compared with the sites discussed above, and at some sites is present in large amounts; it is mostly described as the naked form. From Hochdorf there is one grain of Italian millet (*Setaria italica*).

Two of the remaining sites, both at Aldingen (Halden I and Halden II), present a very different wheat spectrum and at these sites naked wheat is the most important cereal. In the pits of the Schwieberdinger Gruppe (Halden I), in addition to a large amount of well-preserved grains, there were some rachis internodes (Piening 1986a, pp.203–204); they showed the typical characteristics of tetraploid naked wheat (*Triticum turgidum/durum*). Tetraploid naked wheat is also present in the Schussenrieder settlement (Halden II: Piening 1992, fig. 4). At both sites naked wheat outnumbered both einkorn and emmer. In Aldingen Halden II, some well preserved grains and spikelet forks were recovered, which showed very clearly the characteristics of spelt (*Triticum spelta*; Piening 1992, pp.133–136). Naked barley is also present in fairly large amounts.

Other cultivated plants of the Schussenrieder Kultur sites in the Neckar area include pea, flax and opium poppy (table 14.4). Flax and poppy are only present in in large amounts at the site of Hochdorf, and here over 100 grains of parsley (*Petroselinum crispum*) were also found.

In the only settlement (an 'Erdwerk' or earthwork) of the Michelsberger Kultur that has been investigated in our area, which is dated to around 3600 cal BC, naked wheat is the dominant wheat species, followed by emmer and einkorn (table 14.4). In addition to naked wheat grains there were 169 rachis internodes (Stika 1996, pp.35–37), and these clearly showed the characteristics of the tetraploid form. Barley is also present in large amounts and is represented mainly by the naked form. There are many finds of stored peas and interestingly lentil is also present. Flax and opium poppy are rare but this may be due primarily to preservation conditions.

Gathered plants are rare in most of the sites mentioned here (table 14.5). They were found in large amounts, including a high diversity of other taxa, only at Hochdorf and Heilbronn-Klingenberg, the two most thoroughly investigated sites. Hazelnuts and wild apples seem to have been important; a 'store' of carbonised apples was found in Aldingen Halden II (Piening 1992, p.129).

Oberschwaben (Upper Swabia), including the Federsee

There are four settlements with quantifiable data from well stratified contexts (table 14.2, figure 14.6). Two of these (Ehrenstein near Ulm and Alleshausen-Hartöschle at the Federsee) belong the Schussenrieder Kultur and are dated shortly after 4000 cal BC (figure 14.3). In addition, there is a collection of museum material from old excavations (19th century) of settlements

Settlement	Preservation	Representativeness	Gold of pleasure	Hazelnut	Wild apple	Sloe	Acorns	Black/rasp-berry	Elder	Wild strawberry	Rose	Brassica rapa
Gaimersheim (D) Kr. Eichstätt	M	3	.	.	.	.	.	.	.	.	.	.
Vilsbiburg (D) Kr. Landshut	M	5	.	.	.	.	.	.	.	.	.	.
Vilsbiburg (D) Kr. Landshut, Bauplatz König und Lerchenstrasse	M	3	.	.	.	.	.	.	1	.	.	.
Ergolding (D) Fischergasse, Kr. Landshut	M(W)	2	.	1	1	.	.	20	101	.	.	.
Ehrenstein (D) Phasen I-III, Kr. Ulm,	W	4	.	>81	>60	5	.	42699	6	10267	.	.
Alleshausen (D) Hartöschle Kr. Biberach	W	5	.	+	+++	.	.	+	+	+	+	.
Reute (D) Schorrenried Kr. Ravensburg	W	4	.	60	1	.	1	>724	4	12	.	.
Oedenahlen (D) Riedwiesen, Kr. Biberach	W	3	162	303	102	.	1	1598	47	2744	8	33
Aldingen (D) Gem. Remseck, Kr. Ludwigsburg "Halden I"	M	3	.	.	.	.	.	1	.	.	.	.
Hochdorf (D) Gem. Eberdingen, Kr. Ludwigsburg	M	1	.	32	1009	.	.	.	8	6	.	4
Aldingen (D) Gem. Remseck, Kr. Ludwigsburg "Halden II"	M	5	.	1	185	.	.	.	.	.	.	.
Freiberg (D) Geisingen, Kr. Ludwigsburg	M	5	.	.	.	.	.	.	.	.	.	.
Grossachsenheim (D) Holderbüschle Kr. Ludwigsburg	M	5	.	.	.	.	.	.	.	1	.	.
Schlösslesfeld (D) Kr. Ludwigsburg	M	3	.	.	.	.	.	.	.	.	.	.
Heilbronn (D) Klingenberg	M	1	.	41	48	5	.	3	61	1	.	.
Hornstaad (D) Hörnle IA Kr. Konstanz	W	1	10	33	298	1	3	300	1	1359	3	25
Thayngen (CH) Weier, Sch. 16-19, Profil III	W	5	.	3	39	3	.	707	43	3455	.	25
Wallhausen (D) Ziegelhütte, Kr. Konstanz	W	5	.	+	30	.	16	19	+	110	.	.
Sipplingen (D) Osthafen, Pfyner Schichten, Bodenseekreis	W	2	.	29	221	.	.	405	7	1244	.	4
Gachnang (CH) Niederwil	W	2	.	>215	>307	++	>30	16448	3850	12328	.	30849
Hornstaad (D) Hörnle IB Kr. Konstanz	W	5	9	10	716	2	1	676	1	1119	.	25
Zürich (CH) Kleiner Hafner, Schichten 5A+B	W	5	7	68	300	.	1	1065	.	1325	7	36
Egolzwil (CH) 3, Wauwiler Moos	W	2	9	552	164	.	28	5185	57	3405	116	197
Cham (CH) Eslen	W	2	.	53	2668	4	7	1095	8	287	15	4
Zürich (CH) Kleiner Hafner Schichten 4A-C/D	W	3	.	548	86	.	2	166	.	629	1	21
Zürich (CH) Kleiner Hafner Schichten 4E/F	W	2	46	30	1126	2	2	7769	.	11925	264	420
Zürich (CH) Mozartstrasse Schicht 6	W	5	62	362	493	9	13	4443	2	4555	474	58
Zürich (CH) Mozartstrasse Schicht 5 u,o	W	3	131	3672	302	4	9	5836	10	3041	83	90
Zürich (CH) Kan San Schicht 9	W	2	451	895	988	7	761	4451	.	7464	127	75
Seeberg (CH) Burgäschisee-Süd	W	4	.	1000	>25	5	5	2000	6	>50	2	275
Zürich (CH) AKAD/Pressehaus, Schicht J	W	2	209	6329	4101	318	1552	8612	.	9860	680	1339
Zürich (CH) KanSan Schicht 7	W	5	4	951	505	18	1	2301	7	6993	57	385
Zürich (CH) Mozartstrasse Schicht 4B+4A	W	5	3	159	110	20	1	311	.	151	1	22
Risch (CH) Oberrisch, Aabach	M(W)	1	.	1019	405	41	.	128500	27	15433	23	2894
Cham (CH) St. Andreas	M(W)	5	.	.	.	.	.	.	.	.	.	+++
Zürich (CH) Mozartstrasse Schicht 4 u,m,o	W	3	417	4237	2048	117	2	17658	.	4363	173	7654
Zürich (CH) KanSan Schicht 5	W	5	62	203	480	14	.	5677	70	1690	108	681
Twann (CH) US, E1-2	W	5	.	17	824	.	12	423	1	1745	30	.
Concise (CH) Sous Colachoz (EMS)*	W	2	.	49	535	1	45	114	1	357	21	9
Twann (CH) MS (E3-4)	W	5	.	124	1631	1	4	526	1	2312	70	.
Port (CH) Stüdeli, US	W	4	2	17	39	.	.	59	7	66	1	100
Twann (CH) MS (E5 - 5a)	W	5	1	5	261	.	.	353	.	754	23	9
Twann (CH) OS (6-7)	W	5	.	23	238	.	6	243	.	694	67	7
Port (CH) Stüdeli, OS	W	4	.	3	3	.	.	20	.	41	3	9
Concise (CH) Sous Colachoz COT2 (Häuf.-Klass.)	W	5	—no information available—									
Clairvaux (F) V Motte Aux Magnins	W	3	31	118	685	3	2	280	.	1170	226	38

Table 14.5: Gathered plants of the early phases of the late Neolithic (4300–3500 cal BC). Mainly waterlogged plant remains preserved. Concise, Cortaillod moyen: data are concentrations per litre. For cultural designations, please refer to table 14.4.

(dated to around 4000 cal BC) assigned to the Aichbühler and Schussenrieder cultures in the Federsee (Blankenhorn and Hopf 1982). These results are worthy of mention even though, to some degree, they may be less reliable than those from the later analyses. Two of the recently investigated settlements from this region (Oedenahlen in the Federsee and Reute in the Schorrenried) have later dates of around 3700 cal BC and belong to the Pfyn-Altheimer Kultur (figure 14.3). All these settlements are types of 'lake dwelling' and have yielded waterlogged material. In Ehrenstein, however, mainly carbonised grain stores from a burnt layer were analysed.

All four settlements are investigated relatively well. At these settlements, there is a difference in the role of the different wheat species between the Schussenrieder and the Pfyn-Altheimer cultures. Glume wheats (emmer and/or einkorn) are predominant in the former and naked wheat is also present in fairly large amounts (table 14.4). The evidence from the old museum collections is slightly different; naked wheat takes the first place, followed by emmer, and einkorn is rare. But the relevance of this result is hard to judge. The type of naked wheat present in these settlements is probably tetraploid as shown by the rachis internodes (illustrations in: Blankenhorn and Hopf 1982, tab. 2, no. 10; Hopf 1968, tab. 9, no. 6).

Naked wheat is the most common wheat species in the two sites belonging to the Pfyn-Altheim cultural group; ear and rachis-remains of the tetraploid type are present at Oedenahlen (Maier 1995, pp.202–211). Large amounts of emmer and einkorn were also found.

In Ehrenstein, Reute-Schorrenried and also in the old museum collections from the Federsee there are some specimens that have been tentatively identified as spelt (*Triticum spelta*). However, the well-preserved material from Ehrenstein should be reassessed on the basis of more recently formulated morphological criteria for identification because it is not really clear from the illustrations in Hopf's publication (Hopf 1968) if *Triticum spelta* is present; the grains appear to be very similar to those of two-grained einkorn.

Barley is common in all the settlements and is often present in large amounts. A majority of the more precisely identified specimens have been assigned to the naked form.

Pulses seem to have been rare, and pea has been found at only two sites (table 14.4). Flax is present in very large amounts in the two settlements from the Federsee; the exceptionally high quantities could of course also be due to the good preservation in the waterlogged layers. In Oedenahlen, there were also several stores of flax (Maier 1995, pp.216–218). Opium poppy was found at two sites, but the remains were only quantified at Oedenahlen, where the numbers are not very high in comparison with other settlements further to the west (see below). It is not clear why poppy is absent at Ehrenstein but it is possible that there were methodological reasons that had affected its recovery.

Gathered plants were found in very large amounts in all these settlements (hundreds or thousands of specimens; table 14.5). Very well-preserved carbonised apple-halves were recovered in Oedenahlen (Maier 1995: 218-221). Worth mentioning are some special finds, for example, *Trapa natans,* which played an important role as a collected resource in the Federsee and in Reute-Schorrenried (Maier 1995, p.222; Karg 1996) and, at the latter site, quantities of *Physalis alkekengi* (over 1,000), most probably from the remains of human coprolites, which thus suggests its use as a source of vitamin C.

Lake Constance and eastern Switzerland

Archaeobotanical data from six lakeshore or peat bog settlements that date from the first half of the 4th millennium BC are available in this region. Four are situated at Lake Constance and two are in peat bogs of northeastern Switzerland (Gachnang-Niederwil and Thayngen-Weier; figure 14.6, table 14.2). Only three have been investigated well and of these, perhaps the most thoroughly analysed of all the lakeshore settlements is Hornstaad-Hörnle IA (Maier 2001). All the settlements belong to the Pfyner Kultur (including the early Pfyner group of Hornstaad) and they date from ca. 3900 to 3500 cal BC (figure 14.3).

The spectra of wheat species in all these sites are uniform and tetraploid naked wheat is dominant (table 14.4). Large amounts of ears were found at Hornstaad Hörnle IA (see Maier 1996), amongst which several landraces could be identified. Emmer is relatively common in all the settlements, whereas einkorn is present in only three. Barley was ubiquitous and was found in fairly large quantities (at five sites it is the second most frequently occurring cereal after naked wheat, and at one it is the most abundant).

Pulses are rare and only pea is present (table 14.4); it was found in large amounts solely at the settlement of Hornstaad Hörnle IA, mainly in the form of uncarbonised remains of the pods that have been interpreted by the author as debris from domestic cleaning. One can conclude, therefore, that the ripe pea pods were harvested and the seeds used (Maier 2001, pp.73–74, table 30).

Many remains of flax and opium poppy are present (table 14.4); stores of both species were found in Gachnang-Niederwil. Of special interest are finds of species that are most likely to be of Mediterranean origin, including *Apium* sp. (Hornstaad and Gachnang-Niederwil), *Anethum graveolens* and *Petroselinum crispum* (Hornstaad). The specimens of *Anethum* from Hornstaad are amongst the oldest finds of these species in the area (see also below, Central Switzerland).

Gathered plants, including hazelnuts, wild apple, acorns, *Rubus* ssp., *Fragaria vesca* etc., are present in very large amounts (table 14.5). Of special interest are 'stores' of less commonly collected plants, for example, in Hornstaad Hörnle IA there were several accumulations of different species with oil-rich seeds (*Descurainia sophia, Brassica rapa, Galeopsis tetrahit* etc.; Maier 2001, pp.123–127), which indicate that these plants were used regularly. Concentrations of *Brassica rapa* seeds were found in Gachnang-Niederwil and here too there were large densities of *Chenopodium album* seeds, which suggests that they were also collected intentionally. The seeds of this weed contain starch and in historical periods in Europe they were used as a staple food when cereal harvests failed (Maurizio 1927). Many other wild plants were collected as has been demonstrated by the investigations at Hornstaad-Hörnle IA (Maier 2001, p.110 ff.).

Central Switzerland

Archaeobotanical data from 16 lakeshore and peat bog settlements are available in this region (table 14.2, figure 14.6). Four are dated to before 4000 cal BC and are the oldest known lakeshore settlements in the area being discussed. They belong to the Egolzwiler and Early Cortaillod-cultures (figure 14.3) and are situated at Lake Zürich (Kleiner Hafner layers 5 and 4 A-C/D), Wauwiler Moos and at Lake Zug. The remaining 12 settlements date from between 4000 and 3600 cal BC. Those dated up to ca. 3800 cal BC belong to the westerly-influenced classical Cortaillod culture but after this date the settlements are assigned to the Pfyn culture, which spread from the east. The settlements are mainly situated at different locations in Zürich, at Lake Zug and at the Burgäschisee in the canton of Berne, further to the west. Six of these settlements have not been investigated well (those ranked 5 in table 14.2).

In the four earliest sites (dated to before 4000 cal BC), where more than 50 cereal remains were found, tetraploid naked wheat is the most important wheat species (table 14.4), einkorn is always the second-most common species and emmer is rare. Six-rowed barley (probably naked) is frequently present at the Zürich-Kleiner Hafner site in layers 4A–CD (including finds of stored grain). All types of remains (i.e., ears, spikelets, chaff, grains) were found. It has not been possible to account for the fact that identifiable cereal remains were astonishingly rare in the three settlements of the Egolzwiler Kultur even though many samples were analysed (for example, in Egolzwil 3; table 14.4). Cereal stores were found only in a burnt layer at the somewhat later settlement of Zürich-Kleiner Hafner (layers 4A-C/D).

The trend is the same for the settlements of the classical Cortaillod and the Pfyn culture, although cereal remains are commonly present in larger amounts. Naked wheat is predominant in all of the 12 sites, followed by emmer (table 14.4). Emmer is very rare in most settlements of the Pfyn culture, and einkorn is absent in some. Only stores of naked wheat, including whole ears, were common and amongst which different landraces were identified (Jacomet et al. 1989, table 12). Only one store of emmer was found at Zürich-Kleiner Hafner in layers 4E–F. In a majority of cases barley is ranked in second place after naked wheat. Stores of six-rowed barley were also found and these probably represented the naked variety.

Pea was found in all the early settlements dating to before 4000 cal BC and in those of the Egolzwiler Kultur it was present in large amounts (table 14.4). This pulse was present in eight of the 12 later settlements but only twice were there more than 10 specimens present; the importance of pea is therefore unclear. Only one tentatively identified lentil was found at Egolzwil 3.

Finds of flax and opium poppy are widespread and both are commonly present in large amounts (table 14.4), however, flax is much rarer in sites before 4000 cal BC. The highest numbers of flax seeds (including stores of the crop) are recorded in sites of the Pfyn culture. The typical Mediterranean weed of flax, *Silene cretica,* appears for the first time in this period (Brombacher and Jacomet 1997, p.265) and it then appears to become more common as flax-growing intensifies. Poppy is abundant in this region.

The relatively frequent finds of *Apium* sp. and *Anethum graveolens* after 4000 cal BC are remarkable. (table 14.4), as is the presence of *Melissa officinalis,* which was found in layer J at Zürich-AKAD. The oldest finds of these Mediterranean species come from layers that are contemporary with the above-mentioned settlement of Hornstaad Hörnle IA (ca. 3900 cal BC). Gathered plants are present in very large amounts (see table 14.5).

Western Switzerland and French Jura

Only nine settlements from five sites have been investigated from this area (table 14.2, figure 14.6). They are found at the Lakes Biel (Twann, Port) and Neuchâtel (Concise) and one is in the French Jura (Clairvaux, Motte aux Magnins). The investigations of most of these settlements are not considered to be representative; only semi-quantitative data are available from Concise (in one layer based only on four samples; Karg and Märkle 2002). As a consequence, our knowledge of the plant economy of these settlements is not comprehensive.

This is perhaps the main reason why the picture reflected by the cereal data is very heterogeneous (table 14.4). At some sites, particularly in the two layers of the Cortaillod culture in Concise, einkorn seems to be the most important cereal and, for example, there were large quantities in the stores at Port-Stüdeli OS. Worth mentioning is the presence of two-grained einkorn in Port (Brombacher and Jacomet 2003, pp.70–76). Emmer is predominant in Port-Stüdeli OS and in Clairvaux, Motte aux Magnins, whereas tetraploid naked wheat is the most important wheat species in the other settlements (for example, all the layers of Twann and also in Port Stüdeli US).

Barley is present in all the sites and often in large amounts. Stores were found in Port in which ears of both hulled and naked six-rowed barley have been identified.

Pea was relatively common (table 14.5) and in a burnt layer (Ensemble 5) at Twann MS there were large amounts of seeds together with parts of the pod and other remains, which suggests that a store of peas in their pods was carbonised *in situ* (Piening 1981, p.78–79).

Flax and opium poppy are present in high numbers (table 14.5). Of interest are the flax stores from Port Stüdeli that contain many seeds of the weed *Silene cretica*. The finds of capsule fragments (parts of the top) of poppy from the Cortaillod Moyen layer of Concise are also exceptional (Karg and Märkle 2002, p.173). *Anethum graveolens* has been found in two sites. Gathered plants are present in large amounts (table 14.5).

14.4 Discussion

All the settlements mirror a picture of a subsistence economy based on cultivation and gathering (also, of course, animal husbandry and hunting). Cereals are present in all sites and must have been a major source of calories, whereas the role of other cultivated plants is relatively hard to judge because the representaton of these taxa is partly determined by the preservation conditions. As far as can be ascertained, flax seems to have been fairly important from the early Neolithic onwards. Opium poppy must have played a role in the plant economy from the early Neolithic, but is only found in large amounts at waterlogged sites from 4300 cal BC onwards and must have played some role in the plant economy since then. The role of pulses is very unclear; pea is found regularly although it is seldom represented by many specimens and lentil is rare until the end of the 5th millennium cal BC after which time it disappears totally (see below).

In the area discussed, the glume wheats (einkorn and emmer) were the dominant cereals in the early and middle Neolithic (table 14.3) but naked wheat is rare. Hexaploid naked wheat was identified in the LBK well of Kückhoven in the Rhineland (Knörzer 1998) but there is no direct evidence of this type in our area.

The same glume wheat species continue to be the most important wheats in the middle Neolithic in Bavaria and most of the Neckar region. At this time, however, naked wheat becomes much more important, not only in our area but also elsewhere in Europe (see Maier 1998, p.21). The hexaploid type is present at two sites in our area and it is a matter of debate where it originated from prior to its penetration of the western part of Central Europe. Maier (1998) suggests that hexaploid wheat is perhaps a component of the plant suite of the 'Danubian' cultures.

Einkorn and emmer continue to be dominant in lower Bavaria, in most of the Neckar region and also in some places in Upper Swabia during the late Neolithic (figure 14.7). Tetraploid naked wheat appears to have been the main wheat species in the other areas (figure 14.8). This type of naked wheat is by far most important wheat species until 3500 cal BC in a majority of the settlements from Upper Swabia westwards, and it is also the dominant cereal at three sites of the Neckar-region. However, there are a few exceptions to this general trend, for example, at the lakes in western Switzerland and in the French Jura. The present state of research, which is particularly weak in the western-most part of the area, does not allow us to assign a reason for these regional differences.

The origin of the tetraploid naked wheat that reached the Alpine Foreland has been defined more clearly only recently. For over 20 years there have been theories that it must have come from the western Mediterranean, from the sphere of influence of the Cardial/Impressa and the succeeding cultural groups (see figure 14.3; Jacomet and Schlichtherle 1984; Schlichtherle 1997b, p.12-13; Maier 1998, p.211 ff.). However, until a few years ago no rachis remains had been found in the Mediterranean area. New excavations have shown that there are also Cardial lakeshore settlements in the Mediterranean (in central Italy: La Marmotta, see Rottoli in this volume; in NW Spain: La Draga, Bosch i Lloret et al. 2000; Buxó et al. 2000). The remains of ears, spikelets and rachis fragments of tetraploid naked wheat have been identified at these sites (pers. comm. R. Buxo and M. Rottoli). So it now seems much more likely that *Triticum turgidum/durum* penetrated our area from southwest, together with many other influences that are visible in the archaeological artefacts (see introduction).

Of note, the tetraploid naked wheat is not restricted to the sphere of influence of the westerly-based cultural groups, for example, the Egolzwiler and Cortaillod cultures. It spread very rapidly as far as the Federsee region and entered the spheres of influence of more easterly-based cultural groups, above all the Pfyn culture. Here the signs of strong influence from the east include not only the presence of copper but also the smelting furnaces. Conversely, there are also westerly characteristics. I don't yet have a non-problematic explanation as to why tetraploid naked wheat

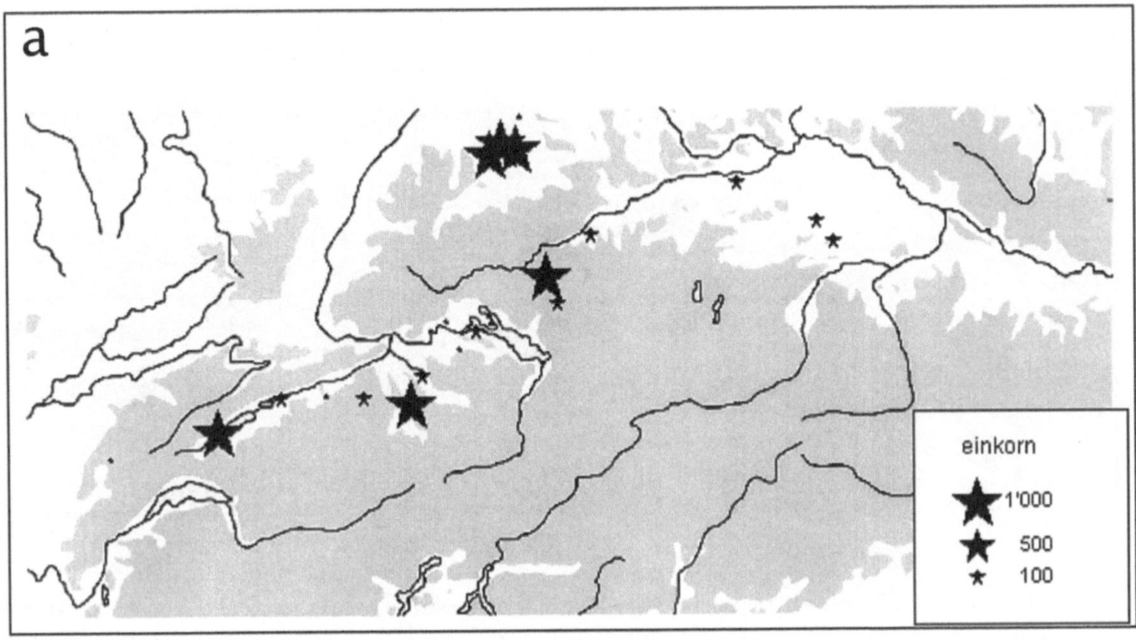

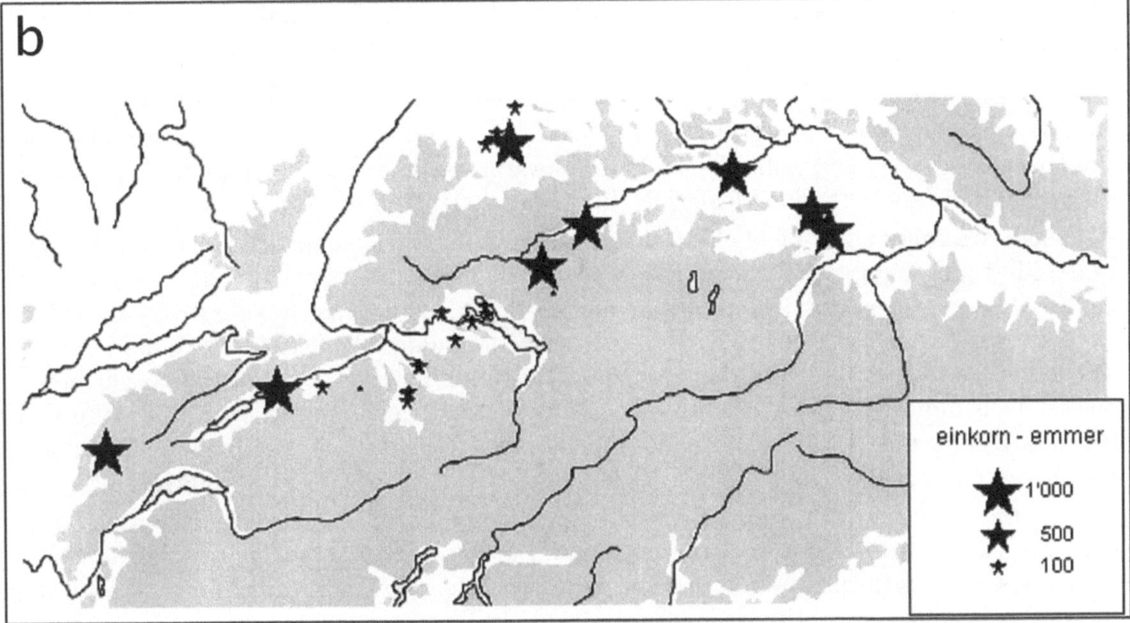

Figure 14.7: (a) Importance of einkorn and (b) einkorn and emmer (mainly emmer) in late neolithic settlements between 4300–3500 cal BC. Logarithmic scale: 1000 signifies the most important wheat species in a site. Map: R. Ebersbach, IPAS, Basel University. For geographical context, see figure 14.1. For details of sites, see tables 14.2 and 14.5.

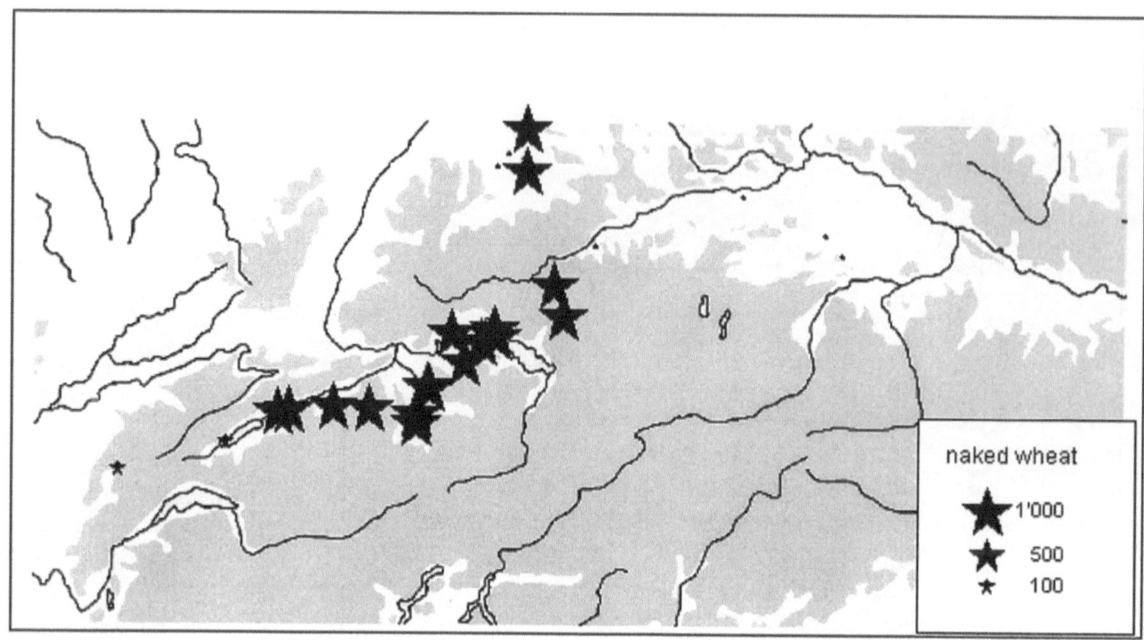

Figure 14.8: Importance of tetraploid naked wheat in late neolithic settlements between 4300–3500 cal BC. Logarithmic scale: 1000 signifies the most important wheat species in a site. Map: R. Ebersbach, IPAS, Basel University. For geographical context, see figure 14.1. For details of sites, see tables 14.2 and 14.5.

did not expand more to the east, but westerly influences in the archaeological material gradually disappear east of upper Swabia. Also, lower Bavaria seems to remain a 'glume wheat area' in the younger phases of the late Neolithic (Bittmann 2001; Küster 1995).

There are other signs of Mediterranean influence in settlements after 4000 cal BC (figure 14.9). Finds of plants with a Mediterranean origin, for example, parsley, dill, celery and lemon balm (*Melissa officinalis*), which appear for the first time around 4000 cal BC, are restricted to the regions where tatraploid naked wheat is the dominant wheat form (Jacomet et al. 1989). The seeds are usually found only in very small amounts, however, to date they occur in several places. Thus indicating either that the species were grown in the region or that their seeds were frequently imported. *Apium* is originally a plant of salty places near the coast and it is possible that it reached our area with the salt trade.

The role of barley is also somewhat surprising; it is very rare in the early Neolithic but later, with the exception of some settlements in lower Bavaria, is commonly found in much larger amounts. One gets the impression that barley, mainly the naked, six-rowed form, at least in part shows the same patterns of spread as tetraploid naked wheat.

Millets were found very rarely: *Panicum miliaceum* occurred three times in settlements of various dates and in different locations (figure 14.10). Interestingly, millets didn't penetrate the main extension area of tetraploid naked wheat, the region of the lake dwellings from western Switzerland to Upper Swabia. Lentil was found in an area more or less coincidental with that occupied by the millets (figure 14.11). Evidence of this pulse is mainly restricted to early and middle neolithic settlements in the East of the area, with the exception of the late Michelsberger site of Heilbronn-Klingenberg. (The find from Egolzwil 3 cannot be identified with certainty.) Millets and lentil seem, therefore, to have behaved in an opposite fashion to that of tetraploid naked wheat and the Mediterranean species.

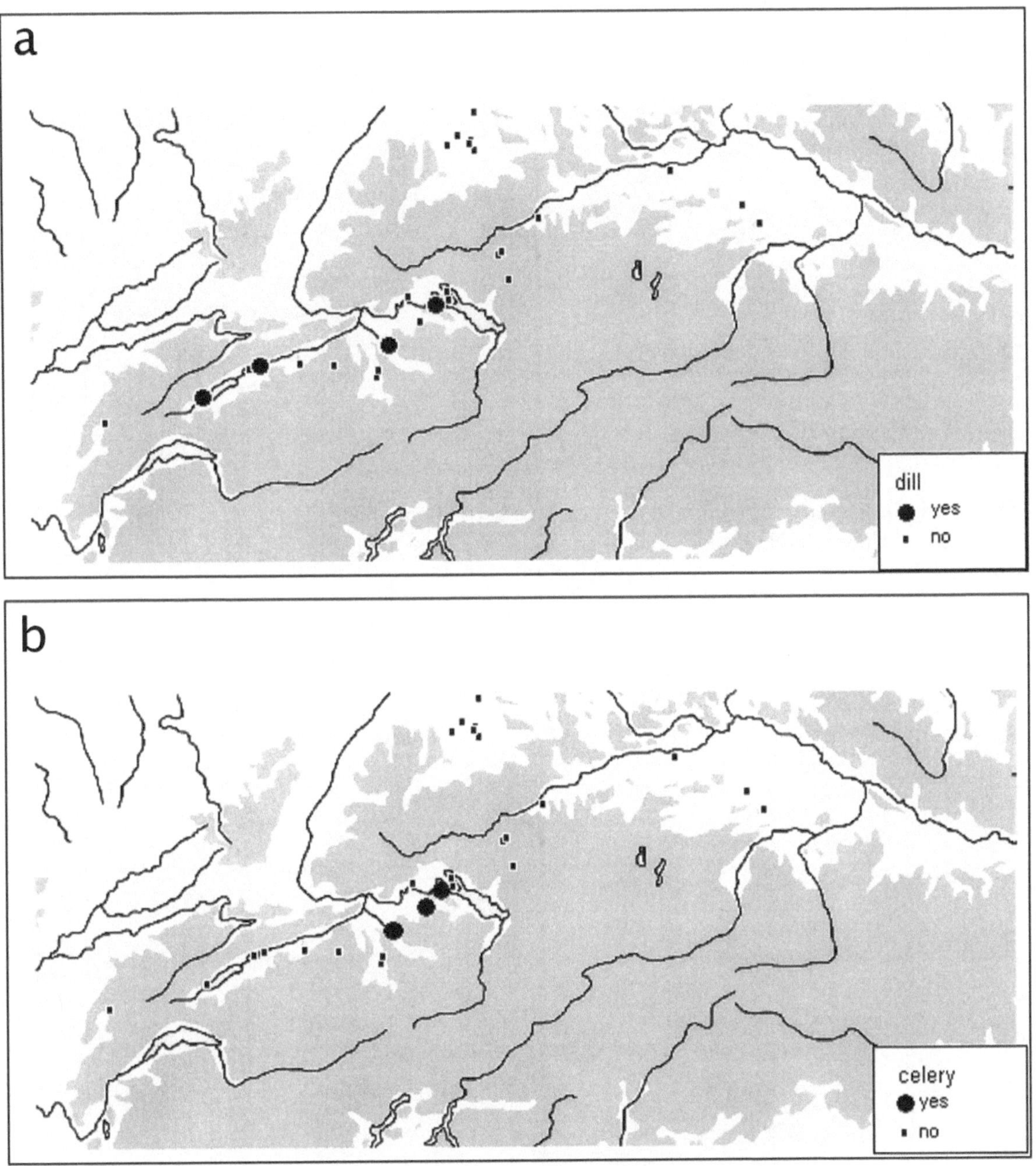

Figure 14.9: Presence of 'Mediterranean species' in late neolithic settlements between 4300–3500 cal BC. (a) Dill (*Anethum graveolens*) and (b) celery (*Apium graveolens*). Map: R. Ebersbach, IPAS, Basel University. For geographical context, see figure 14.1. For details of sites, see tables 14.2 and 14.5.

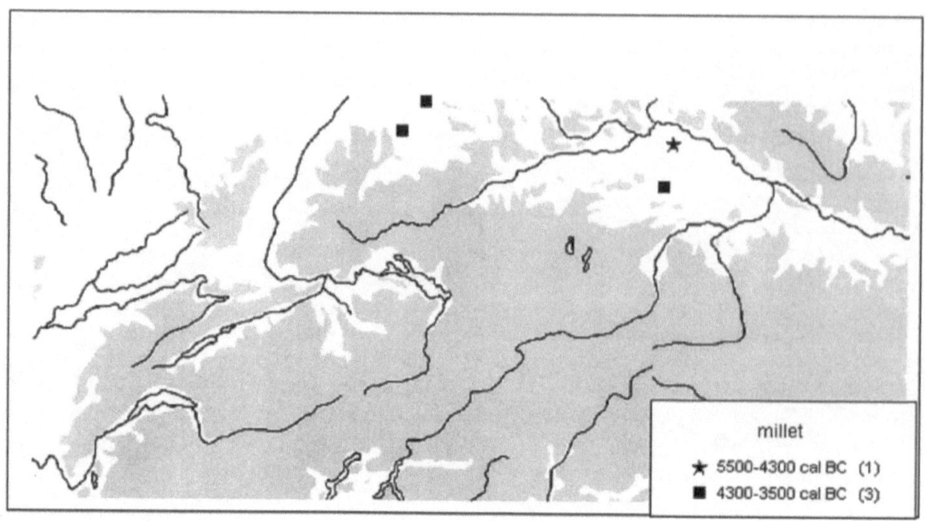

Figure 14.10: Presence of millets in the neolithic settlements north of the Alps. Map: R. Ebersbach, IPAS, Basel University. For geographical context, see figure 14.1. For details of sites, see tables 14.1, 14.3, 14.2 and 14.5.

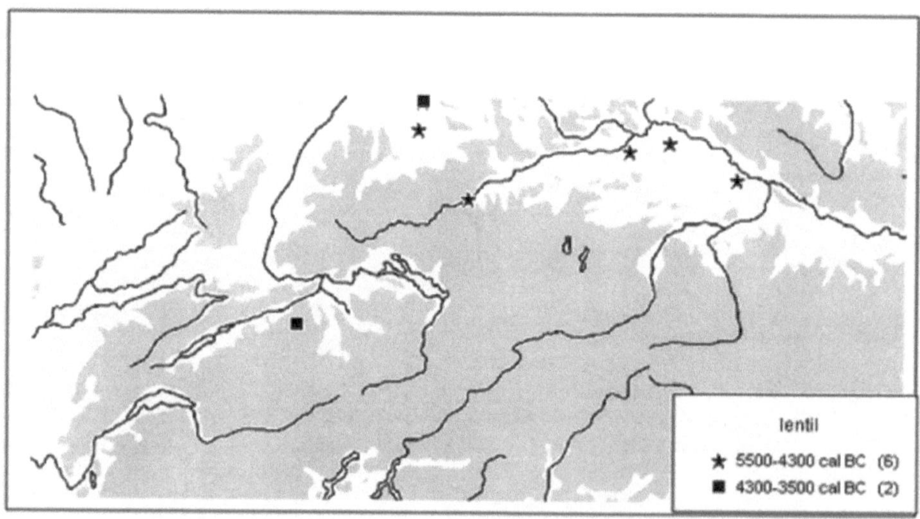

Figure 14.11: Presence of lentil in the neolithic settlements north of the Alps. Map: R. Ebersbach, IPAS, Basel University. For geographical context, see figure 14.1. For details of sites, see tables 14.1, 14.3, 14.2 and 14.5.

At eight sites in the whole area from middle neolithic times onwards, several flat *Triticum* grains with parallel flanks were identified as *Triticum* cf. *spelta*, however, glume bases assigned to the species were only found in two of these. If spelt is proven to be present at all, it has to be considered as being very rare. In my view there is no unequivocal evidence for the species before 3500 cal BC (nor after this date, until the very end of the Neolithic period). We also remain some time away from when we are able to explain where spelt emerged (for new results see, for example, Blatter et al. 2002).

14.5 Conclusion

The history of the cultivated plants in the area north of the Alps shows in an interesting way the changing influences of cultural spheres. It also provides hints that the distribution areas of cultivated plants are not always the same as the areas of archaeologically defined cultures. The factors that determined the decisions about whether or not to grow certain crops are not known. Unfortunately, the state of research at present is such that we are unable to give definitive answers to many important questions, or even yet to investigate suspected patterns in greater detail.

Acknowledgments

I would like to thank several people who gave me information about their unpublished data, including Simone Riehl, Ursula Maier and several people from the Archaeobotany Group in Basel, especially Christoph Brombacher. Renate Ebersbach and Rosemarie Arbogast were a great help in producing the maps. Albert Hafner, Helmut Schichtherle and Christian Jeunesse, Irenäus Matuschik, Bodo Dieckmann and Samuel van Willigen helped me with the compilation of the chronology table. Thanks for inspiring discussions to Marianne Kohler-Schneider and to James Conolly and Sue Colledge for considerably improving my English.

References

Amman, B., T. Bollinger, S. Jacomet, H. Liese-Kleiber, and U. Piening (1981). *Die neolithischen Ufersiedlungen von Twann. Vol. 14: Botanische Untersuchungen.* Bern: Staatlicher Lehrmittelverlag.

Bakels, C. (1978). *Four Linnearbandkeramik Settlements and their Environment: A Palaeoecological Study of Sittard, Stein, Elsoo and Hienheim*, Volume XI of *Analecta Praehistorica Leidensia.* Leiden: Leiden University Press.

Bakels, C. (1984). Pflanzenreste aus Niederbayern: Beobachtungen in rezenten Ausgrabungen. *Berichte der Bayrischen Bodendenkmalpflege 24/25*, 157–166.

Beckmann, M. (2003). *Pollenanalytische Untersuchung der Zeit der Jäger und Sammler und der ersten Bauern an zwei Lokalitäten des Zentralen Schweizer Mittellandes: Umwelt und erste Eingriffe des Menschen in die Vegetation vom Paläolithikum bis zum Jungeneolithikum.* Ph. D. thesis, Universität Bern.

Bertsch, K. and F. Bertsch (1947). *Geschichte unserer Kulturpflanzen.* Stuttgart: Wissenschaftliche Verlagsgesellschaft.

Binz, A. and C. Heitz (1990). *Schul- und Exkursionsflora für die Schweiz* (19th ed.). Basel: Schwabe & Co.

Bittmann, F. (2001). Die jungneolithische Feuchtbodensiedlung Pestenacker, Lkr. Landsberg/Lech: Auswirkungen auf die Landschaft aus botanischer Sicht. In P. Schauer (Ed.), *DFG–Graduiertenkolleg 462. "Paläoökosystemforschung und Geschichte". Beiträge zur Siedlungsarchäologie und zum Lanschaftswandel. Ergebnisse zweier Kolloquien in Regensburg 9.–10. Oktober 2000, 2.–3. November 2000*, Volume 7 of *Regensburger Beiträge zur Prähistorischen Archäologie*, pp. 93–107. Regensburg: Universitätsverlag.

Blankenhorn, B. and M. Hopf (1982). Pflanzenreste aus spätneolithischen Moorsiedlungen des Federseerieds. *Jahrbuch des Römisch-Germanischen Zentralmuseums Mainz 29*, 71–99.

Blatter, R. H. E., S. Jacomet, and A. Schlumbaum (2002). Spelt-specific alleles in HMW glutenin genes from modern and historical European spelt (*Triticum spelta* L.). *Theoretical and Applied Genetics 104*, 329–337.

Bollinger, T. (1994). *Samenanalytische Untersuchung der früh-jungsteinzeitlichen Seeufersiedlung EGOLZWIL 3.* Dissertationes Botanicae 221. Stuttgart: Gebr. Borntraeger Verlagsbuchhandlung, Science Publishers.

Bosch i Lloret, A., C. Sanchez, and J. Tarrus i Galter (2000). *El Poblat Lacustre Neolitic de la Draga. Excavaciones de 1990 a 1998.* Monografies del CASC. Girona: Museu d'Arqueologia de Catalunya.

Brombacher, C. and S. Jacomet (1997). Ackerbau, Sammelwirtschaft und Umwelt: Ergebnisse archäobotanischer Untersuchungen. In J. Schibler, H. Hüster-Plogmann, S. Jacomet, C. Brombacher, E. Gross-Klee, , and A. Rest-Eicher (Eds.), *Ökonomie und Ökologie neolithischer und bronzezeitlicher Ufersiedlungen am Zürichsee. Ergebnisse der Ausgrabungen Mozartstrasse, Kanalisationssanierungen Seefeld, AKAD/Pressehaus und Mythenschloss in Zürich*, Volume 20 of *Monographien der Kantonsarchäologie Zürich*, pp. 220–279. Zürich and Egg: Direktion der öffentlichen Bauten des Kantons Zürich.

Brombacher, C. and S. Jacomet (2003). Ackerbau, Sammelwirtschaft und Umwelt. In H. Zwahlen (Ed.), *Die jungneolithische Siedlung Port-Stüdeli*, Volume 7 of *Ufersiedlungen am Bielersee*, pp. 66–86. Bern: Archäolog. Dienst.

Buxó, R., N. Rovira i Buendia, and C. Saüch (2000). Les Restes Vegetals de Llavors i Fruits. In A. Bosch i Lloret, C. Sanchez, and J. Tarrus i Galter (Eds.), *El Poblat Lacustre Neol;tic de la Draga. Excavaciones de 1990 a 1998*, Volume 2 of *Monografies del CASC*, pp. 129–140. Girona: Museu d'Arqueologia de Catalunya.

Dieckmann, B., J. Hoffstadt, U. Maier, and H. Spatz (1998). Zum Stand der Ausgrabungen auf den "Offwiesen" in Singen, Kr. Konstanz. *Archäologische Ausgrabungen in Baden-Württemberg 1997*, 43–47.

Dieckmann, B., U. Maier, and R. Vogt (1997). Hornstaad-Hörnle, eine der ältesten jungsteinzeitlichen Ufersiedlungen am Bodensee. In H. Schlichtherle (Ed.), *Pfahlbauten rund um die Alpen, Archäologie in Deutschland, Sonderheft*, pp. 15–21. Stuttgart: Konrad Theiss Verlag.

Driehaus, J. (1960). *Die Altheimer Gruppe und das Jungneolithikum in Mitteleuropa.* Mainz: Verlag des Römisch–Germanischen Zentralmuseums.

Erny-Rodmann, C. (1995). *Von der Urlandschaft zur Kulturlandschaft. Pollenanalytische Untersuchungen an drei Uferprofilen aus dem Zürcher "Seefeld" zu anthropogenen Vegetationsveränderungen seit dem Mesolithikum.* Ph. D. thesis, Universität Basel.

Erny-Rodmann, C., E. Gross-Klee, J. N. Haas, S. Jacomet, and H. Zoller (1997). Früher 'human impact' und Ackerbau im Übergangsbereich Spätmesolithikum-Frühneolithikum im schweizerischen Mittelland. *Jahrbuch der Schweizerischen Gesellschaft für Ur- und Frühgeschichte 80*, 27–56.

Fredskild, B. (1997). Seeds and fruits of the neolithic settlement Weier, Switzerland. *Botanisk Tidsskrift 72*, 189–201.

Gregg, S. (1989). Paleo-ethnobotany of the Bandkeramik phases. In C.-J. Kind (Ed.), *Ulm-Eggingen. Die Ausgrabungen 1982 bis 1985 in der bandkeramischen Siedlung und der mittelalterlichen Wüstung*, Volume 34 of *Forschungen und Berichte zur Vor- und Frühgeschichte in Baden-Württemberg*, pp. 367–399. Stuttgart: Konrad Theiss Verlag.

Gronenborn, D. (2003). Migration, acculturation and culture change in western temperate Eurasia, 6500-5000 cal BC. *Documenta Præhistorica 30*, 79–91.

Gross-Klee, E. and S. Hochuli (2002). Die jungsteinzeitliche Doppelaxt von Cham-Eslen. Gesamtbericht über einen einzigartigen Fund aus dem Zugersee. *Tugium 18*, 69–101.

Guilaine, J. (2003). *De la Vague à la Tombe, la Conquête Néolithique de la Méditerranée*. Paris: Seuil.

Hafner, A. (1998). Archäobotanische Untersuchungen in Reute-Schorrenried. In M. Mainberger (Ed.), *Archäologische Untersuchungen in der jungneolithischen Siedlung Reute Schorrenried*, pp. 385–418. Staufen i. Br: Taraqua CAP.

Hafner, A. and P. J. Suter (1997). Entwurf eines neuen Chronologie-Schemas zum Neolithikum des schweizerischen Mittellandes. *Archäologisches Korrespondenzblatt 27*, 549–565.

Hafner, A. and P. J. Suter (2003). Das Neolithikum in der Schweiz. *jungsteinSITE: Informationen zur Neolithikum-Forschung*. `www.jungsteinsite.de` Accessed 2004-11-18.

Hofmann, R. (1986). Die vegetabilischen vor- und frühgeschichtlichen Funde aus Niederbayern und der Oberpfalz südlich der Donau im Rahmen der Siedelgeschichte. Ein Beitrag zur Aussagefähigkeit der Paläo-Ethnobotanik. *Bericht der Bayerischen Bodendenkmalpflege 24/25*, 112–157.

Hopf, M. (1968). Früchte und Samen. In H. Zürn (Ed.), *Das jungsteinzeitliche Dorf Ehrenstein (Kreis Ulm)*, Volume 10/II of *Veröffentlichungen des Staatlichen Amtes für Denkmalpflege Stuttgart, Reihe A*, pp. 7–77. Stuttgart: Verlag Silberburg.

Hopf, M. (1977). Sämereien und Holzkohlenfunde. In J. Lüning and H. Zürn (Eds.), *Die Schussenrieder Siedlung im Schlösslesfeld, Markung Ludwigsburg*, Volume 8 of *Forschungen und Berichte zur Vor- und Frühgeschichte in Baden-Württemberg*, pp. 3–10. Stuttgart: Konrad Theiss Verlag.

Hopf, M. (1981). Die Pflanzenreste aus Schernau, Landkreis Kitzingen. *Materialhefte zur bayerischen Vorgeschichte 44*, 152–160.

Hopf, M. and B. Blankenhorn (1986). Kultur- und Nutzpflanzen aus vor- und frühgeschichtlichen Grabungen Süddeutschlands. *Berichte der Bayrischen Bodendenkmalpflege 24/25*, 76–111.

Hosch, S. and S. Jacomet (2004). Ackerbau und Sammelwirtschaft. In S. Jacomet, U. Leuzinger, and J. Schibler (Eds.), *Arbon Bleiche 3: Wirtschaft und Umwelt*, Volume 12 of *Archäologie im Thurgau*. Frauenfeld: Erziehungsdirektion Thurgau.

Hosch, S. and P. Zibulski (2003). The influence of inconsistent wet-sieving procedures on the macroremains concentration in waterlogged sediments. *Journal of Archaeological Science 30*, 849–857.

Jacomet, S. (1981). Neue Untersuchungen botanischer Grossreste an jungsteinzeitlichen Seeufersiedlungen im Gebiet der Stadt Zürich (Schweiz). *Zeitschrift für Archäologie 15*, 125–140.

Jacomet, S. (1986). Kulturpflanzenfunde aus der neolithischen Seeufersiedlung Cham–St.Andreas. *Jahrbuch der Schweizerischen Gesellschaft für Ur- und Frühgeschichte 69*, 55–62.

Jacomet, S. (2005). Reconstruction of intra-site patterns in neolithic lakeshore settlements: the state of archaeobotanical research and future prospects. In P. Della Casa and M. Trachsel (Eds.), *WES'04–Wetland Economies and Societies. Proceedings of the International Conference in Zurich, 10–13th March 2004. Collectio Archaeologica.* Zürich: Chronos.

Jacomet, S., C. Brombacher, and M. Dick (1989). *Archäobotanik am Zürichsee. Ackerbau, Sammelwirtschaft und Umwelt von neolithischen und bronzezeitlichen Seeufersiedlungen im Raum Zürich.* Zürcher Denkmalpflege, Monographien 7. Zürich: Verlag Orell Füssli.

Jacomet, S. and H. Schlichtherle (1984). Der kleine Pfahlbauweizen Oswald Heer's: Neue Untersuchungen zur Morphologie neolithischer Nacktweizen–Ähren. In W. van Zeist and W. A. Casparie (Eds.), *Plants and Ancient Man*, Studies in Palaeoethnobotany. Proceedings of the Sixth Symposium of the International Work Group for Palaeoethnobotany, Groningen, 30 May–3 June 1983, pp. 153–176. Rotterdam: A. A. Balkema.

Jeunesse, C. and P. Pétrequin (1997). La région de la Trouée de Belfort au Ve millénaire. Evolution des styles céramiques et transformations techniques. In C. Constantin, D. Mordant, and D. Simonin (Eds.), *La Culture de Cerny. Nouvelle Économie, Nouvelle Société au Néolithique*, Actes du Colloque International de Nemours, 9-11 mai 1994, pp. 593–616. Nemours: Mémoires du Musée de Préhistoire de l'Ile de France.

Jörgensen, G. (1975). *Triticum aestivum* s.l. from the neolithic site of Weier in Switzerland. *Folia Quaternaria 46*, 7–21.

Karg, S. (1996). *Aus Pfahlbauers Pflanzenwelt.* Trapa natans*: die Wassernuss.* Stuttgart: Württembergisches Landesmeuseum Stuttgard.

Karg, S. and T. Märkle (2002). Continuity and changes in plant resources during the Neolithic period in western Switzerland. *Vegetation History and Archaeobotany 11*, 169–176.

Knörzer, K.-H. (1998). Botanische Untersuchungen am bandkeramischen Brunnen von Erkelenz-Kückhoven. In H. Koschik (Ed.), *Brunnen der Jungsteinzeit. Internationales Symposium Erkelenz 27. bis 29. Oktober 1997*, Volume 11 of *Materialien zur Bodendenkmalpflege im Rheinland*, pp. 229–246. Köln und Bonn: Landschaftsverband Rheinland. Rheinisches Amt für Bodendenkmalpflege.

Köninger, J., M. Mainberger, H. Schlichtherle, and M. Vosteen (2002). *Schleife, Schlitten, Rad und Wagen*, Volume 3 of *Hemmenhofener Skripte.* Gaienhofen–Hemmenhofen: Landesdenkmalamt Baden-Württemberg.

Kreuz, A. (1990). *Die ersten Bauern Mitteleuropas: Eine archäobotanische Untersuchung zu Umwelt und Landwirtschaft der Ältesten Bandkeramik.* Analecta Praehistorica Leidensia 23. Leiden: Leiden University Press.

Küster, H. (1985). Neolithische Pflanzenreste aus Hochdorf, Gemeinde Eberdingen (Kreis Ludwigsburg). In H. Küster and U. Körber-Grohne (Eds.), *Hochdorf I. Forschungen und Berichte zur Vor- und Frühgeschichte in Baden-Württemberg*, Volume 19, pp. 15–83. Stuttgart: Theiss.

Küster, H. (1989). Pflanzenreste in spätneolithischen Siedlungsgeschichten von Ergolding-Fischergasse, Lkr.Landshut. In K. Schmotz (Ed.), *Vorträge des 7. Niederbayerischen Archäologentages*, pp. 17–27. Rahden, Westfalen: Verlag Marie Leidorg GmbH.

Küster, H. (1991). Mittelneolithische Pflanzenreste aus Künzing-Unternberg. In J. Petsch (Ed.), *Die jungsteinzeitliche Kreisgrabenanlage von Künzing-Unternberg*, Volume 6 of *Archäologische Denkmäler im Landkreis Deggendorf*, pp. 26–31. Deggendorf: Landratsamt Deggendorf.

Küster, H. (1992). Mittelneolithische Pflanzenreste aus Gaimersheim, Landkreis Eichstätt (Oberbayern). *Das Pfostenloch. Beiträge zur Geschichte der Jungsteinzeit 1*, 89–94.

Küster, H. (1995). *Postglaziale Vegetationsgeschichte Südbayerns. Geobotanische Studien zur prähistorischen Landschaftskund.* Berlin: Akademie Verlag.

Lenneis, E. (1995). Altneolithikum: Die Bandkeramik. In J. W. Neugebauer (Ed.), *Jungsteinzeit im Osten Österreichs*, Volume 17 of *Forschungsberichte zur Ur- und Frühgeschichte*, pp. 11–56. St. Pölten, Wien: Niederösterreichisches Pressehaus.

Lundström-Baudais, K. (1989). Les macrorestes végétaux du niveau V de la Motte-aux-Magnins. In P. Pétrequin (Ed.), *Les Sites Littoraux Néolithiques de Clairvaux-les-Lacs (Jura). Tome II: Le Néolithique Moyen*, pp. 417–439. Paris: Éditions de la Maison des Sciences de l'Homme.

Lüning, J. (1996). Erneute Gedanken zur Benennung der neolithischen Perioden. *Germania 74*, 233–237.

Lüning, J. (2000). *Steinzeitliche Bauern in Deutschland. Die Landwirtschaft im Neolithikum*, Volume 58 of *Universitätsforschungen zur prähistorischen Archäologie.* Bonn: Dr. Rudolf Habelt Verlag.

Maier, U. (1995). Moorstratigraphische und paläobotanische Untersuchungen in der jungsteinzeitlichen Moorsiedlung Ödenahlen am Federsee. In *Siedlungsarchäologie im Alpenvorland III. Die neolithische Moorsiedlung Ödenahlen*, Volume 46 of *Forschungen und Berichte zur Vor- und Frühgeschichte in Baden-Württemberg*, pp. 143–253. Stuttgart: Konrad Theiss Verlag.

Maier, U. (1996). Morphological studies of free-threshing wheat ears from a neolithic site in southwest Germany, and the history of the naked wheats. *Vegetation History and Archaeobotany 5*, 39–55.

Maier, U. (1998). Der Nacktweizen aus den neolithischen Ufersiedlungen des nördlichen Alpenvorlandes und seine Bedeutung für unser Bild von der Neolithisierung Mitteleuropas. *Archäologisches Korrespondenzblatt 28*, 205–218.

Maier, U. (2001). Archäobotanische Untersuchungen in der neolithischen Ufersiedlung Hornstaad-Hörnle IA am Bodensee. In U. Maier and R. Vogt (Eds.), *Siedlungsarchäologie im Alpenvorland VI. Botanische und pedologische Untersuchungen zur Ufersiedlung Hornstaad-Hörnle IA*, Forschungen und Berichte zur Vor- und Frühgeschichte in Baden-Württemberg 74, pp. 9–384. Stuttgart: Konrad Theiss Verlag.

Maier, U. (2004). Archäobotanische Untersuchungen in jung- und endneolithischen Moorsiedlungen am Federsee. In J. Köninger and H. Schlichtherle (Eds.), *Ökonomischer und ökologischer Wandel am vorgeschichtlichen Federsee. Archäologische und Naturwissenschaftliche Untersuchungen*, Volume 5 of *Hemmenhofener Skripte*, pp. 71–159. Gaienhofen-Hemmenhofen: Landesdenkmalamt Baden-Württemberg.

Märkle, T. (2000). Die Wildpflanzen der Cortaillod moyen-zeitlichen Besiedlung von Concise-sous-Colachoz, Kt. Waadt, Schweiz. Eine Analyse der botanischen Makroreste. Magisterarbeit, Universität Tübingen.

Martinoli, D. and S. Jacomet (2002). Pflanzenfunde aus Cham-Eslen: Erste Ergebnisse zur Versorgung mit pflanzlichen Nahrungsmitteln. *Tugium 18*, 76–77.

Maurizio, A. (1927). *Die Geschichte unserer Pflanzennahrung von den Urzeiten bis zur Gegenwart.* Berlin: Paul Parey.

Mazurié de Keroualin, K. (2003). *Genèse et diffusion de l'agriculture en Europe. Agriculteurs, Chasseurs, Pasteurs.* Paris: Editions Errance.

Piening, U. (1979). Neolithische Nutz- und Wildpflanzenreste aus Endersbach, Rems-Murr-Kreis, und Ilsfeld, Kreis Heilbronn. *Fundberichte aus Baden-Württemberg 4*, 1–17.

Piening, U. (1981). Die verkohlten Kulturpflanzenreste aus den Proben der Cortaillod- und Horgener Kultur. In B. Ammann, T. Bollinger, S. Jacomet, H. Liese-Kleiber, and U. Piening (Eds.), *Die neolithischen Ufersiedlungen von Twann*, Volume 14, pp. 69–88. Bern: Staatlicher Lehrmittelverlag.

Piening, U. (1982). Botanische Untersuchungen an verkohlten Pflanzenresten aus Nordwürttemberg (Neolithikum bis römische Zeit). *Fundberichte aus Baden-Württemberg 7*, 239–271.

Piening, U. (1986a). Verkohlte Getreidevorräte von Aldingen, Gem. Remseck am Neckar, Kreis Ludwigsburg. *Fundberichte aus Baden-Württemberg 11*, 191–208.

Piening, U. (1986b). Verkohlte Nutz- und Wildpflanzenreste aus Grossachsenheim, Gem. Sachsenheim, Kreis Ludwigsburg. *Fundberichte aus Baden-Württemberg 11*, 177–208.

Piening, U. (1988). Neolithische und hallstattzeitliche Pflanzenreste aus Freiberg-Geisingen (Kreis Ludwigsburg). In H. Küster (Ed.), *Der prähistorische Mensch und seine Umwelt. Festschrift für Udelgard Körber-Grohne zum 65. Geburtstag*, Volume 31 of *Forschungen und Berichte zur Vor- und Frühgeschichte in Baden-Württemberg*, pp. 213–228. Stuttgart: Konrad Theiss Verlag.

Piening, U. (1989). Pflanzenreste aus der bandkeramischen Siedlung von Bietigheim-Bissingen, Kreis Ludwigsburg. *Fundberichte aus Baden-Württemberg 14*, 119–140.

Piening, U. (1992). Nutzpflanzenreste der Schussenrieder Kultur von Aldingen, Kreis Ludwigsburg. *Fundberichte aus Baden-Württemberg 17*, 125–142.

Piening, U. (1998). Die Pflanzenreste aus Gruben der Linearbandkerammik und der Rössener Kultur von Ditzingen, Kr. Ludwigsburg. *Fundberichte aus Baden-Württemberg 22*, 125–160; Supplement.

Robinson, D. E. (2003). Neolithic and bronze age agriculture in southern Scandinavia: recent archaeobotanical evidence from Denmark. *Environmental Archaeology 8*, 145–166.

Rösch, M. (1990). Botanische Untersuchungen in spätneolithischen Ufersiedlungen von Wallhausen und Dingelsdorf am Überlinger See (Kr. Konstanz). In *Siedlungsarchäologie im Alpenvorland II*, Volume 37 of *Forschungsberichte zur Ur- und Frühgeschichte*, pp. 227–266. Stuttgart: Konrad Theiss Verlag.

Rösch, M. (1992). Zwei pflanzenhaltige Gruben der Linearbandkerammik vom Viesenhäuser Hof, Stadt Stuttgart-Mühlhausen. *Archäologische Ausgrabungen in Baden-Württemberg 1991*, 53–56.

Rösch, M. (1998). Botanische Untersuchungen in der bandkeramischen Siedlung. In R. Krause (Ed.), *Die bandkeramischen Siedlungsgrabungen bei Vaihingen an der Enz, Kreis Ludwigsburg (Baden-Württemberg)*, Volume 79 (1998) of *Bericht der Römisch-Germanischen Kommission*, pp. 64–73. Frankfurt: von Zabern.

Ruttkay, E. (1995). Spätneolithikum. In J. W. Neugebauer (Ed.), *Jungsteinzeit im Osten Österreichs*, Forschungsberichte zur Ur- und Frühgeschichte 17, pp. 108–209. St. Pölten and Wien: Verlag Niederösterreichisches Pressehaus.

Sangmeister, E. (Ed.) (1993). *Zeitspuren. Archäologisches aus Baden*, Volume 50 of *Archäologische Nachrichten aus Baden.* Freiburg: Kehrer.

Schlichtherle, H. (1988). Das Jung- und Endneolithikum in Baden-Württemberg. Zum Stand der Forschung aus siedlungsarchäologischer Sicht. Archäologie in Württemberg. In D. Plank (Ed.), *Archäologie in Württemberg, Ergebnisse und Perspektive archäologischer Forschung von der Altsteinzeit bis zur Neuzeit*, pp. 91–110. Stuttgart: Konrad Theiss Verlag.

Schlichtherle, H. (1990a). Aspekte der siedlungsarchäologischen Erforschung von Neolithikum und Bronzezeit im südwestdeutschen Alpenvorland. Siedlungsarchäologische Untersuchungen im Alpenvorland 5. Kolloquium der Deutschen Forschungsgemeinschaft vom 29.-30. März 1990 in Gaienhofen-Hemmenhofen. *Bericht der Römisch-Germanischen Kommission 71*, 208–244.

Schlichtherle, H. (1990b). Siedlungen und Funde jungsteinzeitlicher Kulturgruppen zwischen Bodensee und Federsee. In M. Höneisen (Ed.), *Die Ersten Bauern 2: Einführung, Balkan, angrenzende Regionen der Schweiz*, pp. 135–156. Zürich: Schweizerisches Landesmeuseum.

Schlichtherle, H. (1991). Jungsteinzeitliche und bronzezeitliche Siedlungen im Federseebecken. In J. Hahn and C.-J. Kind (Eds.), *Urgeschichte in Obserschwaben und der mittleren Schwäbishchen Ald. Zum Stand neuerer Untersuchungen der Steinzeit-Archäologie*, pp. 65–69. Stuttgart: Landesdenkmalamt Baden-Württemberg.

Schlichtherle, H. (Ed.) (1997a). *Pfahlbauten rund um die Alpen*, Volume 1997 of *Special Volume 1997 of Archäologie in Deutschland, Sonderheft.* Stuttgart: Konrad Theiss Verlag.

Schlichtherle, H. (1997b). Pfahlbauten rund um die Alpen. In H. Schlichtherle (Ed.), *Pfahlbauten rund um die Alpen*, Special Volume 1997 of Archäologie in Deutschland, Sonderheft, pp. 7–14. Stuttgart: Konrad Theiss Verlag.

Stadler, P. (1995). Ein Beitrag zur Absolutchronologie des Neolithikums in Ostösterreich aufgrund der 14C-Daten. In J. W. Neugebauer (Ed.), *Jungsteinzeit im Osten Österreichs*, Volume 17 of *Forschungsberichte zur Ur- und Frühgeschichte*, pp. 210–224. St. Pölten and Wien: Verlag Niederösterreichisches Pressehaus.

Stika, H.-P. (1991). Die paläoethnobotanische Untersuchung der linearbandkeramischen Siedlung Hilzingen, Kreis Konstanz. *Fundberichte aus Baden-Württemberg 16*, 63–104.

Stika, H.-P. (1996). *Vorgeschichtliche Pflanzenreste aus Heilbronn-Klingenberg. Archäobotanische Untersuchungen zum Michelsberger Erdwerk auf dem Schlossberg (Bandkeramik, Michelsberger Kultur, Späthallstatt/Frühlatène)*, Volume 34 of *Materialhefte zur Archäologie in Baden-Württemberg*. Stuttgart: Konrad Theiss Verlag.

Stöckli, W. E., U. Niffeler, and E. Gross-Klee (1995). *Neolithikum*, Volume II of *Die Schweiz vom Paläolithikum bis zum frühen Mittelalter SPM*. Basel: Schweizerische Gesellschaft für Ur- und Frühgeschichte.

Strahm, C. (1994). Die Anfänge der Metallurgie in Mitteleuropa. *Helvetia Archaologica 25*, 2–39.

Strahm, C. (1997). Chronologie der Pfahlbauten. In H. Schlichtherle (Ed.), *Pfahlbauten rund um die Alpen*, Volume 1997 of *Archäologie in Deutschland, Sonderheft*, pp. 124–126. Stuttgart: Konrad Theiss Verlag.

Suter, P. J. (2002). Vom Spät- zum Endneolithikum. Wandel und Kontinuität um 2700 v. Chr. in Mitteleuropa. *Archäologisches Korrespondenzblatt 32*, 533–541.

van Zeist, W. and A. M. Boekschoten-van Helsdingen (1991). Samen und Früchte aus Niederwil. In H. T. Waterbolk and W. van Zeist (Eds.), *Niederwil. Eine Siedlung der Pfyner Kultur. Band III: Naturwissenschaftliche Untersuchungen, Academica Helvetica*, pp. 49–97. Bern and Stuttgart: Haupt.

Villaret-von Rochow, M. (1967). Frucht– und Samenreste aus der neolithischen Station Seeberg, Burgäschisee–Süd. In J. Boessneck, J.-P. Jequier, and H. R. Stampfli (Eds.), *Seeberg Burgäschisee Süd, Teil 4: Chronologie und Umwelt*, pp. 21–63. Bern: Verlag Stämpfli & Cie.

Villaret-von Rochow, M. (1970). Détermination des céréales du niveau XI. In P. Pétrequin (Ed.), *La Grotte del Baume de Gonvillars*, Annales Litteraires de l'Université de Besançon 107, pp. 127. Paris: Les Belles Lettres.

Villaret-von Rochow, M. (1974). Détermination des céréales du niveau 11. Annexe II. *Bulletin de la Société Préhistoire Française 71*(2), 495–497.

Willerding, U. (1991). Präsenz, Erhaltung und Repräsentanz von Pflanzenresten in archäologischem Fundgut. In W. van Zeist, K. Wasylikowa, and K.-E. Behre (Eds.), *Progress in Old World Palaeoethnobotany: A Retrospective View on the Occassion of 20 Years of the International Work Group for Palaeoethnobotany*, pp. 25–51. Rotterdam: A. A. Balkema.

Zohary, D. and M. Hopf (2000). *Domestication of Plants in the Old World: The Origin and Spread of Cultivated Plants in West Asia, Europe and the Nile Valley* (3rd ed.). Oxford: Oxford University Press.

Chapter 15

Archaeobotanical perspectives on the beginning of agriculture north of the Alps

Angela Kreuz
Landesamt für Denkmalpflege Hessen, Wiesbaden, Germany

15.1 Introduction

The spread of arable and stock farming reached parts of western Central Europe from the middle of the 6th millenium cal BC onwards, where the earliest and contextually unequivocal agricultural finds are of the early Neolithic *Linearbandkeramik* (LBK) culture. The Linearbandkeramik (or, simply *Bandkeramik*) culture is characterized archaeologically by pottery with typical band-like decoration (figure 15.1; e.g., Lüning 1998). The changing decorative style of the pottery enables the differentiation of five phases (LBK I–V) as devised by Meier-Arendt (1966). For comparative archaeobotanical work it is more appropriate to use Meier-Arendt's chronological classification instead of existing local archaeological chronologies, as his phases form the only framework that can be used supra-regionally.

^{14}C dates indicate the whole Bandkeramik cultural complex lasted for about 400 to 500 years (Stöckli 2002). The duration of the earliest phase (LBK I) was probably about half the time-span of the whole culture, from about 5500/5400 to about 5200 cal BC (Stäuble 1995; Stöckli 2002, p.55). Evidence of LBK I has been found in a huge area covering large parts of Europe, between the river Rhine in the west and the western Ukraine in the east, the foreland of the Alps in the south and the foreland of the Harz mountains in the north. In the second part of the culture, from Bandkeramik phase II (Flomborn style) to phase V (LBK II-V), the distribution area became even larger. Bandkeramik finds and settlements are found between the Paris Basin and the Black Sea (Lüning 1998; Lüning 2000).

Apart from the similarity in ceramic styles there also exists a surprising conformity in the landscape types that were selected by the earliest Bandkeramik farmers. They settled on Chernozems developed from loess or fluvial sediments, which were the best soils available at that time (e.g., Kreuz 1990a, 13ff.; Sielmann 1971). Shifting cultivation (for a discussion see Bogaard 2002a, Bogaard 2002b) or manuring of the fields very probably was not necessary due to the high quality of the soils (Lüning 1980). The farmers may have had to alternate yearly the crop species grown on each field, but this was to avoid diseases and pests and not to preserve the soil fertility (see e.g., Pflanzliche Erzeugung 1998).

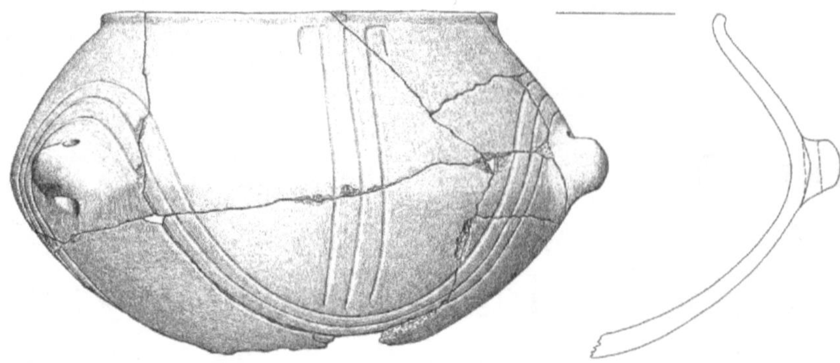

Figure 15.1: Typical pottery of the earliest Bandkeramik phase I: a so-called *Kumpf* with band-like decoration from a settlement near Friedberg-Bruchenbrücken, Hesse (height ca. 24 cm; from Cladders 2001, Tafel 4, 2).

Only those landscapes that today have a warm and especially dry climate, referred to by climatologists as *Trockengebiete*, were settled during LBK I (Kreuz 1990a, 7ff.). These are landscapes with average mean temperatures of about 7–9°C and about ≤600 mm annual precipitation. It was only after Bandkeramik phase I that areas with a more Atlantic climate, like the Lower Rhine area, were also settled (Kreuz 1990a, p.162; Sielmann 1971). High lake levels in other regions could be interpreted as signs of a more humid climate in Central Europe at that time. This might have been caused by precipitation, possibly in the form of heavy rainfall. In addition there are some indications of a cooler (possibly humid) phase around the middle of the 6th millennium BC (for an extended discussion see Beug et al. 1999; Haas et al. 1998; Hormes et al. 2001; Kalis et al. 2003; Kleinmann et al. 2000; Kreuz 1990a, p.8; Magny 1998; Maise 1998; Niggemann et al. 2000; Spurk et al. 2002). However, the available dataset does not allow a reconstruction of the exact climate in our study areas, it can only provide some hints of the assumed macro-climatic conditions during the Bandkeramik period, although it is almost certain that irrigation of the fields was not necessary.

On a certain level there was also a conformity in the form of LBK houses and settlements. Postholes and wall-ditches, as well as pits, are preserved and are well contrasted against the light-coloured soil (figure 15.2). The original surface on which the houses were constructed is more or less eroded (Thiemeyer 1988). The houses in LBK settlements were half-timbered and about 30 metres long by 6 metres wide (Cladders and Stäuble 2003; Lüning 1998; Lüning 2000) and these massive buildings are one of the reasons why it has been assumed that there was a sedentary population. Phosphate analysis of the soil inside and outside Bandkeramik houses revealed that there were no stables inside the buildings; therefore the livestock must have been kept elsewhere (Lienemann 1999; Stäuble and Lüning 1999: 183).

There are changes in the construction between LBK I and the following phases. Figure 15.3 shows Bandkeramik houses and their chronological development according to Cladders and Stäuble (2003, p.494, for further explanation). Typical of the LBK houses of phases I and II are the house-accompanying pits (figures 15.2 and 15.3). For reasons of structural engineering and stability these pits could not have been open after the houses were constructed (Stäuble 1990, p.341, 1997). Therefore all finds inside these pits very probably date to the time before and during the building phase (see below).

In addition there are other pits distributed over the settlement area. These are waste pits,

Figure 15.2: Overview over the excavation of the Bandkeramik site at Friedberg-Bruchenbrücken, Hesse, summer 1985. Postholes, ditches and house-accompanying pits from a building, as well as some single pits, contrast well against the lighter loess soil (picture: Seminar für Vor- und Frühgeschichte, J.W. Goethe-Universität Frankfurt/M.). For the ground plan of the house, see figure 15.3.

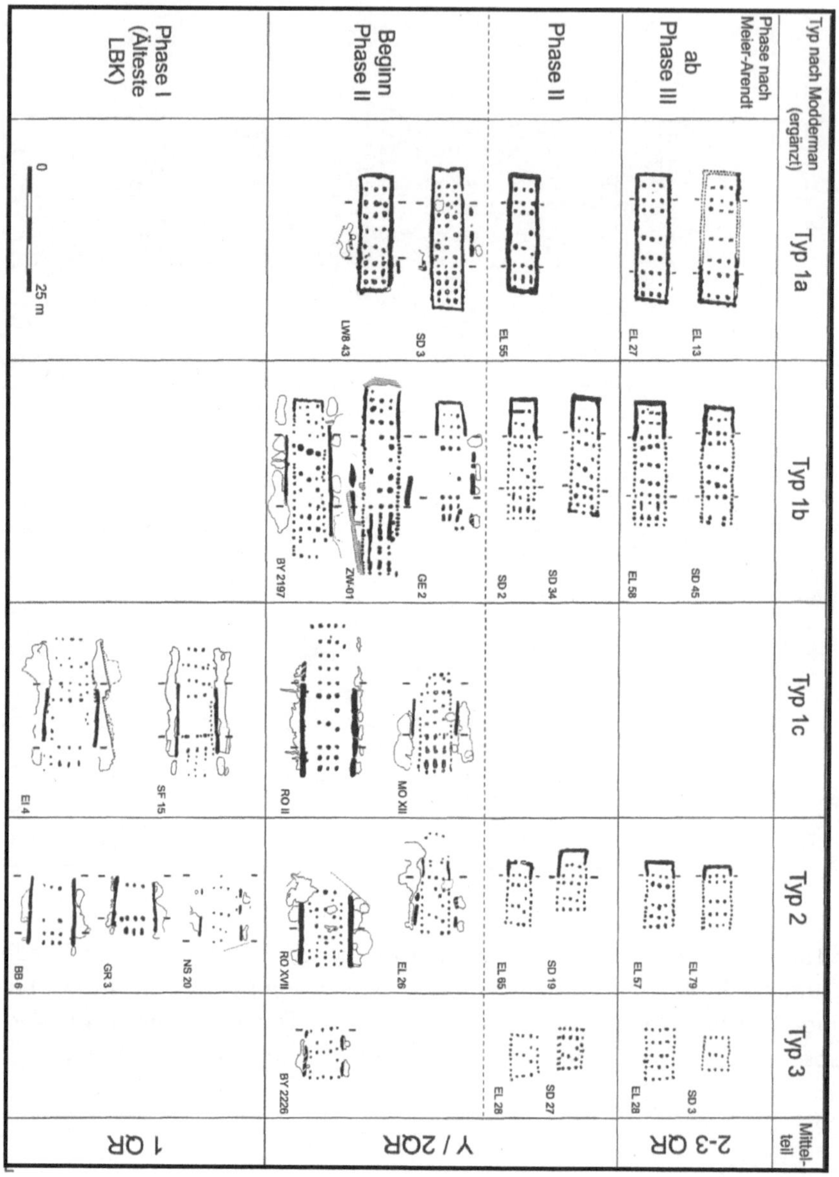

Typ nach Modderman (ergänzt) / Phase nach Meier-Arendt	Typ 1a	Typ 1b	Typ 1c	Typ 2	Typ 3	Mittelteil
ab Phase III	EL 13, EL 27	SD 45, EL 58		EL 79, EL 57	SD 3, EL 28	2-3 QR
Phase II	EL 55	SD 34, SD 2		SD 19, EL 65	SD 27, EL 28	Y / 2QR
Beginn Phase II	SD 3, LW8 43	GE 2, ZW-01, BY 2197	MO XII, RO II	EL 26, RO XVIII	BY 2226	
Phase I (Älteste LBK)	0 – 25 m		SF 15, EI 4	NS 20, GR 3, BB 6		1 QR

Figure 15.3: Typological scheme showing the development of ground plans from Bandkeramik houses of different phases (after Cladders and Stäuble 2003, fig. 2).

which contain mostly mixed settlement waste ('background' or 'noise' after Bakels 1991, p.281) and only very rarely have evidence of refuse from 'single-activities' in form of layers of charred material (Kreuz 1990a, p.125ff., 1990b). Chances of obtaining a representative spectrum of the charred plant remains used inside the settlements are improved by analyzing as many pits as possible from such sites.

The LBK settlement structures and the evidence from graves provide no indication of social ranking and it is likely that a more or less egalitarian social structure existed at that time. Enclosures are rare and only two (at Eitzum and Eilsleben) have curently been been dated to LBK I (Kaufmann 1983; Meyer 2003, p.450; Stäuble 1990). Interaction between settlements is still the subject of discussion (e.g., for further references see Petrasch 2003). Lüning (1998, p.38, footnote 33, 86) estimates a density of 12 people/km^2 for the area of the Aldenhovener Platte (Lower Rhineland, Germany). In general there is a pronounced increase of the population density from phase II onwards (e.g., Saile 1998).

The first farmers settled in landscapes, which in contrast to how they appear today, have been reconstructed as being more or less densely wooded by palynologists and anthracologists (Bakels 1978; Beug 1986, 1992; Kalis 1988; Kreuz 1990a, p.17ff., 1995; Liese-Kleiber 1997; Litt 1990; Schäfer 1996; Schweizer 2001; Stobbe 1996; van Zeist 1967; van Zeist and van der Spoel-Walvius 1980). There are different opinions concerning, for example, the percentages of oak (*Quercus*) and lime (*Tilia*) trees and other tree species, and the kind of woodland cover on the flood plains. In certain pollen-spectra there seems to be an over representation of pollen types resistant to corrosion, like *Pinus*- or *Tilia*-types, but this subject is not of relevance here. In our study areas the forest cover on Chernozems was formed by deciduous woodland. The results of charcoal analysis can be interpreted as managed hedges providing firewood (Groenman-van Waateringe 1970; Kreuz 1988, 1992).

It has to be stressed that the first settlements and fields must have been situated in the woodland rather like islands in the sea. This is also indicated by the low values of anthropogenic indicators in the pollen diagrams (see previous citations). The woodland and its resources must surely have been an essential supplement of early agriculture. In addition we can imagine that the symbolic, ritual or religious aspects of this culture were also strongly influenced by the prevailing presence of woodland at that time. Concerning the question of the Mesolithic/Neolithic transition, it should be mentioned that there is also evidence for pre-LBK openings of the forests in our study areas (e.g., Kreuz et al. 1998; Schäfer 1996; Urz 2000, 2002), but the extent of the impact of this early land use in the Mesolithic was quite different. Jeunesse (2003, p.106) describes that any 'wound' inflicted to the forest by mesolithic people was followed by a period of convalescence destined to erase the traces. All the more, for any indigenous mesolithic population to accept and adopt a neolithic approach to nature presupposes a significant change in the perception of the natural world. That is why it is worth wondering, with Jeunesse, if "*l'établissement de paysages ouverts pérennes n'aient pas été perçus comme une aggression traumatisante par les populations indigenes*" (2003, p.106).

15.2 The archaeobotanical dataset considered

An archaeobotanical dataset from 22 Bandkeramik sites is considered in this paper (figure 15.4). The project was faciliated by funding of the German Research Association (DFG). Table 15.1 presents a general summary of the methodological background of the archaeobotanical work undertaken. A total of 1476 samples from 405 different features have been examined and 414,900

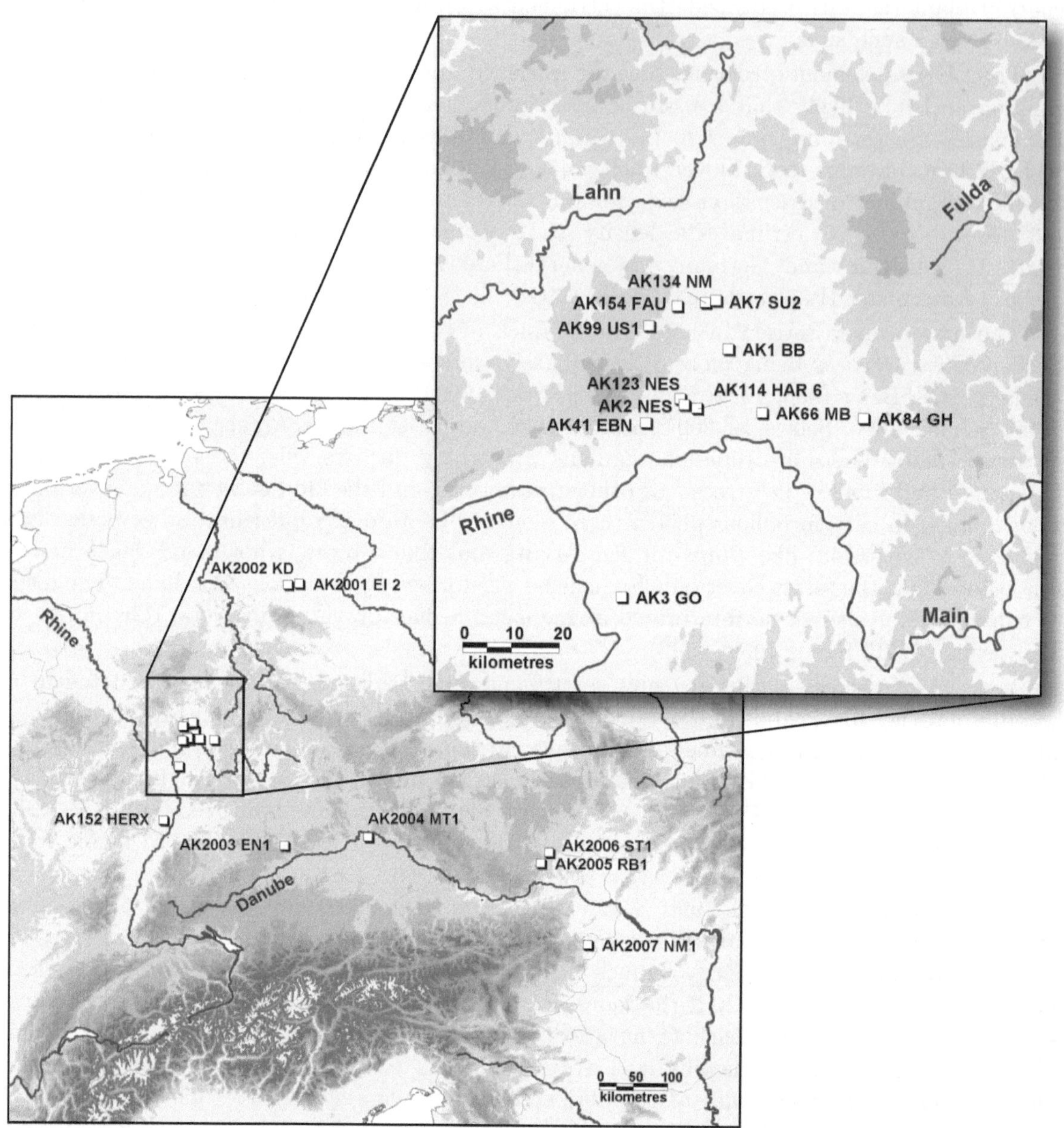

Figure 15.4: Map of sites investigated archaeobotanically. Sites in alphabetical order: EilslebenAK1 BB Bruchenbrücken; AK2001 EI Eitzum; AK2003 EN1 Enkingen; AK41 EBN Eschborn; AK154 FAU Fauerbach; AK84 GH Gelnhausen-Hailer; AK3 GO Goddelau; AK114 HAR Harheim; AK152 HERX Herxheim; AK2002 KD Klein Denkte; AK66 MB Mittelbuchen; AK2004 MT Mintraching; AK2007 NM Neckenmarkt; AK2 and AK123 NES Nieder-Eschbach; AK2005 RB Rosenburg; AK2006 ST1 Strögen; AK99 USI Usingen.

Feature type	Features (n)	Samples (n)	Volume of samples (l)	taxa (n)	Sum of plant remains
LBK I					
pit	154	335	6590.0	51	121908
house-accompanying pit	35	584	10723.5	59	31388
post-hole	1	2	30.0	5	12
ditch	5	30	540.0	11	310
sum	**195**	**951**	**17883.5**	(71)	**153618**
LBK II					
pit	48	81	1111.0	86	112649
house-accompanying pit	4	8	86.0	32	6784
oven	4	5	43.5	31	2143
post-hole	5	9	82.5	21	238
sum	**61**	**103**	**1323.0**	(90)	**121814**
LBK III–V					
pit	134	305	4648.8	126	113296
house-accompanying pit	5	7	83.0	34	2324
post-hole	3	3	26.0	21	162
ditch	7	107	1294.3	80	23686
sum	**149**	**422**	**6052.0**	(134)	**139468**
LBK II-V					
pit	182	386	5759.8	140	225945
house-accompanying pit	9	15	169.0	43	9108
oven	4	5	43.5	31	2143
post-hole	8	12	108.5	31	400
ditch	7	107	1294.3	80	23686
Total	**210**	**525**	**7375.0**	(149)	**261282**

Table 15.1: Archaeobotanical dataset from 22 Bandkeramik sites (identifications by author and Nicole Boenke, Götzis). The sum of the taxa determined is not the numerical sum, but the number of different taxa recorded.

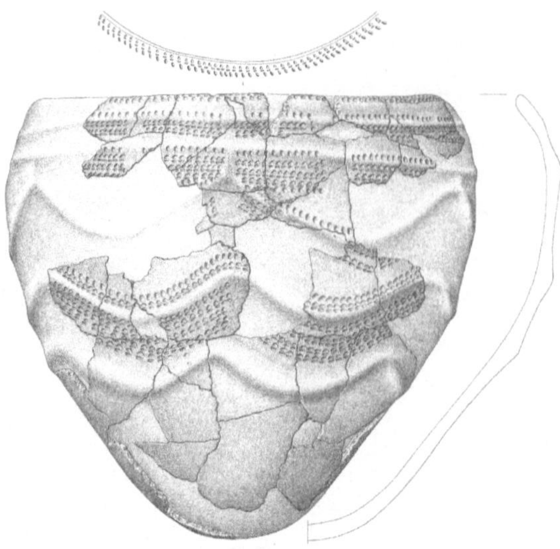

Figure 15.5: Pottery of the so-called La Hoguette style from the site of Dautenheim, Kr. Alzey-Worms, Palatinate (height ca. 45 cm; reproduced with permission from Lüning et al. 1989, fig. 10).

plant remains from 156 plant taxa were identified[1].

Plant remains are preserved exclusively by charring as the sites are situated on well-drained mineral soils. As the density of the charred plant remains is usually quite low the sample volume was never less than ten litres (Kreuz 1990a, 33ff.). It should be stressed that there are comparable numbers of features investigated from the two phases of the Bandkeramik culture. At present there is no archaeological or other reason to assume that taphonomic formation processes during Bandkeramik phases II–V (LBK II–V) were different from those in the earliest Bandkeramik.

As mentioned above, there were several different feature-types at the excavations, e.g., single pits, house-accompanying pits, postholes and ditches. Generally, the fills of postholes and ditches contained no, or at most very few, seeds (see also figure 15.12), and the density of plant remains in these contexts is not comparable with the pits, the final function of which was for the deposition of settlement waste. Most calculations are thus based on samples from pits only. To date there have not been more than 89 samples investigated from 52 pits of Bandkeramik phase II, for this reason the frequency calculations for phases II–V have been summarized (195 pits) in order to provide data comparable to those from phase I (189 pits).

15.3 The Mesolithic/Neolithic transition: fiction and evidence

Sherds of a foreign pottery have been found in some settlements of the western distribution area of the LBK culture. This is La Hoguette pottery, so named after the first place of discovery in France (figure 15.5). The decorative style of this pottery suggests some form of relationship with the Mediterranean Cardial culture (for discussion of Limburg pottery see Jeunesse 1987, p.20;

[1]The results presented here are the outcome of the project: *Archaeobotanical investigations of early, middle and late neolithic landscape development and agricultural systems in Hesse, Germany, and adjacent areas* (applicants: Angela Kreuz in cooperation with Jens Lüning, University Frankfurt/M.). Collaborators have been Nicole Boenke, Götzis, as well as at present Julian Wiethold, Wiesbaden. The data from the sites outside Hesse in northern and southern Germany and southwestern Austria are equivalent to each other, so all data have been included in the following evaluation. The plant remains from eight sites have been identified by Nicole Boenke. Those of the other sites and the charcoal analysis have been done by Angela Kreuz in the course of preceeding projects. New data worked out by Julian Wiethold within the DFG-project mentioned above will be presented elsewhere.

Lüning et al. 1989, p.360). According to Jeunesse, indigenous people of a pre-LBK 'Neolithique autochtone' made this pottery and, furthermore, that mesolithic people introduced cereal cultivation to their broad-spectrum hunting and gathering activities (2003, p.101); he argues, based on palynological data, that they practiced a kind of horticulture in addition to gathering (see also e.g., Gronenborn 1999b, a, 2003; Haas 1996; Lüning 2000, p.5; Schweizer 2001, pp.91-93, 117-119; Sedlmeier 2003).

Until now pure La Hoguette settlements have not been found; even single pits or other features containing only La Hoguette sherds are lacking. In most cases La Hoguette and Bandkeramik sherds have been found together in waste pits of Bandkeramik settlements. So we still do not know what these unique finds represent. In addition, there are several cave or rock-shelter sites that have revealed layers with single La Hoguette sherds but all are problematic with respect to the ^{14}C-dating (due to mixing of material from different layers, or for reasons unknown).

In this context the rock-shelter Bettenroder Berg IX (Reinhausen, Lower Saxony, Germany) should be mentioned, as this is the only site included in the discussion that had charred remains of domesticated plants. One grain of barley (*Hordeum* sp.) and one glume base of emmer (*Triticum dicoccum*), together with mesolithic lithic artefacts have been found (layers 4–6; see Wolf in Grote 1994). Unfortunately, to date, there has been no AMS-dating of these plant remains. As Grote himself mentions, layers in caves and rock-shelters can be problematic due to mixing of material as a result of trampling by people or animals (not to mention rodent activities) within the uppermost 5–15 cm (Grote 1994, part 1, p.64). For example, layer 3 above layers 4–6 contained prehistoric sherds as well as metal artefacts. AMS-dating of the single barley grain as well as the glume base of emmer would be worthwhile in order to prove whether they are really of Bandkeramik (or even late mesolithic date), particularly given that barley was probably a weed in LBK I (see below).

Palynological data are often presented and cited to support theories for early cultivation carried out by late mesolithic groups (e.g. Kalis et al. 2001; Schweizer 2001; Strien and Tillmann 2001; Visset et al. 2002). Single early finds of Cerealia-type pollen grains have been discussed for a long time (e.g., Erny-Rodmann et al. 1997; Haas 1996, p.40–41; Kossack and Schmeidl 1975; Küster 1989; Wiethold 1998, p.128). Inherent methodological problems include the difficulty in differentiating between the morphology of wild grass and cereal pollen grains (e.g., Beug 2004; Firbas 1937; Jacomet and Kreuz 1999, p.169), and contamination with younger material and dating (e.g., *Hartwassereffekt*; see also the critical methodological remarks of Schlütz 2003). It appears that since the work of Erny-Rodmann et al. (1997), which was surely not their intention, there is a new wave of such finds being published and interpreted as 'pre-neolithic cultivation'. Erny-Rodmann et al. (1997) give a critical presentation of their work, however, they still lack suitable archaeological or archaeobotanical explanations for their finds of Cerealia pollen and a single pollen grain of linseed (*Linum usitatissimum*) in sediments dated to the time span between about 6600 and 6100 cal BC (the linseed itself has not been dated). These dates, as the authors themselves cautiously comment, are even older than the Cardial culture from southern France, which is dated to around 5750 cal BC, or the pre-Bandkeramik Karanovo I culture of Bulgaria, which is dated to about 6000–5500 cal BC (Erny-Rodmann et al. 1997, p.42; Marinova 2001 see also the contribution by Jacomet in this volume).

Gronenborn (1997, 1999b, a, 2003) interprets the lithic tradition of the LBK I settlers as evidence for contacts with, or even an origin from, an indigenous mesolithic population (see also Taute 1974 and others cited there). In fact there are some changes in flint production technology, which occur at the beginning of the late Mesolithic, that are still practised by the people of the Bandkeramik culture (see also Kind 1997, p.13; Kind 2003, p.275). Weller (2003, p.106) suggests that one reason for this phenomenon may be the slower speed of the technological changes of flint production due to functional reasons.[2]

Without wanting to ignore the possibility of an adaptation of elements of the Neolithic by local mesolithic groups, it should be stressed that in our study areas there is no convincing evidence to support the hypothesis mentioned above. The models rely largely on botanical finds of insufficient evidence and inadequate contextual affinity, and on lithic artefacts and ^{14}C dates belonging to series which include dates that do not correlate with the established chronological framework of the late Mesolithic/early Neolithic horizon.

Late mesolithic sites contemporary with Bandkeramik settlements have not been found in our study areas and this is not due to lack of exploration by archaeologists (e.g., Gronenborn 1992; Urz 2000, 2002). In the middle of the 6th millennium cal BC, mesolithic groups apparently had almost retreated, for example to the coastal areas of the North Sea and the Baltic Sea, and to southern parts of Germany, south of the Danube (see also the compilation in Kind 1997, e.g., 127–128). Only in these areas were the ecological or other conditions such that they may have supported a hunter-gatherer lifestyle (for theoretical discussion see Kind 2003, p.16ff., 260ff.) and, as a consequence, mesolithic subsistence persisted for a very long time (e.g., Bakker 2003). The introduction of agriculture both caused and required such a fundamental change to the whole socio-economic system that its adoption was likely a 'point of no return' (Kind 2003, p.283). It is hard to imagine that a group of people could have followed two such different subsistence strategies at the same time, to say nothing of maintaining the required technical know-how.

Gronenborn (2003, p.198) has noted that: "...differences between the locals and the immigrants after the contact phase should be archaeologically visible only on a very subtle level." Nevertheless, it seems worthwhile to examine our archaeobotanical data for any signs of 'mesolithic roots' for the Bandkeramik farming-system; for example, high concentrations and frequencies of collected plants (for a critical discussion see Jones 2000; Monk 2000; Stephens; Jones and Rowley-Conwy, this volume). In this regard, of note is the fact that there are only two samples from one pit (AK134 Nieder-Mörlen, feature no. 2333) that show a higher proportion of edible seeds from wild plants. The occurrence of 225 and 154 charred seeds of *Chenopodium album* per litre of sediment in these two samples might suggest that they were gathered deliberately. However, given the large number of samples without such amounts, these finds might also be accidental. In addition these two samples are dated to the younger Bandkeramik (LBK IV/V), and in this context it is of interest that we never find charred roots and tubers (i.e., other collected wild plant food resources) in our samples as our colleagues have on mesolithic or neolithic sites in England or Scandinavia (Greig 1991; Kubiak-Martens 1998).

During all prehistoric periods in our study areas there is evidence that hazelnuts (*Corylus avellana*) were collected, and it is important to investigate whether they contributed more to diets during the early Neolithic than in later prehistoric periods. Nutshells have the advantage of being preserved well by charring:

> Since nutshell is a by-product of consumption, it is likely either to be discarded (possibly by throwing into household fires) or deliberately used as fuel or kindling. Frequent, recurrent burning of small or large quantities of nutshell is therefore to be expected in any community regularly consuming hazelnuts and using fires for cooking, etc. (Jones 2000, p.80).

In all periods concerned, waste pits that contain charred plant remains existed inside the settlements and there is no indication that there were changes in taphonomic processes, which could have resulted in profound quantitative differences in the deposition of hazel nutshells in those pits.

[2] "Sind in der ältesten Bandkeramik noch Elemente des Mesolithikums vertreten, so finden sich solche der jüngeren Phasen der LBK auch noch in der Rössener Kultur, z.T. bis in die Michelsberger Kultur hinein. Die bei der Keramik so leicht fassbaren Veränderungen fanden bei den Steinartefakten wohl nur langsam und in einzelnen Bereichen statt." (See also Lichardus et al. 2000).

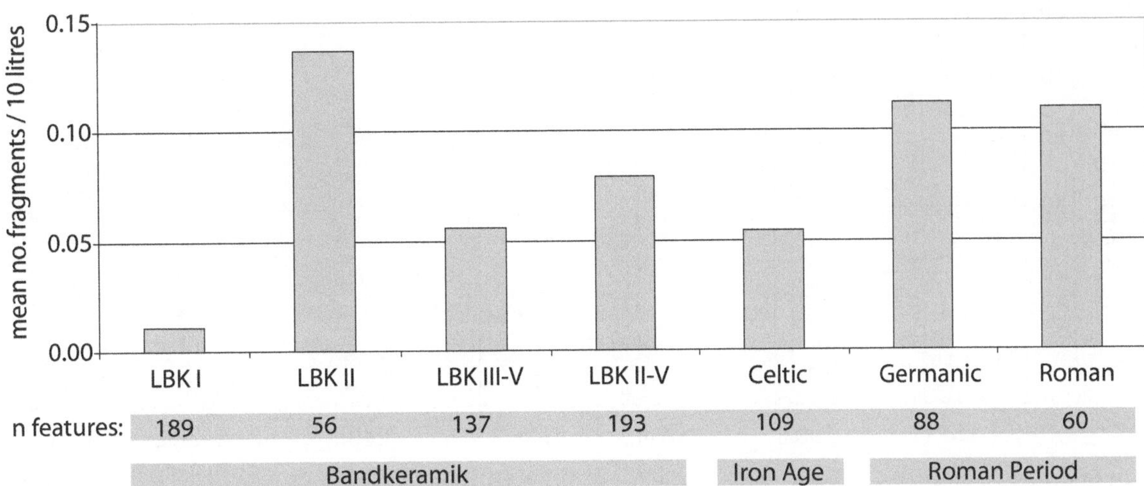

Figure 15.6: Average concentration of charred hazelnut-fragments (*Corylus avellana*) in waste pits from Bandkeramik, Iron Age Celtic and Roman Period (Germanic and Roman) sites. Excluded from the calculations are two extreme values from the sites AK 154 Fauerbach (feature 3/04, LBK III/IV), and AK 134 Nieder-Mörlen (feature 3286, LBK V).

Therefore the average concentrations from pits of different periods have been compared (figure 15.6). The averages were calculated in the following manner: the average concentration of plant remains of a certain species was calculated for every pit, and then the average value of all pits was calculated per period. This enables comparison of the general waste preserved in the settlement pits, with each pit representing a new chance of preservation by deposition.

The average concentrations of hazel nutshell fragments in Bandkeramik pits are compared with those of iron age and Roman periods (data from Kreuz nd). LBK I pits contain the lowest concentrations (figure 15.6). The same trends can be observed in the frequency values (figure 15.7); again the earliest Bandkeramik sites have the lowest value. So for whatever reason, there is no indication for increased gathering of plants in LBK I, which could be interpreted as a 'mesolithic tradition'.

In this context it would be of interest to evaluate critically the existing archaeozoological dataset for the Bandkeramik period, based on absolute and on archaeological dates, as well as on fully quantitative data. The state of research shows clear differences between the spectra of different settlements (e.g., Uerpmann 1997). In contrast to crop cultivation, stock-farming and hunting were thus apparently more *regionally* variable, perhaps arising from adaptation to local ecological conditions.

15.4 LBK crop species: a limited number

In our study areas crops could have been a major component of Bandkeramik nutrition. The cultivated crop species comprise two cereals: einkorn (*Triticum monococcum*) and emmer (*Triticum dicoccum*), and two legumes: pea (*Pisum sativum*) and lentil (*Lens culinaris*), as well as the oil or fibre plant flax (*Linum usitatissimum*). The two cereals could have been autumn or spring sown, whereas the pulses and the oil/fibre plants had to be spring sown (Geisler 1991; Körber-Grohne 1988). Two-grained einkorn occurs regularly in sites of Bandkeramik phases II-V (Kreuz and Boenke 2002).

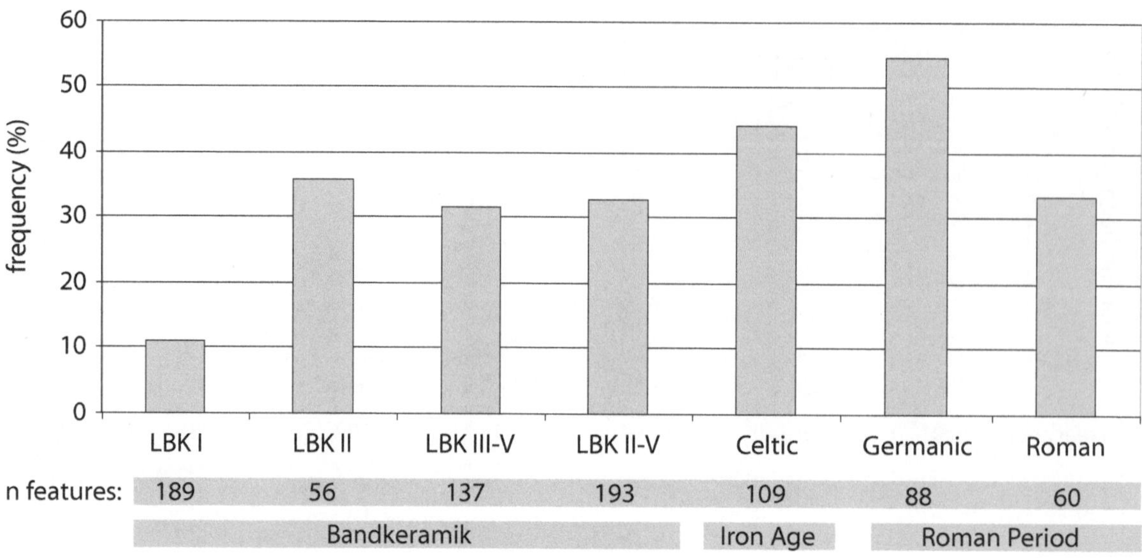

Figure 15.7: Frequency of charred hazelnut-fragments (*Corylus avellana*) in waste pits from Bandkeramik, Iron Age Celtic and Roman times Germanic and Roman sites. Note that there are fewer features from LBK II and III-V investigated than from LBK I, so the increase in frequency may be spurious and thus the overall frequency for LBK II-V is also shown (for average values see figure 15.6).

It is remarkable that poppy (*Papaver somniferum*) is only recorded from LBK II onwards, but not from earliest Bandkeramik sites. Poppy may point to direct or indirect contacts with the western Mediterranean area (Bakels 1982; Kreuz 1990a, p.172). It could have served as a foodstuff as well as a drug in ritual or other contexts.

In contrast to Willerding (1980) and Lüning (2000, p.60), regional differences in crop spectra within the Bandkeramik period have not been detected. Whenever a site has been thoroughly sampled—which due to various reasons often is not the case—it appears that all these species are found, but whether this is true all the time requires further investigation.

The spectrum of Bandkeramik crop species is surprisingly limited. Apparently only some of the crops cultivated in the Near East (Badr et al. 2000; Cappers and Bottema 2002; Colledge 2001; Damania et al. 1998; Jacomet and Kreuz 1999, p.241ff.; Nesbitt 2001, 2002; Nesbitt and Samuel 1996; Nesbitt and O'Hara 1999; Willcox 1998, 1999) or in Southeast Europe (Kroll 1991; Marinova 2001) reached our study areas during the early Neolithic. Solitary finds of species are not interpreted as crops that were grown intentionally, but rather as weeds in the cultivated fields, e.g., barley (*Hordeum* sp.), millet (*Panicum miliaceum*), rye (*Secale cereale*), celtic bean (*Vicia faba*) and bitter vetch (*Vicia ervilia*). Nevertheless, their presence may reflect an introduction of seeds and therefore supra-regional contacts. It is interesting to speculate *why* these useful crops did not become established in our study areas in the early Neolithic. The possibility cannot be ruled out that this was in part due to climatic reasons; free-threshing rye and naked-barley can be infested by pests in wet summers and millet is sensitive to summer rain. *Vicia* species were possibly rejected due to the toxicity of their seeds (Enneking 1995). Alternatively, it cannot be excluded that the early farmers relied on just a few crop species, as their labour resources and technologies did not allow an expansion. This phenomenon is still open to discussion.

Figure 15.8 shows the frequencies of crops based on grains or seeds during LBK I compared with those of the following phases II-V (the frequency of the cereals would be much higher (more than 80%) if the glume bases had been included in the calculations). The frequency of all crop

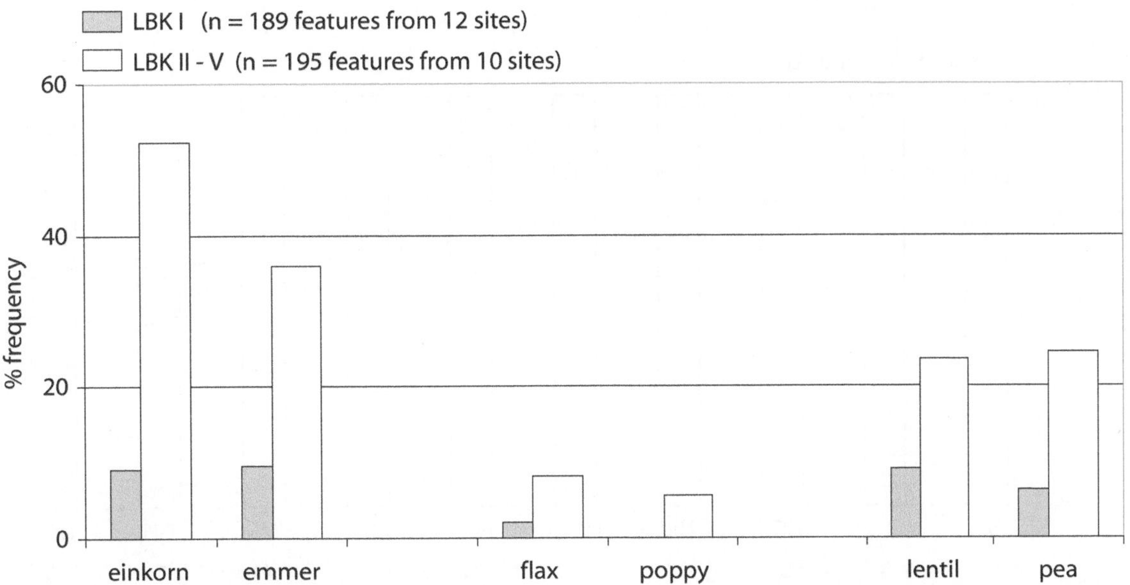

Figure 15.8: Frequency of occurrence (ubiquity) of charred crop-species (grains only) in Bandkeramik pits. The calculations are based on 189 features from twelve earliest Bandkeramik sites of phase I and on 195 features from ten younger Bandkeramik sites from phases II-V (storage finds included). Only species which are interpreted as intentionally grown crops are included.

species increases considerably during the second half of the Bandkeramik culture. Apparently at that time, for whatever reason, charred crop remains are more often deposited in pits, and this does not seem to be biased by sample size, as the number of features investigated is similar: 189 pits in Bandkeramik phase I compared with 195 in phases II-V (table 15.1).

It is questionable which of the two glume wheats was more important. In some literature it is reported that emmer was generally the most important (e.g., Bakels 1991, p.280; Kalis et al. 2003, p.36; Lüning 2000, p.60), which is presumably based largely on supposition. Figure 15.9 presents a comparison of the average relative proportions of charred emmer and einkorn glume bases per site. Only those sites are included at which both cereals have been identified. It appears that einkorn was more important than emmer. Figure 15.10 shows the same calculation based on grains, and the relative proportions are similar to those for grains and glume bases together (this figure is not shown here). The figure 15.10 results are less clear, but it is obvious that most of the sites from the second half of the Bandkeramik culture as well as one from phase I also have higher values of einkorn. The dominance of einkorn corresponds to the situation in the Lower Rhine Area and Bavaria (Knörzer 1991, p.191) and in the foreland of the Harz mountains (for suggestive palynological indications see Beug 1992, pg. 306, fig. 10).

Although our own dataset is too limited to resolve the question it seems worthwhile to focus our attention on this issue. If einkorn really was the dominant cereal it would be quite surprising, as it seems to be the worse choice of the two. The yield of einkorn is almost half of that of emmer (Reynolds 1990; van der Veen and Palmer 1997; bio-farmers pers. comm.) and, in addition, the lower tillering rate of einkorn allows more weeds to grow in the fields compared to emmer. So why should LBK farmers prefer einkorn?

Einkorn passes for a more winter-hardy cereal than emmer (Körber-Grohne 1988), but more importantly, it is the only cereal that keeps standing upright after heavy rainfall, whereas emmer tends to lodge (figure 15.11). Lodging of cereal plants may seriously reduce yields. So in the frequent and heavy rainfalls during Atlantic summers (see above) einkorn would have been the

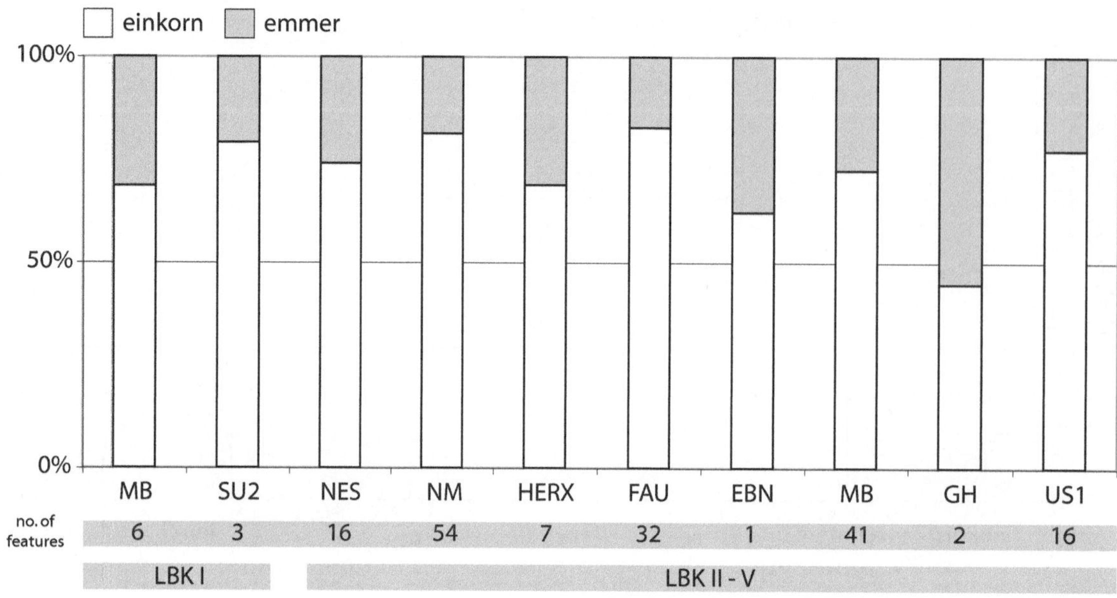

Figure 15.9: Mean proportions of glumes bases of emmer (*Triticum dicoccum*) and einkorn (*Triticum monococcum*) per site. Only those sites are included where both species have been identified. For abbreviations of the sites see figure 15.4.

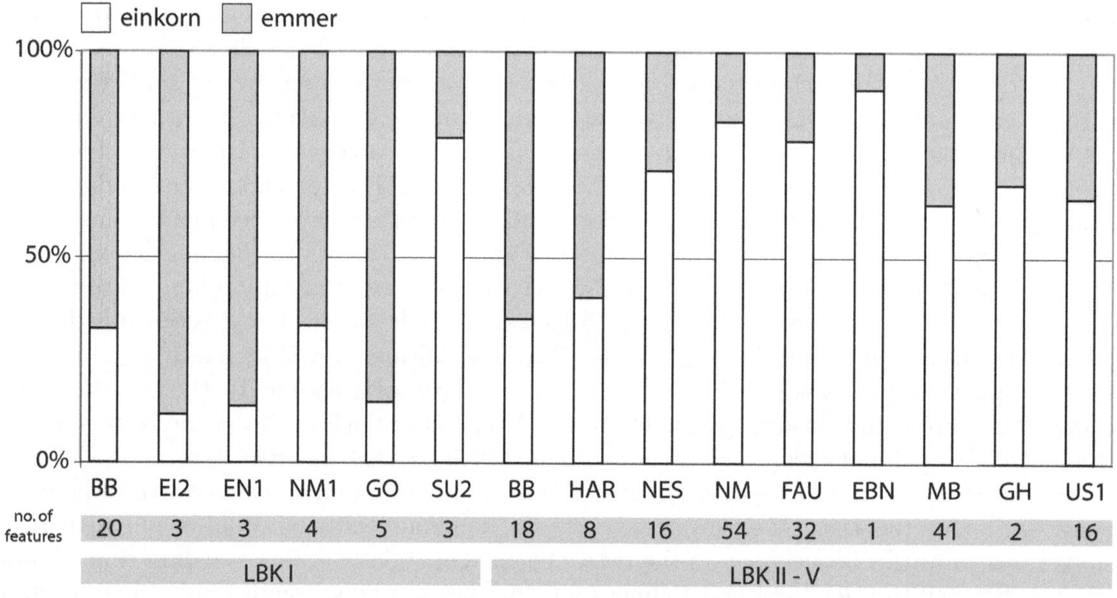

Figure 15.10: Mean proportions of grains of emmer (*Triticum dicoccum*) and einkorn (*Triticum monococcum*) per site. Only those sites are included where both species have been identified. For abbreviations of the sites see figure 15.4.

Figure 15.11: An important advantage of einkorn (*Triticum monococcum*) compared with other wheat crops is its ability to keep upright after heavy rainfall. The picture shows the 'lodging' of emmer plants (*Triticum dicoccum*) in the foreground and einkorn plants still standing on the left in the background (picture: A. Kreuz 1988–15–07, Stuttgart-Hohenheim).

better choice. However, because of possible climatic reasons for einkorn dominance, this issue requires further examination.

Crop remains are always found as mixtures of several species within waste pits but this might be a result of taphonomy and not growing practices in the Bandkeramik period. The question of how to recognize maslins, mixtures and monocrops has been previously discussed, for example by Jones and Halstead (1995). Theoretically, it is possible to grow einkorn and emmer as maslins because of the compatibility of their growth characteristics and their ecological demands. Under certain climatic conditions this might be useful as a strategy for yield risk management.

In this context it is important to differentiate between crops grown for fodder and those grown for human consumption. It might be useful to sow some species together to get good fodder but this mixture would not necessarily represent a good product for human meals. We have no evidence in the Neolithic of the growing of crops for animal fodder but there is no doubt that they were grown for human consumption.

As for the species grown by the LBK farmers, einkorn, emmer and pea could have been sown together at the beginning/middle of March and harvested at the beginning/middle of August. The other crops are less likely to have been grown in combination; lentil is sown as late as the middle of April and is not ripe when the wheats are harvested, so it's likely to have been grown in fields on its own. The same holds true for flax and poppy.

The best evidence for this is provided by the so-called storage finds, which are concentrations of several hundreds of charred crop remains per sample of ten litres volume. To date, three storage finds from two settlements of the younger Bandkeramik (LBK IV/V) exist among the

	Site and Date		
	AK 41 EBN LBK IV	AK 84 GH LBK IV/V	
Feature	2	13	15
Sample number	1	13-1	15-1
Sample volume (l)	32	26.5	17.5
Einkorn	4002	408	884
Emmer	1683	345	559
Pea	415	93	989
Lentil	2	.	.
Pulses cultivated	1478	112	516

Table 15.2: Numbers of crop species from three storage finds of the younger Bandkeramik sites AK 41 Eschborn and AK 84 Gelnhausen-Hailer (determinations Nicole Boenke, Götzis).

data presented here (table 15.2). The results for these three cases support the assumption made above, that emmer, einkorn and pea were grown together. As pea has to be sown in springtime, it is possible that emmer and einkorn were grown as summer crops. Alternatively, the possibility cannot be ruled out that pea was sown in March within the emmer/einkorn fields that had been planted in September (if sown in rows). This question probably cannot be solved. The sowing of glume wheats and pea together would have made it necessary to harvest closer to the ground, possibly requiring the use of sickles. To clarify whether or not this practice was already common from the earliest Bandkeramik we need to examine storage finds dating to this phase.

15.5 Intra-site distribution of plant remains

Crop remains occur in almost all the features investigated. In most cases they might be the waste from crop processing, for example, glume bases and weed seeds from fine sieving (Jones 1984; Hillman 1984). Apparently crop processing took place in every household and this is why the remains are found everywhere in the Bandkeramik settlements. The question then arises as to whether there is any intra-site patterning of the distribution of cereal remains within different feature types.

Figure 15.12 shows the distribution of cereal remains in the three most common Bandkeramik feature-types: pits, house-accompanying pits and postholes. The average concentration of grains in all features are very low, and the lowest are those of the earliest Bandkeramik phase I. The lowest concentrations of glume bases are again recorded for Bandkeramik phase I. The highest values are for the single pits and not the house-accompanying pits. This can be explained in terms of their functions; as mentioned above, the house-accompanying pits very probably did not stay open for a long time.

To estimate the relative importance of these concentrations of glume bases, the LBK values are compared with results for the hulled wheats from iron age and Roman settlements (data from Kreuz nd). Figure 15.13 shows a comparison of the average concentration of glume bases and grains in pits of different periods. Storage finds or concentrations of charred material with several hundreds of grains or chaff per ten litres of sediment have been excluded from the calculations. It is evident that the 'apparently low' values of earliest Bandkeramik phase I are not exceptional when compared with younger prehistoric periods. In contrast, the values of glume bases from

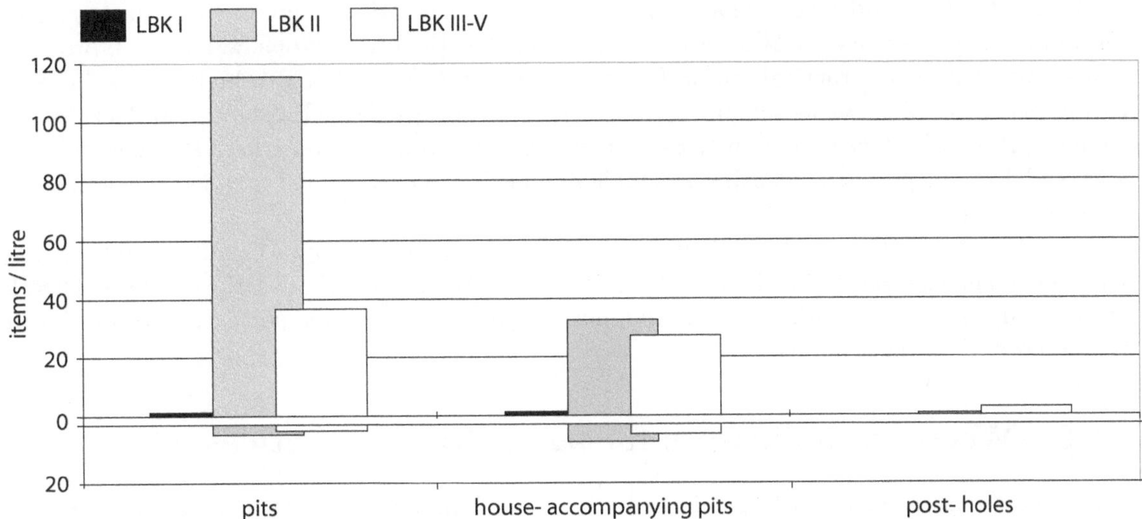

Figure 15.12: Mean number of charred cereal remains by feature type (glume bases upper bars; grains lower bars). See table 15.4 for the number of features in each category.

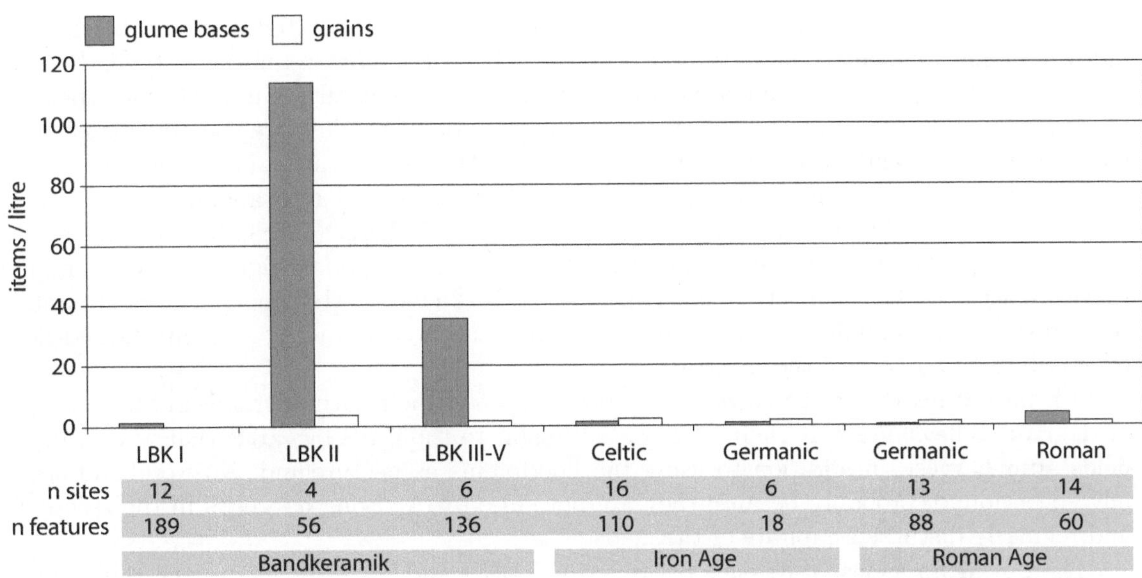

Figure 15.13: Mean number of charred glume bases and cereal grains (Cerealia indet. included) in waste pits from Bandkeramik, Iron Age Celtic and Roman Period Germanic and Roman sites. Storage finds with several hundreds of seeds per ten litres of sediment are excluded from the calculation.

the second half of the Bandkeramik culture are extremely high, especially those of Bandkeramik phase II.

It is difficult to find an explanation for this phenomenon. High amounts of glume bases with low amounts of grain and weed seeds can be explained as the product of a certain crop processing stage, but there is no practical reason for this to be burnt, as chaff is a valuable material, which can be used as fodder or as temper for pottery, etc. (Forbes 1998; Halstead and Tierney 1998; Palmer 1998). Chaff was apparently used in this manner during the earliest Bandkeramik, Iron Age and Roman periods, resulting in smaller quantities entering the archaeological record as charred remains.

In fact, the pottery of LBK I was tempered mainly with chaff (Cladders 2001, p.39ff.). Cladders considers this an ideal reason for the low quantities of chaff in LBK I contexts (Cladders 2001, p.40). However, during the second half of the Bandkeramik culture, pottery is rarely tempered with plant remains.

15.6 Weeds as indicators for agricultural practices

It is interesting to consider potential weed species to examine further the changes between LBK I and subsequent phases. As there are very few finds of stored crops with associated weeds in Bandkeramik sites, it is not always clear how to decide whether or not identified weed species grew alongside crops. Chorological and ecological data for each species are needed to answer this question (see also Kreuz 1990a, p.143ff., 1993, for a discussion of the term weed and for further references).

The Central European flora can be divided into two groups. First, there are those species that came into a region without anthropogenic influence; these are the indigenous species called idiochores (for the terms used see Schroeder 1969). Some of them are able to establish themselves as ruderal plants or weeds in the fields. Such potential weeds are called apophytes. Secondly, there are those species that could only have arrived in and inhabited a region by direct or indirect human influence: the anthropochores, which do not grow as natural stands but only occur in anthropogenically influenced sites. We can expect that all anthropochorous species in the Bandkeramik were introduced during the colonization of the landscapes and this is why in most cases they might represent weeds, and are also likely to indicate a movement of people.

The weed species found at the 22 sites have been separated into apophytes and anthropochores based on Oberdorfer (1990) and Korneck and Sukopp (1988). In total, 48 anthropochores and 17 apophytes have been identified (table 15.3), and in figure 15.14 the number of each calculated for each settlement is presented. It is evident that most species are anthropochores, whereas apophytes, from the natural vegetation surrounding settlements and fields, form only a minor part of the spectra.

It is interesting that today most of the apophytes have their natural stands in the floodplains of the river valleys. Possibly their seeds were brought to the fields by cattle that were using the fields after harvest, in addition to using the floodplains as pastureland. No woodland species have been found, probably because they are not suited to growing as weeds in the steppe-like and regularly disturbed habitats of the fields.

There is again a clear difference between the LBK I and the later half of the Bandkeramik period (LBK II-V), as in the second half there are clearly more species per settlement, especially from phase III–V onwards (figure 15.14, table15.3). Moreover, it is striking that the 'new' weeds of younger Bandkeramik phases are mostly low-growing species with a maximum height of about 40 cm (figure 15.15, table 15.3), and possibly this is indicative of a change in the harvesting technique. According to Hillman (1981) and Reynolds (1985, 1993, p.189) many more weed plants and their seeds can be collected by sickle than by harvesting the ears. Our botanical data have to be statistically analyzed more fully, but our preliminary findings of changes between

Botanical name	Classification	Centre of distribution	LBK I	LBK II	LBK III-V	Found in storage (n)	Life form	Min. height (cm)	Max. height (cm)
Agrostis stolonifera/capillaris	ANT?	.	.	.	7	.	H	20/20	50(-150)/40
Asperula arvensis	ANT	med-smed	.	.	1	.	T	5	25
Atriplex patula/hastata	ANT	euras(subozean)(-smed), circ/euras(-med)	1	4	7	1	T/T	30/30	80/60
Atropa bella-donna	APO	subatl-smed	.	4	.	.	H	60	150
Avena sp.	ANT	((fatua = med))	.	.	1	.	.	.	high
Brassica sp.	ANT	smed/med-atl	.	.	7	7	T	.	high
Bromus cf. *arvensis*	ANT	euras-med	.	31	91	.	T	30	80
Bromus cf. *secalinus*	ANT	(euras)	153	402	1584	835	T	30	80
Bromus cf. *sterilis*	ANT	smed	.	15	166	70	T	30	50
Bromus sterilis	ANT	smed	.	113	82	18	T	30	50
Bromus sterilis/tectorum	ANT	smed/smed-kont	.	56	280	.	T	30/10	low?
Calystegia sepium	APO	euras(subozean)-smed	.	.	1	.	G, Hli	100	300
cf. *Calystegia sepium*	APO	euras(subozean)-smed	.	.	1	.	G, Hli	100	300
Capsella bursa-pastoris	ANT	med(-kont)	.	.	2	.	T	10	30(50)
cf. *Capsella bursa-pastoris*	ANT	med(-kont)	.	.	1	.	T	10	30(50)
Carex muricata agg.	APO	euras(subozean)	.	1	.	.	H	20	80
Carex cf. *muricata* agg.	APO	euras(subozean)	.	1	.	.	H	20	80
Chenopodium album	ANT	no-euras(-med)	919	4780	8415	307	T	10	100
Chenopodium hybridum	ANT	euras(kont)	2	2	4	.	T	30	70
cf. *Digitaria sanguinalis*	ANT	med-smed(-euras), circ	.	.	1	.	T	10	40
Echinochloa crus-galli	ANT	med-smed-euras, circ	1	47	32	9	T	30	80
cf. *Echinochloa crus-galli*	ANT	med-smed-euras, circ	1	3	4	.	T	30	80
Echinochloa/Setaria	ANT	.	.	.	1	.	T	.	medium
Euphrasia/Odontites	ANT	.	.	.	2	.	T	.	low
Festuca/Lolium	ANT	.	.	.	6	.	H/T	.	.
Fragaria cf. *vesca*	APO?	no-euras(subozean)	.	4	11	5	H	5	20
Galium aparine	ANT?	euras(subozean)	.	12	.	.	Tli	50	150
Galium cf. *aparine*	ANT?	euras(subozean)	14	15	134	13	Tli	50	150
Galium aparine/spurium	ANT?	euras(subozean)/smed-euras	5	5	210	7	T/T	50/30	150/100
Galium palustre	APO	no-eurassubozean	1	.	.	.	H	15	40(-80)
Galium cf. *palustre*	APO	no-eurassubozean	1	.	.	.	H	15	40(-80)
Galium spurium	ANT	smed-euras	14	98	805	.	Tli	30	100
Galium cf. *spurium*	ANT	smed-euras	3	2	26	.	Tli	30	100
Galium cf. *verum* agg.	ANT?	euras-smed	.	132	190	.	H	20	70
Hyoscyamus niger	ANT	smed-euras	.	.	2	.	T, H	30	60
Knautia arvensis	ANT	(no-)eurassubozean	.	1	.	.	H	30	80
Lapsana communis	APO	eurassubozean-smed	1	152	443	43	T(H)	30	100
cf. *Lapsana communis*	APO	eurassubozean-smed	.	19	1	.	T(H)	30	100

Table 15.3: Potential weed species from 22 Bandkeramik sites in alphabetical order. Not included are taxa which could not been classified as a weed. ANT=anthropochores; APO=apophytes; ?=very likely, but not completely clarified; *H=Hemicryptophyte* (hibernating near ground surface); *T=Therophyte* (hibernating as seeds); *li*=liana. Centres of distribution, life form and height after Oberdorfer (1990).

Botanical name	Classification	Centre of distribution	LBK I	LBK II	LBK III-V	Found in storage (n)	Life form	Min. height (cm)	Max. height (cm)
Galium aparine	ANT?	euras(subozean)	.	12	.	.	Tli	50	150
Galium cf. *aparine*	ANT?	euras(subozean)	14	15	134	13	Tli	50	150
Galium aparine/spurium	ANT?	euras(subozean)/smed-euras	5	5	210	7	T/T	50/30	150/100
Galium palustre	APO	no-eurassubozean	1	.	.	.	H	15	40(-80)
Galium cf. *palustre*	APO	no-eurassubozean	1	.	.	.	H	15	40(-80)
Galium spurium	ANT	smed-euras	14	98	805	.	Tli	30	100
Galium cf. *spurium*	ANT	smed-euras	3	2	26	.	Tli	30	100
Galium cf. *verum* agg.	ANT?	euras-smed	.	132	190	.	H	20	70
Hyoscyamus niger	ANT	smed-euras	.	.	2	.	T, H	30	60
Knautia arvensis	ANT	(no-)eurassubozean	.	1	.	.	H	30	80
Lapsana communis	APO	eurassubozean-smed	1	152	443	43	T(H)	30	100
cf. *Lapsana communis*	APO	eurassubozean-smed	.	19	1	.	T(H)	30	100
Lotus corniculatus s.str.	ANT?	eurassubozean-smed	.	.	1	.	H	5	30
Lotus cf. *uliginosus*	APO	subatl(-wsmed)	.	1	.	.	H	20	40(-80)
Malva sylvestris	ANT	smed-euras	.	2	.	.	H	20	100
Malva sp.	ANT	.	.	4	10	.	T/H	.	.
cf. *Malva* sp.	ANT	.	.	.	1	.	T/H	.	.
Matricaria perforata	ANT	gemäßkont	1	.	.	.	T	25	80
Medicago lupulina	ANT	euras-smed	.	.	1	.	T, H	10	30
Nepeta cf. *cataria*	ANT	osmed-euraskont	13	.	13	.	H, C	40	100
Phleum pratense s.l.	ANT	no-euras	317	1841	1172	7	H	30	100
cf. *Phleum pratense* s.l.	ANT	no-euras	6	21	36	.	H	30	100
Phleum pratense/Poa annua	ANT	no-euras/no-euras-med	1	15	10	4	H/T,H	30/3	100/20
Picris hieracioides s.l.	ANT?	euras(kont)-smed	1	1	.	.	H, T	30	60(100)
Plantago lanceolata	ANT?	eurassubozean	.	.	2	.	H	10	40
Plantago major s.str.	APO?	no-eurassubozean	.	.	1	.	H	10	30
Plantago major ssp. *intermedia*	APO?	mehr subatl	.	.	1	.	H, T	5	15
Plantago major cf. ssp. *intermedia*	APO?	mehr subatl	.	.	1	.	H, T	5	15
Poa annua	ANT	no-euras-med	1	4	5	2	T, H	3	20
cf. *Poa annua*	ANT	no-euras-med	.	1	1	1	T, H	3	20
Polygonum aviculare agg.	ANT	med-euras-no	2	4	12	.	T	10	50
Polygonum cf. *aviculare* agg.	ANT	med-euras-no	.	.	1	.	T	10	50
Polygonum convolvulus	ANT	(no-)euras	398	663	443	15	T	10	80
Polygonum cf. *convolvulus*	ANT	(no-)euras	3	37	103	.	T	10	80
Polygonum dumetorum	APO	euras(subozean)-smed	70	16	5	.	Tli	100	300
Polygonum lapathifolium agg.	APO	eurassubozean	9	36	15	1	T	30	80
Polygonum cf. *lapathifolium* agg.	APO	eurassubozean	.	3	1	.	T	30	80
Polygonum persicaria	ANT?	euras	.	.	3	.	T	10	60
Rhinanthus cf. *minor*	ANT	no-eurassubozean, circ	.	.	2	.	T	15	30(-50)
Rumex acetosella agg.	(APO?), ANT?	no-euras(subozean)	1	.	.	.	G,H	5	15(-30)
Rumex conglomeratus/sanguineus	APO? /APO	smed(-subatl) /subatl-smed	.	34	2	.	H	30/30	60/60
Rumex crispus/obtusifolius	APO?	eurassubozean-smed /subatl(-smed)	.	.	24	10	H	30/50	120/120
Rumex cf. *crispus/obtusifolius*	APO?	eurassubozean-smed /subatl(-smed)	.	.	3	3	H	30/50	120/120

Table 15.3, cont.

Botanical name	Classification	Centre of distribution	LBK I	LBK II	LBK III-V	Found in storage (n)	Life form	Min. height (cm)	Max. height (cm)
Sambucus ebulus	APO	smed(subatl)	2	.	3	.	H	50	150
Scleranthus annuus s.str.	ANT	(no)eurassubozean-smed	.	.	3	2	T	5	15
Setaria verticillata/viridis	ANT	smed-med/euras-med	7	13	101	.	T/T	5/5	50/50
cf. *Setaria verticillata/viridis*	ANT	smed-med/euras-med	2	.	6	.	T/T	5/5	50/50
cf. *Sherardia arvensis*	ANT	med-smed(-euras)	.	.	1	.	T	5	15
cf. *Sinapis arvensis*	ANT	eurassubozean-smed (origin: med-smed)	1	.	.	.	T	30	60
Solanum nigrum	ANT	smed-euras	2	113	115	2	T	10	80
Solanum cf. *nigrum*	ANT	smed-euras	.	4	1	.	T	10	80
cf. *Solanum nigrum*	ANT	smed-euras	.	4	9	.	T	10	80
Stellaria graminea	ANT	no-eurassubozean	.	.	12	12	H	10	30
Stellaria media agg.	ANT?	no-euras-med	1	2	.	.	T	5	30
Stipa sp.	ANT	kont?	11	.	.	.	H	.	medium
Thlaspi arvense	ANT	euras-smed	1	.	1	.	T	10	30
Torilis arvensis(/japonica)	ANT (/APO)	smed-med (/eurassubozean-smed)	.	.	1	.	.	30/30	80/90(-120)
Trifolium campestre/dubium	ANT	smed-subatl /subatl(smed)	.	.	2	.	T/T	10/10	20/20
Trifolium camp./dub./arv.	ANT?	smed-subatl /subatl(smed)/ eurassubozean-smed	.	1	16	4	T/T/T	10/10/5	20/20 /20(40)
Trifolium medium/pratense	APO /APO?	eurassubozean(smed)/ eurassubozean(-smed)	.	1	.	.	H/H	10/10	20(-50) /30(-40)
Urtica dioica	APO	no-euras	1	3	3	1	H	60	150
Valerianella dentata	ANT	smed-med	.	.	4	.	T	15	35(-50)
Verbascum sp.	ANT?, APO?	.	.	.	3	.	.		.
Veronica arvensis	ANT	eurassubozean(-smed)	1	.	81	.	T	5	20(-30)
Veronica hederifolia s.l.	APO?	eurassubozean-smed	.	.	1	1	T	5	30
Vicia hirsuta	ANT	euras-smed	.	1	6	5	Tli	15	50
Vicia cf. *hirsuta*	ANT	euras-smed	.	1	3	.	Tli	15	50
Vicia hirsuta/tetrasperma	ANT	euras-smed/smed-eurassubozean	2	1	10	7	T/T	15/20	50/60
Vicia tetrasperma	ANT	smed-eurassubozean	1	.	.	.	Tli	20	60

Table 15.3, cont.

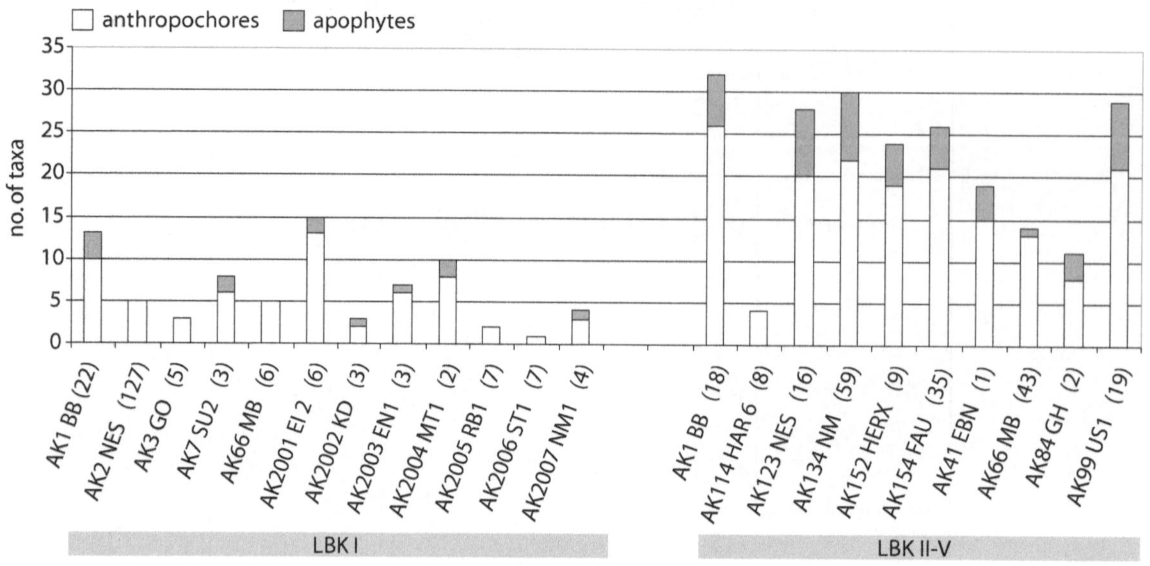

Figure 15.14: Number of species of anthropochores and apophytes per site. Numbers in brackets represent features investigated. For abbreviations of the sites see figure 15.4.

LBK I and later phases find support in the archaeological evidence. For example, flint-working techniques and the quantities of lithic artefacts changed markedly between Bandkeramik phase I and the following phases; in the earliest Bandkeramik settlements the percentage of sickle-blades is lower and they are less standardized (de Grooth 2003, p.402; Gronenborn 1997, p.102 and pers. comm.; Kind 1997, p.140). It is therefore possible that sickles were used less frequently for harvesting cereals during Bandkeramik phase I than in the later phases. This is an important point as experimental work has shown that using flint sickles is three times faster than ear-harvesting (pers. comm. Leonor Peña-Chocarro and Lydia Zapata-Peña, see also Ibáñez et al. 2001). Hand plucking or cutting near the ground are the best ways of harvesting cereals in order to keep the intact straw (Reynolds 1993).

The botanical data are insufficient to confirm that this change in harvesting methods had already occurred at the beginning of Bandkeramik phase II. There is only a slight difference in the weed spectra between phases I and II, whereas there is a distinct difference between phases I and III–V (figure 15.15, table 15.3). Unfortunately there are far fewer pits dated to phase II that have been investigated archaeobotanically in comparison with those dated to phases I or III–V (table 15.1). So it cannot be excluded that the 'lack' of difference between phases I and II is a consequence of the state of research.

Most, but not all, of the potential weeds found are annual species (table 15.3). The question of how the fields were ploughed, hoed or dug cannot be solved due to the lack of archaeological finds (Lüning 2000, p.160ff.). Pigs may have useful for this purpose. It is very probable that the crops were sown in rows and not broadcast, as the latter makes sense only if the soil is extensively ploughed and harrowed (Kreuz nd). The ecological demands of the documented weed species in LBK assemblages indicate no extreme soil conditions. As mentioned above, the Chernozem soils would have have been very well-suited for agriculture.

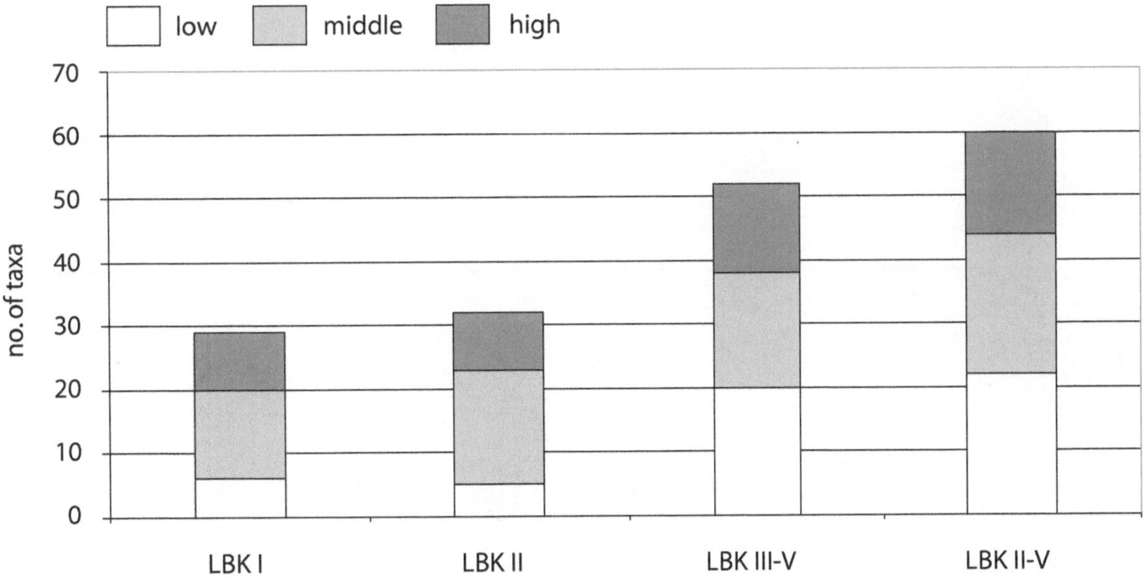

Figure 15.15: Numbers of potential weeds arranged according to their maximum height found at twelve earliest Bandkeramik (LBK I) and ten younger Bandkeramik sites (LBK II-V) (for the numbers of features see table 15.1).

15.7 Where did the strangers come from?

For 45 anthropochores it is possible to define their centre of distribution (*Verbreitungsschwerpunkt*), and their spatial extent (*Areal*) can be suggested using the *Florengebiete* of Oberdorfer (1990; for further explanations see Kreuz 1990a, p.143ff.; Kreuz 1993). Most of the Bandkeramik sites concerned are situated in western Europe (*Florengebiete* b; see table 15.4 and figure 15.16). Of the 45 anthropochorous species found, there are not more than six to nine that have their centre of distribution here, although three to five of these species have a submediterranean distribution pattern. It is on this basis that we may assume that most of the Bandkeramik anthropochorous weed species were introduced with seeds originating outside western Europe.

In fact, the distribution area of most of the anthropochorous species is within the Eurasiatic area, to the east and southeast of western Europe (table 3-4, fig. 16: *Florengebiete* c). The weed species with a mediterranenan (*med*) or submediterranean (*smed*) centre of distribution provide a clear link to the Mediterranean area (table 3-4). Six taxa appear not before Bandkeramik phase II, these are *Bromus* cf. *arvensis*, *Bromus sterilis*, *Malva sylvestris*, *Galium* cf. *verum*, *Trifolium campestre/dubium/arvense* and *Vicia hirsuta*. They might have been transported in the same way as the west Mediterranean poppy (*Papaver somniferum*), which is found from Bandkeramik phase II onwards (see above). Twelve further (sub)mediterranean (*(s-)med*) taxa occur from phases III–V onwards (table 3, 4).

The dataset is too small to confirm the ideas mentioned above. It would be important to compare these preliminary data with the evidence from contemporary sites in Hungary, Czechoslovakia, Poland, Romania and Bulgaria, etc., however, it is considered worth mentioning our botanical results as they concur with the archaeological evidence. Analogies of the ceramics of the Linearbandkeramik culture with those of the late Starçevo and Körös cultures and possibly the early Vinça culture have led the archaeologists to assume that the Bandkeramik culture originated in southeast Central Europe, in the area of western Hungary (e.g., Cladders 2001, p.105; Kind 1997, p.140; Lüning 1998, p.29ff.). This proposition is consistent with our weed

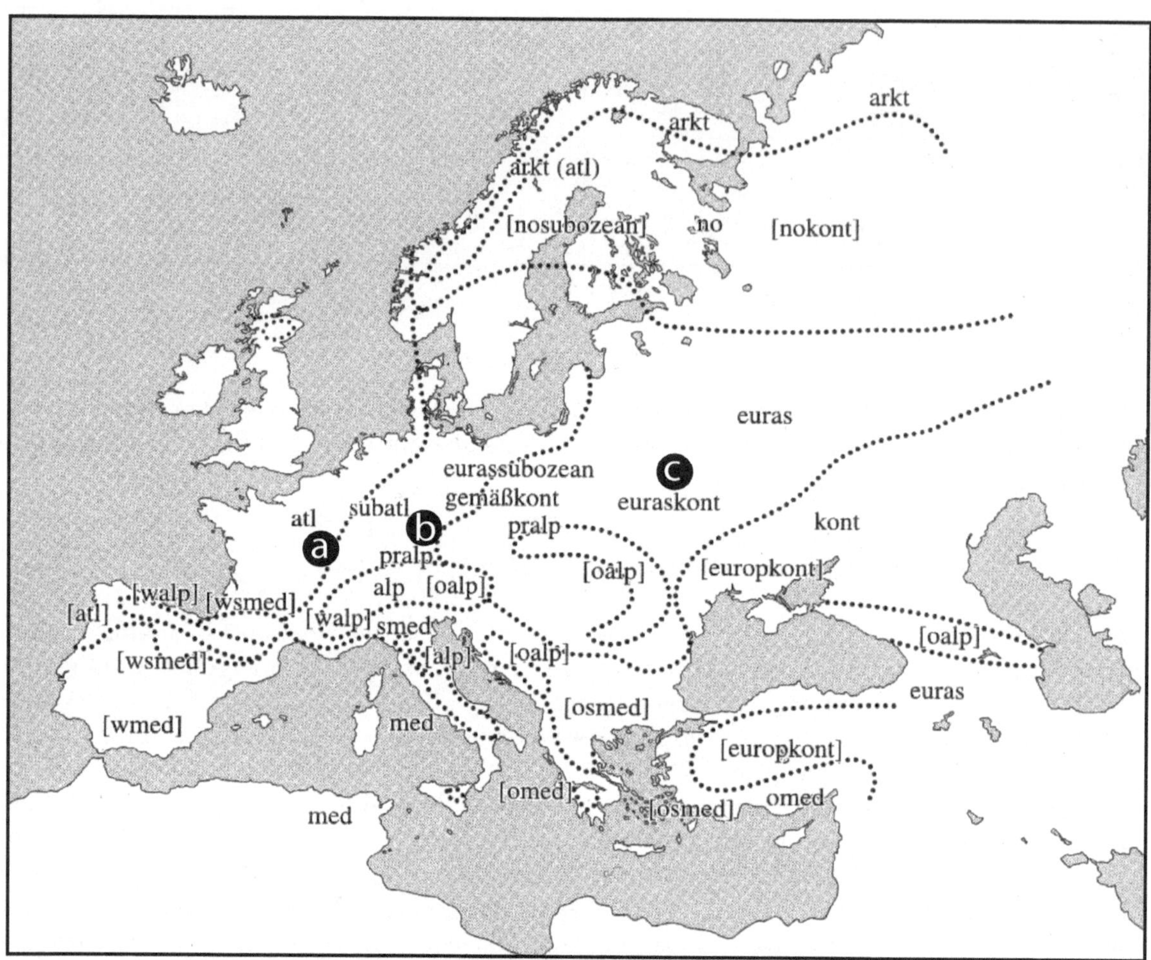

Figure 15.16: *Florengebiete* of Europe (modified after Oberdorfer 1990, fig. 2). Species with *atlantic* (atl), *subatlantic* (subatl) and *eurasiatic-subozeanic* (eurassubozean) centres of distribution are dominant in Area A. Species with *subatlantic* (subatl) and *eurasiatic-subozeanic* (eurassubozean) centres of distribution are dominant in Area B. In Area C, species with *moderate-kontinental* (gemäßkont) and *eurasiatic-kontinental* (euraskont) centres of distribution are dominant.

Botanical name	Classification	Centre of distribution	LBK I	LBK II	LBK III-V	Found in storage (n)
Euras						
Bromus cf. *secalinus*	ANT	(euras)	153	402	1584	835
Chenopodium hybridum	ANT	euras(kont)	2	2	4	.
Galium cf. *aparine*	ANT?	euras(subozean)	14	27	134	13
Phleum pratense s.l.	ANT	no-euras	317	1841	1172	7
cf. *Phleum pratense* s.l.	ANT	no-euras	6	21	36	.
Polygonum convolvulus	ANT	(no-)euras	398	663	443	15
Polygonum cf. *convolvulus*	ANT	(no-)euras	3	37	103	.
Polygonum persicaria	ANT?	euras	.	.	3	.
sum of taxa			4-5	4-5	4-6	
Eurassubozean						
Knautia arvensis	ANT	(no-)eurassubozean	.	1	.	.
Plantago lanceolata	ANT?	eurassubozean	.	.	2	.
Rhinanthus cf. *minor*	ANT	no-eurassubozean, circ	.	.	2	.
Stellaria graminea	ANT	no-eurassubozean	.	.	12	12
sum of taxa				1	2-3	
Eurassubozean-smed						
Lotus corniculatus s. str.	ANT?	eurassubozean-smed	.	.	1	.
Scleranthus annuus s.str.	ANT	(no)eurassubozean-smed	.	.	3	2
Veronica arvensis	ANT	eurassubozean(-smed)	1	.	81	.
Vicia tetrasperma	ANT	smed-eurassubozean	1	.	.	.
Trifolium campestre/dubium	ANT	smed-subatl /subatl(smed)	.	.	2	.
Trifolium camp./dub./arvense	ANT?	smed-subatl /subatl(smed)/eurassubozean-smed	.	1	16	4
sum of taxa			2	0-1	3-4	
Euras-(s)med						
Atriplex patula/hastata	ANT	euras(subozean)(-smed), circ/euras(-med)	1	4	7	1
Bromus cf. *arvensis*	ANT	euras-med	.	31	91	.
Chenopodium album	ANT	no-euras(-med)	919	4780	8415	307
Echinochloa crus-galli	ANT	med-smed-euras, circ	1	47	32	9
cf. *Echinochloa crus-galli*	ANT	med-smed-euras, circ	1	3	4	.
Galium spurium	ANT	smed-euras	14	98	805	.
Galium cf. *spurium*	ANT	smed-euras	3	2	26	.
Galium cf. *verum* agg.	ANT?	euras-smed	.	132	190	.
Hyoscyamus niger	ANT	smed-euras	.	.	2	.
Malva sylvestris	ANT	smed-euras	.	2	.	.
Medicago lupulina	ANT	euras-smed	.	.	1	.
Picris hieracioides s.l.	ANT?	euras(kont)-smed	1	1	.	.

Table 15.4: Numbers of anthropochores from 22 Bandkeramik sites in order of their centres of distribution (after Oberdorfer 1990).

Botanical name	Classification	Centre of distribution	LBK I	LBK II	LBK III-V	Found in storage (n)
euras-(s)med, cont.						
Poa annua	ANT	no-euras-med	1	4	5	2
cf. *Poa annua*	ANT	no-euras-med	.	1	1	1
Polygonum aviculare agg.	ANT	med-euras-no	2	4	12	.
Polygonum cf. *aviculare* agg.	ANT	med-euras-no	.	.	1	.
Solanum nigrum	ANT	smed-euras	2	113	115	2
Solanum cf. *nigrum*	ANT	smed-euras	.	4	1	.
cf. *Solanum nigrum*	ANT	smed-euras	.	4	9	.
Stellaria media agg.	ANT?	no-euras-med	1	2	.	.
Thlaspi arvense	ANT	euras-smed	1	.	1	.
Vicia hirsuta	ANT	euras-smed	.	1	6	5
Vicia cf. *hirsuta*	ANT	euras-smed	.	1	3	.
sum of taxa			8-10	10-13	12-13	
med, med-smed, smed-med						
Asperula arvensis	ANT	med-smed	.	.	1	.
Avena sp.	ANT	((fatua = med))	.	.	1	.
Brassica sp.	ANT	smed/med-atl	.	.	7	7
Bromus sterilis	ANT	smed	.	113	82	18
Bromus cf. *sterilis*	ANT	smed	.	15	166	70
Bromus sterilis/tectorum	ANT	smed/smed-kont	.	56	280	.
Capsella bursa-pastoris	ANT	med(-kont)	.	.	2	.
cf. *Capsella bursa-pastoris*	ANT	med(-kont)	.	.	1	.
cf. *Digitaria sanguinalis*	ANT	med-smed(-euras), circ	.	.	1	.
Setaria verticillata/viridis	ANT	smed-med/euras-med	7	13	101	.
cf. *Setaria verticillata/viridis*	ANT	smed-med/euras-med	2	.	6	.
cf. *Sherardia arvensis*	ANT	med-smed(-euras)	.	.	1	.
cf. *Sinapis arvensis*	ANT	eurassubozean-smed (origin: med-smed)	1	.	.	.
Torilis arvensis(/japonica)	ANT (/APO)	smed-med (/eurassubozean-smed)	.	.	1	.
Valerianella dentata	ANT	smed-med	.	.	4	.
sum of taxa			2	2	9-10	
kont						
Matricaria perforata	ANT	gemäßkont	1	.	.	.
Nepeta cf. *cataria*	ANT	osmed-euraskont	13	.	13	.
Stipa sp.	ANT	kont?	11	.	.	.
sum of taxa			3		1	

Table 15.4, cont.

data.

15.8 Summarizing remarks

Compared with later prehistoric periods the Bandkeramik agricultural system appears relatively restricted as far as the crop species are concerned. The relevance of this phenomenon cannot be discussed without the archaeozoological evidence, which still has to be evaluated.

Interestingly, several important differences between the LBK I and LBK II-V are shown by the archaeological finds and structures, as well as by the botanical macro-remains; the form, temper and decoration of the pottery changes, as well as some details of the house-constructions. The flint-working techniques and the quantities of lithic artefacts also change. During the second half of the Bandkeramik culture a difference in harvesting technique can be assumed and an additional crop, poppy (*Papaver somniferum*), occurs together with 'new' weeds probably of a (sub-)mediterranean origin. New landscape types were settled then and at the same time the population density increased significantly. High concentrations of charred glume bases as well as the occurrence of poppy could be interpreted as arising from changes in ritual practices.

In addition, it is striking that many LBK I sites have no continuity with the later phases; either the settlement ends completely or there is a significant temporal break in occupation, and LBK II settlements were often founded at new sites. Cladders and Stäuble interpret the extent of change in the typology of houses and pottery as being so profound that they could not be adapted by the inhabitants of one and the same settlement:

> *Derartige Änderungen eines mindestens schon zwei Jahrhunderte bestehenden Topos sind in einer traditionellen Gesellschaft mit Sicherheit nicht im Rahmen der gleichen Siedlung möglich. Deshalb muss die Generation der Erneuerer jeweils von der Elternsiedlung fortziehen, die Neuerungen können nur in deutlichem räumlichen Abstand durchgeführt werden.* (Cladders and Stäuble 2003, p.502)

I would like to propose a hypothesis that extends Caldders and Stäuble's propostion. Changes occurred throughout everyday life at the end of the earliest Bandkeramik phase I, yet there is no current evidence for any crisis at this time. It is a well-known fact that subsistence systems are not changed without good reason, thus it is worth questioning if the changes between LBK I and II originated internally from what our evidence suggests was a well-functioning system. Instead, it seems reasonable to hypothesise that at the end of LBK I there was an introduction of new people, and an integration of their traditions with that of the earliest Bandkeramik culture that gave rise to the distinctive LBK II complex. Those sites where it is difficult to separate typologically the ceramics of Bandkeramik phase I from phase II may belong to the period of transition (Cladders 2001, p.66, 107, 115, footnote 147). Further interdisciplinary research is required to gain insight into this complex and fascinating subject.

Acknowledgements

The research project is possible thanks to the funding of the German Research Association DFG, to whom I am very grateful. I would like to thank Maria Cladders and Harald Stäuble (Weixdorf), and Eva Schäfer and Julian Wiethold (Wiesbaden), for critical remarks to the manuscript. Finally I would like to thank Eva Schäfer, who prepared all the graphs and Muazez Elibal, who made the scans, as well as Sue and James for improving my English.

References

Badr, A., K. Müller, R. Schäfer-Pregl, H. El Rabey, S. Effgen, H. H. Ibrahim, C. Pozzi, W. Rohde, and F. Salamini (2000). On the origin and domestication history of barley (*Hordeum vulgare*). *Molecular Biology and Evolution 17*, 499–510.

Bakels, C. (1978). *Four Linnearbandkeramik Settlements and their Environment: A Palaeoecological Study of Sittard, Stein, Elsoo and Hienheim*, Volume XI of *Analecta Praehistorica Leidensia.* Leiden: Leiden University Press.

Bakels, C. (1982). Der Mohn, die Linearbandkerammik und das westliche Mittelmeergebiet. *Archäologisches Korrespondenzblatt 12*, 11–13.

Bakels, C. (1991). Tracing crop processing in the Bandkeramik culture. In J. Renfrew (Ed.), *New Light on Early Farming. Recent Developments in Palaeoethnobotany*, pp. 281–288. Edinburgh: Edinburgh University Press.

Bakker, R. (2003). *The Emergence of Agriculture on the Drenthe Plateau: A Palaeobotanical Study supported by High-Resolution 14C Dating.* Archäologische Berichte 16. Bonn: Die Deutsche Gesellschaft für Ur- und Frühgeschichte (DGUF) e. V.

Beug, H.-J. (1986). Vegetationsgeschichtliche Untersuchungen über das Frühe Neolithikum im Untereichsfeld, Landkreis Göttingen. In K.-E. Behre (Ed.), *Anthropogenic Indicators in Pollen Diagrams*, pp. 115–124. Rotterdam: A. A. Balkema.

Beug, H.-J. (1992). Vegetationsgeschichtliche Untersuchungen über die Besiedlung im Unteren Eichsfeld, Landkreis Göttingen, vom frühen Neolithikum bis zum Mittelalter. . *Neue Ausgrabungen und Forschungen in Niedersachse 20*, 261–339.

Beug, H.-J. (2004). *Leitfaden der Pollenbestimmung für Mitteleuropa und Angrenzende Gebiete.* München: Dr. Friedrich Pfeil.

Beug, H.-J., I. Henrion, and A. Schmüser (1999). *Landschaftsgeschichte im Hochharz. Die Entwicklung der Wälder und Moore seit dem Ende der letzten Eiszeit.* Clausthal-Zellerfeld: Papierflieger Verlag.

Bogaard, A. (2002a). *The Permanence, Intensity and Seasonality of Early Crop Cultivation in western Central Europe.* Ph. D. thesis, Sheffield University.

Bogaard, A. (2002b). Questioning the relevance of shifting cultivation to neolithic farming in the loess belt of Europe: evidence from the Hambach Forest experiment. *Vegetation History and Archaeobotany 11*, 155–168.

Cappers, R. T. J. and S. Bottema (Eds.) (2002). *The Dawn of Farming in the Near East.* Studies in early Near Eastern Production, Subsistence and Environment 6/1999. Berlin: Ex oriente.

Cladders, M. (2001). *Die Tonware der Ältesten Bandkeramik. Untersuchungen zur zeitlichen und räumlichen Gliederung.* Universitätsforschungen zur prähistorischen Archäologie 72. Bonn: Dr. Rudolf Habelt Verlag.

Cladders, M. and H. Stäuble (2003). Das 53. Jahrhundert v. Chr. : Aufbruch und Wandel. In J. Eckert, U. Eisenhauer, and A. Zimmermann (Eds.), *Archäologische Perspektiven. Analysen und Interpretationen im Wandel. Festschrift für Jens Lüning zum 65. Geburtstag*, Internationale Archäologie, Studia honoraria 20, pp. 491–503. Rahden, Westfalen: Marie Leidorf.

Colledge, S. (2001). *Plant Exploitation on Epipalaeolithic and Early Neolithic Sites in the Levant.* BAR International Series 986. Oxford: John and Erica Hedges Ltd.

Damania, A. B., J. Valkoun, J. Willcox, and C. O. Qualset (Eds.) (1998). *The Origins of Agriculture and Crop Domestication. Proceedings of the Harlan Symposium 10-14 May 1997.* Aleppo, Syria: ICARDA.

de Grooth, M. E. T. (2003). They do things differently there. Flint working at the Early Bandkeramik settlement of Geleen-Janskamperveld (The Netherlands). In J. Eckert, U. Eisenhauer, and A. Zimmermann (Eds.), *Archäologische Perspektiven. Analysen und Interpretationen im Wandel. Festschrift für Jens Lüning zum 65. Geburtstag*, Internationale Archäologie, Studia honoraria 20, pp. 401–406. Rahden, Westfalen: Marie Leidorf.

Enneking, D. (1995). *The Toxicity of* Vicia *Species and their Utilisation as Grain Legumes.* Centre for Legumes in Mediterranean Agriculture (CLIMA) Occasional Publication No. 6. Nedlands, W. A.: University of Western Australia.

Erny-Rodmann, C., E. Gross-Klee, J. N. Haas, S. Jacomet, and H. Zoller (1997). Früher 'human impact' und Ackerbau im Übergangsbereich Spätmesolithikum-Frühneolithikum im schweizerischen Mittelland. *Jahrbuch der Schweizerischen Gesellschaft für Ur- und Frühgeschichte 80*, 27–56.

Firbas, H. (1937). Der pollenanalytische Nachweis des Getreidebaus. *Zeitschrift für Botanik 37*, 447–478.

Forbes, H. (1998). European agriculture viewed bottom-side upwards: fodder-provision in a traditional Greek community. In M. Charles, P. Halstead, and G. Jones (Eds.), *Fodder: Archaeological, Historical and Ethnographic Studies*, Environmental Archaeology 1, pp. 19–34. Oxford: Oxbow Books.

Geisler, G. (1991). *Farbatlas landwirtschaftliche Kulturpflanzen.* Stuttgart: Ulmer.

Greig, J. R. A. (1991). The British Isles. In W. van Zeist, K. Wasylikowa, and K.-E. Behre (Eds.), *Progress in Old World Palaeoethnobotany: A Retrospective View on the Occassion of 20 Years of the International Work Group for Palaeoethnobotany*, pp. 299–334. A. A. Balkema.

Groenman-van Waateringe, W. (1970). Hecken im westeuropäischen Frühneolithikum. *Berichten van de Rijksdienst voor het Oudheidkundig Bodemonderzoek 20-21*, 295–299.

Gronenborn, D. (1992). Inventarwerk zu mesolithischen Fundplätzen im Main-Mündungsgebiet. *Berichte der Kommission für Archäologische Landesforschung in Hessen 1*(1990/91), 35–37.

Gronenborn, D. (1997). Die Steinartefakte. In J. Lüning (Ed.), *Ein Siedlungsplatz der Ältesten Bandkeramik in Bruchenbrücken, Stadt Friedberg/Hessen*, Universitätsforschungen zur prähistorischen Archäologie 39, pp. 255–332. Bonn: Dr. Rudolf Habelt Verlag.

Gronenborn, D. (1999a). Ältestbandkeramische Kultur, La Hoguette, Limburg, and. . . what else? Contemplating the Mesolithic-Neolithic transition in southern Central Europe. *Neolithic Studies. Porocilo o Raziskovanju Paleolitika, Neolitika in Eneolitika v Slovenij 25*, 189–202.

Gronenborn, D. (1999b). A variation on a basic theme: the transition to farming in southern Central Europe. *Journal of World Prehistory 13*(2), 123–210.

Gronenborn, D. (2003). Migration, acculturation and culture change in western temperate Eurasia, 6500-5000 cal BC. *Documenta Præhistorica 30*, 79–91.

Grote, K. (1994). *Die Abris im südlichen Leinebergland bei Göttingen. Archäologische Befunde zum Leben unter Felsschutzdächern in urgeschichtlicher Zeit. Part I and II.* Oldenburg: Isensee.

Haas, J. N. (1996). *Pollen and Plant Macrofossil Evidence of Vegetation Change at Wallisellen-Langachermoos (Switzerland) during the Mesolithic-Neolithic Transition 8500 to 6500 Years Ago.* Dissertationes Botanicae 267. Stuttgart: Gebr. Borntraeger Verlagsbuchhandlung, Science Publishers.

Haas, J. N., I. Richoz, W. Tinner, and L. Wick (1998). Synchronous Holocene climatic oscillations recorded on the Swiss Plateau and at timberline in the Alps. *The Holocene 8*(3), 301–309.

Halstead, P. and S. Tierney (1998). Leaf hay: an ethnoarchaeological study in NW Greece. In M. Charles, P. Halstead, and G. Jones (Eds.), *Fodder: Archaeological, Historical and Ethnographic Studies*, Environmental Archaeology 1, pp. 71–80. Oxford: Oxbow Books.

Hillman, G. C. (1981). Reconstructing crop husbandry practices from charred remains of crops. In R. J. Mercer (Ed.), *Farming Practice in British Prehistory*, pp. 123–162. Edinburgh: Edinburgh University Press.

Hillman, G. C. (1984). Interpretation of archaeological plant remains: the application of ethnographic models from Turkey. In W. van Zeist and W. A. Casparie (Eds.), *Plants and Ancient Man*, Studies in Palaeoethnobotany. Proceedings of the Sixth Symposium of the International Work Group for Palaeoethnobotany, Groningen, 30 May–3 June 1983, pp. 1–42. Rotterdam: A. A. Balkema.

Hormes, A., B. U. Müller, and C. Schlüchter (2001). The Alps with little ice: evidence for eight Holocene phases of reduced glacier extent in the central Swiss Alps. *The Holocene 11*(3), 255–265.

Ibáñez, J., J. González Urquijo, L. Peña-Chocarro, L. Zapata, and V. Beugnier (2001). Harvesting without sickles. Neolithic examples from humid mountain areas. In S. Beyries and P. Petrequin (Eds.), *Ethno-Archaeology and its Transfers*, Number 983 in BAR International Series, pp. 23–36. Oxford: Archaeopress.

Jacomet, S. and A. Kreuz (1999). *Archäobotanik. Aufgaben, Methoden und Ergebnisse vegetations—und agrargeschichtlicher Forschung.* UTB für Wissenschaft 8158. Stuttgart: Verlag Eugen Ulmer. With contributions by M. Rösch.

Jeunesse, C. (1987). La céramique de La Hoguette. Un nouvel "élément non-rubané" du néolithique ancien de l'Europe du nord-ouest. *Cahiers Alsaciens d'Archéologie, d'Art et d'Histoire 30*, 3–33.

Jeunesse, C. (2003). Néolithique "initial", néolithique ancien et néolithisation dans l'espace centre-européen: une vision rénovée. *Revue d'Alsace 129*, 97–112.

Jones, G. (1984). Interpretation of archaeological plant remains: ethnographic models from Greece. In W. van Zeist and W. A. Casparie (Eds.), *Plants and Ancient Man*, Studies in Palaeoethnobotany. Proceedings of the Sixth Symposium of the International Work Group for Palaeoethnobotany, Groningen, 30 May–3 June 1983, pp. 43–61. Rotterdam: A. A. Balkema.

Jones, G. (2000). Evaluating the importance of cultivation and collecting in neolithic Britain. In A. S. Fairbairn (Ed.), *Plants in Neolithic Britain and Beyond*, Neolithic Studies Group Seminar Paper 5, pp. 79–84. Oxford: Oxbow Books.

Jones, G. and P. Halstead (1995). Maslins, mixtures and monocrops: on the interpretation of archaeobotanical crop samples of heterogeneous composition. *Journal of Archaeological Science 22*, 103–114.

Kalis, A. J. (1988). Zur Umwelt des frühneolithischen Menschen: ein Beitrag der Pollenanalyse. *Forschungen und Berichte zur Vor- und Frühgeschichte Baden-Württemberg (Festschrift U. Körber-Grohne) 31*, 125–137.

Kalis, A. J., J. Merkt, and J. Wunderlich (2003). Environmental changes during the Holocene climatic optimum in Central Europe: human impact and natural causes. *Quaternary Science Reviews 22*, 33–79.

Kalis, A. J., J. Meurers-Balke, K. van der Borg, A. von den Driesch, W. Rähle, U. Tegtmeier, and H. Thiemeyer (2001). Der La-Houguette-Fundhorizont in der Wilhelma von Stuttgart-Bad Cannstatt. Anthrakologische, archäopalynologische, bodenkundliche, malakozoologische, radiometrische und säugetierkundliche Untersuchungen. In B. Gehlen, M. Heinen, and A. Tillmann (Eds.), *Zeit-Räume. Gedenkschrift für Wolfgang Taute*, Archäologische Berichte 14, pp. 649–671. Bonn: Die Deutsche Gesellschaft für Ur- und Frühgeschichte (DGUF) e. V.

Kaufmann, D. (1983). Die ältestlinienbandkeramischen Funde von Eilsleben, Kr. Wanzleben und der Beginn des Neolithikums im Mittelelbe-Saale Gebiet. *Nachrichten aus Niedersachsens Urgeschichte 52*, 177–202.

Kind, C.-J. (1997). *Die letzten Wildbeuter. Henauhof Nord II und das Endmesolithikum in Baden-Württemberg*, Volume 39 of *Forschungen und Berichte zur Vor- und Frühgeschichte in Baden-Württemberg*. Stuttgart: Konrad Theiss Verlag.

Kind, C.-J. (2003). *Das Mesolithikum in der Talaue des Neckars. Die Fundstellen von Rottenburg Siebenlinden 1 und 3*, Volume 88 of *Forschungen und Berichte zur Vor- und Frühgeschichte in Baden-Württemberg*. Stuttgart: Konrad Theiss Verlag. With contributions by A. M. Miller and J. Hahn.

Kleinmann, A., J. Merkt, and H. Müller (2000). Climatic lake-level changes in German lakes during the Holocene? *Terra Nostra 7*, 55–63.

Knörzer, K.-H. (1991). Deutschland nördlich der Donau. In W. van Zeist, K. Wasylikowa, and K.-E. Behre (Eds.), *Progress in Old World Palaeoethnobotany: A Retrospective View on the Occassion of 20 Years of the International Work Group for Palaeoethnobotany*, pp. 189–206. Rotterdam: A. A. Balkema.

Körber-Grohne, U. (1988). *Nutzpflanzen in Deutschland. Kulturgeschichte und Biologie.* Stuttgart: Konrad Theiss Verlag.

Korneck, D. and H. Sukopp (1988). *Rote Liste der in der Bundesrepublik Deutschland ausgestorbenen, verschollenen und gefährdeten Farn- und Blütenpflanzen und ihre Auswertung für den Arten- und Biotopschutz.* Schriftenreihe Vegetationskunde 19. Bonn-Bad Godesberg: Landwirtschaftsverlag.

Kossack, G. and H. Schmeidl (1975). Vorneolithischer Getreideanbau im Bayerischen Alpenvorland. *Jahresbericht der Bayerischen Bodendenkmalpflege 15/16*, 7–23.

Kreuz, A. (1988). Holzkohle-Funde der ältestbandkeramischen Siedlung Friedberg-Bruchenbrücken: Anzeiger für Brennholz-Auswahl und lebende Hecken? *Forschungen und Berichte zur Vor- und Frühgeschichte Baden-Württemberg (Festschrift U. Körber-Grohne) 31*, 139–153.

Kreuz, A. (1990a). *Die ersten Bauern Mitteleuropas: Eine archäobotanische Untersuchung zu Umwelt und Landwirtschaft der Ältesten Bandkeramik.* Analecta Praehistorica Leidensia 23. Leiden: Leiden University Press.

Kreuz, A. (1990b). Searching for "single-activity refuse" in Linearbandkerammik settlements. In D. E. Robinson (Ed.), *Experimentation and Reconstruction in Environmental Archaeology*, pp. 63–74. Oxford: Oxbow Books.

Kreuz, A. (1992). Charcoal from ten early neolithic settlements in Central Europe and its interpretation in terms of woodland management and wildwood resources. *Bulletin de la Société Botanique de France, Actualités Botaniques 139*, 383–394.

Kreuz, A. (1993). Einheimische oder fremde Pflanzen? Überlegungen zur Herkunft "potentieller Unkräuter" und ihrer Verbreitung zur Zeit der Bandkeramik. In A. J. Kalis and J. Meurers-Balke (Eds.), *7000 Jahre Bäuerliche Landwirtschaft: Entstehung, Erforschung, Erhaltung*, Archaeo-Physika 13, pp. 23–33. Köln: Dr. Rudolf Habelt Verlag.

Kreuz, A. (1995). On-site and off-site data: interpretative tools for a better understanding of early neolithic environments. In H. Kroll and R. Pasternak (Eds.), *Res Archaeobotanicae: 9th Symposium IWGP*, pp. 117–134. Kiel: Oetker-Voges Verlag.

Kreuz, A. (n.d.). Landwirtschaft im Umbruch? Eine archäobotanische Untersuchung zu den Jahrhunderten um Christi Geburt in Hessen und Mainfranken. *Bericht der Römisch-Germanischen Kommission.* In Prep (2005).

Kreuz, A., S. Nolte, and A. Stobbe (1998). Interpretation pflanzlicher Reste aus holozänen Auensedimenten am Beispiel von drei Bohrkernen des Wettertales (Hessen). *Eiszeitalter und Gegenwart 48*, 133–161.

Kroll, H. (1991). Südosteuropa. In W. van Zeist, K. Wasylikowa, and K. E. Behre (Eds.), *Progress in Old World Palaeoethnobotany: A Retrospective View on the Occassion of 20 Years of the International Work Group for Palaeoethnobotany*, pp. 161–178. Rotterdam: A. A. Balkema.

Kubiak-Martens, L. (1998). *The Botanical Component of Hunter-Gatherer Subsistence Strategies in Temperate Europe during the Late Glacial and Early Holocene (Evidence from Selected Archaeological Sites).* Ph. D. thesis, W. Szafer Institute of Botany, Polish Academy of Science, Kraków.

Küster, H. (1989). Pollen analytical evidence for the beginning of agriculture in south Central Europe. In A. Milles, D. Williams, and N. Gardner (Eds.), *The Beginnings of Agriculture*, Symposia of the Association for Environmental Archaeology No. 8. BAR International Series 496, pp. 137–147. Oxford: British Archaeological Reports.

Lichardus, J., I. Gatsov, M. Gurova, and I. K. Illiev (2000). Geometric microliths from the middle neolithic site of Drama-Gerena (southeast Bulgaria) and the problem of mesolithic tradition in southeasten Europe. *Eurasia Antiqua. Zeitschrift für Archäologie Eurasiens 6*, 1–12.

Liese-Kleiber, H. (1997). Erste Pollenanalysen und C-14-Daten aus den mesolithischen Lagerplätzen Henauhof Nord I und II am Federsee. In C.-J. Kind (Ed.), *Die letzten Wildbeuter. Henauhof Nord II und das Endmesolithikum in Baden Württemberg*, Materialhefte zur Archäologie Baden-Württembergs 39, pp. 212–232. Stuttgart: Konrad Theiss Verlag.

Litt, T. (1990). Pollenanalytische Untersuchungen im Allertal bei Eilsleben, Kr. Wanzleben, und ihre Aussagemöglichkeit zur Vegetationsentwicklung während des Frühneolithikums (Vorläufige Mitteilung). *Jahresschrift mitteldeutsche Vorgeschichte 73*, 49–55.

Lüning, J. (1980). Getreideanbau ohne Düngung. *Archäologisches Korrespondenzblatt 10*, 117–122.

Lüning, J. (1998). Frühe bauern in mitteleuropa im 6. und 5. jahrtausend v. chr. *Jahrbuch des Römisch-Germanischen Zentralmuseums Mainz 35*(1), 27–93.

Lüning, J. (2000). *Steinzeitliche Bauern in Deutschland. Die Landwirtschaft im Neolithikum*, Volume 58 of *Universitätsforschungen zur prähistorischen Archäologie*. Bonn: Dr. Rudolf Habelt Verlag.

Lüning, J., U. Kloos, and S. Albert (1989). Westliche Nachbarn der bandkeramischen Kultur: La Hoguette und Limburg. *Germania 67*, 355–393. With contributions by J. Eckert and C. Strien.

Magny, M. (1998). Reconstruction of Holocene lake-level changes in the French Jura: methods and results. In B. Frenzel (Ed.), *Palaeohydrology as Reflected in Lake-Level Changes al Climatic Evidence for Holocene Times*, pp. 67–85. Stuttgart: Gustav Fisher Verlag.

Maise, C. (1998). Vom Einfluß nacheiszeitlicher Klimavariabilität in der Ur- und Frühgeschichte. *Jahrbuch der Schweizerischen Gesellschaft für Ur- und Frühgeschicht 81*, 197–235.

Marinova, E. M. (2001). *Vergleichende paläoethnobotanische Untersuchung zur Vegetationsgeschichte und zur Entwicklung der prähistorischen Landnutzung in Bulgarien*. Ph. D. thesis, Universität Bonn.

Meier-Arendt, W. (1966). *Die Bandkeramische Kultur im Untermaingebiet*. Veröffentlichungen des Amtes für Bodendenkmalpflege im Regierungsbezirk Darmstadt 3. Bonn: Dr. Rudolf Habelt Verlag.

Meyer, M. (2003). Zur formalen Gliederung alt- und mittelneolithischer Einhegungen. In J. Eckert, U. Eisenhauer, and A. Zimmermann (Eds.), *Archäologische Perspektiven. Analysen und Interpretationen im Wandel. Festschrift für Jens Lüning zum 65. Geburtstag*, Internationale Archäologie, Studia honoraria 20, pp. 441–454. Rahden, Westfalen: Marie Leidorf.

Monk, M. (2000). Seeds and soils of discontent: an environmental archaeological contribution to the nature of the early neolithic. In A. Desmond, G. Johnson, M. McCarthy, J. Sheehan, and E. Shee Twohig (Eds.), *New Agendas in Irish Prehistory. Papers in Commemoration of Liz Anderson*, pp. 67–87. Co. Wicklow, Ireland: Wordwell.

Nesbitt, M. (2001). Wheat evolution: integrating archaeological and biological evidence. In P. D. S. Caligari and P. E. Brandham (Eds.), *Wheat Taxonomy: The Legacy of John Percival*, The Linnean Special Issue No 3, pp. 37–59. London: Academic Press.

Nesbitt, M. (2002). When and where did domesticated cereals first occur in Southwest Asia? In R. T. J. Cappers and S. Bottema (Eds.), *The Dawn of Farming in the Near East*, Studies in early Near Eastern Production, Subsistence and Environment 6/1999, pp. 113–132. Berlin: Ex oriente.

Nesbitt, M. and S. O'Hara (1999). Irrigation agriculture in Central Asia: a long-term perspective from Turkmenistan. In C. Gosden and J. Hather (Eds.), *The Prehistory of Food: Appetites for Change*, pp. 103–122. London: Routledge.

Nesbitt, M. and D. Samuel (1996). From staple crop to extinction? The archaeology and history of the hulled wheats. In S. Padulosi, K. Hammer, and J. Heller (Eds.), *Hulled wheats. Promoting the Conservation and Use of Underutilized and Neglected Crops*, pp. 41–100. Rome: IPGRI.

Niggemann, S., A. Mangini, D. K. Richter, and G. Wurth (2000). Holozäne Stalagmiten des Sauerlandes (Deutschland) als Klimaarchive. *Mitteilungen des Verbandes deutscher Höhlen- und Karstforscher 46*, 84–90.

Oberdorfer, E. (1990). *Pflanzensoziologische Exkursionsflora für Deutschland und angrenzende Gebiete.* Stuttgart: Ulmer.

Palmer, C. (1998). The role of fodder in the farming system: a case study from northern Jordan. In M. Charles, P. Halstead, and G. Jones (Eds.), *Fodder: Archaeological, Historical and Ethnographic Studies*, Environmental Archaeology 1, pp. 1–10. Oxford: Oxbow Books.

Petrasch, J. (2003). Zentrale Orte in der Bandkeramik. In J. Eckert, U. Eisenhauer, and A. Zimmermann (Eds.), *Archäologische Perspektiven. Analysen und Interpretationen im Wandel. Festschrift für Jens Lüning zum 65. Geburtstag*, Internationale Archäologie, Studia honoraria 20, pp. 505–513. Rahden, Westfalen: Marie Leidorf.

Pflanzliche Erzeugung (1998). *Band 1. Die Landwirtschaft.* München, Wien, Zürich: BLV Verlagsgesellschaft.

Reynolds, P. (1985). Carbonized seed, crop yield, weed infestation and harvesting techniques of the iron age. In M. Gast, F. Sigaut, and C. Beutler (Eds.), *Les Techniques de Conservation des Graines à Long Terme 3*, pp. 397–407. Paris: C.N.R.S.

Reynolds, P. J. (1990). Ernteerträge der prähistorischen Getreidearten Emmer und Dinkel: "Die ungünstigste Wahl". *Archäologische Informationen 13*, 61–72.

Reynolds, P. J. (1993). Zur Herkunft verkohlter Getreidekörner in urgeschichtlichen Siedlungen: Eine alternative Erklärung. In A. J. Kalis and J. Meurers-Balke (Eds.), *7000 Jahre bäuerliche Landwirtschaft: Entstehung, Erforschung, Erhaltung. Zwanzig Aufsätze zu Ehren von Karl-Heinz Knörzer*, Archaeo-Physika 13, pp. 187–206. Köln: Dr. Rudolf Habelt Verlag.

Saile, T. (1998). *Untersuchungen zur ur- und frühgeschichtlichen Besiedlung der nördlichen Wetterau.* Materialien zur Vor- und Frühgeschichte von Hessen 21. Wiesbaden: Selbstverlag des Landesamtes für Denkmalpflege Hessen.

Schäfer, M. (1996). *Pollenanalysen an Mooren des Hohen Vogelsberges (Hessen) - Beiträge zur Vegetationsgeschichte und anthropogenen Nutzung eines Mittelgebirges.* Dissertationes Botanicae 265. Stuttgart: Gebr. Borntraeger Verlagsbuchhandlung, Science Publishers.

Schlütz, F. (2003). Review of: Schweizer, A. 2001. "Archäopalynologische Untersuchungen zur Neolithisierung der nördlichen Wetterau, Hessen". Dissertationes Botanicae 350. Berlin, Stuttgart: J. Cramer. *Tuexenia 23*, 431–433.

Schweizer, A. (2001). *Archäopalynologische Untersuchungen zur Neolithisierung der nördlichen Wetterau/Hessen.* Dissertationes Botanicae 350. Stuttgart: Gebr. Borntraeger Verlagsbuchhandlung, Science Publishers.

Sedlmeier, J. (2003). Neue Erkenntnisse zum Neolithikum in der Nordwestschweiz. *Archäologie der Schweiz 26*(4), 2–14.

Sielmann, B. (1971). Die frühneolithische Besiedlung Mitteleuropas. Die Anfänge des Neolithikums vom Orient bis Nordeuropa. *Fundamenta A 3*(Va), 1–65.

Spurk, M., H. H. Leuschner, M. G. L. Baillie, K. R. Briffa, and M. Friedrich (2002). Depositional frequency of German subfossis oaks: climatically and non-climatically induced fluctuations in the Holocene. *The Holocene 12*, 707–715.

Stäuble, H. (1990). Die ältestbandkeramische Grabenanlage in Eitzum, Ldkr. Wolfenbüttel. Überlegungen zur Verfüllung und Interpretation von Befunden. *Jahresschrift mitteldeutsche Vorgeschichte 73*, 331–344.

Stäuble, H. (1995). Radiocarbon dates of the earliest Neolithic in Central Europe. In G. T. Cook, D. D. Harkness, B. F. Miller, and E. M. Scott (Eds.), *Proceedings of the 15th International C-14 Conference*, pp. 227–327. Radiocarbon 37.

Stäuble, H. (1997). Häuser, Gruben und Fundverteilung. In J. Lüning (Ed.), *Ein Siedlungsplatz der Ältesten Bandkeramik in Bruchenbrücken, Stadt Friedberg/Hessen*, pp. 17–150. Bonn: Dr. Rudolf Habelt Verlag.

Stobbe, A. (1996). *Die holozäne Vegetationsgeschichte der nördlichen Wetterau: Paläoökologische Untersuchungen unter besonderer Berücksichtigung anthropogener Einflüsse.* Dissertationes Botanicae 260. Stuttgart: Gebr. Borntraeger Verlagsbuchhandlung, Science Publishers.

Stöckli, W. (2002). *Absolute und relative Chronologie des Früh- und Mittelneolithikums in Westdeutschland (Rheinland und Rhein-Main-Gebiet).* Basler Hefte zur Archäologie 1. Basel: Archäologie Verlag.

Strien, H.-C. and A. Tillmann (2001). Die La-Hoguette-Fundstelle von Stuttgart-Bad Cannstatt: Archäologie. In B. Gehlen, M. Heinen, and A. Tillmann (Eds.), *Zeit-Räume. Gedenkschrift für Wolfgang Taut*, Archäologische Berichte 14, pp. 673–681. Die Deutsche Gesellschaft für Ur- und Frühgeschichte (DGUF) e. V.

Taute, W. (1974). Neolithische mikrolithen und andere neolithische silexartefakte aus süddeutschland und österreich. *Archäologische Informationen 2/3*(1973/4), 71–125.

Thiemeyer, H. (1988). *Bodenerosion und holozäne Dellenentwicklung in hessischen Lößgebieten.* Frankfurt: Rhein-Mainische Forschung 105.

Uerpmann, H.-P. (1997). Die Tierknochenfunde. In J. Lüning (Ed.), *Ein Siedlungsplatz der Ältesten Bandkeramik in Bruchenbrücken, Stadt Friedberg/Hessen*, Universitätsforschungen zur prähistorischen Archäologie 39, pp. 333–348. Bonn: Dr. Rudolf Habelt Verlag.

Urz, R. (2000). Begraben unter Auelehm: Frühmesolithische Siedlungsspuren im Mittleren Lahntal. *Archäologisches Korrespondenzblatt 30*, 33–43.

Urz, R. (2002). Archäobotanische Untersuchungen zur Veränderung der Flusslandschaft im mittleren Lahntal (Hessen) in prähistorischer Zeit. *Archäologisches Korrespondenzblatt 32*, 169–186.

van der Veen, M. and C. Palmer (1997). Environmental factors and the yield potential of ancient wheat crops. *Journal of Archaeological Science 24*(2), 163–182.

van Zeist, W. (1967). Palynologische Untersuchungen eines Torfprofils bei Sittard. *Palaeohistoria 6/7*, 19–24.

van Zeist, W. and M. R. van der Spoel-Walvius (1980). A palynological study of the late-glacial and the postglacial in the Paris basin. *Palaeohistoria 22*, 67–109.

Visset, L., A.-L. Cyprien, N. Carcaud, A. Ouguerram, D. Barbier, and J. Bernard (2002). Les prémices d'une agriculture diversifiée à la fin du Mésolithique dans le Val de Loire (Loire armoricaine, France). *Comptes Rendus Palevol 1*, 51–58.

Weller, U. (2003). *Steingeräte der Linearbandkerammik im Leinetal zwischen Hannover und Northeim. Eine technologisch-archäologische Analyse.* Beiträge zur Archäologie in Niedersachsen 4. Rahden, Westfalen: Marie Leidorf.

Wiethold, J. (1998). *Studien zur jüngeren postglazialen Vegetations- und Siedlungsgeschichte im östlichen Schleswig-Holstein.* Universitätsforschungen zur prähistorischen Archäologie 45. Bonn: Dr. Rudolf Habelt Verlag.

Willcox, G. (1998). Archaeobotanical Evidence for the Beginnings of Agriculture in Southwest Asia. In A. B. Damania, J. Valkoun, J. Willcox, and C. O. Qualset (Eds.), *The Origins of Agriculture and Crop Domestication. Proceedings of the Harlan Symposium 10-14 May 1997*, pp. 25–38. Aleppo: ICARDA.

Willcox, G. (1999). Agrarian change and the beginnings of cultivation in the Near East: evidence from wild progenitors, experimental cultivation and archaeobotanical data. In C. Gosden and J. Hather (Eds.), *The Prehistory of Food: Appetites for Change*, pp. 478–500. London: Routledge.

Willerding, U. (1980). Zum Ackerbau der Bandkeramiker. In T. Krüger and H.-G. Stephan (Eds.), *Beiträge zur Archäologie Nordwestdeutschlands und Mitteleuropas*, Volume 16, pp. 421–457. Hildesheim: Lax.

Chapter 16

Early farming in Slovakia: an archaeobotanical perspective

Mária Hajnalová
Slovak Academy of Sciences, Nitra, Slovakia

16.1 Introduction

Early Neolithic Linearbandkeramik (LBK) groups entered the territory of Slovakia from the south via the adjacent Hungarian plains over and/or along the Danube and Tisza rivers. In the east, farming appeared with the Alföldi Linear Pottery culture (='eastern Linear Pottery culture' or 'Alföld Bandkeramik') which occupied the area originally populated by the Körös culture, and which also spread further north beyond these boundaries (Kalicz and Makkay 1977; Baldia 2003). In the west it was introduced with the early Linear Pottery culture (='Early Transdanubian Bandkeramik'), which derived from the Starčevo-Çris culture (Virág and Kalicz 2001; Pavúk 2003).

Chronological studies are based mostly on ceramic analyses, supplemented by only a few ^{14}C dates, and there is no consensus about the date of the spread of farming or whether it spread first to the east or west or if in fact the process was concurrent in the two areas (Pavúk 2003; Šiška 1989; Baldia 2003; Parkinson 1999). Additional absolute, contextually sensitive and uncontaminated dates are required before the chronology of the Neolithic in the eastern Central Europe area can be established.

In recent decades there have been more arguments in favour of the idea that mesolithic populations played an important role in the development of the above mentioned neolithic cultures (see, for example, Makkay 1996; Bánffy 2004; Pavúk 2003). However, on the basis mainly of analyses of material culture (e.g., ceramics) it is argued that the Neolithic in the two lowland regions developed independently (but possibly with some minor interregional influences) (figure 16.1) and, as such, lasted until the Baden culture of the middle Aeneolithic. The environmental evidence available to reconstruct the economy (e.g., crop and/or animal husbandry) is incomplete and is, therefore, inadequate for the purpose of defining the similarities and differences in the development of farming in the two regions.

For all periods there is rather a large discrepancy between the numbers of excavated sites and those sites with archaeobotanical data. All neolithic sites in the territory of Slovakia (figure 16.2) are located on dry soils with unfavourable conditions for the preservation of plant macro-remains. The five most important of the 53 sites have botanical data which derive from identifications of plant remains that have been recovered by flotation as well as those based on imprints in daub and/or ceramics. Of equal value are the data from another eight sites at which the studied material was obtained by flotation. Limited but important information (as will be

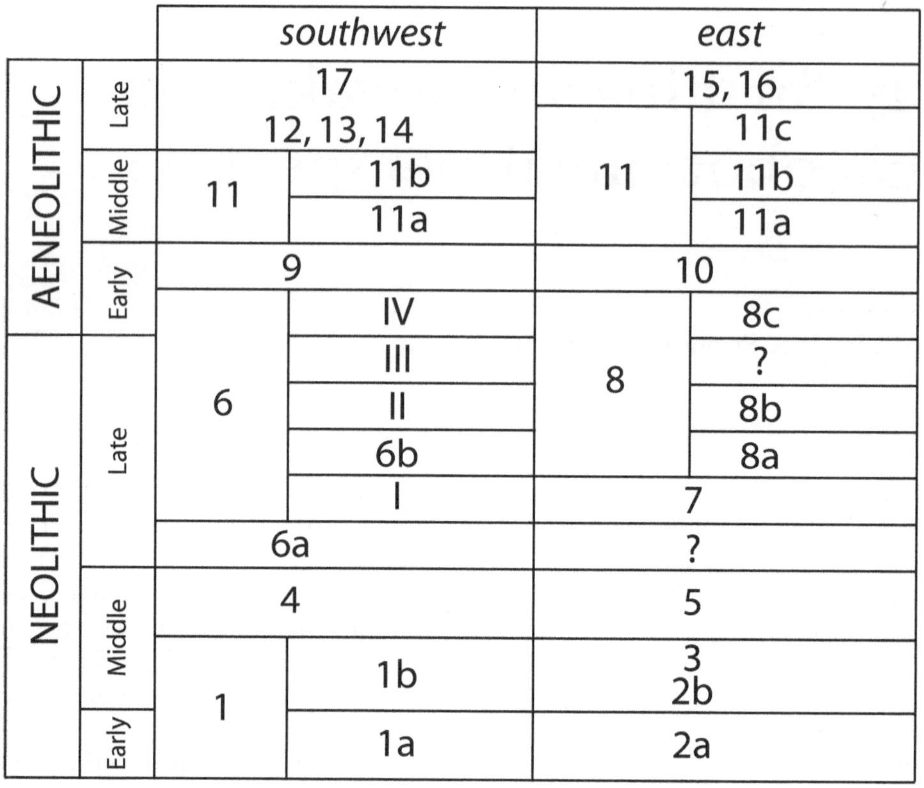

Period	Phase	southwest		east	
AENEOLITHIC	Late	17 12, 13, 14		15, 16	
				11	11c
	Middle	11	11b		11b
			11a		11a
	Early	9		10	
		6	IV	8	8c
NEOLITHIC	Late		III		?
			II		8b
			6b		8a
			I	7	
		6a		?	
	Middle	4		5	
		1	1b	3 2b	
	Early		1a	2a	

Figure 16.1: Simplified relative chronology scheme of the Neolithic and Aeneolithic in Slovakia (J. Pavúk personal communication). Legend: 1—Linear Pottery culture (LBK), 1a—early LBK, 1b—later LBK (Notenkopf culture), 2a—Proto-Linear Pottery (Méhtelek group), 2b—early eastern (=Alföldi) Linear pottery (Kopčany group), 3—later eastern Linear Pottery (Tiszadob and Raškovce groups), 4—Želiezovce group, 5—Bükk culture, 6a—Protolengyel phase, 6—Lengyel culture (I—Svodín and Nitriansky Hrádok; 6b—Santovka phase; II—Pečeňady, III—Moravany-Brodzany, IV—Ludanice group), 7—Stroke-ornamented pottery culture and earlier Tisza culture, 8—Tisza culture (syn. Polgár complex) (8a—Čičarovce I; 8b—Cszöshalom-Oborín I; 8c—Tiszapolgár-Bodrogkeresztúrphases), 9—Bajč-Retz group (Furchenstichkeramik), 10—Lažňany group, 11—Baden culture, 11a—Boleráz group, 11b—Classical Baden phase, 11c—late Baden phase, 12—Bošáca group, 13—Kostolac group, 14—Jevišovice culture, 15—Nýrszeg-Zatín culture, 16—East-Slovakian mounds culture, 17—Kosihy-Čaka culture.

shown later) comes from 16 sites where the plant remains originate exclusively from imprints in daub and/or ceramics. At 17 sites only charcoal fragments visible to the naked eye were collected and, as a consequence, these data contribute little, if anything, towards the reconstruction of crop husbandry. More detailed information is presented and discussed for the sites with rich assemblages; for the others, however, only site names and/or names of crops are listed.

Excavation of most of the neolithic sites was undertaken in the 1960s and 1970s, when archaeobotanical research in Slovakia was just being established and flotation was rarely conducted in the field. The information presented here, unless stated otherwise, is taken from the personal databases of the author and Dr. Eva Hajnalová (both of which contain unpublished data).

16.2 Results

16.2.1 Early Neolithic

The early farming sites in the territory of Slovakia are connected with the most fertile regions of the southwestern and southeastern or eastern lowland zones, which are a part of the larger system of inner Carpathian basins. Mountains separate the two regions and high ranges form borders to the northeast and northwest; both have either flat or rolling landscapes, the latter being dominated by loess dunes.

The early Linear Pottery culture (LBK), or Transdanubian Bandkeramik, is the earliest phase of Linear ceramic culture in southwest Slovakia and is represented to date by 15 sites that have either been excavated and/or identified by field walking; plant remains have been studied at only four sites. Imprints in daub were studied at Nitra-Dolné Krškany and grains of emmer (*Triticum dicoccum*) and indeterminate chaff fragments were identified. Hand-collected charcoal fragments were recorded from the sites of Borovce-Chríb and Borovce-Rakovická ulička. A sample (total volume 100 litres) from a single context at Blatné-Štrky (the fill of a large soil extraction pit) was processed by flotation and has produced over 6,300 plant macro-remains. The main constituents of the sample are emmer and einkorn (*Triticum monococcum*) and in the ratio of two emmer grains to one einkorn grain. As many as one hundred grains are tentatively identified as spelt wheat (*Triticum* cf. *spelta*). Chaff remains together with weed seeds make up to four percent of the sample (Hajnalová 1988). Unfortunately, the sample is from a secondary context, therefore it is not possible to say whether it represents a post-depositional amalgamation of the two cereals (i.e., emmer and einkorn) that were originally grown independently, or a mixture of cereals cultivated in the same fields (a maslin).

Botanical data for the early phase of the Neolithic in eastern Slovakia (the 'eastern Linear Pottery culture', or 'Alföldi Linear Pottery culture'; or 'Protolinear Pottery stage' in local terminology; Šiška 1989, p.58) are known from a single site (Košice-Červený Rak), and originate exclusively from imprints of chaff and grain in daub. They produced imprints of both emmer and einkorn, and cultivated legumes are represented by pea.

16.2.2 Middle Neolithic

The middle Neolithic period in Slovakia is characterised by the evolution of the Linear Pottery culture and the appearance of the so called Notenkopfkeramik in the west and, later, the eastern Linear Pottery culture (Tiszadob and Raškovce groups) in the east. Two discrete cultural phenomenon emerge in the second half of the middle Neolithic: in the west, the Želiezovce group and in the east, the Bükk culture.

Plant remains were studied from four sites dated to the later Linear Pottery culture (i.e., the later phase of 'Transdanubian Bandkeramik', or 'Notenkopfkeramik'); only wood charcoal fragments were recovered from Borovce-Popelkových; imprints identified as indeterminate wheat grains (*Triticum* sp.) were found at Nitra-Dolné Krškany; charcoal fragments and 20 grains of

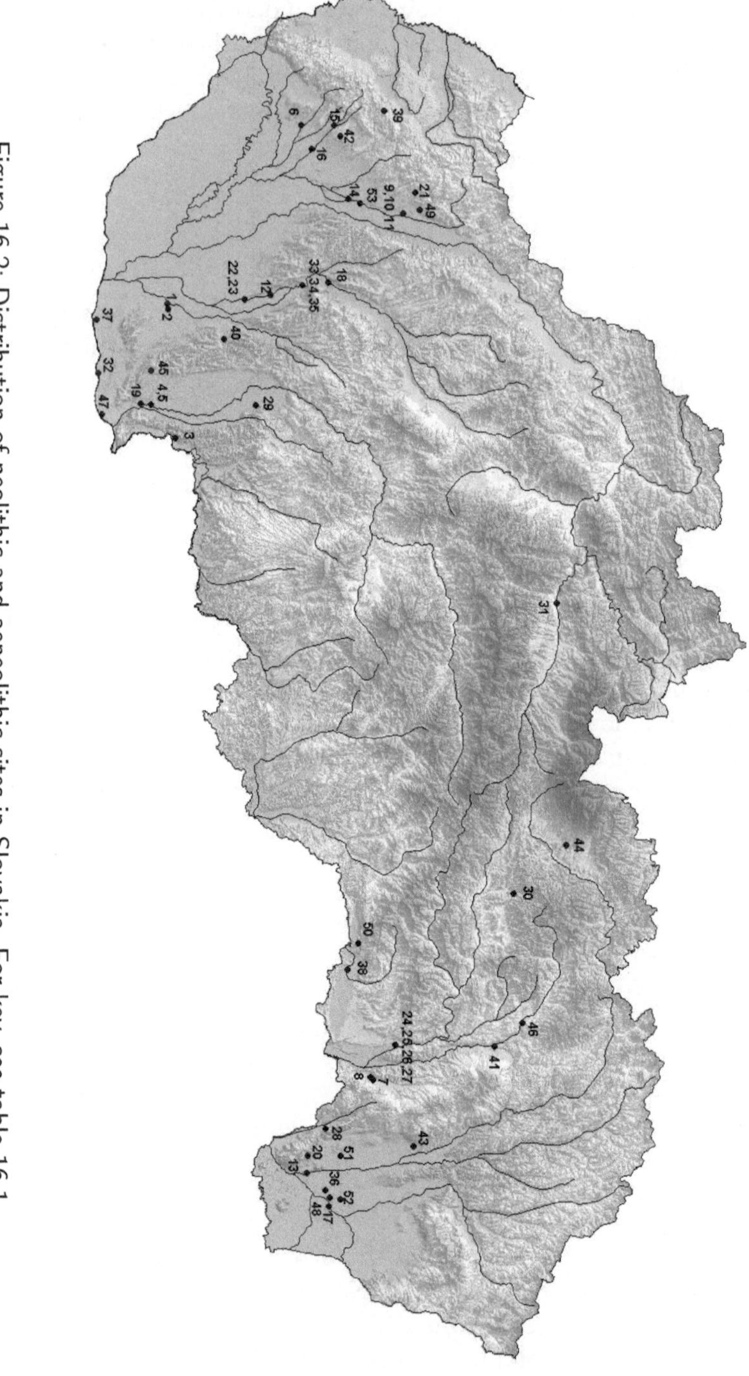

Figure 16.2: Distribution of neolithic and aeneolithic sites in Slovakia. For key, see table 16.1.

Map no.	Site name	Reference
1	Bajč–Medzi kanálmi	Cheben (2000); Cheben and Ruttkay (1991)
2	Bajč–Vlkanovo	Nevizánsky and Točík (1984)
3	Bielovce –Telek	Fusek (1986)
4	Bíňa–Berek	Pavúk and Šiška (1980)
5	Bíňa–Cénapart	Cheben (1981)
6	Blatné–Štrky	Pavúk (1978, 1980)
7	Blažice	Pástor (1965)
8	Bohdanovce–Pod Rákošským lesom	Pástor (1965)
9	Borovce–Chríb	Staššíková-Štukovská (1988)
10	Borovce–Rakovická ulička	Staššíková-Štukovská (1987)
11	Borovce–Popelkova parcela	Staššíková-Štukovská (1987)
12	Branč–Arkuš I	Cheben et al. (1992)
13	Brehov–Pod Veľkým vrchom	Horváthová (2003)
14	Bučany–Kopanice	Bujna and Romsaueer (1981)
15	Budmerice–Gidra	Pavúk (1981)
16	Cífer-Pác–Nad mlynom	Kolník (1980)
17	Ižkovce–Predná hora	Hajnalová (2003)
18	Jelšovce–H.d. JRD	Bátora (1987, 1988)
19	Kamenín–Kiskukoricás	Nevizánsky (1980)
20	Kašov–Čepegov	Bánesz (unpublished ms.)
21	Kočín-Lančár–Dielce	Němejcová-Pavúková (1988, 1987)
22	Komjatice–Kňazová jama	Točík (1980)
23	Komjatice–Tomášové	Točík (1981)
24	Košice–Červený rak	Kaminská and Novák (2002); Šiška (1992)
25	Košice–Barca III	Hájek (1957)
26	Košice–Galgovec I	Kaminská (2001b)
27	Košice–Galgovec III	Kaminská (1999)
28	Lastovce	Hajnalová (1977a)
29	Levice–Pod Krížnym vrchom	Oždáni (1975)
30	Levoča–Nad Podkovou	Staššíková (unpublished ms.)
31	Lisková–Liskovská Jaskyňa	Struhár (1998)
32	Mužla-Čenkov–Vilmakert	Cheben and Hajnalová (1997); Kuzma et al. (1983); Kuzma (1982)
33	Nitra–Šindolka	Chropovský et al. (1987)
34	Nitra-Chrenová–stavba Shell	Březinová (2003); Benediková and Hajnalová (2003)
35	Nitra-Dolné Krškany–Dvor OSP	Hajnalová (1976)
36	Oborín–Kapustniská	Vizdal (1970); Hajnalová (1976)
37	Patince–Teplica	Cheben (1988)
38	Peder–Kenderkert	Pavúk and Šiška (1971)
39	Plavecký Mikuláš–Dzeravá Skala	Kaminská et al. (nd); Hajnalová and Hajnalová (nd)
40	Podhájska–Nad bytovkami	Rejholcová (unpublished ms.)
41	Prešov - Šarišské Lúky	Šiška (1976)
42	Ružindol-Borová–Celiny	Němejcová-Pavúková (1973)
43	Sečovská Polianka–Božčická hrádza	Budinský-Krička (unpublished ms.)
44	Stráne Pod Tatrami–Pod kamenným vrchom	Soják (2000)
45	Svodín–Busahegy	Němejcová-Pavúková (1995)
46	Šarišské Michaľany–Fedelemka	Šiška (1995)
47	Štúrovo–JCP	Pavúk (1994)
48	Veľké Raškovce–Odvodňovací kanál	Vizdal (1973)
49	Vrbové–Beňovský kúria	Klčo (1997)
50	Zádiel-Včeláre–Kostrová jaskyňa	Bárta (unpublished ms.)
51	Zemplínske Hradište–Konopianky	Chovanec (1988)
52	Zemplínske Kopčany–JRD	Šiška (1974)
53	Žlkovce–Vaniga	Pavúk (1992)

Table 16.1: Key to sites in figure 16.1.

emmer were handpicked by the excavator at Blatné-Štrky, and one floated sample from Štúrovo-JCP produced 220 emmer grains and 60 einkorn grains.

Further evolution of the eastern Linear Pottery culture (Alföldi Linear Pottery culture) in eastern Slovakia (and northeastern Hungary) is characterized by the development of spatially discrete local ceramic groups (or stages) e.g. Barca III, Kopčany and the later development of the distinctive Tiszadob and Raškovce groups (Kalicz and Makkay 1977; Šiška 1989), which were connected with occupation in new geographic areas, e.g., the hilly forest zones. Very typical of these stages of eastern Linear Pottery culture are ceramics decorated with various lines, often in combination with painted black lines over a white background and, at a later date, red and yellow painted ornamentation. By the end of the last decade of the twentieth century there were 70 known sites that belonged to the eastern Linear Pottery culture (Šiška 1995, p.14). The botanical data from the later phase of this complex in eastern Slovakia are based on a study of imprints in over 13,000 ceramic sherds and in a comparable number of daub fragments from the same sites (Hajnalová 1977a). The remains of both emmer and einkorn were found at Zemplínske Kopčany, Košice–Barca III, Lastovce, Prešov-Šarišské lúky, Zemplínske Kopčany, Veľké Raškovce and Bohdanovce; emmer was identified at Blažice; barley was found at Zemplínske Kopčany, Veľké Raškovce; pea (*Pisum sativum*) is recorded from Prešov-Šarišské lúky; tentative identifications of millet (cf. *Panicum miliaceum*) were made at Sečovská Polianka; and indeterminate wheat grains were found at Peder-Kenderkert (Hajnalová 1989, p.6).

There are two Tiszadob group sites (Šarišské Michaľany and Košice-Galgovec I) from which only a few plant macro-remains have been obtained by flotation of very small samples (volumes of 20 litres at the first site and 15.1 litres at the second). In total there are 50 items, which include grains and chaff of einkorn, emmer and naked barley (*Hordeum vulgare* var. *coeleste*), and a few weed seeds (Hajnalová 1993; unpublished). Two radiocarbon dates (Ki-8919: 7350±110 bp and Ki-9820:7220±110 bp) for a Tiszadob culture context at Košice-Galgovec (eastern Slovakia), excavated by Kaminská (1999, 2001a), give a date range that spans from 6500 cal BC to 5800 cal BC (with 95.4% probability). However, extremely early age determinations measured on samples of botanically undetermined charred wood such as these from Košice-Galgovec, may well represent an 'old-wood effect' suggesting that the actual date is probably much later. These two dates can be viewed as a terminus post quem for the settlement of the Tiszadob group at Košice-Galgovec I (Peter Barta pers. comm.).

As has already been mentioned, the spread of farming to new areas was characteristic of this phase, for example, into hilly areas in the west, or further to the north and also penetrating the relatively high elevated basins of Spiš and Poprad in the Carpathians. In comparison to other basins in the territory of Slovakia, these two basins have more favourable climatic and edaphic conditions, as is documented by the presence of thermophilous oak forests and the fertile soils on loessic substrata. It is striking that even though they are adjacent to the territory of the eastern Linear Pottery culture with no profound geographical barriers, the ceramic material of the more distant and geographically isolated western Linear Pottery ('Notenkopf') culture is prevalent.

The evolution and development of this important region, which is on the south-north distribution route between Hungary and Poland with relatively easily accessible mountain passes over the Carpathians, is not well understood (see the discussion below). A single soil sample of 0.5 litres has been floated from a site at Stráne pod Tatrami (Hajnalová and Mihályiová 1998) and this produced a few grains and chaff remains of einkorn and a grain of rye (*Secale cereale*).

The Želiezovce group of southwest Slovakia has its core area of distribution along the lower Hron, Nitra, Žitava and Váh rivers. From this core it spreads to Transdanubia (northwest Hungary) and the eastern part of Lower Austria. Originally it was thought to be the last stage of the Linear Pottery culture in the region described, however, recently it has been seen as an independent culture complex. The decoration of ceramic vessels with yellow and red 'painted' lines (similar to the LBK) and use of clay animal statues are characteristic of the group (Točík

1970, p.43). The number of sites (including the ones that have been excavated) is much higher in comparison with those from previous periods.

Plant remains have been studied from Želiezovce group contexts at 10 sites. The types of remains studied (e.g., charcoal, imprints, charred seeds and chaff), the number of samples taken and/or the volume of the floated samples vary significantly between individual site assemblages. The widespread occurrence of emmer and einkorn is recorded, however, it is impossible to state whether emmer predominated over einkorn or vice-versa; but it is clear that both wheats were cultivated as individual crops (see details below). It is interesting to note that emmer grains prevail in number and frequency over einkorn grains in the charred material, while just the opposite is true for chaff remains. However, emmer chaff remains were found more often (in 10 out of 18 samples) than einkorn chaff (in two out of 18 samples) in the daub. Spelt, free-threshing wheat and barley (both hulled and naked six- or four-row varieties) were less common and were found in smaller quantities, and of the pulses, lentil occurred more frequently than pea.

Only charcoals were collected and submitted for analyses at the two sites of Bielovce-Telek and Nitra-Šindolka. Plant imprints in ceramics and daub fragments have been studied from the site of Cífer-Pác and revealed a few grains and chaff remains of barley, emmer and einkorn, and a single pea. Samples obtained exclusively by flotation were studied from the sites of Štúrovo-JCP and Patince-Teplica. The latter is one of the sites that has been sampled most extensively for archaeobotanical remains and it is rather disappointing, therefore, that an assemblage of less than 300 cultivated cereal remains (mostly chaff) of emmer, einkorn, free-threshing wheat (*Triticum aestivum* s.l.) and also barley (cf.) was the product of processing almost 1,000 litres of soil from 100 samples (30 different contexts) by flotation.

Special attention has been paid to a weed-rich assemblage from one sampled pit that was originally tentatively defined as having been used for the storage of gathered goosefoot seeds (*Chenopodium album* aggr.). However, the fill of the context has now been reinterpreted as the waste product from cleaning cereals, on the basis that 206 chaff remains of emmer and 27 indeterminate wheat grains were found in association with the large number (over 700) of goosefoot seeds (Hajnalová and Hajnalová 2004).

In an assemblage that comprised over 1,300 remains, which were recovered from samples (total volume of 100 litres) taken from three pits at Štúrovo-JCP, emmer grains greatly outnumbered those of einkorn and free-threshing wheat, but there was virtually no chaff.

From the remaining five sites (Bajč-Medzi kanálmi, Blatné-Štrky, Jelšovce-JRD, Mužla-Čenkov-Vilmakert and Bíňa-Cénapart), floated material as well as imprints were studied. At Bajč-Medzi kanálmi, over 34,000 mostly glume wheat grains were obtained from a single in situ context of an oven (Cheben and Hajnalová 1997). The average density of grains in the assemblage was 638 per litre of sediment. A majority of these charred wheat grains (over 55%) could not be identified due to bad preservation (for example, because of deformation of the kernels). Over half the kernels that could be assigned to species belong to one- and two-seeded einkorn (62%), with fewer kernels of emmer (22%) and spelt (cf.; 11%). The remaining items (5%) consist of grains of free-threshing wheat and naked four- or six-row barley, chaff remains and weed seeds (Cheben and Hajnalová 1997). Also identified at the site were imprints in daub of chaff remains of emmer, which were found more frequently and in larger quantities than those of einkorn; imprints of barley kernels and possibly glumes were also present but only in very small numbers.

Emmer outnumbered other wheat species (including einkorn, cf. spelt and free-threshing wheat) in an assemblage of over 3,000 remains, which consisted almost exclusively of grains, recovered from seven samples (186 litres of soil) at Blatné-Štrky; a few seeds of lentil (*Lens culinaris*) and pea were also found (Hajnalová 1989, p.6).

Eighty of the 112 cereal remains that were recovered from seven samples (70 litres of soil) at Mužla-Čenkov-Vilmakert belonged to grains of barley; single grains/seeds of einkorn, emmer, broomcorn millet, pea and rose hip (cf. *Rosa* sp.) were also identified. Chaff and grains of emmer and einkorn were found in daub (Cheben and Hajnalová 1997).

Two samples with total volume of two litres were taken at Bíňa-Cénapart (Hajnalová 1989). In the first sample (795 grains, density 530 remains per litre of sediment) emmer grains outnumber einkorn grains in the ratio 27 to 1, whereas in the second sample (193 remains, density 96.5 items per litre of sediment) the proportions are reversed and einkorn grains are slightly more common than emmer grains in the ratio 1.4 to 1 (for 48 grains). Moreover, the second sample also contains chaff (112 items), 30 seeds of field brome (*Bromus arvensis*) and 11 seeds of corn cockle (*Agrostemma githago*). A single sample of 20 litres (density 4.1 items per litre of sediment) at Jelšovce-JRD has yielded grains and chaff of einkorn and naked four- or six-row barley.

It has been recently pointed out by geneticists and molecular biologists (Yan et al. 2003) that there is a strong support for a 'European' spelt secondary origin hypotheses. This suggests that 'European' spelt is a result of hybridisation between cultivated emmer (*Triticum dicoccum*) and club wheat (*Triticum compactum*). It has been stressed out by Akeret (2005), after his discovery of numerous spelt finds from a Bell Beaker culture site in Switzerland, that it would be useful to address the question of distribution and the earliest spelt cultivation in east-central Europe. For this it is necessary both to reexamine and reassess the tentatively assigned finds of spelt and free-threshing wheat finds from east-central Europe (e.g., those form Želiezovce group but also those from the earlier periods).

The Bükk culture, which developed from the Tiszadob culture and other contemporary regional cultures of northern Hungary and eastern Slovakia, is a well-known phenomenon of the Tisza region. In comparison to the previous eastern Linear Pottery culture, the Bükk culture is characterised by higher settlement density (e.g., 150 sites in the territory of Slovakia; Šiška 1995, 14), the colonisation of more elevated, hilly regions (up to 600 m), the use of caves and, most importantly, the manufacture of very fine, black (or dark brown) ceramics with highly superficial linear ornamentation.

Plant remains are known from five sites belonging to this culture. Only charcoal fragments were collected at three of the sites, Veľké Raškovce, Kašov and Zádiel-Včeláre. Sampling and recovery of plant remains by flotation has been conducted at the site of Šarišské Michaľany-Fedelemka (Hajnalová 1993), where 20 contexts (330 litres of soil) produced almost 2,000 plant remains. Hulled and naked barley, emmer and einkorn were the most numerous of the cultivated crops and there were fewer finds of spelt wheat, free threshing wheat, broomcorn millet, rye, pea, lentil and flax (*Linum usitatissimum*). Seeds of wild plants were relatively rare, however, over 30 different taxa were present. Within this group ten pips/stones/seeds of apple (*Malus* sp.), wild cherry (cf. *Cerasus avium*), plum (*Prunus* sp.), raspberry/blackberry (*Rubus* sp.) and water chestnut, *Trapa natans*) were present (Hajnalová and Hajnalová 2004).

A unique discovery was made at site of Zemplínske Hradište-Konopianky, where a lump of charred flax seeds, within which a single lentil was embedded, was found in a small ceramic vessel that had been placed in a settlement pit. Unfortunately, these are the only plant remains recovered from the site. The imprints in ceramic sherds were studied from the site of Zemplínske Kopčany and grains of both emmer and einkorn together with seeds of pea (cf.) were recorded.

16.2.3 Late Neolithic

The late Neolithic in the western part of Slovakia is associated with the Lengyel culture complex which, as it developed, spread from the core area of southern Hungary, southwest Slovakia, and northwest Austria as far as Poland, Bohemia and Germany. The main defining characteristics are painted pottery, clay zoo- and anthropomorphic statutes and fortified settlements (Točík 1970, p.116).

The development and chronology of the Lengyel culture in Slovakia is still being discussed (Lichardus and Vladár 2003; Pavúk 2000, 2004). However, according to the latest works (Pavúk 2000, 2004) it is seen as representing a long period of continuity, interrupted by several evolutionary phases. Five distinctive phases have been described and classified mainly on the basis of the development of the shape and ornamentation of ceramic vessels, rarely on other artefacts or stratified contexts: Protolengyel (including the Bíňa-Bicske and Lužianky stages), Lengyel I (including the Svodín I and Svodín II stages, Nitriansky Hrádok—-contemporary with MOG Ia in Austria and Moravian Painted Pottery culture Ia in Moravia), Santovka stage—contemporary with MOG Ib in Austria and Moravian Painted Pottery culture Ib in Moravia; Lengyell II (Pečeňady—contemporary with MOG II, and Moravian Painted Pottery culture II), Lengyel III (Moravany–Brodzany stages) and finally Lengyel IV (i.e., 'Epilengyel' or Ludanice) including the Nitra and Ludanice stages.

The late Neolithic in the eastern part of Slovakia begins with the Stroke-Ornamented Pottery culture ('Stichbandkeramik') and the Tisza culture of northeast Hungary (contemporary with Lengyel I), and develops into the Tisza-Polgár culture complex. As in the Lengyel culture, there are also several stages in the Tisza-Polgár culture: Čičarovce-Csőshalom I (contemporary with the Santovka stage), Csőshalom II-Oborín I (contemporary with Lengyel II), Oborín II (contemporary with Lengyel III), and Tiszapolgár-Bodrogkereszúr (contemporary with Lengyel IV). The Lengyel IV (Epilengyel) and Tiszapolgár-Bodrogkerestúr cultures are seen as the beginning of the Aeneolithic period (Pavúk 2000, cf.) in Central Europe.

The number of sites that have been excavated as well as sampled for plant remains vary for each phase (or stage) of the late Neolithic, and at some there was no recovery of archaeobotanical material.

Plant remains were collected at four of the excavated sites dated to the Lengyel culture (Lengyel I). An imprint of an emmer grain was found in one sample of daub at Svodín-Busahegy, together with five grains of emmer, one grain of einkorn, 119 wild plant seeds and charcoal fragments that were found in six samples (total volume 60 litres), which had been taken from a storage pit and processed by flotation (Hajnalová 1986). Only individual charcoal fragments were studied from two of the other sites, Ružindol-Borová-Celiny and Bučany-Kopanice, and charcoal fragments as well as a single acorn (*Quercus* sp.) were found at Kočín-Lančár-Dielce.

Veľké Raškovce is the only site dated to the Stroke-Ornamented Pottery culture from which the plant remains have been studied. All botanical data, however, come from imprints in 750 ceramic sherds and a few daub fragments, in which spikelets and chaff of emmer and einkorn are represented (Hajnalová 1977a).

Various wood charcoals and a collection of almost 400 remains of crops and wild plants were found in 37 floated samples (total volume 115 litres) at the site of Žlkovce-Vaniga, dated to the Lengyel II phase. A piece of hazelnut shell (*Corylus avellana*) was identified amongst the gathered fruits. Most of the cereal items were only described as indeterminate grain and chaff fragments due to the bad preservation of the material. The remaining identifiable grains and chaff of emmer, einkorn, barley and spelt (cf.), and a few grains of rye, are impossible to interpret in terms of their economic significance. Rather striking was a higher frequency of rye grains and rachis internodes of barley (the latter were also identified from imprints in one sample of daub). Individual charcoal fragments were collected and analysed at Bučany-Kopanice, another site dated to the Lengyel II phase.

Imprints of emmer spikelets and indeterminate cereal grains were found in a few of the 750 ceramic sherds that were studied at the Csőshalom II-Oborín I culture site of Oborín-Kapustniská in eastern Slovakia (Hajnalová 1977a).

No sites of Lengyel III phase have been sampled for plant macro-remains.

16.2.4 Early Aeneolithic

As mentioned above, the Aeneolithic period, which is characterized by the production of new ceramic forms and ornamentation connected with the spread of copper artefacts (e.g., copper axes and battleaxes), starts with the Epilengyel (Lengyel IV) in the west (Pavúk 2004) and the formation of the Tiszapolgár-Bodrogkerestúr culture in eastern Slovakia.

To date, scarce botanical information has been gathered about early aeneolithic sites. Only individual charcoal fragments were recovered and analysed at the Lengyel IV sites of Budmerice-Gidra and Ružindol-Borová-Celiny. Imprints in daub (including some fragments with white painting) were studied at Jelšovce-JRD (Ludanice group), but only einkorn and indeterminate chaff remains were identified. The finds from Dzeravá Skala cave also belong to this culture, and here the remains of barley, emmer, spelt, pea, lentil, flax and indeterminate grains were found in a pit and within the surrounding sediment. Gathered plants are represented by hazelnuts, wild pear (*Pyrus pyraster*), and winter/bladder cherry (*Physalis alkekengi*). Individual charcoal fragments and one acorn were collected at Bučany-Kopanice, dated to the Bajč-Retz culture, which represents the latest stage of the early Aeneolithic in southwest Slovakia.

An assemblage comprising individually collected charcoal fragments and grains of einkorn and emmer, found in association with human bones at the cave of Liskovská jaskyňa in northern Slovakia, is of the same date (Lengyel IV or Bajč-Retz period). The accumulation of charred material was situated below a layer of stones on which deliberately arranged human bones had been laid; the feature has been interpreted as a ceremonial grave (Struhár 1998).

A remarkable and unique find of more than 1,000 awn fragments of an undetermined feather grass (*Stipa* sp.) was collected by the excavator at Komjatice-Tomášové (Hajnalová 1989; Bieniek 2002). Eight oat grains (*Avena* sp.) and a glume base of spelt were found with the awns.

16.2.5 Middle Aeneolithic

The Baden culture is one of the few cultures that connected not only the eastern and western parts of the Carpathian basin, but also spread further beyond the boundaries of the basin and joined the areas from the former Yugoslavia to southern Poland and from Transylvania to Germany. The question of its origin and development are still under discussion; according to Němejcová-Pavúková (1970, p.184), the central (core) area of Lengyel culture played the most important role in its formation. Firstly with the development of the Boleráz group, which later in the classical Baden culture phase spread further east to the area of the Bodrogkerestúr culture (east Slovakia and Hungary) and west to the area of the Funnel Beaker culture (Bohemia).

Only the charcoals were collected and analysed from the site of Komjatice-Kňazova jama, dated to the earlier phase (Boleráz group). Plant remains were collected from 11 sites of the later, classical phase. Flotation and analyses of imprints in ceramics and daub have been undertaken at three of these sites: Jelšovce-JRD, Svodín-Busahegy and Mužla-Čenkov. Eighty-six samples with a total volume of soil of 838 litres were floated at Jelšovce-JRD; charcoal, cultivated crops, gathered fruits and wild plants were recovered. It was not possible to identify 350 of the 812 cereal grains further, due to bad preservation. On the basis of numbers and frequency of occurrence, most of the remaining grains belong to einkorn, emmer and spelt (cf.), with far fewer grains of free-threshing wheat (cf.), barley and broomcorn millet. Chaff remains (162) were almost exclusively of einkorn, with infrequent finds of emmer, spelt (cf.) and indeterminate fragments. A single seed of a small-seeded pulse was tentatively identified as bitter vetch (*Vicia* cf. *ervilia*). There was also one pea, and single finds of two gathered fruits: cornelian cherry (*Cornus mas*) and hazelnut, and 86 seeds of wild plants that belong to 17 different taxa. Indeterminate chaff fragments were identified from imprints in daub. At Svodín-Busahegy five storage pits were sampled and 130 litres of soil were floated. Charcoal, over 200 grains of emmer, einkorn, barley (both hulled and naked six-or four-row barley) and spelt (cf.), over 100 seeds of broomcorn millet, a single seed of lentil, a chaff fragment of emmer/einkorn, and over 200 seeds of wild plants that represented 22 taxa were recovered from the samples. Imprints of chaff remains of emmer and einkorn and an imprint of an einkorn grain were found in daub (Hajnalová 1986).

Two samples (20 litres) from Mužla-Čenkov-Vilmakert produced only a single grain of einkorn. However, a greater range of taxa, which includes spikelets of emmer and einkorn, grains of barley, free threshing wheat and indeterminate wheat, has been identified as imprints in daub (Cheben and Hajnalová 1997).

At the sites of Brehov-Pod Veľkým vrchom and Šarišské Michaľany-Fedelemka, plant remains were recovered by flotation. The remains of chaff and possibly grains of the 'new' glume wheat type (*Triticum* cf. *timopheevii* (Jones 1998), einkorn chaff, indeterminate grains, and charcoal were found in five small samples with a total volume of 69 litres from the site of Brehov-Pod Veľkým vrchom. Six cornelian cherry seeds and one acorn were included in the 31 seeds that represented six wild plant taxa.

At Šarišské Michaľany-Fedelemka, 70 litres of soil from two pits produced single finds of lentil, emmer, spelt (cf.) and free-threshing wheat, indeterminate grains and 12 seeds of four wild plant taxa (Hajnalová 1993). Three thousand charred flax seeds were found in a pot placed within an oven at Levice-Pod krížnym vrchom. The daub and plaster from this particular oven was also studied and contained imprints of both chaff and grains of emmer and einkorn (Hajnalová 1977b).

Imprints in daub of indeterminate wheat grains and chaff fragments are the only botanical data from the site at Podhájska-Nad bytovkami. Individually collected charcoals were analysed at Bajč-Vlkanovo, Branč-Arkuš, Bielovce-Telek, Ružindol-Borová-Celiny, Svodín-Busahegy and Levoča-Nad podkovou. In addition to the individually collected charcoal fragments at Kamenín-Kiskukoricás, the excavator also took a two litre sample that comprised 300 charred grains of einkorn, 29 grains of emmer and over 650 seeds of various brome species (Hajnalová 1989). This sample was recently interpreted as a residue of stored grain. Seeds of brome, which is thought to have been a 'tolerated' weed, were impossible to eliminate in the crop processing sequence because they are similar in size and shape to cereal grains (Hajnalová and Hajnalová 2004).

16.2.6 Late Aeneolithic

A typical phenomenon of late aeneolithic settlements in the territory of Slovakia is occupation, often also fortification, of gentle slopes, hilltops or naturally protected sites. Four cultural groups have been assigned to this period based on ceramic analyses (the Bošáca, Kostolac, Jevišovice and Kosihy-Čaka groups); all are rather local (with minor influences beyond their boundaries) and relatively short-lived. Sites of the Bošáca group are situated on the slopes of mountains on the northern borders of the southwest Slovakian plain. The settlements of this cultural group, in contrast to the preceding Baden culture, occupied hilly regions almost to the exclusion of all other localities. To date, charcoal analysed from Kočín-Lančár-Dielce, together with imprints of chaff of einkorn, emmer, spelt and indeterminate grain in a single sample from Nitra-Šindolka, are the only archaeobotanical data for the Bošáca group. No plant remains were collected at sites of the Kostolac group, and the sites of this group found in southwest Slovakia represent the northernmost boundary of its distribution. Charcoal remains from Kočín-Lančár-Dielce, which is on the eastern boundary of the Jevišovice culture (its core is in southern Moravia and northeastern Austria), represent the only botanical data for this group in Slovakia. However, a relatively rich assemblage assigned to the Jevišovice culture was recovered in Austria (Kohler-Schneider, this volume). The developmental stages of the early farming period culminate with the Kosihy-Čaka culture, which towards the end is subsumed within the incoming bronze age cultures. The settlement sites are situated in the lowlands or hilly zones of southwest Slovakia. There is little botanical information for this culture; charcoal and imprints of emmer spikelets and fragments of indeterminate grains come from the sites of Jelšovce-JRD and Mužla-Čenkov-Vilmakert.

16.3 Conclusions and Summary

The beginning of farming in Slovakia is closely connected with development of the Neolithic on the Hungarian plain in the 6th millennium BC. The valleys and plains along the Danube and Tisza rivers represent the main corridors via which farming spread further east, north and west in east-central Europe; beyond the plains of southwest and eastern Slovakia it involved crossing the mountain ranges of the Carpathian arch.

Between the two lowland areas of southwest and east Slovakia that were colonized at an early date there is elevated mountainous terrain, within which there are very few neolithic sites. Those present are connected with the use of caves and are dated not earlier than the later Neolithic or Aeneolithic. However, there is the question of whether the sparse evidence for neolithic occupation in this region is apparent rather than real, and is due to the concentration of research in the lowland and comparative lack of investigation of the highland areas. Nevertheless, the independent development of the east and west regions since the earliest Neolithic until the middle Aeneolithic, with only minor influence and/or communication from external sources as suggested by ceramic types and decorative styles, has been proposed. At present there are not sufficient data to reconstruct crop and animal husbandry regimes, to estimate the role of gathered plants and hunted wild animals in the subsistence economy, or to gain insight into the social life (etc.), for example, information that would either support or dismiss the hypothesis of independent development of the two regions.

As shown previously, the archaeobotanical data are rather fragmentary and the tendency has been to over-interpret on the basis that the 'unique' finds are representative of a period or culture. What we can confidently say about the production and consumption of plant resources is summarised below.

The cultivated crops of the earliest stages of the Neolithic in Slovakia were einkorn and emmer. Although the present evidence seems to support the idea that these cereals were grown as individual crops, the possibility can't be ruled out that they were cultivated as a mixture (maslin). However, it should be stressed that finds comprising a mixture of einkorn and emmer are always from secondary contexts, whereby the combination of the two might be a result of various taphonomic processes. Arguments that favour the preference of either einkorn or emmer during certain periods or in certain regions are also problematic; for example, when results of analyses of charred macro-remains and imprints in daub from the same sites (or sites of the same region and archaeological culture) are compared, they are often contradictory. Both emmer and einkorn were cultivated throughout the Neolithic and Aeneolithic periods, but their numbers decrease towards the end of the Neolithic and the beginning of the Bronze Age.

Only a single barley grain imprint is known to date to the early Neolithic, however, numbers increase in the middle Neolithic and shortly afterward it seems to become an established crop. It is presumed that naked six-row barley appeared earlier (i.e., in the later phase of the Linear Pottery culture) than the hulled six-row variety in Slovakia. However, this may be due to the fact that while naked grains can be easily identified, earlier grains of the hulled variety are usually deformed and cannot be identified with certainty. Barley that has been identified with certainty as the hulled variety is known from the later Aeneolithic.

Free-threshing wheat grains (no chaff remains have been found and so it is not certain whether macaroni wheat or bread wheat is present) frequently occur on neolithic sites but in very small numbers; numbers increase in the late Aeneolithic period. Glumes and grains of spelt are also sporadically present from the middle Neolithic. At one middle aeneolithic site from eastern Slovakia (Brehov-Pod Veľkým vrchom), chaff fragments were cautiously assigned to a new-glume wheat type (*Triticum* cf. *timopheevi*), but the identifications are not definite. The rarity of free threshing wheats and spelt and the absence of bulk samples of stored grain of these cereals suggests that during the Neolithic they were probably tolerated field weeds.

Broomcorn millet is the only other cereal species present in the Neolithic. A single find (a ceramic imprint) is dated to the early Neolithic and later sporadic finds occur during the middle Neolithic. At one middle aeneolithic site (Svodín-Busahegy) grains of broomcorn millet were found in such large numbers and at such high frequencies that suggest it was cultivated as an individual crop.

Two species of pulses were cultivated. The first evidence of pea is in the form of ceramic imprints found at the earliest neolithic site in eastern Slovakia, and the earliest lentil is recorded in the middle Neolithic; both crops are present infrequently in later contexts. The low numbers are thought to be due to under-representation as a result of taphonomic processes; pulses come into (direct) contact with fire during processing and preparation for consumption less frequently than cereals and by comparison, therefore, the chances of being preserved by charring are greatly reduced. Only one seed of a small-seeded vetch from a middle neolithic site was tentatively identified as bitter vetch (*Vicia* cf. *ervilia*).

Despite the presence of archaeological artefacts connected with textile production from the earliest neolithic, flax seeds from middle neolithic contexts in eastern Slovakia are so far the earliest evidence for its cultivation. Finds of flax are even more rare than those of pulses. The use of flax for purposes other than fibre production, e.g., as a medicine or oil, is suggested by the recovery *in situ* of a ceramic vessel full of charred flax seeds placed inside an oven.

The role of plant gathering during the development of agriculture is another issue that needs to be addressed. As mentioned in the introduction, it has been argued that local mesolithic populations played an important part in the diffusion of agriculture in east-central Europe. Therefore, it might be useful to compare the food resources of the two. Unfortunately, data about mesolithic plant foods gathered and consumed by foragers in southwest or southeast Slovakia are virtually non-existent, as no mesolithic sites have yet been found in these areas. Charred hazelnuts (*Corylus avellana*), raspberry (cf. *Rubus idaeus*, *Rubus* sp.), elderberry (*Sambucus nigra*), goosefoot (*Chenopodium album*) and grass seeds (Poaceae) published by Pokorný (2003, pp.272–273), recovered at the site of Jezevčí převis in northwest Bohemia, provide the nearest information on mesolithic plant use (Svoboda 2003). Most of the plants gathered by the first farmers in Slovakia are never found in high frequencies and belong to low energy fruits, e.g. cornelian cherry, wild pear, plum and rosehip. Plant foods of higher calorific value, e.g., hazelnut and water chestnut, are much rarer. The gathering for consumption of other wild plants also represented by their seeds cannot be ruled out, but there is little evidence to date to support this. Moreover, there are alternative (equally likely) explanations for the presence of wild taxa, as has been shown for the bulk find of goosefoot seeds that also contained chaff remains and small grains, which were evidence that it was possibly a crop processing residue rather than a store of collected wild foods.

Assemblages of wild plants can be excellent tools with which to reconstruct agricultural practices, as has been pointed out by Jones (1984, 1987), and more recently by Bogaard (2002). However, to address questions such as intensity of cultivation, sowing season, or specific location of fields (e.g. in clearings or on alluvial terraces) it is necessary that the archaeobotanical samples fulfil certain requirements. For example, they should ideally comprise weed-rich samples of unmixed origin to enable the identification of the crop processing sequence from which they derive. In the Slovakian Neolithic/Aeneolithic there are very few samples of unequivocally unmixed origin and most are cleaned crop stores with only a few weeds (of similar shape and size as the cultivars), which are therefore not suitable for this type of analysis.

With the exception of the above-mentioned gathered fruits and *Rumex* cf. *crispus*, all the wild plants identified to species level are annuals. Sixteen species (*Agrostemma githago*, *Asperula arvensis*, *Bromus arvensis*, *Bromus secalinus*, *Bromus sterilis*, *Bromus tectorum*, *Chenopodium* cf. *murale*, *Echinochloa crus-galli*, *Fallopia convolvulus*, *Galium* cf. *tricornutum*, *Persicaria* cf. *lapathifolia*, *Setaria glauca*, *Setaria verticillata/viridis*, *Solanum nigrum* and *Stachys annua*) are considered to be archaeophytes of east-central Europe, which probably came into this area as weeds of the cultivated crops. There is not yet a consensus about whether the six other species (*Chenopodium album*, *Chenopodium glaucum*, *Chenopodium hybridum*, *Chenopodium polyspermum*, *Polygonum aviculare* and *Stachys arvensis*) are local apophytes or archaeophytes, but they are almost exclusively weeds or ruderal plants. Local apophytes represent a group of collected edible fruits and medicinal herbs (e.g., *Sambucus nigra*, *Sambucus ebulus*, *Solanum dulcamara* and *Chelidonium majus*) or herbaceous plants of forest clearings or woodland edges (e.g., *Cerastium arvense*, *Galium aparine*, *Lycopus europaeus*, *Rumex crispus*, *Silene vulgaris* and *Veronica hederifolia*). The last two groups are today considered to be plants characteristic of synanthropic stands (i.e., associated with humans and their dwelling places). Both winter and summer annuals are present among the weeds and potential weeds. Three quarters of the wild species are tall plants of similar height to einkorn or emmer (even though their heights can vary considerably according to different environmental and climatic conditions). The remaining species are of medium height and only two apophytes are low-growing plants. In the assemblages of stored crops the seeds are almost exclusively of tall plants. This might suggest that cereals were harvested high on the culm by sickles, or that ears were plucked by hand in the field or cut from reaped straw at the settlement.

The shift from the Neolithic to the Aeneolithic is characterised by the introduction of track animals (cattle) for ploughing. This method of working the soil replaces the smaller-scale cultivation typical of 'garden' type agriculture of the earlier neolithic period, and enables increased production. It is appropriate to find out whether it is possible to recognise this change using archaeobotanical evidence. The increased area of arable farming might result in an increase of field weeds in crop-rich samples, both in terms of greater numbers of species and also specimens represented. There is one site dating to the earliest neolithic phase that contained a store of cleaned grain with only a few weed species, whereas, for some of the later periods there are a number of sites with rich weed assemblages (four middle neolithic sites, one late neolithic and four late aeneolithic sites). The most commonly occurring weed species remain the same from the early-middle Neolithic to the late Aeneolithic.

Alternatively, the possibility of recognizing the change from a 'garden' type (i.e., intensive) to an extensive system of production using archaeobotanical evidence is questionable. In the former system, where small plots of cereals and pulses are cultivated, there would be only shallow disturbance of the soil as a result of light tillage methods; however, intensive weeding would prevent infestation by annual or perennial weed species. When animal power is available, larger plots with deeper disturbed soil could be farmed. With the use of animals to pull ploughs not only would it have been possible to cultivate larger plots but also the disturbance to the soil would have been considerably greater. Deeper ploughing would eliminate the perennial species but effective weeding would not have been possible in the larger plots; hence, this may have resulted in an increase of annual weeds. However, if the crop harvesting method did not change (i.e., by cutting high on the culm or by ear plucking), it is likely that the same range of weed species would be represented in assemblages of stored crops. Consequently, plant remains alone are probably not an ideal source of information with which to investigate the change from intensive to extensive methods of crop production.

Acknowledgments

My greatest thanks go to E. Hajnalová for her endless support, advice and encouragement, as well as for her permission to use unpublished data. I am particularly grateful to M. Kohler-Schneider,

S. Riehl and J. Pavúk for their consultations, suggestions and discussions. Special thanks go to M. Loughton for his comments on the manuscript and his corrections to the language and to J. Mihályová and M. Bartík for technical help. I wish to thank the reviewers for their comments and above all the editors of this volume for their patience and for giving me the opportunity to publish this paper.

References

Akeret, Ö. (2005). Plant remains from a Bell Beaker site in Switzerland, and the beginnings of *Triticum spelta* (spelt) cultivation in Europe. *Vegetational History and Archaeobotany 14*(4), 279–286.

Baldia, M. O. (2003). Starčevo-Körös-Çris. Version 2.13. `http://www.comp-archaeology.org/Starcevo-Koros-Cris.htm` Accessed 2005-04-08.

Bánffy, E. (2004). *The 6th Millennium BC Boundary in Western Transdanubia and its Role in the Central European Neolithic Transition (The Szentgyörgy-Pityerdomb Settlement).* Varia Archaeologica Hungarica 15. Budapest: Régeszeti Intézet, Magyár Tudományos Akadémiai.

Bátora, J. (1987). Piata sezóna záchranného výskumu v Jelšovciach. *Archeologické výskumy a nálezy na Slovensku 1986*, 32–33.

Bátora, J. (1988). Záverečná sezóna záchranného výskumu v Jelšovciach. *Archeologické výskumy a nálezy na Slovensku 1987*, 33–34.

Benediková, L. and M. Hajnalová (2003). Rastlinné zvyšky a mazanica (1996-2000). In G. Březinová (Ed.), *Nitra-Chrenová. Archeologické výskumy na plochách stavenísk Shell a Baumax*, Archaeologica Slovaca Monographie. Catalogi. Tomus IX, pp. 74–84, 144–146. Nitra: Archeologického ústav SAV.

Bieniek, A. (2002). Archaeobotanical analysis of some early neolithic settlements in the Kujawy region, central Poland, with potential plant gathering activities emphasised. In S. Jacomet, G. Jones, M. Charles, and F. Bittmann (Eds.), *Archaeology of Plants: Current Research in Archaeobotany. Proceedings of the 12th IWGP Symposium, Sheffield 2001*, Vegetation History and Archaeobotany 11, pp. 33–40. London: Springer.

Bogaard, A. (2002). Questioning the relevance of shifting cultivation to neolithic farming in the loess belt of Europe: evidence from the Hambach Forest experiment. *Vegetation History and Archaeobotany 11*, 155–168.

Bujna, J. and P. Romsaueer (1981). Tretia sezóna výskumu v Bučanoch. *Archeologické výskumy a nálezy na Slovensku 1980*, 55–57.

Březinová, G. (Ed.) (2003). *Nitra-Chrenová. Archeologické výskumy na plochách stavenísk Shell a Baumax.* Archaeologica Slovaca Monographie. Catalogi. Tomus IX. Nitra: Archeologický ústav SAV.

Cheben, I. (1981). Druhá sezóna výskumu v Bíni. *Archeologické výskumy a nálezy na Slovensku 1980*, 88–90.

Cheben, I. (1988). Ukončenie výskumu v Patinciach. *Archeologické výskumy a nálezy na Slovensku 1987*, 54–55.

Cheben, I. (2000). *Bajč–Eine Siedlung der Želiezovce-Gruppe.* Bonn: Dr. Rudolf Habelt Verlag.

Cheben, I. and E. Hajnalová (1997). Neolitische und äneolitische Öfen in der Slowakei aus der Sicht des Archäologen und Archäobotanikers. *Archaeologia Austriaca 81*, 41–52.

Cheben, I. and M. Ruttkay (1991). Ukončenie výskumu pieskovej duny v Bajči. *Archeologické výskumy a nálezy na Slovensku 1989*, 41–42.

Cheben, I., M. Ruttkay, and J. Ruttkayová (1992). Záchranné výskumy na trase výstavby ropovodu v okrese Nitra. *Archeologické výskumy a nálezy na Slovensku 1991*, 64–65.

Chovanec, J. (1988). Sídlisko ľudu s bukovohorskou kultúrou v Zemplínskom Hradišti. *Archeologické výskumy a nálezy na Slovensku v roku 1987*, 57–58.

Chropovský, B., J. Hečková, and G. Fusek (1987). Ukončenie záchranného výskumu v Nitre na Šindolke. *Archeologické výskumy a nálezy na Slovensku v roku 1986*, 52–53.

Fusek, G. (1986). Záchranný výskum v Bielovciach. *Archeologické výskumy a nálezy na Slovensku 1985*, 83–84.

Hájek, L. (1957). Nová skupina páskové keramiky na východním Slovensku. *Archeologické rozhledy 9*(3–8), 33–36.

Hajnalová, E. (1976). Prehľad botanických nálezov z výskumov v r. 1975. *Archeologické výskumy a nálezy na Slovensku v roku 1975*, 95–100.

Hajnalová, E. (1977a). Odtlačky kultúrnych rastlín z neolitu na východnom Slovensku. *Archeologické rozhledy 29*, 121–136.

Hajnalová, E. (1977b). Zyhoľnatené rastlinné zvyšky v eneolitickej nádobe z Levíc. *Slovenská archeológia 25*, 7–12.

Hajnalová, E. (1986). Paläobotanische reste aus Svodín. *Slovenská Archeológia 34*, 177–184.

Hajnalová, E. (1988). Analýza a hodnotenie rastlinných zvyškov v neolitických objektoch v blatnom poloha Štrky. *Študijné zvesti 24*, 11–28.

Hajnalová, E. (1989). Katalóg zvyškov semien a plodov v archeologických nálezoch na slovensku. In E. Hajnalová (Ed.), *Súčasné Poznatky z Archeobotaniky Na Slovensku*, Acta Interdisciplinaria Archaelogica IV, pp. 3–192. Nitra: Archeologický ústav SAV.

Hajnalová, E. (1993). Praveké osídenie lokality Sarišské Michaľany dokumentované rastlinnými zvyškami. *Východoslovenský pravek 4*, 49–65.

Hajnalová, E. and M. Hajnalová (2004). Zbierané rastliny ako zdroj potravy v praveku strednej Európy a ich archeobotanické nálezy na Slovensku. In V. Janák and S. Stuchlik (Eds.), *Otázky neolitu a eneolitu našich zemí*, Acta Archaeologica Opaviensia, pp. 33–47. Opava: University of Opava.

Hajnalová, E. and J. Mihályiová (1998). Archeobotanické nálezy v roku 1996. *Archeologické výskumy a nálezy na Slovensku v roku 1996*, 61–67.

Hajnalová, M. (2003). Finds of cultivated crops and utilised wood dated to Aeneolthic and Hallstatt period at a site of Ižkovce-Predná Hora. *Východoslovenský pravek 6*, 161–162.

Hajnalová, M. and E. Hajnalová (n.d.). The plant macro-remains from Dzeravá Skala cave: the environment and plant foods. In Ľ. Kaminská, J. K. Kozlowski, and J. Svoboda (Eds.), *Dzerravá Skala: Upper Plesitocene Cave Sequence in Western Carpathians*. Kraków: Polska Akademia Umjejentnosc.

Horváthová, E. (2003). Záchranný výskum v Brehove. *Archeologické výskumy a nálezy na Slovensku 2002*, 54–55.

Jones, G. (1984). Interpretation of archaeological plant remains: ethnographic models from Greece. In W. van Zeist and W. A. Casparie (Eds.), *Plants and Ancient Man*, Studies in Palaeoethnobotany. Proceedings of the Sixth Symposium of the International Work Group for Palaeoethnobotany, Groningen, 30 May–3 June 1983, pp. 43–61. Rotterdam: A. A. Balkema.

Jones, G. (1987). A statistical approach to the archaeological identification of crop processing. *Journal of Archaeological Science 14*, 311–323.

Jones, G. (1998). Wheat grain identification: why bother? *Environmental Archaeology 22*, 29–34.

Kalicz, N. and J. Makkay (1977). *Die Linienbandkeramik in der grossen Ungarischen Tiefebene*. Studia Archaeolgica. Budapest: Akademiai Kiadó.

Kaminská, Ĺ. (1999). Záchranný výskum na preložke cesty v Košiciach. *Archeologické výskumy a nálezy na Slovensku v roku 1997*, 93–94.

Kaminská, Ĺ. (2001a). Archeobotanické nálezy v roku 2000. *Archeologické výskumy a nálezy na Slovensku v roku 2000*, 96–97.

Kaminská, Ĺ. (2001b). Záchranné výskumy v Košiciach. *Archeologické výskumy a nálezy na Slovensku v roku 2000*, 96–97.

Kaminská, Ľ., J. K. Kozlowski, and J. Svoboda (Eds.) (n.d.). *Dzerravá Skala: Upper Plesitocene Cave Sequence in Western Carpathians*. Kraków: Polska Akademia Umjejentnosc. In press.

Kaminská, Ĺ. and M. Novák (2002). Sídliskové nálezy bukovohorskej kultúry v polohe Košice-Červený rak. *Archeologické výskumy a nálezy na Slovensku v roku 2001*, 82–83.

Klčo, M. (1997). Archeologický výskum na nádvorí kúrie—rodný dom M.A. Beňovského vo Vrbovom. *Balneologický spravodaj 35*, 49–170.

Kolník, T. (1980). Výskum v Cíferi-Páci v roku 1978. *Archeologické výskumy a nálezy na Slovensku v roku 1978*, 142–155.

Kuzma, I. (1982). Druhá etapa výskumu v Mužli-Čenkove. *Archeologické výskumy a nálezy na Slovensku v roku 1981*, 171–175.

Kuzma, I., O. Oždáni, and M. Hanuliak (1983). Tretia sezóna výskumu v Mužli-Čenkove. *Archeologické výskumy a nálezy na Slovensku v roku 1982*, 143–148.

Lichardus, J. and J. Vladár (2003). Gliederung der Lengyel-kultur inder Slowekei ein rückblick nach Viercig Jahren. *Slovenská Archeológia LI/2*, 195–216.

Makkay, J. (1996). Theories about the origin, the distribution and the end of the Körös culture. In L. Tálas (Ed.), *At the Fringes of Three Worlds: Hunter-Gatherers and Farmers in the Middle Tisza Valley*, pp. 35–49. Szolnok, Hungary: Damjanich Múzeum Press.

Nevizánsky, G. (1980). Záchranný výskum v Kameníne. *Archeologické výskumy a nálezy na Slovensku 1978*, 187–188.

Nevizánsky, G. and A. Točík (1984). Predbežné výsledky predstihového záchranného výskumu v Bajči-Vlkanove. *Archeologické výskumy a nálezy na Slovensku 1983*, 156–158.

Němejcová-Pavúková, V. (1970). Kultúra s kanelovanou keramikou. In A. Točík (Ed.), *Slovensko v mladšej dobe kamennej*, pp. 182–206. Bratislava: Archeologického ústavu SAV.

Němejcová-Pavúková, V. (1973). Správa o ukončení výskumu v Ružindole. *Archeologické výskumy a nálezy na Slovensku 1972*, 102–104.

Němejcová-Pavúková, V. (1987). Záchranný výskum v Kočíne. *Archeologické výskumy a nálezy na Slovensku 1986*, 75–76.

Němejcová-Pavúková, V. (1988). Záchranný výskum v Kočíne. *Archeologické výskumy a nálezy na Slovensku 1987*, 94–96.

Němejcová-Pavúková, V. (1995). *Svodín. Zwei Kreisgrabenanlagen der Lengyel-Kultur*, Volume II of *Studia Archaeologica et Mediaevalia*. Bratislava: Univerzita Komenského.

Oždáni, O. (1975). Nové sídliskové nálezy z Levíc. *Archeologické výskumy a nálezy na Slovensku v roku 1974*, 71–72.

Parkinson, W. A. (1999). *The Social Organization of Early Copper Age Tribes, on the Great Hungarian Plain.* Ph. D. thesis, University of Michigan. Ann Arbor.

Pástor, J. (1965). Blažice, Bohdanovce i Hraničná pod Koszycami (vykopaliska v latach 1963-1964). *Acta Archaeologica Carpathica VII*, 87–91.

Pavúk, J. (1978). Výskum neolitického sídliska v Blatnom. *Archeologické výskumy a nálezy na Slovensku 1977*, 192–195.

Pavúk, J. (1980). Výskum neolitického sídliska v Blatnom. *Archeologické výskumy a nálezy na Slovensku 1979*, 160–165.

Pavúk, J. (1981). Sídlisko lengyelskej kultúry v Budmericiach. *Archeologické výskumy a nálezy na Slovensku 1980*, 220–222.

Pavúk, J. (1992). Sídlisko lengyelskej kultúry v Žlkovciach ohradené palisádami. *Archeologické Rozhledy 44*, 3–9.

Pavúk, J. (1994). *Ein Siedlungsplatz der kultur mit Linearkeramik und der Želiezovce Gruppe.* Archaeologica Slovaca Monographie. Communicationes Instituti Archaeologici Nitriensis Academiae Scientarum Slovacae. Nitra: Archeologický ústav SAV.

Pavúk, J. (2000). Das Epilengyel/Lengyel IV als kulturhistorische einheit. *Slovenská Archeológia XLVIII*(1), 1–26.

Pavúk, J. (2003). Stará lineárna keramika na Slovensku a neolitizácia strednej Európy. In M. Lutovský (Ed.), *Questions in the Neolithic and Eneolithic 2003. Papers From the 22nd Working Meeting of Researchers into the Neolithic and Eneolithic. Český Brod–Kounice, Sept. 23rd-26th, 2003*, pp. 11–28. Prague: Ústav archeologické památkové péče středních Čech.

Pavúk, J. (2004). Kommentar zu einem Rückblick nach vierzig Jahren auf die Gliederung der Lengyel-Kultur. *Slovenská Archeológia LII*(1), 139–160.

Pavúk, J. and S. Šiška (1980). Neolit a eneolit. *Slovenská archeológia 28*, 137–156.

Pavúk, J. and S. Šiška (1971). Neolitické a eneolitické osídlenie Slovenska. *Slovenská Archeológia 19*, 319–364.

Pokorný, P. (2003). Rostlinné makrozbytky. In J. Svoboda (Ed.), *Mezolit Severních Čech*, pp. 272–273. Brno: Archeologický ústav Brno, Národní Park České Švýcarsko, Oblastní Muzeum Děčín.

Šiška, S. (1974). Abdeckung von Siedlungen und einem Gräberfeld aus der jungeren Steinzeit in Kopčany, Kr. Michalovce. *Archeologické rozhledy 26*, 4–14.

Šiška, S. (1976). Sídlisko z mladšej doby kamennej v Prešove–Šarišských lúkach. *Slovenská archeológia 24*, 83–117.

Šiška, S. (1989). *Kultúra s východnou lineárnou keramikou na Slovensku.* Bratislava: Veda.

Šiška, S. (1992). Protoliearnaja keramika iz vostočnej Slovakii i jej otnoščnie k jugo-vostočnej Jevrope. *Studia Praehistorica 11–12*, 35–40.

Šiška, S. (1995). *Dokument o Spoločnosti Mladšej Doby Kamennej: Šarišské Michaľany.* Bratislava: Veda.

Soják, M. (2000). Nálezy z prieskumov a záchranných exploatácii na Spiši. *Archeologické výskumy a nálezy na Slovensku 1999*, 115–120.

Staššíková-Štukovská, D. (1987). Archeologický výskum v Borovciach. *Balneohistorica Slovaca 26*, 158–169.

Staššíková-Štukovská, D. (1988). Sídliskové nálezy a hrob z Boroviec. *Študijne zvesti Archeologického ústavu SAV 24*, 173–190.

Struhár, V. (1998). Eneolitický kolektívny hrob z jaskyne pri Liskovej, okres Ružomberok. In I. Kuzma (Ed.), *Otázky neolitu a eneolitu našich krajín*, pp. 203–216. Nitra: Archeologický ústavu SAV.

Svoboda, J. (Ed.) (2003). *Mezolit Severních Čech.* Brno: Archeologický ústav Brno, Národní Park České Švýcarsko, Oblastní Muzeum Děčín.

Točík, A. (Ed.) (1970). *Slovensko v Mladšej Dobe Kamennej.* Bratislava: Vydavateľstvo Slovenskej Akadémie Vied.

Točík, A. (1980). Pokračovanie záchranného výskumu v polohe Kňazová jama v Komjaticiach. *Archeologické výskumy a nálezy na Slovensku 1979*, 215–229.

Točík, A. (1981). Záverečná správa zo záchranného výskumu v Komjaticiach v roku 1977 a 1979. *Študijne zvesti Archeologického ústavu SAV 20*, 139–157.

Virág, Z. M. and J. Kalicz (2001). Neuere Siedlungsfunde der fruhneolithischen Starčevo-Kultur aus Sudwestungarn. In B. Ginter (Ed.), *Problems of the Stone Age in the Old World. Jubilee Book Dedicated to Professor Janusz K. Kozlowski*, pp. 268–279. Kraków: Jagelonian University, Institute of Archaeology.

Vizdal, J. (1970). Neskoroneolitické nálezy z Oborína. *Slovenská Archeológia 18*, 217–227.

Vizdal, J. (1973). *Zemplín v Mladšej Dobe Kamennej.* Košice: Východoslovenské vydavateľstvo.

Yan, Y., S. L. Hsam, J. Z. Yu, Y. Jiang, I. Ohtsuka, and F. J. Zeller (2003). HMW and LMW glutenin alleles among putative tetraploid and hexaploid European spelt wheat (*Triticum spelta* L.) progenitors. *Theoretical and Applied Genetics 107*, 1321–1330.

Chapter 17

Early neolithic agriculture in south Poland as reconstructed from archaeobotanical plant remains

Maria Lityńska-Zając
Polish Academy of Science, Kraków, Poland

17.1 Introduction

In a survey of plants recovered from archaeological sites in Poland, K. Wasylikowa discussed the history of plant cultivation and the role of some gathered, edible plants wild plants (Wasylikowa et al. 1991). Another review, published in 1997, was based on data from 27 sites belonging to different cultural and chronological units in the loess uplands of western Małopolska (Little Poland; Lityńska-Zając 1997a). The results discussed in this paper come from selected archaeobotanical sites—those that yielded fairly rich plant remains—situated in southern Poland (Appendix 1, figure 17.1).

The records of plant material from archaeological sites in Poland are unequal with respect both to chronology and geography. Most data is derived from sites dating to the early Middle Ages and the Roman period. Particularly intensive studies from different urban areas have provided numerous plant remains dating to the Medieval period. There is also a significant amount of data from the Neolithic and Lusatian culture sites, while the bronze age sites have yielded the poorest record of plant remains. The geographical distribution of sites shows two concentrations, in Wielkopolska (Great Poland) and Małopolska (Little Poland), simply because the most intensive archaeological investigations have been conducted in these areas.

In this paper an attempt has been made to reconstruct the nature of cereal cultivation, taking into account the cultural differences, in the earliest Neolithic of southern Poland. The data used in the analysis are derived from a relational database that includes all published archaeobotanical information from Poland (Lityńska-Zając 1999). The plant materials from settlements at Donosy, Kraków-Prądnik Czerwony and Gwoździec were only partly described, because botanical and/or archeological work is still in progress (Appendix 1).

Macrofossil analysis provides fairly comprehensive information about the species that were cultivated in prehistoric times. The reconstruction of the role played by any individual species in the agricultural economy of different cultures is, however, much more difficult. In order to gain better insight into this aspect, two methods of comparison are used in the interpretation presented below; these are: (i) taxon abundance, i.e., the number of specimens of a taxon present at a site (tables 17.1–17.4); and (ii) taxon frequency, expressed as the number of sites in which

Figure 17.1: Polish Neolithic sites discussed in this paper. 1= Bronocice (Małoposka), 2= Donosy site no. 3 (Małopolska), 3= Gniechowice site no. 2a (Dolnośląskie), 4= Gwoździec site no. 2 (Małopolska), 5= Iwanowice-Klin (Małopolska), 6= Kazimierza Mała site no. 1 (Małopolska), 7= Kraków-Mogiła site no. 62 (Małopolska), 8= Kraków-Pleszów site no. 17 (Małopolska), 9= Kraków-Zesławice site no. III (Małopolska), 10= Kraków-Prądnik Czerwony (Małopolska), 11= Niedźwiedź site no. 1 (Małopolska), 12= Olszanica (Małopolska), 13= Stary Zamek site no. 2 (Dolnośląskie), 14= Strachów site no. 2 (Dolnośląskie), 15= Tomaszowice site no. 1 (Małopolska).

a taxon was found (table 17.5). For taxon frequency, an assumption is made that both large deposits and single finds are of equal rank.

17.2 Description of the material

The botanical material was collected from various features such as houses, pits, postholes and hearths. Different numbers of samples were studied from each site but they had all been taken from features with well-defined chronological and cultural affiliation. The numbers and types of plant remains recovered from the separate sites were not equal; for example, only plant impressions in daub and potsherds were present at Gniechowice and Stary Zamek, while at Gwoździec, Kazimierza Mała and Kraków-Mogiła charred remains were examined.

Charred cereal remains include caryopses, spike fragments (e.g., glumes, paleas, lemmas, spikelet forks and rachis internodes) and stem fragments. In a majority of cases a small number of specimens were scattered in different features. The only large concentrations of charred specimens were found in pits 416 at Kraków-Mogiła and 32 at Kraków-Prądnik Czerwony. Daub fragments contained impressions of cereal grains and vegetative plant parts, as well as scattered imprints of seeds and fruits of wild plants, some of which had charred specimens preserved inside.

The results that will be discussed in this paper come from sites of the Linear Pottery culture, dated from between 5440/5380 and 4580 cal BC, the Lengyel culture, dating from 4580–3640/3610 cal BC, and the Funnel Beaker culture, with a date range of 3970–3060 cal BC (Kruk and Milisauskas 1999).

17.3 Results

The Linear Pottery culture is represented by six sites (table 17.1) situated in the regions of Małopolska (Little Poland) and Dolny Śląsk (Silesia) (figure 17.1). The greatest number of identified plant remains from this chronological phase come from the sites at Gwoździec and Strachów and the smallest number are those in Kazimierza Mała. The greatest diversity of species is noted at Olszanica. Hulled wheats dominated the cereal remains in the Linear Pottery culture sites. Most of the remains are of emmer wheat (*Triticum dicoccum*), usually accompanied by less abundant remains of einkorn wheat (*Triticum monococcum*). Barley (*Hordeum vulgare*) is less abundant than the wheats.

On the sites at Strachów and Olszanica a few specimens represent spelt (*Triticum spelta*) and another of the hulled wheats, and these are the oldest remains of this species from Poland. At present it is not possible to interpret the agricultural significance of this wheat in the early Neolithic; more finds are needed in order to confirm whether or not it was cultivated. Imprints of a spike fragment and grains, as well as charred grains of rye (*Secale cereale*), were found at Olszanica (Giżbert 1961). The only other cultivated plant species from the Linear Pottery culture was flax (*Linum usitatissimum*), which was recovered from the site in Gwoździec.

The sites described have provided very little information about weeds because preservation of plant material was mainly in the form of imprints on daub. There are a few seeds or fruits of wild herbaceous plants from Gwoździec, including *Chenopodium album*, *Bromus* sp., *Echinochloa crus-galli* and *Setaria pumila*, which would have grown in fields. The last two species are weeds characteristic of root-crop cultivation; they have been found more frequently in Polish Neolithic sites than in the other parts of Central Europe (Willerding 1986). Their spread must have been associated with that of the most common cereals, emmer and einkorn wheat (Wasylikowa et al. 2002).

Cereals from the Lengyel culture are represented at six sites (table 17.2) located in Małopolska (Little Poland) and Dolny Śląsk (Silesia). The greatest number of specimens was found at site

Taxon	Plant part	Gwoździec ch	Gwoździec i	Kazimierza Mała ch	Olszanica ch	Olszanica i	Strachów i	Gniechowice i	Stary Zamek i
Cerealia indet.	chaff	1	.	.	+	.	+	+	+
Cerealia indet.	c	173	.	14	11	.	38, 1⋆	8	2
Hordeum vulgare	c	3	.	.	7	8	18	4	6⋆
Hordeum vulgare	sp fr	.	.	.	.	4	1	20	.
Panicum miliaceum	c	.	.	.	127	5	.	.	.
Triticum aestivum s.l.	c	.	.	.	16	.	.	.	.
Triticum dicoccum + cf.	c	278	.	.	1	15	53, 5⋆	.	1
Triticum dicoccum + cf.	sp fr	7	36	.	.	1	40, 254⋆	40⋆	134
Triticum monococcum	c	9	.	.	.	.	1⋆	.	2
Triticum monococcum	sp fr	14	.	.	.	.	1, 4⋆	7⋆	115
Triticum spelta + cf.	c	.	.	.	2	1	1⋆	.	.
T. dicoccum/monococcum	c	50	.	4	.	.	10, 7⋆	1, 28⋆	.
T. dicoccum/monococcum	sp fr	39	7	.	5	.	32, 4⋆	.	182
Triticum sp.	c	1	.	.	10	4	.	4	.
Triticum sp.	sp fr	.	.	.	.	4	.	44	.
Secale cereale	c	.	.	.	1	.	.	.	.
Secale cereale	sp	.	.	.	.	1	.	.	.
Linum usitatissimum	s	1	.	.	.	.	.	.	.

Table 17.1: Plant remains from the Linear Pottery culture sites in south Poland. Type of remains: c=caryopsis, s=seed, sp=fragments of spike, spikelet, glumes and spikelet forks; state of preservation: ch=charred, i=imprints, ⋆=charred remains inside the imprints; number of specimens: + = present, +++ = abundant.

no. 62 at Kraków-Mogiła and the poorest material came from the site in Strachów. At Kraków-Mogiła, as in the Linear Pottery culture sites, emmer was the predominant cereal with a small admixture of einkorn and a considerable number of caryopses and spike fragments of six-row naked barley (*Hordeum vulgare* var. *nudum*). A few remains of rye (*Secale cereale*) were found at Kraków-Zesławice. At Iwanowice-Klin there are a number of crushed wheat caryopses, possibly an indication of the preparation of groats.

A few specimens of wild herbaceous plants have been preserved in the form of imprints in daub. Several field and ruderal weeds were identified in feature 8 at Iwanowice-Klin, including *Agrostemma githago*, *Chenopodium album*, *Papaver rhoeas*, *Polygonum aviculare*, *Fallopia convolvulus* and *Lithospermum arvense*. One sample contained a single fruit of *Lithospermum officinale*, which was probably collected from natural habitats and used for medicinal purposes or as a source of dye.

The richest finds of wild herbs are in the samples of ancient grain from the site at Kraków-Mogiła (Gluza 1984). They include plants common in several present-day communities such as cultivated fields, meadows and xerothermic grasslands (table 17.3). According to Gluza (1984), the presence of a large number of perennial species (for instance from the genus *Bromus*) was most probably connected with primitive methods of tillage, which did not destroy plant roots and were typical of garden-type agriculture (Kruk 1993).

Plant material from six Funnel Beaker culture settlements in Małopolska (Little Poland) and Dolny Śląsk (Silesia) has also been described (table 17.4). The most diverse material was preserved at the site at Zawarża. Abundant plant remains were found in a pit at Kraków-Prądnik Czerwony, which contained numerous charred cereal grains and a small number of spikelet forks and glume fragments of emmer wheat (*Triticum dicoccum*), with an admixture of caryopses and spikelet fragments of einkorn (*Triticum monococcum*). In the samples analysed thus far, caryopses of emmer constitute 90% of the total grains. Single seeds and fruits of weeds were also found in the samples, namely *Bromus secalinus*, *Agrostemma githago*, *Fallopia convolvulus*, *Rumex crispus*, *Galium aparine* and *Phleum pratense*. The most interesting of these was *Phleum pratense*; this species, which is associated with present-day meadow communities from the class Molinio-Arrhenatheretea, is often found in archaeological samples of cereals together with other typical weeds. In the material from the site at Niedźwiedź there are imprints of caryopses of naked barley (*Hordeum vulgare* var. *nudum*). From the site at Kraków-Mogiła most of the remains are of emmer (*Triticum dicoccum*). One charred caryopsis and one impression of bread wheat (*Triticum aestivum*) were identified at the sites of Bronocice and Zawarża. A few imprints of spelt (*Triticum spelta*) grains were described from Zawarża. Two lentils (*Lens culinaris*) and one flax seed (*Linum usitatissimum*) were also found at Bronocice.

Large numbers of wild herbaceous plants were found at the site of Bronocice. Most of them were weeds characteristic of cereal fields (*Bromus arvensis*, *Bromus secalinus*, *Festuca arundinacea*, *Polygonum* spp., *Rumex* spp. and undetermined Caryophyllaceae); these were present in different features. The increased number of weeds was probably due to a change in farming methods, i.e., the start of extensive slash-and-burn cultivation in the period of the Funnel Beaker culture development (Kruk 1993), which could have resulted in the establishment of typical segetal communities.

Plant gathering was an important part of the economy in the Neolithic. In the sites described in this paper only a few food plants were found that were collected in the forests. Pieces of hazelnut shells (*Corylus avellana*) have been found in the samples from Kraków-Mogiła and Kraków-Pleszów. The most interesting find was recovered from site 2 at Gwoździec; samples contained fruit fragments of wild apple (*Malus sylvestris*) and pip fragments determined as *Malus* sp. (Bieniek and Lityńska-Zając 2001).

Taxon	Plant part	Iwanowice-Klin ch	Iwanowice-Klin i	Kraków-Mogiła ch	Kraków-Mogiła i	Kraków-Pleszów ch	Kraków-Pleszów i	Kraków-Zesławice ch	Kraków-Zesławice i	Strachów i	Tomaszowice i
Cerealia indet.	chaff	.	.	10	.	.	.	.	.	+	+
Cerealia indet.	c	+	.	.	.	8	.	.	.	+	1⋆
Hordeum vulgare	c	.	.	1194	1	.	.	.	7	5	.
Hordeum vulgare	sp fr	.	.	30	5	.	.	.	6	.	6, 1⋆
Panicum miliaceum	c	.	4	.	.	.	.	.	.	.	.
T. dicoccum + cf.	c	59	.	1670	.	.	.	.	.	.	.
T. dicoccum + cf.	sp fr	.	.	86	27	.	1	.	8	3	83, 11⋆
T. monococcum	c	30	.	450	1	1	.	.	.	.	.
T. monococcum	sp fr	.	.	28	19	.	1	.	6	2	27
T. dicoccum/ monococcum	c	1247	.	539	2	.	13	.	168, 11⋆	3	.
T. dicoccum/ monococcum	sp fr	28	.	2256	223	.	41	.	.	.	98
Triticum sp.	c	76	.	23cm^3	1	1	.	1	.	.	11
Triticum sp.	sp fr	.	.	.	.	.	2	16	.	.	148, 7⋆
Secale cereale	c	.	.	.	.	.	+⋆	.	.	.	.
Secale cereale	sp	.	.	.	.	.	1	.	.	.	.

Table 17.2: Abundance of plant remains from the Lengyel culture sites in south Poland. See table 17.1 for codes.

Habitat/Taxon	no. of specimens
Weeds and ruderal plants	
Bromus arvensis	$68cm^3$
Chenopodium album	457
Melandrium album	38
Polygonum persicaria	7
Ruderal plants	
Bromus sterilis	63
Chenopodium urbicum	9
Linaria cf. *vulgaris*	1
Weeds of root-crops and summer cereals	
Chenopodium polyspermum	121
Digitaria ischaemum	1
Digitaria cf. *sanguinalis*	4
Echinochloa crus-galli	12
Polygonum cf. *minus*	3
Setaria viridis/verticillata	10
Solanum nigrum	26
Weeds of cereals and root-crops	
Fallopia convolvulus	2475
Viola tricolor/arvensis	2
Moist and fresh meadows	
Bromus racemosus	$27cm^3$
Centaurea jacea	1
Galium cf. *boreale*	31
Galium verum/mollugo	20
Lotus cf. *corniculatus*	1
Trifolium hybridum	2
Trifolium repens	3
Xerothermic grassland	
Ajuga genevensis/reptans	2
Centaurea scabiosa	2
Plantago media	1
Silene vulgaris	2
Trifolium cf. *montanum*	67
Others	
Polygonum aviculare	2
Potentilla reptans	7
Polygonum lapathifolium subsp. *lapathifolium*	48
Sambucus ebulus	6
Melandrium rubrum	6

Table 17.3: List of wild herbaceous species from the Lengyel culture pit at site Kraków-Mogiła (Gluza 1984)

Taxon	Plant part	Bronocice ch	Bronocice i	Donosy ch	Donosy i	Kraków-Prądnik Czerwony ch	Niedźwiedź i	Strachów i	Zawarża i
Cerealia indet.	chaff	.	+	7	+	.	+	+, +⋆	+, 4⋆
Cerealia indet.	c	⋆	+	8	.	.	+, 16⋆	+, +⋆	+, 2⋆
Hordeum vulgare	c	22	.	2	4	.	165, 2⋆	2	3⋆
Hordeum vulgare	sp fr	.	.	.	.	.	1	.	.
Triticum aestivum	c	1	.	.	.	.	.	.	1⋆
T. dicoccum + cf.	c	68	48, 2⋆	1	22	+++	2, 38⋆	8, 4⋆	16, 3⋆
T. dicoccum + cf.	sp fr	.	.	5	19	+	38, 120⋆	5, 86⋆	9, 7⋆
T. monococcum	c	15	2	.	2	+	1, 5⋆	1⋆	10
T. monococcum	sp fr	.	.	.	.	+	6⋆	7⋆	5⋆
T. spelta + cf.	c	.	.	.	.	.	.	.	2⋆
T. spelta + cf.	sp fr	.	.	.	.	.	.	.	3⋆
T. dicoccum/monococcum	c	11	4	117	3	.	1	2	.
T. dicoccum/monococcum	sp fr	.	.	1	100	.	69⋆	.	.
Triticum sp.	c	50	8	.	.	.	23	.	8
Triticum sp.	sp fr	.	.	.	.	.	.	.	1
Secale cereale	c	.	.	.	.	.	.	.	1
Lens culinaris	s	2	.	.	.	.	.	.	.
Linum usitatissimum	s	1	.	.	.	.	.	.	.

Table 17.4: Abundance of plant remains from the Funnel Beaker culture sites in south Poland. See table 17.1 for codes.

Taxon	Number of sites LCP	LC	FBC
Cerealia indet.	6	5	5
Hordeum vulgare	5	4	5
Panicum miliaceum	1	1	.
Triticum aestivum s.l.	1	1	2
T. dicoccum + cf.	5	6	6
T. monococcum	4	3	6
T. spelta + cf.	2	.	1
T. dicoccum/monococcum	6	6	4
Triticum sp.	4	4	3
Secale cereale	1	1	1
Lens culinaris	.	.	1
Linum usitatissimum	1	.	1

Table 17.5: The number of sites from different cultures of the early Neolithic that have evidence of various crop remains. LCP=Linear Pottery culture, LC=Lengyel culture, FBC=Funnel Beaker culture.

17.4 Conclusions

The taxonomic composition of cultivated plant assemblages from the earliest neolithic sites in southern Poland shows no differences between the three cultures discussed here (table 17.5). On this basis and in agreement with the results obtained from sites of the same age in the other parts of Poland, it may be concluded that emmer (*Triticum dicoccum*) was the main crop of the Linear Pottery, Lengyel and Funnel Beaker cultures and that the second most common was einkorn (*Triticum monococcum*) or perhaps barley (*Hordeum vulgare*).

Large deposits of emmer and einkorn grains have been found on a few settlements, for example, the Lengyel culture site 62 at Kraków-Mogiła (Gluza 1984) and, most importantly, the Funnel Beaker culture site at Kraków-Prądnik Czerwony (Rook and Nowak 1993). This indicates that mixtures of emmer and einkorn were sown in south Poland, as in many other parts of Europe. The two wheats have similar edaphic requirements and similar life cycles, which allowed them to be grown together, but emmer had a dominant character in cultivations. However, at the Funnel Beaker site of Ćmielów in Sandomierska Upland (Klichowska 1975), a find of a large number of grains consisting almost entirely of emmer indicates that this wheat was monocropped.

The role of barley in neolithic agriculture in the area discussed is difficult to assess. Barley remains have often been described from the neolithic settlements in Poland (Wasylikowa et al. 1991) but only rarely have they been found in large quantities. Larger accumulations were discovered, for example, at Piotrowice Wielkie Linear Pottery culture, Śląsk (Silesia) (Klichowska 1969) and in the site discussed here at Kraków-Mogiła. Barley was probably sown as a separate crop. Other cereals played a minor role; spelt (*Triticum spelta*), possible bread/club wheat (*Triticum aestivum* s.l.) and rye (*Secale cereale*) are rarely found and always in small quantities and they probably occurred in the fields as unintentional admixtures rather than as deliberately sown cereals. Common millet (*Panicum miliaceum*), which is present in low frequencies, was possibly cultivated on a small scale.

A large proportion of the cereal specimens were not identified (Cerealia indet.). They were the remnants of leaves, straw or spikelet fragments and strongly damaged or distorted fragments of caryopses, parts of which were preserved inside burnt clay. Plentiful plant admixtures in daub indicate that threshing remains were added as temper to clay that was used for construction purposes.

On the basis of the archaeobotanical data it may be suggested that the other cultivated plants played a small role in the early Neolithic of southern Poland. They were less frequent than cereals and appeared as single specimens. There is evidence for flax (*Linum usitatissimum*) cultivation as early as the Linear Pottery culture, and lentil (*Lens culinaris*) cultivation in the Funnel Beaker culture.

Appendix 1: sites with archaeobotanical data referred to in this paper

1. Bronocice (Małoposka): charred plant remains and impressions in daub from the different Funnel Beaker culture features (Lityńska-Zając 1997a).

2. Donosy site no. 3 (Małopolska): charred plant remains and impressions in daub from the different Funnel Beaker culture features (Lityńska-Zając nda).

3. Gniechowice site no. 2a (Dolnośląskie): plant impressions in daub and potsherds from the Linear Pottery culture features (Gluza 1994).

4. Gwoździec site no. 2 (Małopolska): plant materials from features of the earliest part of the Linear Pottery culture (Bieniek and Lityńska-Zając 2001) and new material identified by Lityńska-Zając (ndb).

5. Iwanowice-Klin (Małopolska): charred plant remains and imprints in daub from one Lengyel culture pit (Lityńska-Zając 1990).

6. Kazimierza Mała site no. 1 (Małopolska): charred plant remains from one Linear Pottery culture pit, identified by Lityńska (nd).

7. Kraków-Mogiła site no. 62 (Małopolska): charred plant remains and imprints in daub from one Lengyel culture pit (Gluza 1984).

8. Kraków-Pleszów site no. 17 (Małopolska): charred plant remains and impressions in daub from the different Lengyel culture features (Gluza et al. 1988).

9. Kraków-Zesławice site no. III (Małopolska): plant impressions in daub from different features, Lengyel culture (Gluza 1970).

10. Kraków-Prądnik Czerwony (Małopolska): cereals from one pit, Funnel Beaker culture (Lityńska-Zając 1997a).

11. Niedźwiedź site no. 1 (Małopolska): impressions in daub from the different Funnel Beaker culture features Burchard and Lityńska-Zając 2002).

12. Olszanica (Małopolska): charred plant material and impressions in daub and ceramics, from the different Linear Pottery culture features (det. Gibert, in Kozłowski and Kulczycka 1961; Ford 1986; Lityńska 1986b, a).

13. Stary Zamek site no. 2 (Dolnośląskie): plant impressions in daub and potsherds from the Linear Pottery culture features (Gluza 1994).

14. Strachów site no. 2 (Dolnośląskie): plant impressions in daub from features of the Linear Pottery, Lengyel and Funnel Beaker cultures (Lityńska-Zając 1997b).

15. Tomaszowice site no. 1 (Małopolska): imprints in daub from different Lengyel culture features (Gluza 1986).

16. Zawarża site no. 1 (Świętokrzyskie): imprints in daub from different Funnel Beaker culture features (Lityńska-Zając 2002).

References

Bieniek, A. and M. Lityńska-Zając (2001). New finds of *Malus sylvestris* Mill. (wild apple) from neolithic sites in Poland. *Vegetation History and Archaeobotany 10*, 105–106.

Burchard, B. and M. Lityńska-Zając (2002). Plant remains from the Funnel Beaker Culture site at Niedźwiedź, Słomniki commune, Małopolska province. *Acta Palaeobotanica 41*(2), 171–176.

Ford, R. I. (1986). Archaeobotanical data from Olszanica. In S. Milisauskas (Ed.), *Early Neolithic Settlement and Society at Olszanica*, Memoirs of the Museum of Anthropology 19, pp. 263. Ann Arbor: University of Michigan.

Giżbert, W. (1961). Wyniki analizy botanicznej szczštków i odcisków roślin na polepie z paleniska przy jamie 1, ze stanowiska w Olszanicy, pow. Kraków. (Annex 1), in J. K. Kozłowski and A. Kulczycka "Materiały kultury starszej ceramiki wstęgowej z Olszanicy, pow. Kraków". *Materiały Archeologiczne 3*, 29–43.

Gluza, I. (1970). Odciski roślin z osady neolitycznej i z wczesnej epoki brązu, odkrytej na stan. III w Nowej Hucie-Zesławicach. *Materiały Archeologiczne 11*, 127–132.

Gluza, I. (1984). Neolithic cereals and weed from the locality of the Lengyel Culture at Nowa Huta–Mogiła near Cracow. *Acta Palaeobotanica 23*(2), 123–184.

Gluza, I. (1986). Ślady neolitycznego rolnictwa w obiektach kultury lendzielskiej z Tomaszowic, gm. Wielka Wieś, woj. Kraków. *Materiały Archeologiczne 23*, 225–231.

Gluza, I. (1994). Wstępne wyniki badań odcisków roślinnych ze stanowisk kultury ceramiki wstęgowej rytej w Gniechowicach i Starym Zamku, woj. wrocławskie. *Polish Botanical Studies: Guidebook Series 11*, 55–69.

Gluza, I., Z. Tomczyńska, and K. Wasylikowa (1988). Uwagi o użytkowaniu drewna w neolicie na podstawie analizy węgli drzewnych ze stanowisk archeologicznych w Krakowie-Nowej Hucie. *Materiały Archeologiczne Nowej Huty 12*, 1–19.

Klichowska, M. (1969). Analiza botaniczna materiałów z osady neolitycznej z Pietrowic Wielkich, pow. Racibórz. *Sprawozdania Archeologiczne 20*, 413–416.

Klichowska, M. (1975). Najstarsze zboża z wykopalisk Polskich. *Archeologia Polski 20*(1), 83–143.

Kozłowski, J. K. and A. Kulczycka (1961). Materiały kultury starszej ceramiki wstęgowej z Olszanicy, pow. Kraków. *Materiały Archeologiczne 3*, 29–43.

Kruk, J. (1993). Rozwój społeczno-gospodarczy i zmiany środowiska przyrodniczego wyżyn lessowych w neolicie (4800-1800 bc). *Sprawozdania Archeologiczne 45*, 7–17.

Kruk, J. and S. Milisauskas (1999). *Rozkwit i upadek społeczeństw rolniczych neolitu. The Rise and Fall of Neolithic Societies.* Kraków: Instytut Archeologii i Etnologii Polskiej Akademii Nauk. [In Polish with an abbreviated English text.].

Lityńska, M. (1986a). Plant remains identified through analysis of imprints in ceramics. In S. Milisauskas (Ed.), *Early Neolithic Settlement and Society at Olszanica*, Memoirs of the Museum of Anthropology 19, pp. 264. Ann Arbor: University of Michigan.

Lityńska, M. (1986b). Plant remains identified through analysis of imprints in daub. In S. Milisauskas (Ed.), *Early Neolithic Settlement and Society at Olszanica*, Memoirs of the Museum of Anthropology 19, pp. 264. Ann Arbor: University of Michigan.

Lityńska, M. (nd). Plant remains from the Linear Pottery Culture at Kazimierza Mała. Unpublished ms.

Lityńska-Zając, M. (1990). Zboża i chwasty z neolitycznego stanowiska Iwanowice-Klin, woj. Kraków. *Sprawozdania Archeologiczne 42*, 105–108.

Lityńska-Zając, M. (1997a). Środowisko i uprawa roślin w czasach pra- i wczesnohistorycznych. In K. Tunia (Ed.), *Z archeologii Małopolski. Historia i stan badań zachodniomałopolskiej wyżyny lessowej*, pp. 473–497. Kraków: Polska Akademia Nauk. Instytut Archeologii i Etnologii PAN Oddział w Krakowie.

Lityńska-Zając, M. (1997b). Wyniki badań odcisków roślinnych z neolitycznego stanowiska w Strachowie, woj. Wrocław. In A. Kulczycka-Leciejewiczowa (Ed.), *Strachów. Osiedla neolitycznych rolników na Śląsku*, pp. 267–277. Wrocław: Instytut Archeologii i Etnologii PAN.

Lityńska-Zając, M. (1999). Komputerowa baza danych makroskopowych szczątków roślinnych z wykopalisk archeologicznych. *Polish Botanical Studies: Guidebook Series 23*, 333–342.

Lityńska-Zając, M. (2002). Odciski roślinne na polepie z osady kultury pucharów lejkowatych w Zawarży. In A. Kulczycka-Leciejewiczowa (Ed.), *Zawarża. Osiedle neolityczne w południowopolskiej strefie lessowej*, pp. 129–134. Wrocław: Instytut Archeologii i Etnologii PAN.

Lityńska-Zając, M. (nda). Charred plant remains and impressions in daub from the Funnel Beaker culture site at Donosy. Unpublished ms.

Lityńska-Zając, M. (ndb). Gwoździec site no 2. Plant materials from the Linear Pottery Culture features. Unpublished ms.

Rook, E. and M. Nowak (1993). Sprawozdanie z badań wielokulturowego stanowiska w Krakowie—Prądniku Czerwonym, w latach 1990 i 1991. *Sprawozdania Archeologiczne 45*, 35–71.

Wasylikowa, K., M. Cârciumaru, E. Hajnalová, B. P. Hartyányi, G. Pashkevich, and Z. V. Yanushevich (1991). East-Central Europe. In W. van Zeist, K. Wasylikowa, and K.-E. Behre (Eds.), *Progress in Old World Palaeoethnobotany: A Retrospective View on the Occasion of 20 Years of the International Work Group for Palaeoethnobotany*, pp. 207–239. Rotterdam: A. A. Balkema.

Wasylikowa, K., M. Lityńska-Zając, A. Bieniek, and I. Gluza (2002). Archeobotaniczne badania nad trawami. In L. Frey (Ed.), *Polska Księga Traw*, pp. 39–52. Kraków: Instytut Botaniki im. W. Szafera PAN.

Willerding, U. (1986). *Zur Geschichte der Unkräuter Mitteleuropas.* Gttinger Schriften zur Vor-und Frühgeschichte, Band 22. Neumunster: Karl Wachholtz.

Chapter 18

Neolithic plant husbandry in the Kujawy region of central Poland

Aldona Bieniek
W. Szafer Institute of Botany, Kraków, Poland

18.1 Introduction

The archaeology surrounding the town of Brześć Kujawski in the Kujawy region of central Poland has been studied since the 1930s (Jażdżewski 1938; Grygiel 1986; Grygiel and Bogucki 1997). Archaeobotanical research started in the 1980s at Brześć Kujawski site 4 (Wasylikowa in Grygiel 1986) and was also undertaken at other archaeological sites during excavations conducted by P. Bogucki and R. Grygiel (Osłonki 1, Guźlin 2, Konary 1, Konary 1a, Miechowice 4, Miechowice 4a, Smólsk 4, Wolica Nowa 1, Zagajewice 1: Bogucki and Grygiel 1983, Grygiel and Bogucki 1997). Archaeobotanical material from these sites has been studied by K. Wasylikowa, Z. Tomczyńska and A. Bieniek (Nalepka et al. 1998; Bieniek 1999, 2002, 2003a, b).

The Kujawy region is the driest part of the country with less than 500 mm of precipitation per year (Kondracki 2000, p.138; Szafer 1972, p.47) and it includes areas of steppe-like and halophytic vegetation (Szafer 1972, pp.44–50, 68–71; Matuszkiewicz et al. 1995). In the 6th millennium cal BC it was probably warmer and relatively drier in Central and Eastern Europe than today (Ralska-Jasieiczowa and Starkel 1999, p.176–179; Starkel 1995, p.38). The study area is covered mainly with black earth and brown or grey-brown podzolic soils (Dobrzański et al. 1972) and is bordered by the Płock Basin, which is covered mainly with sands. The region was under ice during the last glacial period. Despite the fact that the Kujawy region differs geographically from the southern Polish uplands, which are covered with loess, it was also occupied by early Danubian farmers from the beginning of the Neolithic period in Poland (from about 6670±70 bp Lodz-1177, i.e. 5650–5480 cal BC, Nalepka nd). On the basis of several radiocarbon dates it is supposed that the first indigenous neolithic culture (the Funnel Beaker culture or TRB) could have developed in the Kujawy region or in the area between the lower Elbe and the middle Vistula (Midgley 1992, pp.227–228; Nowak 2001, p.590). The neolithisation process in the Polish lowlands was long and complicated and was completed in about 3500 BC, but in the Kujawy region near-contemporary occurrence of the mesolithic tribes, the Lengyel-Polgar groups, and the early Funnel Beaker farmers has been observed (Nowak 2001, fig.4).

The studied archaeobotanical material comes from nine archaeological sites located in the vicinity of the town of Brześć Kujawski (figure 18.1) and from features dated to the Linear Pottery culture (ca. 5500–5000 cal BC, Guźlin 2, Miechowice 4, Smólsk 4, Wolica Nowa 1 and Zagajewice 1) the Lengyel culture (ca. 4400–4000 cal BC, Guźlin 2, Konary 1, Konary 1a, Miechowice 4, Miechowice 4a and Osłonki 1) and the Funnel Beaker culture (Wolica Nowa 1). In the neighbourhood of Brześć Kujawski the Linear Pottery culture (LBK) was usually followed by intensive Lengyel settlement; this was observed at the sites of Guźlin 2, Miechowice 4 and Zagajewice 1 (Bieniek 2002). The settlement at Smólsk was similar and the LBK was followed by the Lengyel culture, but no archaeobotanical material was taken from the features dating to the latter period (Grygiel, pers. comm.). At site 1 of Wolica Nowa, apart from evidence of LBK occupation there was also a TRB settlement. At four archaeological sites only features dated to the Lengyel culture were discovered (Osłonki 1, Konary 1, Konary 1a and Miechowice 4a).

Plant material from features dated to the Linear Pottery, Lengyel and Funnel Beaker cultures (TRB) will be presented and compared in this paper.

18.2 Material and methods

Most of the samples were taken from archaeological features in the way described by Jones (1991, p.55) as 'purposive or judgement sampling'. The charred archaeobotanical material was usually recovered in the same way from soil samples processed either in the field or in a laboratory. Soil samples of known volume were mixed with water and the floating fraction was poured through sieves with mesh sizes of 0.2 and 0.5 mm. At the sites of Osłonki 1, Miechowice 4 and Miechowice 4a, soil samples of unknown volume were sieved in the field using very coarse meshes. The material from these samples is marked in table 18.1 by an asterisk (*) and has not been included in the graphs.

Dried material was sorted and identified under a binocular microscope with ×6 to ×70 magnification. For the examination of some fruits and seeds an incident light metallurgical microscope (×25 to ×200) was used when higher magnifications were required. Identification was undertaken with the aid of the seed reference collection of the Department of Palaeobotany and the herbarium of the W. Szafer Institute of Botany Polish Academy of Sciences. Plant nomenclature follows Mirek et al. (2002).

Only the charred plant remains have been studied fully. Uncharred macro-remains were extracted and stored separately, and tentative identifications have been made but these taxa do not form part of this study. The results of charcoal analysis and imprints in daub will be presented in another paper.

18.3 Results

Lists of taxa found in the samples are presented in table 18.1.

18.3.1 Cultivated plants

A wide range of cultivated plants was recorded in samples from LBK sites, including mainly the remains of glume wheats: einkorn (*Triticum monococcum*), emmer (*Triticum dicoccum*) and the 'new' type of glume wheat (figure 18.2a) (Jones et al. 2000; Kohler-Schneider 2003), but the most commonly occurring species was einkorn (table 18.1).

After thorough identification of the glume wheat remains dated to the Lengyel culture (figure 18.3), it is clear, particularly from the sum of the chaff items, that the 'new' type was dominant at two sites and that the proportions of einkorn were greater at five. Emmer was also present but never reached more than 35% of the identified chaff material (whole spikelet bases were rare in the studied material). It could be suggested on the basis of the composition of the remains that the wheats were sown as a mixture. Hulled barley (*Hordeum vulgare*) was present in the material dated to LBK and the Lengyel cultures, but was not found in the TRB. Bread wheat (*Triticum aestivum* s.l.) was only identified in the material dated to the Lengyel culture, and was always represented by very few specimens. The identifications of grains of millet (cf.

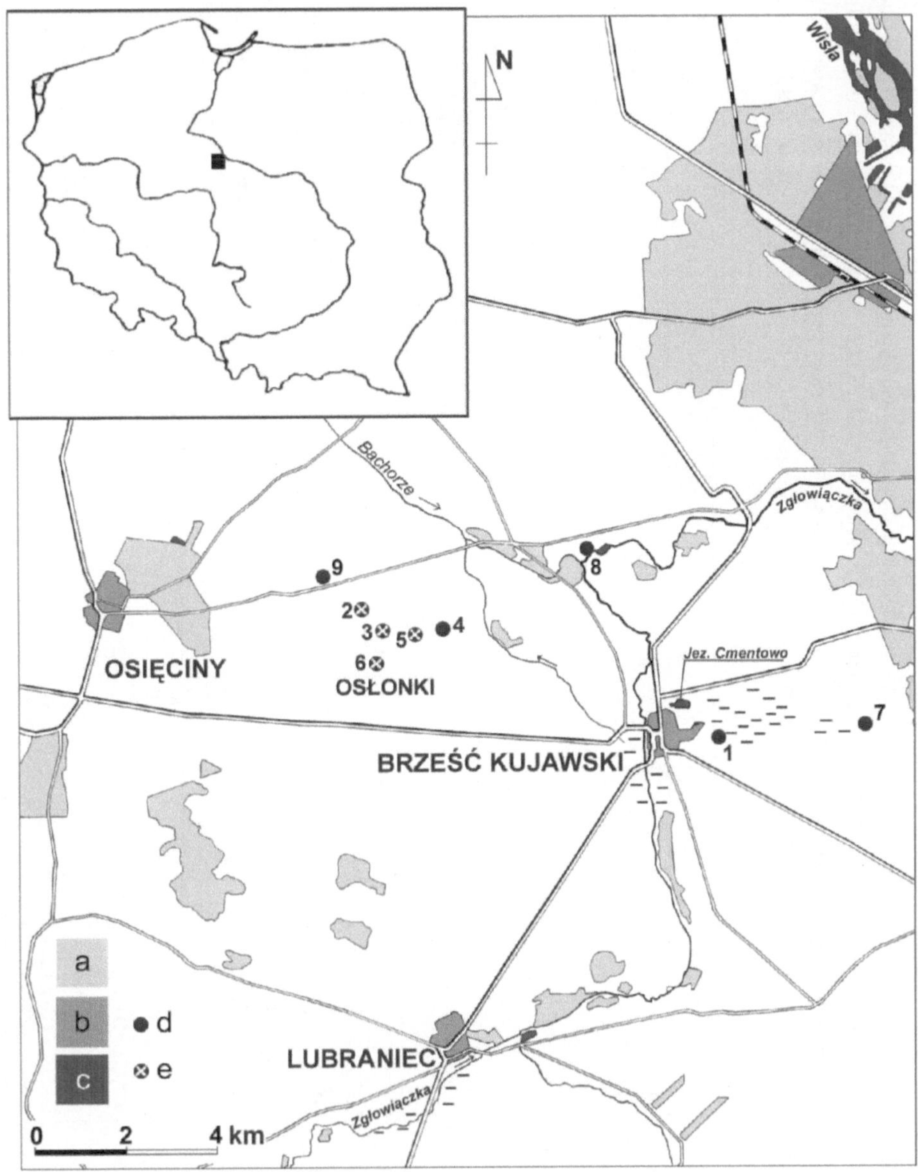

Figure 18.1: Location of sites mentioned in the text. 1. Guźlin 2 (G2); 2. Konary 1 (K1); 3. Konary 1a (K1a); 4. Miechowice 4 (M4); 5. Miechowice 4a (M4a); 6. Osłonki 1 (O1); 7. Smólsk 4 (S4); 8. Wolica Nowa 1 (WN1); 9. Zagajewice 1 (Z1); a. trees; b. buildings; c. water; d. sites with LBK features; e. sites with only the Lengyel culture material. Drawn by J.W. Wieser.

Taxon	Type of remains	LBK G2	LBK M4	LBK S4	LBK WN1	LBK Z1	L G2	L K1	L K1a	L M4	L M4a	L O1	L Z1	TRB WN1
Cultivated plants														
Hordeum vulgare, hulled	c	1	.	.	1	.	.	.	.	.	.	55*	.	.
Triticum aestivum s.l. (and cf.)	r	.	.	.	.	.	.	.	.	1	.	.	.	.
	c	.	.	.	.	.	.	.	.	1	2+3*	17+1*	.	.
Triticum monococcum type	c	.	1*	.	3	.	.	.	.	4+1*	1+1*	114+292*	2	4
	sb	.	23	.	10	.	.	1	2	41	21	18	9	6
	gb	.	122	.	20	.	2	8	6	54	32	20	18	51
Triticum dicoccum type	c	.	2*	.	4	.	.	.	.	40+20*	5+1*	257+1089*	.	8
	sb	.	2	.	5	.	.	.	.	.	.	.	3	.
	gb	.	29	.	7	.	.	2	2	104	12	31	14	6
Triticum 'new' type	sb	.	8	1	2	.	.	2	8	342	13	13	2	.
	gb	.	3	.	2	1	.	2	4	30	8	10	4	6
Triticum cf. 'new' type . terminal	sb	.	.	.	.	.	.	.	.	14	1	1	.	.
T. dicoccum/monococcum	c	.	.	.	.	.	.	.	.	.	.	249+1000*	.	.
Triticum sp. glume wheat	sb	3	11	.	8	.	.	6	3	494	18	51	10	20
	gb	8	99	1	23	.	25	35	23	2093	307	106	45	70
Triticum sp.	c	.	.	.	6	7	.	1	2	23+66*	4+4*	59+26*	1	12
cf. *Panicum miliaceum*	c	.	.	.	1	.	.	.	.	.	.	.	.	.
cf. *Secale cereale*	c	.	.	1	.	.	.	.	.	.	.	.	.	.
Cerealia indet.	c	5	7	6	32	14	5	10	3	104+69*	46+3*	35+43*	1	27
	gb	.	.	.	.	.	.	1	.	.	.	.	.	.
Linum usitatissimum	s	.	1	1	1	1	.	.	.	.	.	.	.	.
Papaver somniferum	s	.	.	1	.	.	.	.	.	.	.	.	.	8
cf. *Pisum sativum*	s	.	1	.	1	3	.	.	.	.	1	1	.	1
Wild plants														
Avena sp.	c	.	.	.	.	.	.	.	.	.	.	1	.	.
	a	1	.	.	.	.	.	.	.	.	.	.	.	.
Bromus hordeaceus	c	.	.	.	.	.	.	.	.	4	.	.	.	.
Bromus racemosus/arvensis	c	.	.	.	.	.	.	.	.	8	8+13*	5+1*	4	.
Bromus sterilis/tectorum	c	.	.	.	.	.	.	.	.	.	3	1	.	.
Bromus sp.	c	.	.	.	.	.	.	1	.	12	14 +4*	11	.	.
Campanula sp.	s	.	.	.	.	.	.	.	.	.	6	.	.	.
Chenopodium album type	s	.	50	5	24	9	.	2	15	4	47	13	.	26
Chenopodium sp.	s	1	.	.	.	.	.	5	.	.	.	.	.	2
Corylus avellana	f	.	.	.	.	.	.	.	.	.	1	.	.	.

Table 18.1: Charred plant remains (excluding charcoal and remains from daub) found at sites in the Kujawy region. Abbreviations: LBK: the Linear Pottery culture, L=the Lengyel culture, TRB=the Funnel Beaker culture, G=Guźlin, K=Konary, M=Miechowice, O=Osłonki, S=Smólsk, WN=Wolica Nowa, Z=Zagajewice, a=awn fragments, c=caryopses, f=fruits, gb=glume bases, s=seeds, sb=spikelet bases, *= specimens obtained from soils samples of unknown volume processed in the field using sieves with very coarse meshes (ca. 3 mm mesh size).

Taxa	Type of remains	LBK					L							TRB
		G2	M4	S4	WN1	Z1	G2	K1	K1a	M4	M4a	O1	Z1	WN1
Wild plants, cont.														
Echinochloa crus-galli	c	.	1	2	2	2	.	.	.	.	4	.	.	.
Fallopia convolvulus	f	2	11	34	9	10	.	4	2	23+1*	33+9*	7+2*	.	12
Fragaria vesca	f	.	.	.	.	.	.	.	.	.	1	1	.	.
Fragaria/Potentilla	f	.	1	.	1	.	.	.	.	.	2	.	.	8
Galium aparine	f	.	.	.	.	.	.	.	.	1	.	.	.	.
Galium spurium	f	.	.	.	.	.	.	.	.	9	3	.	.	.
Galium verum type	f	.	.	.	.	.	.	.	.	2	15	2	.	.
Hierochlöe cf. *australis*	c	.	.	.	.	.	.	.	.	.	15	2	.	.
Juncus sp.	f	.	.	.	.	.	.	.	1	.	.	.	.	.
Lamium amplexicaule/purpureum		.	.	.	.	.	.	.	.	.	.	.	1	.
cf. *Malus sylvestris*	s	.	.	.	1	.	.	.	.	.	.	.	.	.
Malva sp.	s	.	.	.	1	.	.	.	.	.	.	.	.	3
cf. *Mentha* sp.	f	.	.	.	.	.	.	.	.	.	1	.	.	2
Origanum vulgare	f	.	.	.	.	.	.	.	.	.	1	.	.	.
Phleum pratense	c	.	.	.	3	.	.	.	.	.	.	.	.	8
Physalis alkekengi	s	.	.	.	1	.	.	.	.	.	.	.	.	.
Poa annua type	c	.	.	.	.	.	.	.	1	.	.	.	.	.
Poa nemoralis type	c	.	.	.	.	.	.	.	.	.	2	.	.	.
Polycnemum arvense	s	.	.	.	2	.	.	.	.	.	.	.	.	1
Polygonum aviculare	f	.	.	.	.	.	.	.	.	.	2	.	.	.
Polygonum lapathifolium/minus/persicaria	f	2	2	11	6	5	.	.	.	.	.	.	.	10
Polygonum sp.	f	.	.	.	2	.	.	.	.	.	.	.	.	1
Prunus/Cerasus	f	.	.	.	1	.	.	.	.	.	.	.	.	.
Rumex acetosella	f	.	.	.	.	.	.	.	.	.	.	.	.	1
Rumex sp.	f	.	.	3	.	.	.	.	.	.	.	.	.	1
Schoenoplectus tabernaemontani	f	.	.	.	.	1	.	.	.	.	2	.	.	.
Scleranthus sp.	f	.	.	.	.	.	.	.	.	.	.	.	.	3
Setaria viridis/verticillata	c	.	.	1	3	.	1	.	.	.	3	1	.	1
Silene sp.	s	.	1	1	5	.	.	1	.	.	.	.	.	.
Solanum nigrum	s	.	.	1	1	.	.	.	.	.	.	.	.	.
Stipa pennata s.l.	c	.	.	.	.	.	.	6	20	.	4+3*	2	.	.
	a	1	.	.	.	.	.	519	696	122	761	504	.	.
Trifolieae indet.	s	1	.	.	1	.	.	.	.	1	2	.	.	9
cf. *Vaccinium vitis-idaea*	f	.	1	.	.	.	.	.	.	.	.	.	.	.
Verbascum sp.	s	.	.	.	1	.	.	.	.	.	2	.	.	.
cf. *Veronica / Plantago*	s	.	.	.	.	.	.	.	3	.	.	.	.	2
Vicia cf. *tetrasperma*	s	.	.	.	.	.	.	.	.	.	1	.	.	.
Boraginaceae type	f	.	.	.	1	.	.	.	.	.	.	.	.	1
Caryophyllaceae indet.	s	.	.	.	1	.	.	.	.	1	.	.	.	.
Chenopodiaceae indet.	s	.	.	.	.	.	.	.	.	.	1	.	.	.
Cyperaceae indet.	f	.	.	.	.	.	.	.	.	.	3	.	.	1
Lamiaceae indet.	f	.	.	.	.	.	.	1	.	.	1	.	.	1
Poaceae indet. (small)	c	1	1	1	.	.	.	.	1	4	7	3	.	27
Poaceae indet. (large)	c	4	.	2	10	2	.	6	.	14	10+5*	1	1	11
Ranunculaceae indet.	f	.	.	.	.	.	.	.	.	.	.	.	1	.
Rubiaceae indet.	f	.	.	1	.	.	.	.	.	.	.	.	.	.
Solanaceae indet.	s	.	.	.	1	.	.	.	.	.	.	.	.	.
no. of features		2	2	3	2	2	1	13	3	7	12	15	2	3
no. of samples		2	2*	3	22	6	1	20	6	7*	26*	24*	3	12
volume of soil [dm^3]		16	16*	49	189	56	8	20	62	64*	130*	80*	89	92

Table 18.1, cont.

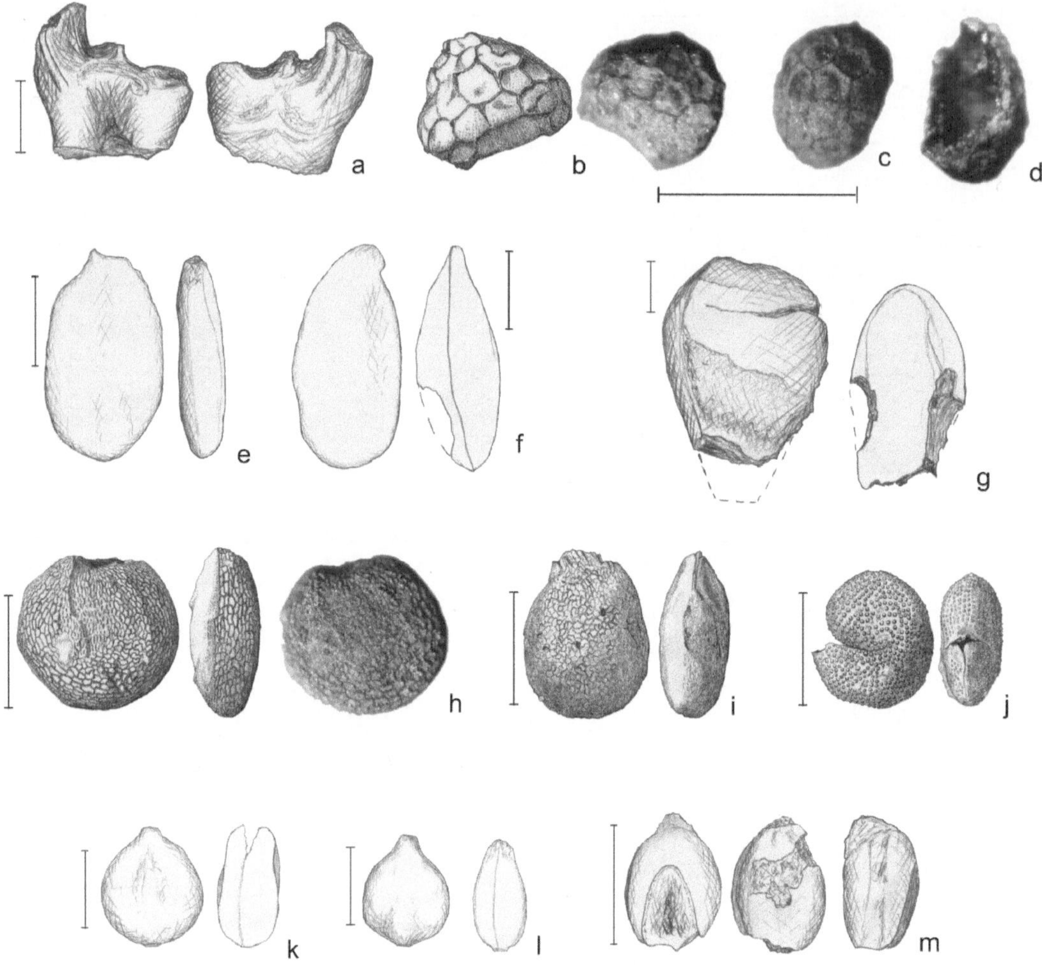

Figure 18.2: Charred plant macroremains. a. *Triticum* "new" type, spikelet base; b-d. *Papaver somniferum*; e, f. *Linum usitatissimum*; g. cf. *Malus*; h. *Physalis alkekengi*; i. *Solanum nigrum*, j. *Polycnemum arvense*; k, l. *Polygonum lapathifolium* s.l./ *minus*/*persicaria*; m. *Setaria viridis*/*verticillata*. Scale bars: 1 mm; c and d are dated to the Funnel Beaker culture; others dated to the LBK. Items b and h–j drawn by J.W. Wieser.

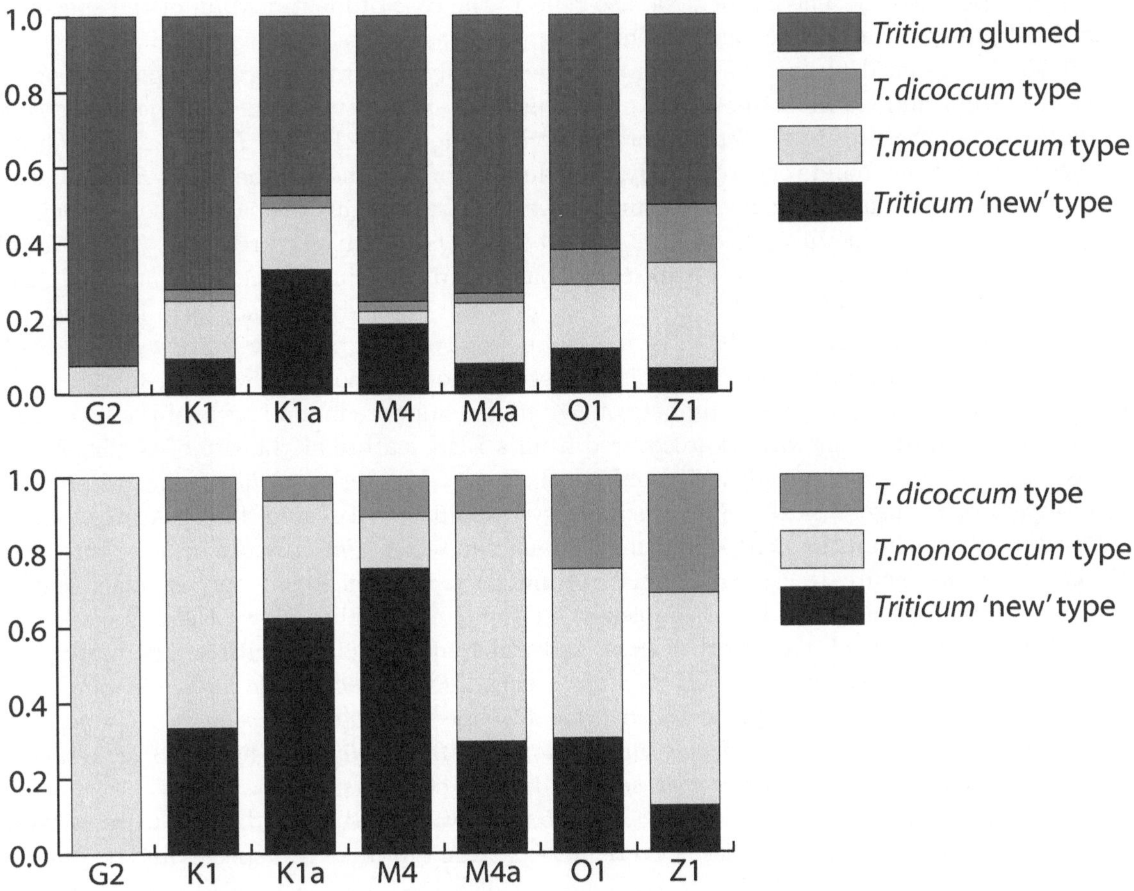

Figure 18.3: Lengyel culture wheat chaff composition. Upper: total number of glume bases; lower: identified glume bases only. Spikelet bases are counted as two glume bases in both graphs. See figure 1 for definition of site abbreviations.

Panicum miliaceum) and rye (cf. *Secale cereale*) in the LBK were uncertain because of their poor preservation.

In addition to the cereals, peas (cf. *Pisum sativum*, present only as fragments and without testa) were probably also cultivated by neolithic settlers in the vicinity of Brześć Kujawski. Flax (*Linum usitatissimum*) (figure 18.2e, 18.2f) was found only in the LBK material. Poppy seeds (*Papaver somniferum*) (figure 18.2b–d) have been recorded at two sites (Smólsk 4 and Wolica Nowa 1). The find of this taxon in the feature dated to the Linear Pottery culture (Smólsk site 4) is very important because such early evidence has been recorded only in the west of Poland (Bakels 1992; Knörzer 1974, 1980, 1998; Kreuz 1990; Wasylikowa et al. 1991, table 7). It should be noted that poppy found at Zesławice was incorrectly reported by Zohary and Hopf (2000, after Giżbert 1960b) as being dated to the LBK; the site is dated to the Radial Pottery culture (3500–2900 cal BC) (Giżbert 1960a, p.350).

18.3.2 Wild plants

The composition of wild plants (table 18.1) differs slightly at each site and in each culture. This is in part connected with the type of the archaeobotanical material that comes from different archaeological features, but may also be due to the fact that the numbers of samples taken from the sites varied greatly. The list of taxa also reflects the ease of identification of the specimens; some well-preserved and were easily identifiable, whereas others were damaged and therefore could not be assigned to species or even to genus/family. Nevertheless, some plants occurred in each culture and either at most (e.g., *Chenopodium album* type and *Fallopia convolvulus*) or only some of the sites (e.g., *Setaria viridis/verticillata*, figure 18.2m, *Trifolium* sp.). Grasses (Poaceae) have been found very frequently, including grains assigned to the genus *Bromus*, which were identified only in the Lengyel culture, however, larger grains probably also belonging to this genus were present in each culture. *Echinochloa crus-galli* was preserved in four of the five LBK sites, while *Hierochloe* cf. *australis* was found only in the Lengyel material (some damaged specimens similar to this grass, listed as small-grained Poaceae, have been recorded in the material from other periods). The most numerous of the grasses was *Stipa pennata* (s.l.), represented mainly by awn fragments. This species was closely associated with the features dated to the Lengyel culture (very high numbers of awn fragments were found in five of the seven sites) but a single awn fragment was also recovered from a LBK feature at the site of Guźlin 2.

Galium species have been identified only in the Lengyel material. *Polygonum lapathifolium/minus/persicaria* (figure 18.2k, 18.2l) was preserved in all the LBK and TRB features, but was completely absent from the Lengyel culture. Solanaceae seeds (*Physalis alkekengi*, figure 18.2h, *Solanum nigrum*, figure 18.2i, and Solanaceae indet.) have been found only in LBK material from two sites, Wolica Nowa 1 and Smólsk 4. At the site of Wolica Nowa there is evidence of several plants that are absent at other sites, and which are associated with sandy, more or less acid soils (e.g., *Phleum pratense*, *Polycnemum arvense*, *Rumex acetosella* and *Scleranthus* sp.). Many of the taxa listed are represented by very few specimens (e.g., *Avena* sp., *Corylus avellana*, *Juncus* sp., *Lamium amplexicaule/purpureum*, cf. *Malus sylvestris*, figure 18.2g, *Origanum vulgare*, *Physalis alkekengi*, *Rumex acetosella* and *Vicia* cf. *tetrasperma*).

One seed of *Physalis alkekengi* (1.4×1.3×0.55 mm) was found in the LBK feature at Wolica Nowa site 1. The seed-coat has a distinct reticulate pattern typical of this species and is composed of bigger cells than those of *Solanum dulcamara*, which has a very similarly sculptured testa. The cells are unstructured in the centre of the seed but they form very regular rows at the margins; these were clearly visible in the charred specimen (figure 18.2h). This is the first find of *Physalis alkekengi* in the Neolithic of the present-day territory of Poland. It was quite common in archaeobotanical material dated to the late Neolithic (3800–3600 cal BC) from Lake Zurich, where it was found in densities of more than 10 specimens per 1 dm^3 of sediment (Brombacher and Jacomet 1997, p.280, table 43).

18.3.3 Comparison of the different cultures

For the purposes of comparison the plant remains have been divided into large groups, for example, glume bases (i.e., one spikelet base is counted as two glume bases), cereal grains, seeds of other cultivated plants, grains of wild grasses, *Chenopodium* seeds, *Fallopia convolvulus* fruits, *Polygonum* spp. fruits and seeds of other wild plants (awn fragments have been omitted from the calculations). In figure 18.4 the number of remains in each of the groups is represented as a percentage of the total found in each culture. The first graph (figure 18.4a) includes chaff remains, fruits and seeds, but for clarity, in the second graph chaff remains have been excluded (figure 18.4b). The material dated to the Lengyel culture is dominated by cereals (chaff and grains), however, the high values represented in figure 18.4 are more a reflection of the rich contents of a single storage pit discovered at the Osłonki site, and so in the third graph the remains from this feature have been omitted from the totals (figure 18.4c).

Wheat chaff is the most abundant plant type in each of the cultures, but in the Lengyel material it represents almost 80 % of the total number of the remains (figure 18.4a). Comparison of only fruits and grains/seeds without chaff clearly shows that the percentage of cereal grains in the Lengyel culture is far higher than in the other cultures (figure 18.4b, 18.4c), and of note

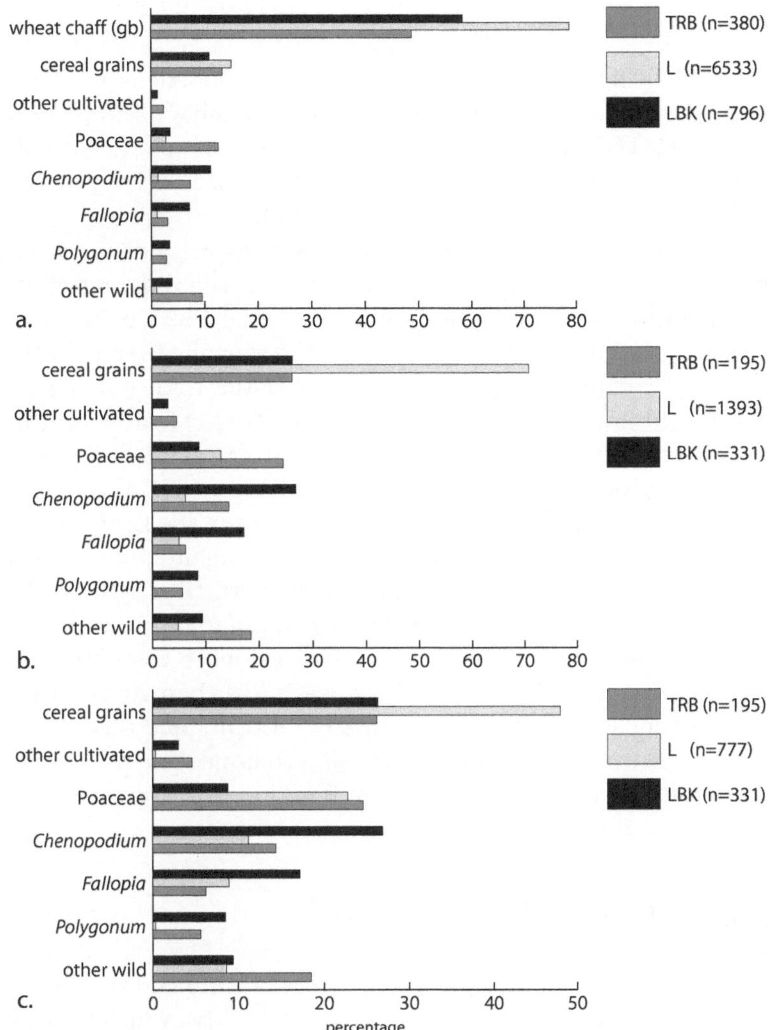

Figure 18.4: Composition of plant macro-remains (excluding charcoal and awn fragments) in the studied cultures. Graph a: including chaff remains; b: only fruits and seeds, c: excluding remains from the storage pit from Osłonki. Data are given as a percentage of total number of specimens found in features dated to each culture.

also for the Lengyel material is the near absence of other cultivated plants (with the exception of two peas). The proportions of wild grasses increase with time from less than 10 % in LBK to 25 % in TRB. *Chenopodium* seeds and *Fallopia convolvulus* fruits are abundant in all the studied material, but they are significantly more numerous in LBK. Fruits of the genus *Polygonum* are present in very low numbers in the Lengyel culture (only two fruits of *Polygonum aviculare* have been found) and are more numerous in the LBK (ca. 8%) and TRB (ca. 6%).

18.3.4 Comparison of the sites

The graphs presented in figure 18.5 are based on the proportions of the different groups of plant types in the separate sites, ordered by culture, to enable a more detailed comparison of taxonomic composition. The charts include two separate bars for the site of Osłonki, one with the data from the storage pit included in the calculations of relative proportions and a second with these excluded from the calculations (marked by an asterisk).

The figure 18.5a graph shows that chaff comprises a majority of the remains in the Lengyel sites (always more that 50%). In the LBK material from the different sites the proportion of chaff varies greatly, from ca. 2% at the Zagajewice site 1 to ca. 82% at the Miechowice site 4, but at four of the five sites chaff forms less than 50% of the remains. In the TRB material from the Wolica Nowa site chaff comprises approximately half of the total number of the remains. The comparison of fruits and grains/seeds (after having omitted cereal chaff from the calculations, figure 18.5b) shows very different proportions of cereal grains at each site, with no apparent trend according to the cultures, but in the LBK and TRB material these never reach more than 40%. Cereal grains constitute more than 65% of the remains in the Lengyel material at three sites (Guźlin 2, Miechowice 4 and Osłonki 1). Other cultivated plants, i.e., *Pisum sativum* (present in all three cultures), *Papaver somniferum* (present in both the LBK and TRB cultures), and *Linum usitatissimum* (only identified in LBK features), are found in very low proportions at all sites. Wild grasses and cereals together usually comprise more than 50% of the remains (with the exception of LBK Miechowice 4 and Smólsk 4, each less than 20%); in the Lengyel material from four of the sites they constitute more than 80% of fruits and seeds (Guźlin 2, Miechowice 4, Osłonki 1 and Zagajewice 1). Fruits of *Polygonum* are noticeable in every LBK and TRB bar. *Chenopodium* and *Fallopia convolvulus* have been found at every site, with the exception of the Lengyel material from Zagajewice.

18.4 Discussion and final remarks

The archaeological evidence dated to the Linear Pottery culture and the Funnel Beaker culture is rather poor, especially when compared with Lengyel settlements, which have many archaeological features including houses and even a fortification ditch (Osłonki 1, Grygiel and Bogucki 1997). Nevertheless, it has been possible to gain some information about the earliest plant husbandry in the region.

Evidence in the form of well-preserved cereal chaff and grains is proof that the cultivation of wheat was common from the beginning of the Neolithic in the Kujawy region. The composition of the wheat remains is very similar at each of the sites studied: the neolithic settlers cultivated mainly einkorn (*Triticum monococcum*), the 'new' type of glume wheat and emmer (*Triticum dicoccum*), probably together as a mixed crop. If we assume that the 'new' wheat type is equivalent to *Triticum timopheevi*, the mixture of crops would have been similar to Georgian 'zanduri', which consists mainly of *Triticum monococcum* and *Triticum timopheevi* (Hanelt and Institute of Plant Genetics and Crop Plant Research 2001, p.2567).

The only obvious difference in the overall composition of the cereals is the presence of bread wheat in the Lengyel material, but it is possible that the large numbers of samples taken from

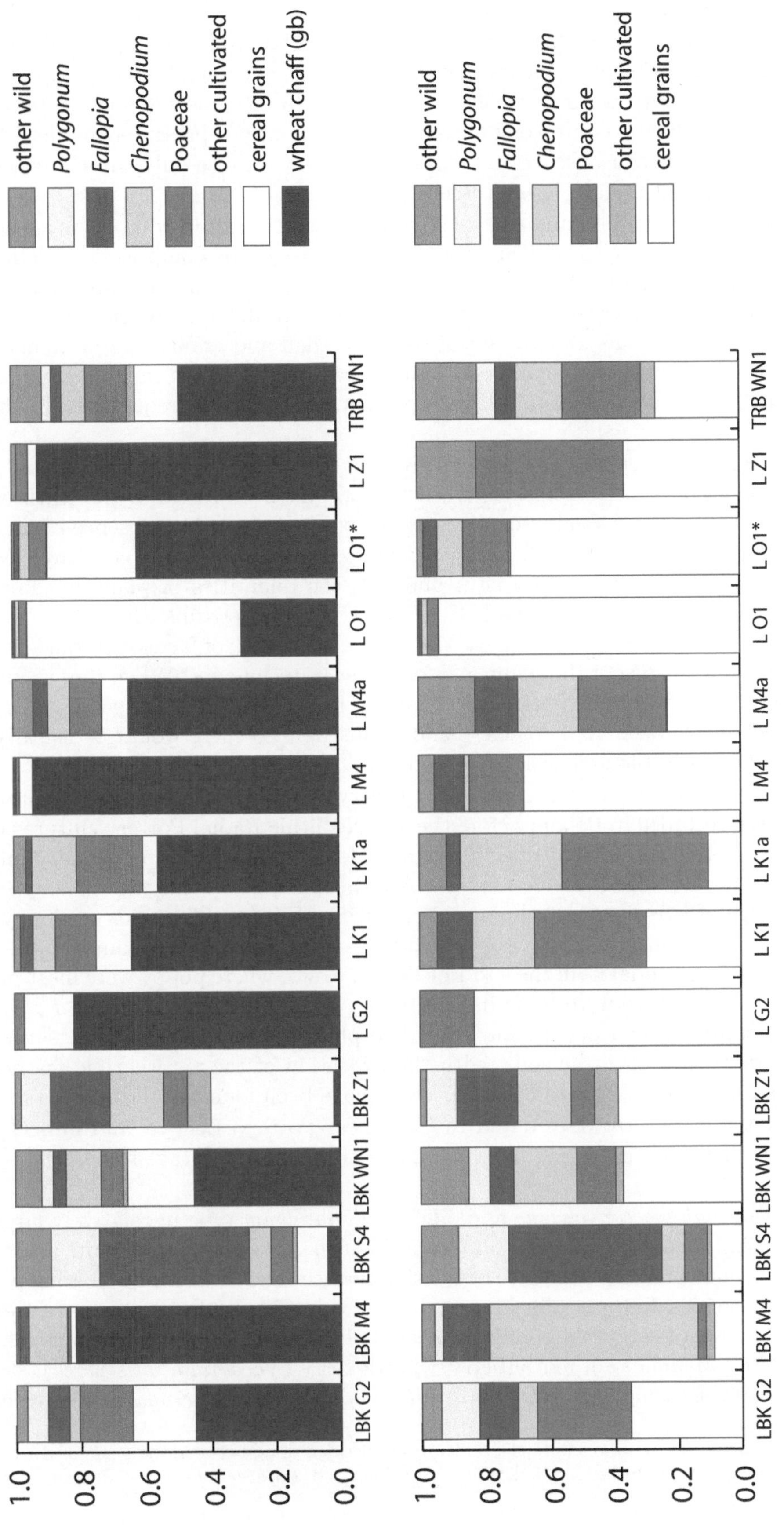

Figure 18.5: Composition of plant macro-remains (excluding charcoal and awns) at separate sites. Graph a (upper) fruits and seeds including chaff remains; graph b (lower) fruits and seeds only.

this period meant that there was better overall representation of taxa than in the other cultures. The lack of free-threshing cereal rachis (naked wheats and barley) could be accounted for by their under-representation in archaeobotanical material, as a result of removal during crop processing and/or due to the fact that they are less likely to survive (in recognisable form) after charring (Boardman and Jones 1990). Grains of hulled barley are poorly represented in the LBK and the Lengyel material and their absence in the TRB could be because only three archaeological features dated to this culture have been studied (table 18.1).

When we compare only the fruit and seed composition, misinterpretation is quite likely. In the LBK, at the Miechowice 4 and Smólsk 1 sites, cereal grains comprise less than 10% of the assemblage, which suggests that crop cultivation was of little significance in LBK plant husbandry (figure 18.5b). However, the picture is completely different when we compare the presence of chaff remains (figure 18.5a). Cereal remains (chaff and grains) are prominent in the Lengyel material if the sites are amalgamated (figure 18.4) and also when the individual sites are compared (figure 18.5). Cereal remains are also present, but in lower frequencies, in the other cultures.

Pea was cultivated by settlers of each culture but was always represented by only a few fragments of seeds. Flax is interesting because it has not yet been found in the features dated to the Lengyel and the Funnel Beaker cultures at the studied sites. The absence of flax in the Lengyel features could not be explained purely by chance because there was good archaeological evidence for the culture at the studied sites and also the quantities of plant remains of the same date were much greater than in the LBK (table 18.1). It is possible that it was no longer cultivated in that microregion after the LBK. Alternatively, the lack of flax seeds could have been caused by a change in the use of the plant and/or agrarian techniques. When flax is cultivated for its fibres rather than for its oil, the plants should be harvested before the seeds ripen (at the white seeds stage) (Kaznowski 1951, p.604)—and if this were so there would be less likelihood of finding charred seeds in the archaeological material.

The discovery of poppy in the LBK and TRB is very important for this area because it has not previously been recorded in Poland before the late Neolithic Radial Pottery culture (Giżbert 1960a). Poppy was taken into cultivation in western Mediterranean Europe (Zohary and Hopf 2000). It was not known in the Near East from where the 'package' of cultivated plants together with the knowledge of agricultural techniques spread into Europe. The finds of poppy in the Kujawy region possibly suggest contact of the settlers with the western groups of the Linear Pottery culture. In the material from the site 1 of Wolica Nowa, where poppy were most probably cultivated in the TRB,[1] there were some indicators of poor, sandy and acid soils *Polycnemum arvense*, *Rumex acetosella* and *Scleranthus* sp.). The plant material preserved at seven of the other sites is more typical of fertile soils, with the exception of the xerothermic feather grass (*Stipa pennata* s.l.). No typical weeds of winter crops have been found in the studied material.

The plant material from the sites dated to the Linear Pottery, Lengyel and Funnel Beaker cultures indicates that plant gathering must still have played an important role in the economy of the settlers.

The presence of *Fallopia convolvulus* and *Chenopodium album* type in relatively large numbers at each site (e.g., *Fallopia* comprises almost 50 % of the total number of fruits and seeds at Smólsk 4) could be interpreted as evidence of the significance that the role of gathering played in the subsistence economies. In the studied material dated to the Linear Pottery culture there were also some tentative identifications of familiar collected plants; for example, crab apple (cf. *Malus sylvestris*), mint (cf. *Mentha* sp.), red bilberry/whortleberry (*Vaccinium vitis-idaea*), sloe/wild cherry (*Prunus/Cerasus*), mallow (*Malva* sp.) and wild strawberry (*Fragaria/Potentilla*). No

[1]The doubt expressed in this case refers to the difficulty in distinguishing between the seeds of the cultivated species (*Papaver somniferum*) and the wild/weedy species/subspecies (*Papaver setigerum*/ *Papaver somniferum* subsp. *setigerum*).

traces of hazelnut shells *(Corylus avellana)* have been discovered in the material dated to the LBK and TRB and only one fragment has been found in the Lengyel material (table 18.1). The find of *Physalis alkekengi* (figure 18.2h), a species that is not a native of the flora of Poland (Mirek et. al. 2002), is the first in the Neolithic of Poland. Its berries are edible and it can also be used as a spice or for medicinal purposes (e.g., it is a diuretic and a remedy for rheumatism) (Podbielkowski 1989, p.226; Broda and Mowszowicz 2000, pp.563–564). Nowadays it is used mainly as a decorative plant in Poland.

Comparison of the wild plant taxa shows an increase in the proportions of grasses with time, indicating that large parts of the area were being deforested because of intensive exploitation for wood. In addition, the presence of *Stipa pennata* suggests that the area could have been partly covered by dry, steppe-like grasslands. Certain archaeological features, for example a well dated to the Stroke-Ornamented Ware culture at Konary I (Grygiel 2002), could indicate a lack of fresh water in the neighbourhood, possibly because of eutrophication and erosion of the surrounding area from deforestation (Gąsiorowski and Nalepka nd; Nalepka nd; Nowaczyk et al. 2002).

It could be concluded, therefore, that the first allochthonous farmers in the Kujawy region used similar cultivated plants as other LBK tribes. The exception is their use of poppy, which could be evidence of their contact with western groups. The presence of the 'new' type of glume wheat along with einkorn and emmer has been attested in the material dated to the Linear Pottery and Lengyel cultures as well as the Funnel Beaker culture. Glume wheats are the most common remains in all the studied material—einkorn, the 'new' type of glume wheat and emmer are always found together, suggesting that they were cultivated as a mixed crop. Bread wheat has only been found in the Lengyel features. Flax has been recorded in the Linear Pottery culture and its absence in later periods was probably due to the fact that it was no longer cultivated or that there were changes in the use of the plant (from its use for oil to exploitation for its fibres). Hulled barley and possibly pea were also cultivated by the settlers.

Many of the wild plants found at the sites are now common weeds; the presence of some (e.g., *Fallopia convolvulus* and *Chenopodium album* type) was significant, especially in LBK and TRB material, suggesting that their edible fruits and seeds could have been collected. Gathered plants are represented by a few specimens of *Corylus avellana*, cf. *Malus sylvestris*, cf. *Cerasus/Prunus*, *Fragaria vesca* and cf. *Vaccinium vitis-idaea*. In fact most of the wild plants found in the material potentially have some useful properties.

The Lengyel culture differs from the other cultures because of the dominance of crop taxa, very few other cultivated plants and an abundance of feather grass (*Stipa pennata* s.l.) remains. Despite the fact that feather grass could also have been collected (Bieniek 2002) the relatively high numbers of grains/awns during the Lengyel culture suggests that the area was strongly deforested by this date. The increase in presence of other wild grasses supports this suggestion.

There are some differences in the material from the sites studied; Wolica Nowa 1 (TRB) and Smólsk 4 (LBK) differ from the other sites by the presence of *Papaver somniferum*. Some of the plants that prefer sandy soils have been found only at the Wolica Nowa site (e.g., *Phleum pratense*, *Polycnemum arvense*, *Rumex acetosella* and *Scleranthus* sp.). Wolica Nowa was the only site in the studied micro-region where there was a TRB settlement.

Acknowledgements

I would like to thank my tutor Professor Krystyna Wasylikowa for her care and comments on the first draft of the paper. I am very grateful to Professor Stefanie Jacomet for the *Physalis alkekengi* identification. I am also obliged to thank Dr. Hab. R. Grygiel for archaeobotanical material and necessary archaeological data regarding chronology.

I should also like to thank the Foundation for Polish Science (FNP) for the metallurgical microscope NIKON Eclipse ME600. Much of this work has been done thanks to the State

Committee for Scientific Research grants 6 PO4F 041 15 and 6 PO4F 026 18 and the W. Szafer Institute of Botany of the Polish Academy of Sciences.

References

Bakels, C. (1992). Fruits and seeds from the Linearbandkerammik settlement at Meindling, Germany, with special reference to *Papaver somniferum*. *Analecta Praehistorica Leidensia 25*, 55–68.

Bieniek, A. (1999). Bread wheat (*Triticum aestivum* s.l.) and feather grass (*Stipa* sp.) in the early Neolithic in the Kujawy region [in Polish with English summary]. *Polish Botanical Studies, Guidebook Series 23*, 89–106.

Bieniek, A. (2002). Archaeobotanical analysis of some early neolithic settlements in the Kujawy region, central Poland, with potential plant gathering activities emphasised. In S. Jacomet, G. Jones, M. Charles, and F. Bittmann (Eds.), *Archaeology of Plants: Current Research in Archaeobotany. Proceedings of the 12th IWGP Symposium, Sheffield 2001*, Vegetation History and Archaeobotany 11, pp. 33–40. London: Springer.

Bieniek, A. (2003a). *Gospodarka rolna ludności kultur naddunajskich w Polsce świetle analizy szczątków roślinnych ze stanowisk archeologicznych na Kujawach [Danubian cultures and their plants found in the Kujawy region]*. Ph. D. thesis, Instytut Botaniki im. W. Szafera, Polska Akademia Nauk, Kraków, Poland.

Bieniek, A. (2003b). Small-seeded grasses from the early neolithic sites in Kujawy, central Poland [in Polish with English summary]. *Botanical Guidebooks 26*, 249–266.

Boardman, S. and G. Jones (1990). Experiments on the effects of charring on cereal plant components. *Journal of Archaeological Science 17*, 1–11.

Broda, B. and J. Mowszowicz (2000). *Przewodnik do oznaczania roślin leczniczych, trujących i użytkowych [Guidebook to Identification of Medical, Toxic and Useful Plants]*. Warszawa: Panstwowy Zaklad Wydawnictw Lekarskich.

Brombacher, C. and S. Jacomet (1997). Sammelwirttschaft: Die Nutzung wildwachsender pflanzlicher Ressourcen zur Gewinnung vor Nahrung und Rohstoffen. In J. Schibler, H. Hüster-Plogmann, S. Jacomet, C. Brombacher, E. Gross-Klee, and A. Rast-Eicher (Eds.), *Ökonomie und Ökologie neolitischer und bronzenzeitlicher Ufersiedlungen am Zürichsee*, pp. 277–285. Zürich: Monographien der Kantonsarchäologie Zürich 20.

Dobrzański, B., J. Siuta, M. Strzemski, T. Witek, and S. Zawadzki (Eds.) (1972). *Polska Mapa Gleb*. Warszawa: Komitet Gleboznawstwa i Chemii Rolnej PAN, Instytut Uprawy Nawożenia i Gleboznawstwa, Polskie Towarzystwo Gleboznawcze. Wydawnictwa Geologiczne.

Gąsiorowski, M. and D. Nalepka (n.d.). Reconstruction of paleoenvironment of fossil lake in Oslonki (Kujawy, Poland) based on cladoceran and pollen analyses [in Polish with English summary]. *Prace i Materiały Muzeum Archeologicznego i Etnograficznego w Łodzi, Seria Archeologiczna*. In Press.

Giżbert, W. (1960a). Nowe stanowisko kopalne maku (*Papaver somniferum* L.) na ziemiach polskich [Nouvelles stations fossiles de pavot (*Papaver somniferum* L.]. *Materiały Archeologiczne 2*, 349–354.

Giżbert, W. (1960b). Studium porównawcze nad ziarnami żyta kopalnego [A comparative study on excavated grains of rye]. *Archeologia Polski 5*(1), 6–90.

Grygiel, R. (1986). The household cluster as a representation of the fundamental social unit of the Brześć Kujawski Group of the Lengyel culture in Polish Lowlands. *Prace i Materiały Muzeum Archeologicznego i Etnograficznego w Łodzi, Seria Archeologiczna 31*, 41–334.

Grygiel, R. (2002). A well of the Stroke-Ornamented Ware culture from Konary near Brzesc Kujawski (Poland). *Archeologické Rozhledy 54*, 106–113.

Grygiel, R. and P. Bogucki (1997). Early farmers in north-central Europe: 1989-1994 excavations at Osłonki, Poland. *Journal of Field Archaeology 24*(2), 161–179.

Hanelt, P. and Institute of Plant Genetics and Crop Plant Research (Eds.) (2001). *Mansfeld's Encyclopedia of Agricultural and Horticultural Crops (Except Ornamentals)*, Volume 5. Berlin: Springer-Verlag.

Jażdżewski, K. (1938). Gräberfelder der bandkeramischen Kultur und die mit ihnen verbundenen Siedlungsspuren in Brześć Kujawski [in Polish with German summary]. *Wiadomości Archeologiczne 15*, 1–105.

Jones, G., S. Valamoti, and M. Charles (2000). Early crop diversity: a 'new' glume wheat from northern Greece. *Vegetation History and Archaeobotany 9*, 133–146.

Jones, M. (1991). Sampling in paleoethnobotany. In W. van Zeist, K. Wasylikowa, and K.-E. Behre (Eds.), *Progress in Old World Palaeoethnobotany: A Retrospective View on the Occassion of 20 Years of the International Work Group for Palaeoethnobotany*, pp. 53–62. Rotterdam: A. A. Balkema.

Kaznowski, L. (1951). Rośliny włókniste. In A. Listowski (Ed.), *Szczegółowa uprawa roślin*, pp. 589–623. Warszawa: Państwowe Wydawnictwa Rolnicze i Leśne.

Knörzer, K.-H. (1974). Bandkeramische Pflenzenfunde von Bedburg-Garsdorf, Kreis Bergheim/Erft. *Rheinische Ausgrabungen 15*, 173–192.

Knörzer, K.-H. (1980). Pflanzliche Großreste des Siedlungsplatzes Wanlo (Mönchengladbach), Naturwissenschafliche Beiträge zur Archäologie. *Archaeo-Physika 7*, 7–20.

Knörzer, K.-H. (1998). Botanische Untersuchungen am bandkeramischen Brunnen von Erkelenz-Kückhoven. In H. Koschik (Ed.), *Brunnen der Jungsteinzeit. Internationales Symposium Erkelenz 27. bis 29. Oktober 1997*, Volume 11 of *Materialien zur Bodendenkmalpflege im Rheinland*, pp. 229–246. Köln und Bonn: Landschaftsverband Rheinland. Rheinisches Amt für Bodendenkmalpflege.

Kohler-Schneider, M. (2003). Klima und Vegetation während des Endneolithikums im Raum Dunkelsteiner Wald – Östliches Alpenvorland. *Archäologie Österreichs 14*(2), 49–52.

Kondracki, J. (2000). *Geografia regionalna Polski.* Warszawa: Państwowe Wydawnictwo Naukowe.

Kreuz, A. (1990). *Die ersten Bauern Mitteleuropas: Eine archäobotanische Untersuchung zu Umwelt und Landwirtschaft der Ältesten Bandkeramik.* Analecta Praehistorica Leidensia 23. Leiden: Leiden University Press.

Matuszkiewicz, W., J. B. Faliński, A. S. Kostrowicki, J. M. Matuszkiewicz, R. Olaczek, and T. Wojterski (Eds.) (1995). *Potencjalna roślinność naturalna Polski. Mapa przeglądowa 1:300000. Arkusz 5: Pojezierze Wielkopolskie i Pojezierze Chełmińsko—Dobrzyńskie.* Warszawa: Instytut Geografii i Przestrzennego Zagospodarowania PAN.

Midgley, M. (1992). *TRB Culture. The First Farmers of the North European Plain.* Edinburgh: Edinburgh University Press.

Mirek, Z., A. Piękoś-Mirkowa, A. Zając, and M. Zając (2002). *Flowering Plants and Pteridophytes of Poland. A Checklist.* Biodiversity of Poland 1. Kraków: W. Szafer Institute of Botany, Polish Academy of Sciences.

Nalepka, D. (n.d.). Vegetation and its changes in the neighbourhood of archaeological site at Osłonki (Kujawy region) in the light of pollen analysis of sediments from a small mire [in Polish with English summary]. *Prace i Materiały Muzeum Archeologicznego i Etnograficznego w Łodzi, Seria Archeologiczna.* In Press.

Nalepka, D., K. Wasylikowa, Z. Tomczyńska, and A. Bieniek (1998). The vegetation of the Kuyavia region (central Poland) and the use of plants during the Lengyel culture settlement: a preliminary report [in Polish with English summary]. *Prace i Materiały Muzeum Archeologicznego i Etnograficznego w Łodzi, Seria Archeologiczna 39*, 139–174.

Nowaczyk, B., D. Nalepka, and I. Okuniewska-Nowaczyk (2002). The role of prehistoric man in the formation and deposits on selected areas of the Wielkopolska-Kujawy Lowland [in Polish with English summary]. *Geographia. Studia et Dissertationes 25*, 34–60.

Nowak, M. (2001). The second phase of neolithization in east-central Europe. *Antiquity 75*, 582–592.

Podbielkowski, Z. (1989). *Słownik roślin użytkowych [Dictionary of Usable Plants].* Warszawa: Państwowe Wydawnictwa Rolnicze i Leśne.

Ralska-Jasieiczowa, M. and L. Starkel (1999). Zmiany klimatu i stosunków wodnych w holocenie. In L. Starkel (Ed.), *Geografia Polski, Środowisko Przyrodnicze*, pp. 175–180. Warszawa: Państwowe Wydawnictwa Naukowe.

Starkel, L. (1995). Reconstruction of hydrological changes between 7000 and 3000 BP in the upper and middle Vistula River Basin, Poland. *The Holocene 5*(1), 34–42.

Szafer, W. (1972). Szata roślinna Polski niżowej. In W. Szafer and K. Zarzycki (Eds.), *Szata roślinna Polski Niżowej*, pp. 476–478. Warszawa: Państwowe Wydawnictwo Naukowe.

Wasylikowa, K., M. Cârciumaru, E. Hajnalová, B. P. Hartyányi, G. Pashkevich, and Z. V. Yanushevich (1991). East-Central Europe. In W. van Zeist, K. Wasylikowa, and K.-E. Behre (Eds.), *Progress in Old World Palaeoethnobotany: A Retrospective View on the Occasion of 20 Years of the International Work Group for Palaeoethnobotany*, pp. 207–239. Rotterdam: A. A. Balkema.

Zohary, D. and M. Hopf (2000). *Domestication of Plants in the Old World: The Origin and Spread of Cultivated Plants in West Asia, Europe and the Nile Valley* (3rd ed.). Oxford: Oxford University Press.

Chapter 19

Nature or culture? Cereal crops raised by neolithic farmers on Dutch loess soils

Corrie Bakels
University of Leiden, The Netherlands

19.1 Introduction

There is a long-standing debate about whether or not the lives of people, and especially those who lived in prehistoric or early historic times, were substantially influenced by their environment. Some researchers, for instance, consider that neolithic farmers were continually struggling with nature; impoverished soils, encroaching weeds and worsening climates were their lot. To what extent did 'nature' control the activities of these farmers? The answers may be as diverse as the regions farmed and probably there is no single answer. I will content myself on this occasion with considering a specific aspect of the economy: the cereals grown during the early and middle Neolithic in a very restricted area, i.e., the loess area in the southeastern part of the Netherlands. This area measures approximately 25 km by 25 km and lies between the towns of Sittard, Maastricht and Heerlen (figure 19.1).

19.2 The first farmers

The first farmers belonged to the Linearbandkeramik culture, which in the area under consideration has been divided into two phases, Phases I and II. Phase I lasted from 5300 cal BC to 5100 cal BC, and Phase II from 5100 cal BC to 4900 cal BC. After this period of time, the farmers seem to have left the area, but in the regions immediately to the east the practice of growing crops continued. Farmers moved into the area again around 4650 cal BC. They belonged to the Rössen culture, a tradition with strong roots in the Linearbandkeramik. This study ends with their successors, who were the bearers of the Michelsberg culture, which lasted from 4400–3800 cal BC.

Eighteen sites in total have been investigated archaeobotanically: three sites of the Linearbandkeramik I, 11 of the Linearbandkeramik II, one Rössen, Netherlands Rössen site and three belonging to the Michelsberg culture (Bakels 2003). All the plant remains were charred, with cereals being the dominant crop plants. One would have expected there to have been continuity, or at most a decline or rise in the relative importance of certain species, because during the entire period there was no substantial influx of new people. The core of the rural population

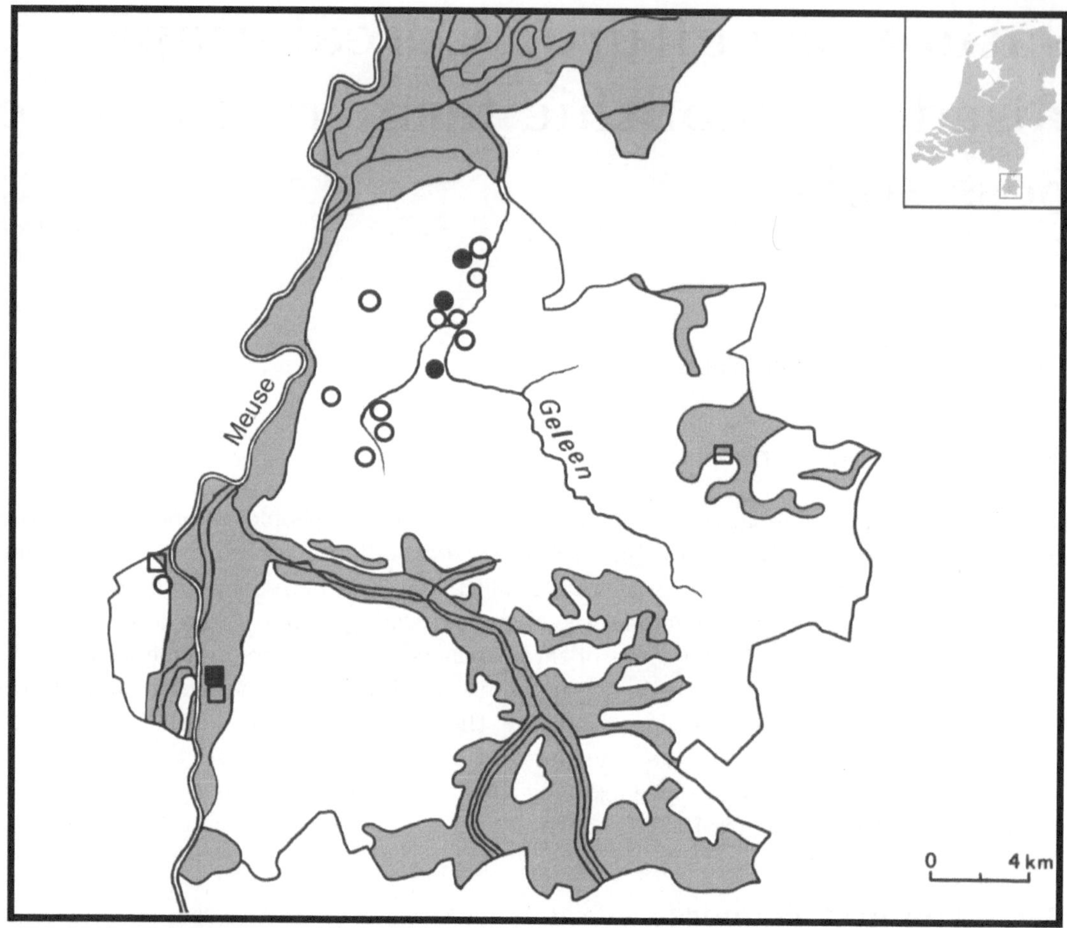

Figure 19.1: The southeastern part of the Netherlands with the location of the sites. Black dots: Linearbandkeramik culture I; open circles: Linearbandkeramik culture II; black square: Rössen culture; open squares: Michelsberg culture. Within the boundaries of the Netherlands loess soils are indicated in white, other soil types in grey.

Cereal taxa	LBK I (3)	LBK II (11)	Rössen (1)	Michelsb. (3)
T. dicoccum	D	D	P	P
T. monococcum	D	D	P	s
Triticum hexaploid naked	.	.	D	.
Triticum tetraploid naked	.	.	.	D
Hordeum vulgare var. *nudum*	.	s	D	P

Table 19.1: Cereal taxa documented on LBK I and II, Rössen and Michelsberg sites. Number of sites noted in brackets. D = dominant, P = present, s = single grains.

remained the same; its material culture was transformed gradually due to internal processes and some influence of ideas adopted from outside the region. Moreover, farmers have the reputation of being slow to adopt changes. It is a surprise, therefore, that the expected continuity is absent, at least between the Linearbandkeramik and Rössen cultures, and between the Rössen and Michelsberg cultures (table 19.1).

The cereal crop of the Linearbandkeramik culture (phases I and II) consists of a mixture of emmer (*Triticum dicoccum*) and einkorn (*Triticum monococcum*). A third species and potential cereal represented at this time is naked barley (*Hordeum vulgare* var. *nudum*); at one site two grains were found, and at a second just a single grain was identified. It would seem likely that the barley was not grown intentionally, but was a weed in the wheat fields. At the Rössen site, however, there was a dominance of a naked wheat with compact kernels, and naked barley; emmer and einkorn were present but are less abundant. Rachis remains of the naked wheat were characteristic of a hexaploid species, presumably in this instance a compact variety of *Triticum aestivum*. Only one site dating to this period has been investigated, but the contrast with the 14 Linearbandkeramik sites is remarkable. A similar trend has been recognised on the Rössen sites of neighbouring Germany (Knörzer and Gerlach in Knörzer et al. 1999, p. 80) where naked barley is also one of the dominant crops. According to U. Körber-Grohne (1988, p.30) naked wheat as a main crop does not occur in Central Europe before the Rössen period. A very small number of Linearbandkeramik sites have revealed evidence of naked wheat, but it seems never to have been a 'true' crop. The Michelsberg sites differ again; at these naked wheat with compact grains predominates, but the rachis remains of this wheat are characteristic of a tetraploid species (Bakels 2003), possibly *Triticum turgidum* or *Triticum durum*. Emmer and naked barley are also present, but einkorn has become scarce, if indeed the einkorn grains have been correctly assigned and are not emmer grains from one-grained spikelets.

What are the reasons for these differences in the choice of crops? The most obvious explanations that spring to mind are changes in the location of the fields or the properties of the soil. Changes in climate may also have influenced which crops were grown. To consider the latter, the period under review falls within the Atlantic optimum, when the climate is known to have been rather stable. Several small disturbances are known, the Piora oscillation in the final phase of the Atlantic being the most important, but these are not considered to have been severe enough to trigger a change in crops. The alternative explanation relating to location and/or soil may, therefore, be appropriate for the periods in question.

The location of the Rössen site does indeed differ from those of the Linearbandkeramik sites (figure 19.1). The site is situated on the lower terrace of the river Meuse, not strictly within the area of the loess soils. The terrain forms a higher part of the terrace and is covered with a river loam, which is mainly derived from loess and has similar properties. As far as could be ascertained, it was never flooded during the Rössen occupation. In contrast, the Rössen sites in nearby Germany are situated on loess soils, just as their Linearbankeramik predecessors were.

Weed taxa	LBK I (3)	LBK II (11)	Rössen (1)	Michelsb. (2)
Bromus secalinus type	100	91	.	.
Chenopodium album	100	100	100	50
Fallopia convolvulus	67	100	100	50
Phleum sp.	67	73	100	.
Echinochloa crus-galli	67	27	.	.
Persicaria lapathifolia	67	9	100	.
Lapsana communis	33	82	100	50
Bromus sterilis/tectorum	33	55	100	50
Vicia hirsuta/tetrasperma	33	64	100	50
Galium aparine	33	36	.	50
Rumex sp.	33	55	.	50
Persicaria maculosa	33	46	.	.
Chenopodium polyspermum	33	18	.	.
Galium spurium	33	18	.	.
Setaria verticillata/viridis	33	27	.	.
Silene sp.	.	27	.	50
Galium cruciata	.	18	.	.
Rumex cf. *sanguineus*	.	18	100	.
Stachys arvensis/sylvatica	.	18	.	.

Table 19.2: The percentage presence of field weeds occurring more than once in LBK I and II, and Rössen and Michelsberg sites. Number of sites used for calculations noted in brackets.

We should not draw the conclusion, therefore, that the farmers of the Rössen culture chose a completely different location in which to live and grow their crops. One of the three Michelsberg sites is in exactly the same position as the Rössen site, while the other two lie on loess, and one of these, to the west of the river Meuse, is situated on the same terrain as a Linearbandkeramik site. Location, therefore, seems not to provide an adequate reason for changes in the crops.

Differences in the properties of the soil may be the factors underlying the impetus for the changes. A.J. Kalis, J. Meurers-Balke and J. Richter (Richter 1997, pp.8, 35) attribute the disappearance of the Linearbandkeramik in the Netherlands and a contemporaneous, temporary decline in the population in adjacent Germany to over-population, failure of crops and overall crisis at the end of the culture. The population decline can be seen in pollen diagrams: a rise in the pollen percentages of hazel, ash, oak and lime points to a regeneration of the forests, following on the abandonment of settled areas and fields (Richter 1997, p.35). Over-exploitation of the arable land and the resultant exhaustion of the soil may indeed be one aspect of the crisis, but it remains questionable as to whether this is the cause of a change in crops in the succeeding cultures. The locations of the later settlements are not quite the same and the soils surrounding them may not have become depleted. There is no continuous occupation in exactly the same location; where two sites are on the same terrain, i.e., the Linearbandkeramik and Michelsberg sites west of the Meuse, the time difference between the cultures is about 1,000 years. A comparison of the arable weeds associated with the cereals may elucidate to some extent the condition of the soils.

In table 19.2 the percentage presence of those taxa occurring more than once are given for the two phases of the Linearbandkeramik culture, and the Rössen and Michelsberg cultures. The frequencies are based on presence or absence in sites. For the Michelsberg only two sites are included, because the samples from the third site were processed such that they could not be compared in this way. The table is arranged according to the frequencies of taxa in the Linearbandkeramik I sites. The presence or absence of taxa ranked lower than *Persicaria maculosa* may be coincidental, but in the upper part of the list this is unlikely. For taxa higher in the list, the ranking remains more or less the same and the commonly occurring taxa remain the same for all four cultures. There are, however, three taxa that do not adhere to this general trend. The first is *Bromus secalinus*-type (which comprises *Bromus secalinus*, *Bromus hordeaceus* and *Bromus arvensis*); the absence of this taxon in a Rössen or Michelsberg context may be fortuitous, but another possibility is that it is not really a weed, but a Linearbandkeramik crop plant, as suggested by K.-H. Knörzer (Knörzer and Gerlach in Knörzer et al. 1999, p.76). If this were the case *Bromus secalinus*-type should not appear in the table. The other two exceptions are *Echinochloa crus-galli* and *Persicaria maculosa*, which are not suspected of being primitive crop plants. Both species require soils rich in nutrients; similarly the *Bromus* species referred to previously also prefer nutrient-rich soils. It follows that the absence of the three taxa from the younger fields may indicate an impoverishment of the soils. However, the evidence of the other species does not support this conclusion. *Chenopodium album* and *Lapsana communis*, indicators of richer soils, are present and in high frequencies across all four cultures. The sole indicator of poorer soil conditions, *Setaria viridis*, has not yet been detected after the Linearbandkeramik culture; however, there is a possibility that the seeds may belong to *Setaria verticillata*, which prefers richer soils. Therefore, on the basis of the evidence of the weed taxa, there are no obvious indications for changes in the condition of the soil.

If natural causes are not the reasons determining the choice of certain cereals, then cultural preference is the obvious alternative. Cereals today constitute a large part of the daily provisions, this was also the case in the past, and food is part of culture. The changes seen in the early and middle Neolithic of the southeastern part of the Netherlands should probably not be attributed to the environment. Which cereals were preferred was probably a cultural trait, as it is today. The changes may have been triggered by cultural influences from outside. Nature obviously imposes constraints, but crop plants are as much a part of culture as, for instance, pottery.

References

Bakels, C. (2003). Die neolithischen Weizenarten des südlimburgischen Lössgebiets in den Niederlanden. In J. Eckert, U. Eisenhauer, and A. Zimmermann (Eds.), *Archäologische Perspektiven. Analysen und Interpretationen im Wandel. Festschrift für Jens Lüning zum 65. Geburtstag*, Internationale Archäologie, Studia honoraria 20, pp. 225–232. Rahden, Westfalen: Marie Leidorf.

Knörzer, K.-H., R. Gerlach, J. Meurers-Balke, A. J. Kalis, U. Tegtmeier, W. D. Becker, and A. Jürgens (1999). *Pflanzenspuren. Archäobotanik im Rheinland: Agrarlandschaft und Nutzpflanzen im Wandel der Zeiten.* Köln: Rheinland-Verlag.

Körber-Grohne, U. (1988). *Nutzpflanzen in Deutschland. Kulturgeschichte und Biologie.* Stuttgart: Konrad Theiss Verlag.

Richter, J. (1997). *Neolithikum. Geschichtlicher Atlas der Rheinlande Beiheft II/2. 1 - II/2. 2.* Köln: Rheinland-Verlag.

Chapter 20

The plant remains from the Neolithic Funnel Beaker site of Wangels in Holsatia, northern Germany[1]

Helmut Kroll
Christian-Albrechts Universität zu Kiel, Germany

20.1 Introduction

Recent excavations at Wangels, on the border of the Oldenburg channel in Eastern Holsatia (54.17 N 10.52 E), unearthed rich archaeological deposits dating to the Mesolithic/early Neolithic and the middle Neolithic of the Funnel Beaker culture. Only the archaeobotanical material from the middle neolithic layer is presented here; the mesolithic layer had no cultivated plants. Two different types of samples were submitted for analysis: handpicked material from the wet-sieving process (i.e., for the recovery of bones, shells, microliths, sherds and beads), and unsieved deposits. Carbonised and uncarbonised (waterlogged) plant material was preserved in the middle neolithic layer. The handpicked samples comprised some specimens as small as cereal grains although in many cases only items as large as hazelnuts had been extracted. The finds from Wangels are perfectly preserved, not only the flint and ceramics but also the bones and plant remains. Wangels is not a settlement site, but a place visited at the water's edge, so that the cultural layers have become mixed as a result of trampling by people and domestic livestock. The continuously rising sea level of the Baltic has sealed the layer until now. For the first time, from the rich array of plant species represented at Wangels, it is possible to throw some light on the developed nature of the neolithic Funnel Beaker culture in northern Central Europe.

20.2 The plant remains

The preservation at the site is similar to that of the pile dwellings in Central Europe and so, for the first time, it has been possible to compare the plant taxa found on a Funnel Beaker site with those of the Alpine Neolithic pile dwelling sites. Until now, there has been little known about the Funnel Beaker economy, and evidence has been based only on very poorly preserved carbonised plant remains from dry mineral soils (Behre 1979; Behre and Kucan 1994). For example, at Rastorf, a megalithic site in Holsatia, which was excavated recently, there are only small numbers of grains of naked barley and bread wheat, and very few grains and spikelet bases

[1]This article is a slightly shortened English version of a article published in German by Kroll (2001).

	Sample no.											
Taxon	1	2	3	4	5	6	7	8	9	10	11	12
Hordeum vulgare var. *nudum*	6	.	2	1	2	2	1	1	1	1	.	.
Triticum dicoccum	.	.	.	1	.	.	.	1	.	.	.	1
T. dicoccum, spikelet base	.	.	1	.	2	.	1	.	.	.	.	1
T. aestivum/*T. durum*	2	.	.	1	.	2	.	.	.	.	2	.
Cerealia indet.	1	1	.	1	3	.	1	1	.	3	.	1
Corylus avellana	2	1	1	1	1	1	1	1	1	1	1	1
Tilia sp., unripe	.	.	3	.	.	1	2	.	1	1	.	2
Galium aparine	.	.	.	.	1	1	.	.	1	1	.	1
Rubus fruticosus	.	.	.	.	1	.	.	.	1	.	.	.
Vicia type	.	.	1	.	.	.	.	.	.	.	.	.
Lamiaceae	.	.	.	.	.	.	.	.	1	.	.	.
cf. *Daucus carota*	.	.	.	.	.	.	.	.	.	.	.	1
Total	11	2	8	5	10	7	6	4	6	7	3	8

Table 20.1: Numbers of carbonized plant remains from a megalithic grave and house context (Rastorf, Kreis Plön, Megalithic grave LA 6c, excavations 2000).

of emmer, together with remains of hazel, lime and bramble, but no other records of cultivated plants (table 20.1).

The number and diversity of finds at Wangels is much greater. The taxa identified in 12 samples and in the handpicked material from the middle neolithic layer are listed in table 20.2. As at Rastorf, naked barley, emmer and bread wheat are present, but there is no einkorn. Of note are the large numbers of glumes and rachis fragments of emmer and barley, indicating, therefore, that either a consumer site (settlement) or a threshing area is located nearby. The site is just in the vicinity of a permanent Funnel Beaker settlement on the dry shore.

Opium poppy (*Papaver somniferum*) is the only cultivated plant in the Neolithic of western European origin and the earliest finds are from sites in the lower Rhine. Until now neolithic opium poppy was associated with Linear Pottery sites and with sites on the loess soils in Central Europe (Bakels 1982; Knörzer et al. 1999). Opium poppy is absent from the records in many areas between the neolithic Linear Pottery period and the Middle Ages, however it is present on many Urnfield sites. Poppy seeds are very small and tough, and are often found in a carbonised and an uncarbonised state. Providing the meshes used for retrieval are small, poppy seeds will be recovered and, because the cell patterns on the seed coats are very characteristic, they are easily identifiable. The presence of poppy seeds at Wangels is an indication that their absence at other sites is due to poor preservation on most and methodological inadequacies on others. Opium poppy at Wangels was probably cultivated mainly for nutritional purposes but, in addition, possibly also for medicinal use. There were no finds of linseed (*Linum usitatissimum*) at Wangels. At most sites, linseed is a more valuable oil plant than opium poppy. And flax fibres were second only in importance to lime bast. If flax had been of economic significance at the site, there would have been some record of its presence in the form of seeds or capsule fragments.

The fruits and nuts are as one would expect to find at any time in Central Europe; for example, hazel, bramble, raspberry, wild strawberry, wild rose, sloe, and hawthorn. Guelder rose (*Viburnum opulus*) is also considered by some to be a useful plant, although others doubt its economic value. The numbers of seeds at Wangels are equivalent to the numbers found in most prehistoric contexts. The numbers of crab apple or cultivated apple pips (*Malus* sp.), however, are exceptional for this period and the quantities are more typical of those found on medieval

Taxon	Sample no.												
	1	9	7	2	11	3	10	5	6	12	8	4	hand
Cereals													
Triticum dicoccum carb.	2	1	1	.	3	1	1	2	1	4	1	.	54
T. dicoccum, spikelet base	102	516	691	259	658	16	75	54	92	121	136	.	.
T. dicoccum, spikelet base, carb.	56	73	50	54	52	4	61	15	4	27	14	.	.
Hordeum vulgare var. *nudum* carb	9	10	6	7	6	.	3	1	.	2	5	.	139
Hordeum sp., rachis, carb.	5	10	21	11	16	2	6	.	1	5	7	.	.
Hordeum sp., rachis	.	.	1	.	.	.	.	.	.	2	.	.	.
Triticum aestivum carb.	.	1	1	.	.	.	.	.	.	.	.	.	8
T. aestivum, rachis, carb.	3	.	1	.	1	.	1	.	.	.	.	.	.
Cerealia indet.	2	5	.	4	1	.	4	1	1	3	3	.	.
Oil plants													
Papaver somniferum	1	1	1	.	.	.	8	1	.	.	9	.	.
P. somniferum carb.	.	.	.	1	.	.	.	.	1	.	.	.	.
Fruits and nuts													
Corylus avellana	5	1	3	5	3	2	2	3	1	4	1	.	9
C. avellana carb.	.	.	.	.	1	.	.	.	.	.	.	.	.
Rubus idaeus	26	37	31	43	59	3	13	9	3	21	16	.	.
R. fruticosus	6	25	14	21	30	3	3	5	3	16	4	.	.
R. fruticosus carb.	.	.	1	.	.	.	.	.	.	.	.	.	.
Fragaria vesca	5	43	22	17	9	3	6	1	3	2	9	.	.
Malus sp.	3	19	13	13	15	.	9	3	.	15	4	.	111
Malus sp., fruit	.	.	.	.	.	.	.	.	.	.	.	.	1
Rosa sp.	.	5	1	.	1	.	1	1	.	.	.	.	3
Crataegus monogyna	.	.	.	1	.	1	.	.	.	.	.	.	39
Crataegus laevigata	.	.	.	.	.	.	.	1	.	.	.	.	.
Prunus sp.	.	1	.	.	.	.	.	.	.	.	.	.	.
Viburnum opulus	.	.	.	.	.	.	.	.	.	.	.	.	1

Table 20.2: Numbers of plant remains from the middle neolithic layer (Wangels, Kreis Ostholstein, site LA 505, excavation 1996). Uncarbonised seeds if not specified; carb: carbonised; hand: handpicked seeds.

	Sample no.												
Taxon	1	9	7	2	11	3	10	5	6	12	8	4	hand
Weeds													
Chenopodium album	107	232	132	208	294	62	153	42	76	137	181	.	2
Ch. album carb	1	1	1	.	.	1	.	.	.	.	.	.	.
Ch. glaucum/Ch. rubrum	74	288	35	96	69	245	240	52	78	149	263	1	.
Atriplex patula type	19	97	25	49	32	19	22	33	49	51	49	.	7
A. prostrata type	16	21	33	37	22	31	15	18	31	7	32	.	3
Polygonum convolvulus	6	15	7	.	6	2	7	3	4	11	4	.	2
Stellaria media	2	10	4	8	1	1	6	2	3	2	5	.	.
Polygonum aviculare	1	3	3	7	5	1	2	5	2	7	4	.	.
Urtica dioica	7	15	9	.	21	1	5	5	4	2	17	.	.
Viola sp.	3	24	5	5	2	2	2	.	3	5	2	.	.
Brassica nigra	6	2	4	7	2	1	9	.	2	7	2	.	.
Chenopodium ficifolium	2	1	6	3	4	1	12	.	1	1	2	.	.
Polygonum persicaria type	.	17	1	4	.	6	5	1	15	1	3	.	.
P. persicaria type carb.	.	.	.	.	.	1	.	.	.	.	.	.	.
Silene type	5	1	4	3	1	.	.	2	2	1	.	.	.
Rumex crispus type	.	2	1	4	3	.	1	3	.	2	1	.	.
Carex hirta	4	14	.	12	6	.	3	1	.	6	.	.	.
Cirsium sp.	2	3	.	2	4	6	.	.	2	3	.	.	.
Ranunculus repens	2	2	1	.	5	.	.	1	.	2	1	.	.
Plantago major	2	9	.	.	8	1	12	.	.	26	.	.	.
Descurainia sophia	.	.	1	2	.	.	12	2	.	9	9	.	.
D. sophia carb	.	.	.	.	.	.	.	.	.	.	1	.	.
Polygonum lapathifolium	.	1	2	.	2	.	2	.	1	.	1	.	.
Valerianella dentata	1	1	.	1	1	.	.	1	.	.	2	.	.
Phleum sp.	.	.	.	2	.	.	.	.	.	.	.	.	.
Phleum sp. carb.	7	.	2	4	1	.	.	1	.	.	.	.	.
Ranunculus acris type	.	.	1	2	.	1	1	.	6	.	.	.	.
Rumex acetosella	3	2	.	.	.	2	1	.	.	.	2	.	.
Bromus secalinus/arvensis	1	2	.	.	3	.	.	.	.	.	.	.	.
B. secalinus/arvensis carb.	.	.	.	.	.	.	.	.	1	1	.	.	1

Table 20.2, cont.

Taxon	Sample no. 1	9	7	2	11	3	10	5	6	12	8	4	hand
Weeds, cont.													
Carex vulpina type	1	2	.	2	.	.	.	.	.	2	.	.	.
Lapsana communis	3	1	2	.	1	.	.	.	.	.	.	.	.
Solanum nigrum	1	1	1	.	.	.	.	.	.	1	.	.	.
Poa sp.	.	.	.	1	1	16	.	.	.	.	.	.	.
Cerastium sp.	.	1	1	.	.	.	.	.	.	8	.	.	.
Sonchus asper	3	.	1	.	1	.	.	.	.	.	.	.	.
Hypericum perforatum	1	.	2	.	.	.	.	1	.	.	.	.	.
Galeopsis sp., large seeds	.	.	.	1	.	.	.	.	.	.	1	.	1
Agrostis type	.	.	.	.	.	.	1	1	.	.	.	.	.
Daucus carota	.	1	1	.	.	.	.	.	.	.	.	.	.
Leontodon autumnalis	.	1	.	.	.	1	.	.	.	.	.	.	.
Vicia type carb.	1	.	.	.	.	.	.	1	.	.	.	.	.
Galium aparine	.	.	.	.	.	.	.	.	.	.	.	.	2
Lolium sp., small seeds	.	.	.	.	.	.	.	.	.	2	.	.	.
Saponaria officinalis	.	.	2	.	.	.	.	.	.	.	.	.	.
Alliaria officinalis	.	.	.	.	.	1	.	.	.	.	.	.	.
Brassica sp., large seeds	.	.	.	.	.	.	.	.	.	1	.	.	.
Carduus sp.	.	.	.	.	.	.	.	.	1	.	.	.	.
Dianthus sp.	.	.	.	1	.	.	.	.	.	.	.	.	.
Myosotis sp.	.	.	.	1	.	.	.	.	.	.	.	.	.
Sonchus arvensis	.	.	.	.	.	.	1	.	.	.	.	.	.
Tilia sp.	.	.	.	.	.	.	.	.	.	1	.	.	.

Table 20.2, cont.

Taxon	Sample no. 1	9	7	2	11	3	10	5	6	12	8	4	hand
Wet-loving plants													
Characeae, Oogonium	54	1120	552	143	335	5508	1305	735	764	513	1212	170	.
Alnus glutinosa	7	14	6	8	7	44	12	11	17	4	5	.	50
A. glutinosa carb.	.	.	.	.	.	.	.	1	.	.	.	.	.
A. glutinosa, catkin	.	.	.	.	.	.	.	.	.	.	.	.	275
Zannichellia palustris	2	29	10	11	17	25	18	9	27	7	32	2	.
Schoenoplectus lacustris	11	18	9	7	53	3	12	2	10	17	14	.	14
Sch. lacustris carb.	1	.	.	.	.	.	.	.	.	.	.	.	.
Juncus bufonius	93	202	26	18	16	57	321	129	11	582	132	.	.
Mentha sp.	17	146	19	34	53	105	69	16	26	64	89	.	.
Potamogeton sp.	.	33	16	35	16	22	10	12	16	23	15	.	1
Potamogeton sp. carb.	.	.	.	.	.	1	.	.	.	.	.	.	.
Batrachium spp.	.	8	7	3	1	34	6	6	1	11	1	.	.
Montia type	3	1	4	1	8	.	1	1	2	.	.	.	.
Solanum dulcamara	2	4	2	3	3	.	3	1	.	1	.	.	.
Lythrum salicaria	.	16	.	1	.	32	1	1	2	.	25	.	.
Phragmites australis	1	.	.	1	.	21	.	1	.	.	1	.	.
Lycopus europaeus	1	.	1	3	.	.	.	1	2	.	.	.	.
Betula sp.	.	.	.	1	.	2	1	1	4	.	.	.	.
Polygonum hydropiper	.	1	2	.	.	.	.	.	.	1	2	.	.
Rumex hydrolapathum	1	.	.	3	.	.	.	.	1	1	.	.	.
Eupatorium cannabinum	.	.	.	.	.	2	1	1	1	.	.	.	.
Stellaria palustris type	2	.	4	.	.	.	.	.	1	.	.	.	.
Epilobium hirsutum	.	.	.	3	.	.	.	1	1	.	.	.	.
Juncus effusus type	1	.	.	.	.	16	.	.	.	.	.	.	.
Rorippa type	.	.	.	.	.	1	3	.	.	.	.	.	.
Eleocharis palustris	.	1	.	.	1	.	.	.	.	.	.	.	.
Typha sp.	1	.	.	.	.	.	1	.	.	.	.	.	.
Nuphar lutea	.	.	.	.	.	.	.	.	.	.	.	.	21
Iris pseudacorus	.	.	.	.	.	.	.	.	.	.	.	.	19
I. pseudacorus carb.	.	1	.	.	.	.	.	.	.	.	.	.	.
Lychnis flos-cuculi	.	.	.	.	.	.	.	.	1	.	.	.	.
Apium sp.	.	.	1	.	.	.	.	.	.	.	.	.	.
Carex disticha	.	.	.	.	.	1	.	.	.	.	.	.	.

Table 20.2, cont.

Taxon	Sample no.												
	1	9	7	2	11	3	10	5	6	12	8	4	hand
Wet-loving plants, cont.													
Ceratophyllum demersum	.	.	.	.	.	.	.	.	.	.	.	.	1
Potentilla anserina	.	.	.	.	.	1	.	.	.	.	.	.	.
Salix sp, fruit	.	.	.	.	.	.	.	.	.	.	.	.	1
Stachys palustris	.	.	.	.	1	.	.	.	.	.	.	.	.
Veronica sp., flat seeds	.	.	.	.	.	.	.	.	.	.	1	.	.
Saltmarsh plants													
Ruppia maritima	11	43	8	17	5	92	22	13	26	19	18	3	2
R. maritima carb	.	.	.	.	.	1	.	.	.	.	.	.	.
Najas marina	10	30	13	11	9	3	9	8	4	17	9	.	.
Juncus gerardii	.	34	4	2	16	192	48	14	3	.	57	2	.
Bolboschoenus maritimus	4	6	13	2	1	2	2	.	3	2	1	.	.
B. maritimus carb	.	.	.	.	1	.	.	.	.	.	.	.	.
Suaeda maritima	.	2	.	2	1	4	.	1	1	1	1	.	.
Aster tripolium	.	1	1	.	1	.	.	.	2	.	1	1	.
Glaux maritima	.	.	.	2	.	19	.	.	1	.	.	1	.
Samolus valerandi	4	2	.	2	1	.	.	.	.	.	.	.	.
Salicornia europaea	.	.	.	.	.	2	.	2	2	.	.	.	.
Spergularia salina	.	.	.	.	.	1	.	.	.	.	.	.	.
Unspecified													
Carex: *Eucarex* spp.	1	4	9	5	2	6	2	3	7	4	7	.	1
Poaceae	5	.	.	2	1	1	1	2	2	1	2	.	.
Poaceae carb	.	.	4	.	.	.	1	.	.	.	.	.	.
Carex: *Vignea* spp.	2	.	.	.	1	.	2	.	.	.	.	.	.
Brassicaceae	2	.	.	.	.	.	1	2	.	.	.	.	.
Brassicaceae, fruit	1	.	.	.	.	.	.	.	.	.	.	.	.
Asteraceae	2	.	.	5	.	.	.	.	.	.	.	.	.
Trifolium type	1	.	.	.	.	.	.	1	.	.	.	.	.
Polygonaceae	.	.	.	.	.	.	.	.	.	.	.	.	5
Lamiaceae	.	.	.	2	.	.	.	.	.	.	.	.	.
Chenopodiaceae	.	.	.	.	.	.	.	.	.	.	1	.	.
Geraniaceae	.	.	.	1	.	.	.	.	.	.	.	.	.
Rubiaceae	.	.	.	.	.	1	.	.	.	.	.	.	.
Sum of all taxa in table 20.2	746	3234	1856	1226	1900	6636	2557	1247	1329	1935	2417	180	773

Table 20.2, cont.

sites. This is very unusual for collected fruits, perhaps indicating that at this time the activities of humans (e.g., woodland clearance) and cattle had resulted in more space for the growth of light-loving species like apple or sloe. Cattle and game, however, love apple twigs and bark, and small trees are soon destroyed if animals are allowed easy access to them. The young trees have to be protected until they are strong enough to withstand cattle and so as to guarantee a good crop of fruit, and for this reason it is proposed that Wangels neolithic people had 'orchards'. These apples from Wangels connect the north to the south: the pile dwellers of the alpine region had already some large fruit sports (Schweingruber 1979). Prior to this, there were no records of carbonised apples in Holsatia earlier than the Bronze Age (Hopf 1973).

The weeds also show connections with Linear Pottery sites. Knörzer (1971) created the term *Bromo-Lapsanetum praehistoricum* to characterise the neolithic weed association of the Linearbandkeramik culture. At Wangels, *Bromus secalinus*, *Lapsana communis*, *Polygonum convolvulus*, *Valerianella dentata*, and *Solanum nigrum* have also been identified. Most 'modern' weeds are absent from the list of taxa. In most other sites of the Funnel Beaker culture we only have records of *Chenopodium* spp. and *Polygonum* spp.; these species are also common in the Wangels assemblage.

There are some species that add culture and colour, for example, *Brassica nigra* (Black mustard) is a cultivated or useful plant; *Descurainia sophia* is a well-known remedy for all purposes and its seeds contain oil like those of *Camelina*; *Hypericum perforatum* is also a widely used herb; and similarly *Alliaria officinalis Saponaria officinalis*, large-seeded *Brassica* sp., and *Daucus carota*. None of these plants are common in a wet and brackish environment and so it could be concluded that they were all deliberately collected for food or as medicines. It could also be concluded that shelter was given to the stands of these plants and as such this would have been the first step towards cultivation. The water plants and the marsh plants are typical of a wet, salty environment.

References

Bakels, C. (1982). Der Mohn, die Linearbandkerammik und das westliche Mittelmeergebiet. *Archäologisches Korrespondenzblatt 12*, 11–13.

Behre, K.-E. (1979). Die natürliche Umwelt der Trichterbecherkultur. In H. Schirning (Ed.), *Großsteingräber in Niedersachsen*, pp. 199–202. Hildesheim: Lax.

Behre, K.-E. and D. Kucan (1994). Die Geschichte der Kulturlandschaft und des Ackerbaus in der Siedlungskammer Flögeln, Niedersachsen, seit der Jungsteinzeit. *Probleme der Küstenforschung 21*, 1–227.

Hopf, M. (1973). Getreide, Äpfel, Eicheln. *Offa 30*, 200–204.

Knörzer, K. H. (1971). Urgeschichtliche Unkräuter im Rheinland ein Beitrag zur Entstehungsgeschichte der Segetalgesellschaften. *Vegetatio 23*, 89–110.

Knörzer, K.-H., R. Gerlach, J. Meurers-Balke, A. J. Kalis, U. Tegtmeier, W. D. Becker, and A. Jürgens (1999). *Pflanzenspuren. Archäobotanik im Rheinland: Agrarlandschaft und Nutzpflanzen im Wandel der Zeiten.* Köln: Rheinland-Verlag.

Kroll, H. (2001). Der Mohn, die Trichterbecherkultur und das südwestliche Ostseegebiet. Zu den Pflanzenfunden aus der mittelneolithischen Fundschicht von Wangels, Kr. Ostholstein. In R. Kelm (Ed.), *Zurück zur Steinzeitlandschaft: Archäologische und ökologische Forschung zur jungsteinzeitlichen Kulturlandschaft und ihrer Nutzung in Nordwestdeutschland*, pp. 70–76. Heide: Albersdorfer Forschungen zur Archäologie und Umweltgeschichte 2.

Schweingruber, F. (1979). Wildäpfel und prähistorische Äpfel. In U. Körber-Grohne (Ed.), *Festschrift Maria Hopf zum 65. Geburtstag am 14. September 1979*. Köln: Rheinland-Verlag.

Chapter 21

Exploitation of plant resources in the Mesolithic and Neolithic of southern Scandinavia: from gathering to harvesting

David Earle Robinson
English Heritage, Portsmouth, UK

21.1 Introduction

This paper is a review of the archaeobotanical evidence for the exploitation of plant resources in the mesolithic and neolithic economies of southern Scandinavia. It spans the transition from a total reliance on collecting and gathering, to the advent and establishment of arable agriculture. The area covered is that represented by present-day Denmark and southern Sweden (figure 21.1). A summary of the local mesolithic and neolithic chronology is given in table 21.1.

21.2 The Mesolithic

The Mesolithic of southern Scandinavia was a time of dependence on collected and gathered plant resources, involving the exploitation of the vegetation of natural and, with time, of humanly-modified habitats within the landscape. It has also been suggested that actual cultivation took place. During the Mesolithic there was the first appearance and subsequent increase of a number of weed and ruderal plant species; and Göransson (1988), for example, has proposed mesolithic 'garden cultivation' as an explanation for early (i.e., pre-elm decline) finds of cereal pollen in southern Swedish pollen diagrams. However logical this may seem, in the light of centuries of contact between the mesolithic population and farming communities south of the Baltic Sea (e.g., on the loess areas of northern Germany), there is no clear archaeobotanical evidence for the presence of crop products and still less for their cultivation. Accordingly, mesolithic cultivation, as envisaged by Goransson, still seems unlikely.

As such, we must conclude that plant resources in the Mesolithic were exclusively collected from wild plants, although some modification and/or management of the vegetation certainly took place. Archaeobotanical evidence for collection from the wild, typically charred or waterlogged wood, bark, hazelnut shells, acorns and concentrations of fruit stones and pips, has been routinely recorded over the years as hand-picked finds recovered during archaeological investigations (Jensen 1985 and Regnell 1998, summarised in table 21.2). Systematic sampling

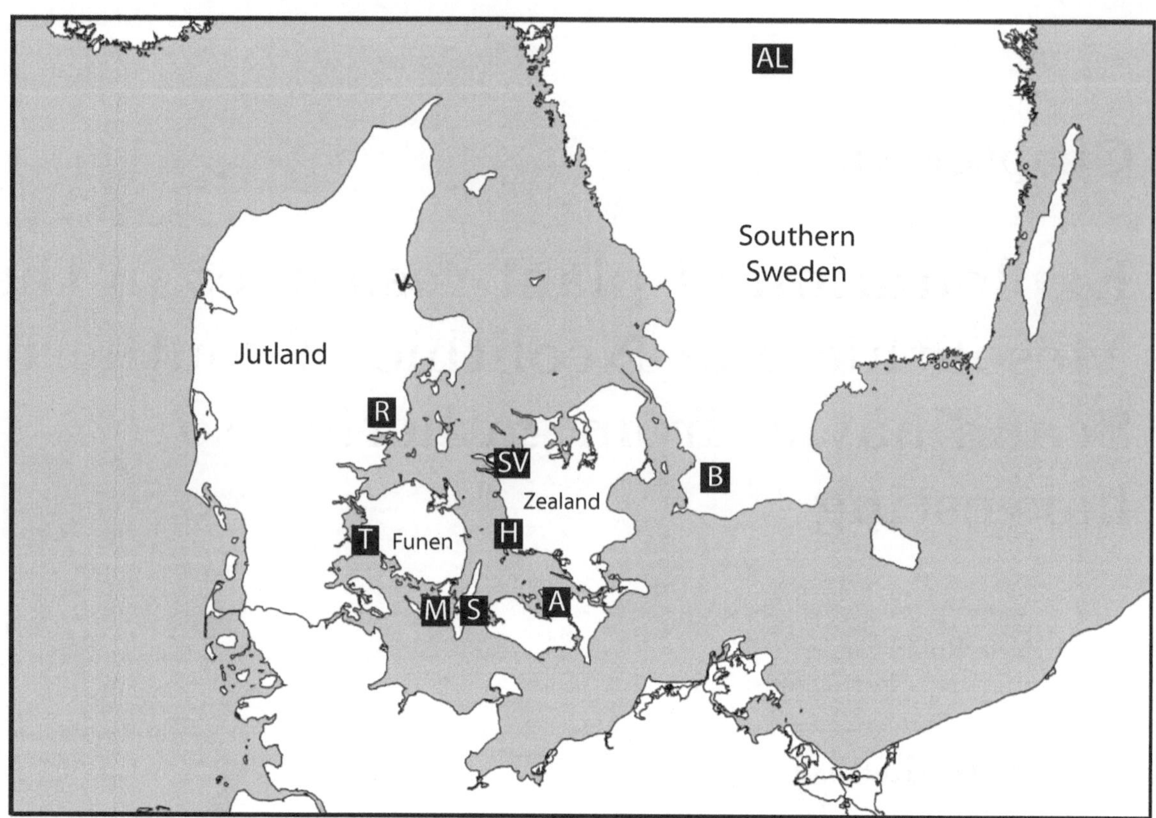

Figure 21.1: Southern Scandinavia showing locations mentioned in the text: A= Argus; M= Møllegabet II; H= Halsskov; R= Ringkloster; B= Bökeberg; T= Tybrind Vig; SV= Saltbæk Vig; S= Spodsbjerg; AL = Alvastra.

Cultural group	Date range
Mesolithic	
Maglemose culture	9000–6400 cal BC
Kongemose culture	6400–5400 cal BC
Ertebølle culture	5400–3900 cal BC
Neolithic	
Funnel Beaker culture (Early and Middle Neolithic)	3900–2800 cal BC
Single Grave/Battle Axe culture	2800–2400 cal BC
Late Neolithic (incl. Beaker) culture	2400–1700 cal BC

Table 21.1: The Mesolithic–Neolithic chronology of southern Scandinavia in calendar years (after Hvass and Storgaard 1993).

Taxon	Mesolithic	Neolithic
Allium ursinum	.	.
Apium graveolens	1	.
Beta vulgaris ssp. *maritima*	1	.
Cladium mariscus	4	1
Conopodium majus	1	.
Cornus sanguinea	4	2
Corylus avellana	22	26
Crataegus spp.	3	2
Daucus carota	.	1
Fragaria vesca	1	5
Glyceria spp.	2	3
Iris pseudacorus/sp.	2	.
Malus/*Pyrus*	1	33
Mentha sp.	2	3
Nuphar lutea/*pumilla*	7	2
Phragmites australis	2	.
Prunus padus	.	1
Prunus spinosa	1	2
Quercus sp.	6	4
Ranunculus ficaria	.	2
Rosa sp.	1	1
Rubus caesius	1	.
Rubus fruticosus/idaeus	8	15
Rubus saxatilis	1	.
Sorbus aucuparia	3	2
Vicia angustifolium	1	1
Viscum album	.	2
Vitis sp.	.	2

Table 21.2: Summary of potentially collected plants from the Mesolithic and Neolithic, excluding obvious arable weeds and ruderals. Numbers refer to number of sites at which remains of a particular taxon have been recorded. Based on Regnell (1998) and more recent studies mentioned in the text.

and analysis of mesolithic deposits specifically for plant remains, as an integrated part of the excavation procedure, has been less common. Fortunately, this situation has changed in recent times and the number of systematic investigations of plant remains from southern Scandinavian Mesolithic sites is now probably greater per unit area than anywhere else in Europe, although the coverage is still far from optimal.

Mesolithic settlement sites in the region are almost exclusively littoral in character and are located either on the contemporary shores of inland lakes and watercourses or on brackish fjords and the marine coast. Many existed and were active over very long periods and continued in use in the Neolithic and later periods, leading to the formation of extensive refuse deposits, so-called *køkkenmøddinger* (kitchen middens). These multi-period middens are a distinct feature of coasts to the northeast of a line running diagonally northwest-southeast through the region. Here there has been upheaval of the land relative to the sea, and the sites were accessible throughout prehistoric and more recent times. The sites lie above the water table and are generally well-drained. Preservation of plant material is largely the result of carbonisation and, to a lesser extent, mineralisation.

Other sites, comprising variously dwellings, hearths, graves and more modest refuse layers, were occupied over a shorter period or periods. In many cases occupation was curtailed by short- or long-term marine transgression, particularly in the southwestern half of the region, or by changes in lake water level. Most marine coastal sites in the south and west of the region now lie submerged under varying depths (several to many metres) of water, depending on their age and location. Typically, submersion has led to the creation of exceptional conditions for the preservation of uncarbonised organic material, and circumstances generally suggest that inundation occurred relatively rapidly and with minimal erosion. These well-preserved submerged sites provide a broad range of information that has largely been lost from dry-land sites. They are, however, now under increasing threat from erosion, shipping, dredging and aggregate extraction.

Sweden and Denmark are quite rightly famous for the large number of mesolithic sites that have been investigated archaeologically over the years, both on land and, more recently, below water (Robinson and Harild 2002). As mentioned above, finds of plant material made during excavation have been routinely recorded at the great majority of sites, whereas integrated systematic archaeobotanical investigations are much less common. The oldest site to be studied in this way is a submerged Kongemose site on the Argus ground, which lies off the southern Danish islands of Lolland and Falster (Fischer et al. 1987, nd). The settlement was occupied for a relatively brief period around 5700 cal BC and the remains now lie at depth of between 4 and 6 m. Sampling was restricted to a well-preserved hearth, and plant remains comprised very modest numbers of carbonised and/or uncarbonised remains of *Corylus avellana*[1], *Rubus idaeus/fruticosus*, *Ruppia maritima*, *Zannichellia* sp. and *Zostera* sp., as well as quantities of charcoal. Remains of fish, birds and both terrestrial and marine mammals were much more abundant. Human remains were also recovered, and stable isotope analyses revealed interesting differences in the terrestrial and marine origins of the food eaten by male and female humans and by dogs.

Much more information is available from the subsequent Ertebølle period, with detailed analyses from five sites, four of which are in Denmark and one in Sweden (figure 21.1 and table 21.3). These comprise three sheltered marine coastal sites in Denmark: Halsskov (Kubiak-Martens 2002; Robinson 1997; Robinson and Harild 2002), Møllegabet II (Robinson, cited in Grøn and Skaarup 1993 and Robinson and Harild 2002) and Tybrind Vig (Kubiak-Martens 1999), and two inland lake sites, one Danish: Ringkloster (Jørgensen unpub., cited in Andersen 1995 and Robinson and Harild 2002) and one Swedish: Bökeberg (Regnell et al. 1995). There has also been further work on material from Møllegabet II and from a site on Saltbaek Vig.

[1] Nomenclature for wild plant species follows Hansen (1981).

This had not been published in full prior to completion of this paper (Mason et al. 2002; Mason 2004).

A summary of the plant remains identified from these sites is given in table 21.3; both charred and uncharred remains are included. Table 21.3 gives an impression of the great range of plant species present at these sites, many of which were exploited. Many more were doubtless also a part of daily life, but it is always difficult to demonstrate deliberate human use of commonly occurring plants. There is also evidence for the character of the local vegetation, in particular plants that benefit from human activities, i.e., those requiring light open habitats and, most importantly, ruderal plants which thrive on the disturbed refuse dumps and nutrient rich soils in and around settlements.

21.2.1 Plant resources

The main plant resources likely to have been used at the sites are outlined briefly below. Much greater detail is given in the individual publications cited above.

Root vegetables

One of the most exciting developments in recent years has been the use of electron microscopy for the identification of vegetative plant tissue preserved in archaeological contexts. This has become possible primarily due to the development of extensive comparative reference collections. In particular, the work of Kubiak-Martens (1999, 2002) at the Ertebølle sites of Tybrind Vig and Halsskov has revealed the presence of charred roots of *Beta vulgaris* ssp. *maritima*, bulbs of *Allium* cf. *ursinum*, tubers of *Conopodium majus* and fragments of unidentified dicotyledonous root. These are all sources of starch and, in the case of *Allium ursinum*, possibly also flavouring. Kubiak-Martens suggests that these starchy plant parts were prepared by being roasted in pit ovens. The exploitation of vegetative plant material has previously been assumed to be important in the Mesolithic, but the concrete evidence for this has proved more or less 'invisible' to conventional archaebotanical techniques. These are important results and this avenue of research will no doubt provide further evidence in the future thereby adding to the already increasing body of data.

Greens

We know (or assume) that plants such as *Allium ursinum*, *Apium graveolens*, *Atriplex* spp., *Beta vulgaris*, *Brassica* spp., *Cakile maritima*, *Chenopodium* spp. (especially *Chenopodium album*), *Mentha* sp., *Phragmites australis* and other members of the Poaceae, *Rumex* spp., *Stellaria media*, *Taraxacum vulgare* and *Urtica dioica* have all been used as green vegetables and herbs in the past. Although it is difficult to find direct evidence for their exploitation, it seems likely that these taxa were also used in the Mesolithic and, similarly, many other plants, which also provided flavour, bulk and vitamins.

Nuts, seeds, fruits and berries

Nuts of *Corylus avellana* figure prominently at almost all of the sites as do acorns (*Quercus* sp.). Both would have provided valuable storable food (carbohydrate) resources, although acorns require treatment to remove toxins and tannins before they are edible. Caryopses of *Glyceria plicata*, *Glyceria maxima* and*Glyceria fluitans* seem also to have been eaten and the same may well have been the case for seeds and fruits of *Atriplex* spp., *Brassica* spp., *Chenopodium* spp., *Iris* spp., *Polygonum* spp., *Nuphar* spp., *Scirpus* spp. and *Vicia* spp. These primarily provide starch and, in some cases, fats and oils.

Taxon	Plant part	Halsskov	Møllegabet	Ringkloster	Bökeberg	Tybrind Vig
Allium cf. *ursinum*	Bulb/bulb frags.	+⋆	.	.	.	.
Alnus sp.	Fruit	.	.	++	+++++	.
Alnus sp.	Female catkin	.	.	+	+++	.
Apium graveolens	Fruit	.	.	.	+	.
Apium inundatum	Fruit	.	.	.	+	.
Atriplex hastata/patula	Seed	.	.	.	.	++++
Atriplex prostrata/patula	Seed	++⋆	.	.	.	.
Atriplex litoralis	Seed	.	.	.	.	+++
Atriplex sp.	Seed	++/++⋆	.	.	.	.
Beta vulgaris	Root parenchyma	.	.	.	.	++⋆
Betula pubescens	Fruit	.	.	.	+++++	.
Brassica campestris	Seed	+	.	.	.	.
Briza media (cf.)	Fruit	+⋆	.	.	.	.
Bromus sterilis (cf.)	Fruit	+⋆	.	.	.	.
Cakile maritima	Capsule	.	.	.	.	+
Carex sp.	Nutlet	+/+⋆	.	.	.	.
Carex distigmata type	Nutlet	.	.	.	++	.
Carex lasiocarpa type	Nutlet	.	.	.	+	.
Carex rostrata type	Nutlet	.	.	.	++	.
Ceratophyllum demersum	Fruit	.	.	.	++	.
Chara sp.	Oospore	.	.	.	+++	.
Chenopodium album	Seed	++/+⋆	+	.	.	+++
Cicuta virosa	Fruit	.	.	.	+	.
Cladium mariscus	Fruit	.	.	.	++++	.
Conopodium majus	Tuber frags	++⋆	.	.	.	.
Cornus sanguineus	Fruitstones	.	++	++	+	.
Corylus avellana	Nuts/nutshell	+/++⋆	++	.	++	++⋆
Crataegus monogyna/oxycantha	Fruitstones	.	+++	+++/+⋆	.	++
Crataegus monogyna/oxycantha	Fruitstones & epidermis	.	+	.	.	.

Table 21.3: Plant macrofossil remains from recent systematic investigations of Ertebølle (late Mesolithic) sites in southern Scandinavia shown on a relative abundance scale from present + to extremely abundant + + + + + (⋆=charred).

Taxon	Plant part	Halsskov	Møllegabet	Ringkloster	Bökeberg	Tybrind Vig
Dicotyledon	Secondary root	+⋆	.	.	.	.
Equisetum sp.	Node	.	.	.	+	.
Eupatorium cannabinum	Achene	.	.	.	++	.
Fragaria vesca	Seed	.	.	.	.	+
Frangula alnus	Fruit	.	.	.	+	.
Galium aparine	Fruit	++⋆	.	.	.	.
Galium sp.	Fruit	+⋆	.	.	+	.
Glyceria fluitans	Fruit	.	.	.	.	+⋆
Glyceria plicata (cf.)	Fruit	+⋆	.	.	.	.
Iris pseudacorus	Seed	.	++	.	.	.
Lemna sp.	Seed	.	.	.	+	.
Malus sylvestris	Seed	.	.	.	.	++
Melica uniflora (cf.)	Fruit	+⋆	.	.	.	.
Mentha sp.	Nutlet	.	.	.	+	.
Menyanthes trifoliata	Seed	.	.	+	+	.
Monocotyledon	Swollen internode	+⋆	.	.	.	.
Myriophyllum cf. *verticilliatum*	Mericarp	.	.	.	+	.
Najas marina	Fruit	.	.	.	++	.
Nuphar luteum	Seed	.	.	+++	+++	.
Nuphar pumilla	Seed	++⋆	.	.	.	.
Nyphaea alba	Seed	.	.	+++	+++	.
Phragmites australis	Stem frags	++⋆	.	.	++	.
Poa type	Fruit	.	.	.	+	.
Poaceae	Fruit	+⋆	.	.	++	.
Polygonum aviculare	Fruit	+	.	.	.	+
Polygonum dumetorum	Fruit	+	.	.	.	.
Potentilla palustris	Achene	.	.	.	+	.
Potomogeton sp.	Fruitstone	.	.	+	++	.
Prunus spinosa	Fruitstone	.	.	.	+	.

Table 2, cont.

Taxon	Plant part	Halsskov	Møllegabet	Ringkloster	Bökeberg	Tybrind Vig
Quercus sp.	Acorn parenchyma	++/++⋆	++	.	.	++++/+⋆
Quercus sp.	Acorn epidermis	++/++⋆	++	.	++	++++/+⋆
Quercus sp.	Cupule/cupule frag.	.	++	.	.	.
Rosa sp.	Achene	.	.	.	.	++
Rubiaceae	Fruit	.	.	.	+	.
Rubus caesius	Achene	.	.	.	.	++
Rubus idaeus	Achene	+++	.	.	++	++
Rumex sp.	Fruit	+⋆	.	.	.	.
Rumex crispus	Fruit	+⋆	.	.	.	++/+⋆
Ruppia maritima	Fruit	++++	.	.	.	.
Scirpus lacustris/maritimus	Fruit	.	.	.	+++	.
Scirpus maritimus	Fruit	.	.	.	.	+
Silene cf. *vulgaris*	Seed	.	.	.	+	.
Silene sp.	Seed	.	.	.	+	.
Solanum dulcamara	Seed	.	.	+	+	.
Solanum nigrum	Seed	+/+⋆	.	.	.	.
Sorbus aucuparia	Achene	.	.	.	++	+
Sparganium sp.	Fruit	.	.	++	.	.
Stellaria media	Seed	+	.	.	.	.
Suaeda maritima	Seed	+	.	.	.	.
Taraxacum vulgare	Achene	.	.	.	+	.
Tilia sp.	Fruit	.	.	+++	+++++	.
Typha sp.	Fruit	.	.	.	++	.
Urtica dioica	Seed	.	.	.	+++	+
Viburnum opulis	Seed	.	.	.	.	++
Vicia cf. *palustris*	Seed	.	.	.	+	.
Vicia sativa ssp. *angustifolium*	Seed	+++⋆	.	.	.	.
Viola sp.	Seed	+⋆	.	.	.	.
Burnt organic (plant) material	.	++⋆	.	.	.	.

Table 2, cont.

Fruits and berries, including *Cornus sanguinea*, *Crataegus* spp., *Fragaria vesca*, *Malus/Pyrus*, *Prunus spinosa*, *Rosa* sp., *Rubus caesius*, *Rubus idaeus/fruticosus* and *Viburnum opulis*, provide sugars and flavouring and would similarly have been of great importance at certain times of year.

Non-food use

Plant resources have also had a variety of important non-food uses. For example, there is evidence from Bökeberg to suggest the use of the sedge *Cladium mariscus* as thatch. Several other species would also have been suitable for thatching and additional construction purposes, as well as for basketry and even for clothing. A rather strange usage of a plant material has also been identified at Bökeberg, where pine resin (*Pinus sylvestris*) had apparently served as chewing gum. A piece of resin was recovered that had clear human tooth marks and pollen analysis indicated that the resin came from *Pinus* (Regnell et al. 1995).

Environment

One of the most striking features of the results of these archaeobotanical analyses is the range and abundance of plant species characteristic of soils that are open, disturbed and/or rich in nutrients—i.e., weedy and ruderal species. The activities and settlements of mesolithic populations clearly had an effect on the local vegetation, a phenomenon that is further demonstrated by pollen analysis, e.g., from Bökeberg (Regnell et al. 1995). This is a trend that accelerates in the course of the Neolithic period.

21.3 The Neolithic

The systematic investigations of plant remains at late mesolithic sites have revealed the breadth, depth and importance of resources exploited by southern Scandinavian hunter-gatherers. The introduction of agriculture represented a major innovation. Arable agriculture apparently reached southern Scandinavia as a complete 'package', comprising five cereals: einkorn (*Triticum monococcum*), emmer (*Triticum dicoccum*), bread wheat (*Triticum aestivum* s.l.), naked barley (*Hordeum vulgare* var. *nudum*) and hulled barley (*Hordeum vulgare* var. *vulgare*). Small fields were cultivated in areas of forest that had been cleared by slash and burn methods. After only a few harvests the natural fertility of the soil, augmented by the nutrients from the wood ash, would have become depleted resulting in abandonment of the plots and prompting the creation of new fields elsewhere. This shifting form of agriculture apparently continued in various forms throughout the Neolithic and it is in the Bronze Age that there are the first signs of more permanent arable fields (Robinson 2000, 2003). The extent, nature and effect of settlement, cultivation and animal husbandry varied from region to region. Pollen analysis has revealed that northwestern Denmark was more or less treeless by the end of the Neolithic, whereas much of eastern Jutland maintained its forests for a further 1,500 years; the situation in southern Jutland lay somewhere in between (Robinson 2003).

In a recent overview of arable agriculture attempts have been made to standardise the available archaeobotanical data resulting from analyses carried out over a period of almost a century (Robinson 2003). This has involved each species present at a site being scored on a scale of 1–5 according to its perceived abundance at that site: 1=present; 2=few remains; 3=many remains; 4=major component; 5=dominant. The scores for each species from each period have then been summed and expressed as a percentage of the total scores for that period. This revealed a series of marked changes in the course of the Neolithic (figure 21.2), which can be summarised briefly as follows: the earliest agriculture was based on *Triticum dicoccum* and *Hordeum vulgare* var. *nudum*, with minor contributions from *Triticum aestivum* s.l. and *Hordeum vulgare* var. *vulgare*, and possibly augmented later by *Triticum spelta* (as suggested by a tentative identification only

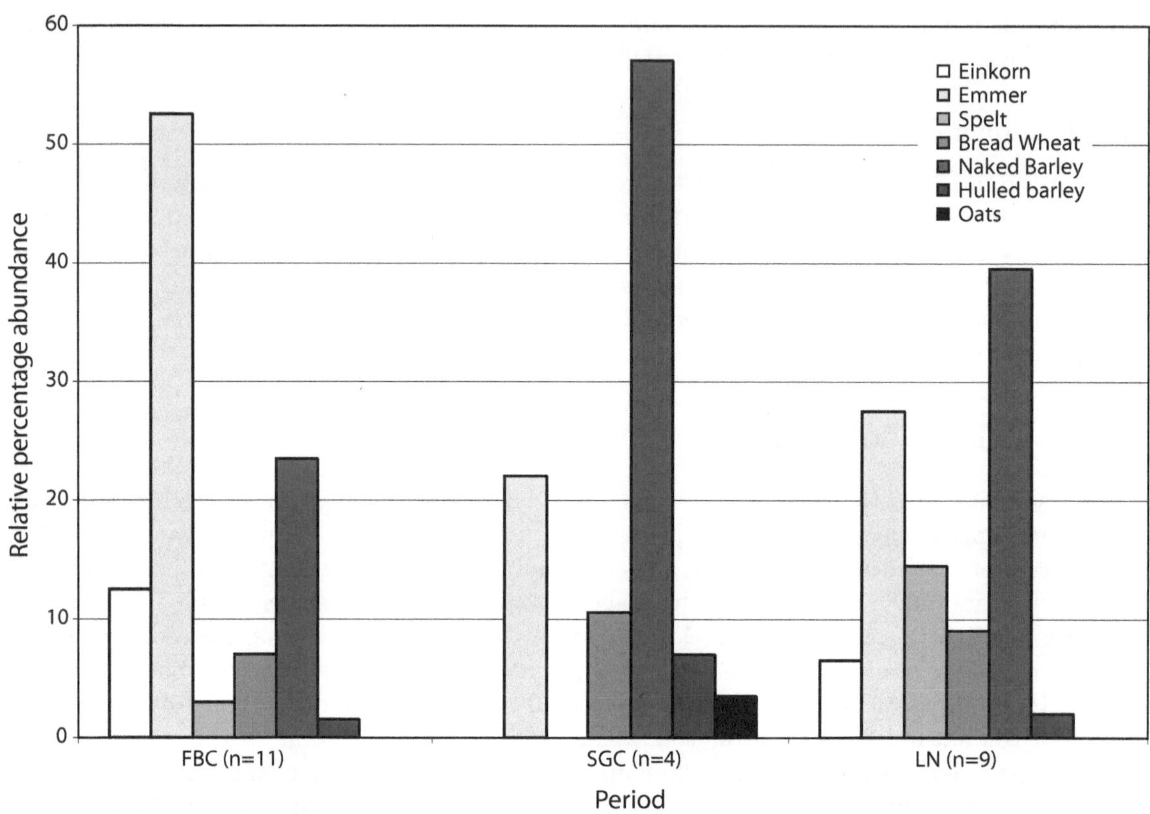

Figure 21.2: Summary of the relative percentage abundance of crop plant species at Danish Neolithic sites: Funnel Beaker culture (FBC); Single Grave culture (SGC); Late Neolithic including Bell Beaker culture (LN). N = number of sites from each period for which suitable data are available (after Robinson 2003).

from Spodsbjerg (Robinson 1998)). This situation persisted virtually unchanged until the end of the middle Neolithic. In the Single Grave Period there was a marked decline in the prominence of *Triticum* species, in favour of *Hordeum vulgare* (mostly of the naked form), before yet another change in the late Neolithic. The latter period saw the introduction of *Triticum spelta* and *Panicum miliaceum* and an increase in importance of *Triticum aestivum* s.l.. The first recorded *Avena* grains are from the Single Grave Period and late Neolithic, but these are more likely to represent weeds rather than a cultivated crop. The first *Panicum* grains appear in the late Neolithic, although cultivation of this crop appears to have been much more prominent in the Bronze Age. There are no neolithic finds of *Linum usitatissimum* and there is a recent record of an imprint of *Camelina sativa* from a neolithic pot (Koch 2002), however, this seems unlikely to be from a cultivated plant.

The 'purity' (i.e., absence of seeds of arable weeds and ruderals) of grain finds from the Scandinavian Neolithic has often been remarked upon and has been the subject of much discussion (Jørgensen 1977; Rowley-Conwy 1978; Robinson and Kempfner 1987; Robinson 2000). Several possible explanations have been proposed, including harvesting techniques that involved plucking or cutting single ears, harvesting of the ears high on the straw, or careful sorting of the crop prior to threshing. It may simply have been that a substantial weed flora did not, or could not, develop under the neolithic regime of shifting agriculture. A clear and dramatic contrast is apparent in the quantity and diversity of weeds later in the Bronze Age, when permanent fields had become established and manuring with animal dung and household waste was an integral part of arable agriculture (Robinson 2000, 2003).

21.3.1 Collected and gathered plant resources in the Neolithic

The importance of collected and gathered resources is rather more difficult to evaluate for the Neolithic because of the addition into the landscape of areas of cultivated land, abandoned fields and secondary woodland and, as a consequence, of the increased likelihood of the representation of 'useful' weed and ruderal species. Table 21.2 has been produced on the basis of Regnell's catalogue published in 1998 (Regnell 1998), supplemented by data from Danish Mesolithic sites published after this date and from the early neolithic layers of the kitchen midden at Visborg (see below). It shows the number of mesolithic and neolithic sites at which various potentially useful wild plants have been recorded. Potential arable weeds/ruderals, for example, *Polygonum*, *Brassica* and *Chenopodium* species etc., have been excluded although these clearly may have grown in, and been collected from, non-arable habitats. Accordingly, this table gives but an approximate indication of the plants represented at settlement sites dated to the two periods.

It is perhaps rather surprising that there is so little difference between the two sets of data; *Corylus avellana* is clearly the most common find in both periods and plants such as *Allium ursinum*, *Mentha* sp., *Cornus sanguinea*, *Glyceria* spp., *Quercus* sp., *Crataegus* spp. and *Sorbus aucuparia* show little change in the frequency of occurrence. There seems to be a reduction in the incidence of *Nuphar* spp. and *Iris* spp. remains, which may reflect the fact that cereals had replaced them as a major source of starch. Conversely, the absence in the neolithic record of vegetative remains of *Beta vulgaris* ssp. *maritima*, *Conopodium majus* and *Phragmites australis* is not surprising, given that the new methods of detecting and identifying vegetative plant remains described above have yet to be applied to material from Scandinavian sites of this date. There are major increases in the incidence of *Malus/Pyrus* sp. and *Rubus fruticosus/idaeus* in the Neolithic and also the first finds of *Daucus carota*, *Prunus padus*, *Ranunculus ficaria*, *Viscum album* (see below) and *Vitis* sp.

It must be remembered, however, that in many cases these are very small data sets and, more importantly, we are not comparing like with like; the characteristics of the sites investigated, the nature of contexts sampled and the material analysed differs greatly in the Neolithic and the Mesolithic. Fortunately, this is not exclusively so—many mesolithic settlement sites (especially shell middens) show continuation of use and deposition into the Neolithic and later times. As yet no mesolithic kitchen midden layers have been subjected to the kind of systematic and rigorous analysis as that carried out on the material from Halsskov and Tybrind Vig. However, carbonised remains from the extensive (early Neolithic) Funnel Beaker culture kitchen midden at Visborg on Mariager Fjord have been investigated (Robinson et al. nd). This site contains primarily animal remains, including terrestrial game animals, birds, sea mammals, fish and shellfish, with enormous numbers of oysters, cockles and mussels. They show that the users of the midden were primarily hunters and fishers. The occupants also extended their gathering activities to incorporate plant resources; small quantities of charred plant remains were found associated with hearths embedded within the midden deposits. A summary of the plant remains recovered by processing more than 400 litres of hearth deposits is given in table 21.4. There are the remains of fruits (*Malus/Pyrus*), wild grains (e.g., *Glyceria plicata*) and possible sources of greens, for example, *Chenopodium album*, *Rumex acetosella* and *Mentha* sp. A total of $4\frac{1}{2}$ charred cereal grains were recovered, together with a few possible arable weeds and ruderals such as *Plantago major*, *Galium aparine* and *Veronica hederifolia*, although the latter probably grew on and around the refuse dump. Three of the cereal grains were identified; there was one grain each of *Hordeum vulgare*, cf. *Triticum aestivum* and *Triticum dicoccum*. A small number of grain and grain fragments were unidentifiable. The state of preservation of the grains was poor and as a consequence the *Hordeum vulgare* grain could not be assigned to either the hulled (var. *vulgare*) or the naked (var. *nudum*) variety; the identification of the grain of *Triticum aestivum* is uncertain for the same reason. All of these cereal species have previously been recorded from early neolithic contexts in Denmark and Sweden. Whether the cereals were grown by the users of the midden is open to question. However no cereal processing waste was found, prompting the logical conclusion that the cereals were grown and processed elsewhere.

Perhaps the most remarkable and intriguing finds at Visborg were the charred remains of mistletoe (*Viscum album*) present in several hearth samples. The characteristic seeds were either whole or fragmented. There were also fragments tentatively identified as epidermis of *Viscum* berries (several showed the typical terminal stigmatic scar) as well as possible whole and fragmented female inflorescences. Remains of *Viscum album* may have been present incidentally on branches of arboreal hosts brought to the site, as the plant was a regular component of Danish forests at this time (e.g., Troels-Smith 1960). If this were the case, the *Viscum* must have been on green wood intended for use as a raw material for tools and structures, etc., rather than on dead wood collected for fuel. The latter would be unlikely to retain intact *Viscum* plants complete with berries and seeds. Why were the berries brought on to site, if indeed they were intentionally imported? There are three obvious possibilities: for animal fodder, for medicinal use, or for purposes relating to mythology, superstition, religion or cult.

In the light of this, it is worth noting that are no published archaeobotanical finds of *Viscum album* from an archaeological context from the Scandinavia Mesolithic. Uncarbonised *Viscum* remains have previously been found in bog deposits from the Atlantic period, for example, a leaf from Taarbæk in northeastern Zealand (Jensen 1985) and a twig from Dyrholmen, Djursland (Andersen et al. 1983), but the Visborg finds represent the first evidence of seeds and berries and also the first carbonised material.

The use of *Viscum* as animal fodder in historical and recent times in Scandinavia is well documented (e.g., Brøndegaard 1980), several authors have also discussed its possible use as animal feed in antiquity, notably Troels-Smith (1960) and more recently Akeret and Rentzel (2001). However, one of the most fascinating accounts of the use of *Viscum* and of the broader role of plants in the Neolithic is by Göransson (2002). He gives an inspired description and interpretation of the plant remains and other assorted finds from the Alvastra pile dwelling, a waterlogged site from the middle Neolithic, which he quite clearly interprets as a byre. In contrast with most neolithic sites included in this review, the preservation of organic material

Taxon	Total	No. samples where present
Cultivated species		
Hordeum vulgare	1	1
cf. *Triticum aestivum* s.l.	1	1
Triticum dicoccum	1	1
Unidentified cereals	1.5+3F	3
Potentially useful wild species		
Pyrus/Malus	2	2
Viscum album	6+10F	.
cf. *Viscum album* (berry epidermis frag.)	3F	.
cf. *Viscum album* (bud)	1	7
cf. *Viscum album* (budscale)	9F	.
Glyceria plicata	5	2
Arable weeds and ruderals		
Aphanes cf. *arvensis*	1	1
Plantago major	3	2
Plantago lanceolata	1	1
Rumex acetosella	2	2
Veronica hederifolia	5	4
Other and broader habitats		
Galium sp.	1	1
Juncus sp.	7	2
cf. *Luzula* sp.	1	1
Mentha sp.	2	2
Moehringia trinerva	0.5	1
Poaceae	4+F	4
Poaceae (node)	1	1
Trifolium sp.	1,5	2

Table 21.4: Summary of carbonized macrofossils from the early neolithic hearths at Visborg. F=fragment.

Remains	Total	No. samples where present
Diverse		
Burnt bark (frag.)		few
Charcoal (frag.)		few to many
Bud/budscale		few
Carbonised plant tissue (frag.)		many
Unidentified seeds	33	.
Unidentified capsule	0.5+2F	.
Cenococcum sp.		some
Mollusc shells		many
Oyster shells		few to many
Burnt amber (frag.)	5	.
Burnt bone (frag.)		some
Small bone	2	.
Modern (uncarb.) seeds		some
Green glass (frag.)	1	.
Pottery		few
Chalk (frag.)		few to many

Table 21.4, cont.

at Alvastra is exceptional, and serves to underline the information that potentially has been lost from dry land sites such as the Visborg kitchen midden and many neolithic settlement sites. The Alvastra material also has species in common with the Visborg material, including carbonised caryopses of *Glyceria* spp. and pips of *Malus sylvestris*. Göransson deals in depth with the customs, practices and superstitions associated with both ancient and also more recent agriculture and, in particular, animal husbandry. In doing so he makes a compelling argument, if any were needed, of how the advent of agriculture was much more than the introduction of crop plants and domesticated animals; it represented not only a completely new way of life, but also a fundamental change in thought and practice, not least in the way plants were perceived, valued and used.

It may well be that the abundant remains of *Viscum album* found in the Visborg hearths are a much more revealing indication of the superimposition of neolithic society on a mesolithic way of life than the few carbonised cereal grains that hitherto have commanded so much attention.

References

Akeret, Ö. and P. Rentzel (2001). Micromorphology and plant macrofossil analysis of cattle dung from the neolithic lake shore settlement of Arbon Bleiche. *Geoarchaeology 16*(6), 687–700.

Andersen, S. H. (1994-1995). Ringkloster. Ertebølle trappers and wild boar hunters in eastern Jutland. *Journal of Danish Archaeology 12*, 13–59.

Andersen, S. T., P. Mortensen, O. Olsen, A. Steensberg, and J. Troek-Smitt (1983). Kommissionen til udforskning af Danmarkr ældste landbrug og de samtidige naturforhold. *Kongelige Danske Videnskabernes Selskab 1982/83*, 95–97.

Brøndegaard, V. J. (1978-1980). *Folk og Flora*, Volume 1-4. Copenhagen: Rosenkilde and Bagger.

Fischer, A., U. Møhl, P. Bennike, H. Tauber, C. Malmros, J. Schou Hansen, and P. Smed (1987). *Argus-grunden: En Undersøisk Boplads fra Jægerstenalderen.* Fortidsminder og Kulturhistorie Antikvariske Studier 8. Copenhagen: Skov- og Naturstyrelsen.

Fischer, A., M. Richards, J. Olsen, D. E. Robinson, P. Bennike, L. Kubiak-Martens, and J. Heinemeier (n.d.). *Mesolithic Menu. Food Remains and Stable Isotopes in Bones from the Submerged Settlement on the Argus Bank, Denmark.* In prep.

Göransson, H. (1988). Can exchange during Mesolithic time be evidenced by pollen analysis? In B. Hårdh, L. Larsson, D. Olausson, and R. Petre (Eds.), *Trade and Exchange in Prehistory. Studies in Honour of Berta Stjernquist*, Acta Arch Lundensia 8:16, pp. 233–248. Lund: Publications of the Archaeological Institute.

Göransson, H. (2002). Alvastra pile dwelling—a 5000-year-old byre? In K. Viklund (Ed.), *Nordic Archaeobotany. NAG 2000 in Umeå*, pp. 67–84. Umeå: University of Umeå.

Grøn, O. and J. Skaarup (1993). Møllegabet II. A submerged mesolithic site and a boat burial from Ærø. *Journal of Danish Archaeology 10*, 38–50.

Hansen, K. (1981). *Dansk Feltflora.* Copenhagen: Glydendal.

Hvass, S. and B. Storgaard (1993). *Digging into the Past. 25 Years of Archaeology in Denmark.* Aarhus: Jutland Archaeology Society, Aarhus University.

Jensen, H.-A. (1985). *Catalogue of Late- and Post-Glacial Macrofossils of Spermatophyta from Denmark, Schleswig, Scania, Halland, and Blekinge dated 13,000 B. P. to 1536 A. D.*, Volume 6 of *Geologiske Undersøgelser Serie A.* København: Reitzels forlag.

Jørgensen, G. (1977). Et kornfund fra Sarup. *KUML 1976*, 47–64.

Koch, E. (2002). Strange holes—a case study concerning making and identifying casts of imprints of plant material in neolithic pots. In K. Viklund (Ed.), *Nordic Archaeobotany. NAG 2000 in Umeå*, pp. 143–160. Umeå: University of Umeå.

Kubiak-Martens, L. (1999). The plant food component of the diet at the late mesolithic (Ertebølle) settlement at Tybrind Vig, Denmark. *Vegetation History and Archaeobotany 8*, 117–127.

Kubiak-Martens, L. (2002). New evidence for the use of root foods in pre-agrarian subsistence recovered from the late mesolithic site at Halsskov, Denmark. *Vegetation History and Archaeobotany 11*, 23–31.

Mason, S. (2004). Archaeobotanical analysis from Møllegabet II. In J. Skaarup and O. Grøn (Eds.), *Møllegabet II: A Submerged Ertebølle Dwelling*, British Archaeological Reports S1328. Oxford: Archaeopress.

Mason, S., J. Hather, and G. Hillman (2002). The archaeobotany of European hunter-gatherers: some preliminary investigations. In S. Mason and J. Hather (Eds.), *Hunter-Gatherer Archaeobotany: Perspectives from the Northern Temperate Zone*, pp. 188–196. London: University College London, Institute of Archaeology.

Regnell, M. (1998). Archaeobotanical finds from the Stone Age of the Nordic countries. A catalogue of plant remains from archaeological contexts. *Lundqua Report 36*, 1–14. Lund University.

Regnell, M., M.-J. Gaillard, T. S. Bartholin, and P. Karsten (1995). Reconstruction of environment and history of plant use during the late Mesolithic (Ertebølle culture) at the inland settlement of Bökeberg III, southern Sweden. *Vegetation History and Archaeobotany 4*, 67–91.

Robinson, D. E. (1997). Ancient plant remains from the Halsskov site. In L. Pedersen, A. Fischer, and B. Aaby (Eds.), *The Danish Storebælt since the Ice Age: Man, Sea and Forest*, pp. 196–200. Copenhagen: A/S Storebælt Fixed Link and The National Forestry and Nature Agency.

Robinson, D. E. (1998). Plantemakrofossiler fra Spodsbjergbopladsen. In *Spodsbjerg—en Yngre Stenalders Boplads på Langeland*, pp. 175–189. Rudkøbing: Langelands Museum.

Robinson, D. E. (2000). Det slesvigske agerbrug i yngre stenalder og bronzealder. Arkæobotanikkens udsagn. In P. Ethelberg, E. Jørgensen, D. Meier, and D. Robinson (Eds.), *Det Sønderjyske Landbrugs Historie. Sten- og Bronzealder*, pp. 281–298. Haderslev: Haderslev Museum and Historisk Samfund for Sønderjylland.

Robinson, D. E. (2003). Neolithic and bronze age agriculture in southern Scandinavia: recent archaeobotanical evidence from Denmark. *Environmental Archaeology 8*, 145–166.

Robinson, D. E. and J. A. Harild (2002). Archaeobotany of an early Ertebølle (Late Mesolithic) site at Halsskov, Zealand, Denmark. In S. Mason and J. Hather (Eds.), *Hunter-Gatherer Archaeobotany: Perspectives from the Northern Temperate Zone*, pp. 84–95. London: University College London, Institute of Archaeology.

Robinson, D. E. and D. Kempfner (1987). Carbonised grain from Mortens Sande 2. *Journal of Danish Archaeology 6*, 125–129.

Robinson, D. E., E. Koch, and S. H. Anderson (n.d.). Cereals, sweetgrass and mstletoe: carbonised plant remains from an early neolithic kitchen midden at Visborg, Mariager Fjord, Denmark. In Prep.

Rowley-Conwy, P. (1978). Forkullet korn fra Lindebjerg. *KUML 1978*, 159–171.

Troels-Smith, J. (1960). Ivy, mistletoe and elm: climate indicators, fodder plants: a contribution to the interpretation of the pollen zone border VII-VIII. *Danmarks Geologiske Undersøgelser 4*(4), 1–32.

Chapter 22

Reconsidering the evidence: towards an understanding of the social contexts of subsistence production in neolithic Britain

Chris J. Stevens
Wessex Archaeology, Salisbury, UK

22.1 Introduction—no conflict: competing patterns of subsistence

Going as far back as the 4th century BC to the writings of Aristotle, and later revived by the antiquarians of the enlightenment, we find elements of subsistence placed in conflict with one another (cf. Bradley 1978, p.29). Cultures became pigeonholed into seemingly opposing subsistence groups in many of these and later works, for example, as either pastoralists and cultivators or hunter-gatherers and agriculturalists. Inherent in these categorisations are assumptions concerning the degree of sedentism, so that pastoralists and hunter-gatherers are considered to be mobile, whereas cultivators and agriculturalists are thought to be sedentary.

Such associations have played an important role in theories concerning transitions to agriculture. Early social evolutionists saw key evolutionary stages as characterised by certain modes of subsistence. Hence Coleridge (1836, p.327), Nilsson (1868) and later Frederick Jackson Turner (1893) proposed an evolutionary sequence from hunting and gathering, passing through pastoralism to agriculture. Within the work of the diffusionists we find similar themes of conflict, but here migrating sedentary agriculturalists are thought to have replaced mobile hunter-gatherers, which are later substituted by pastoralists (Childe 1940).

Various authors have continued to embrace one of two models for the spread of agriculture and the emergence of neolithic society in Britain (cf. Thorpe 1996, p.100). Hawkes and Hawkes (1943), Clark (1952), Smith (1974, p.103), Megaw and Simpson (1979), Barker and Webley (1978), Mercer (1981), and Dennell (1983, p.174) promote a model for the spread of agriculture whereby mobile hunter-gatherers are replaced by sedentary agriculturalists. The alternative model favoured by Hawkes and Hawkes (1958), Troels-Smith (1960), Field et al. (1964), Bradley (1987), Pryor (1988), Entwhistle and Grant (1989), Thomas (1991, pp.18–20), Thomas (1996), Barrett (1994, pp.139–141), Richards (1996), and Richmond (1999, p.23) sees mobile hunter-gatherers replaced by semi-mobile pastoralists with little or no cereal cultivation.

Thomas (1999, p.13) rightly challenges these associations, highlighting how our expectations and prejudices provide a pre-conceptual framework into which the evidence is slotted. Thus we conceive that because neolithic people had cereals and domesticated animals they must have also been sedentary (cf. Whittle 1996, p.6).

22.2 Debating the evidence

Given Thomas' warnings against preconceived associations it is surprising that he seems determined to diminish the importance of cereals to fit the model that promotes a mobile pastoralist society with a greater dependence on wild plant foods (Thomas 1991, pp.22–24). This, more so than the other model, has gained the most approval in the last two decades. Notably, a paper by Moffet, Robinson and Straker (1989) has been cited as providing evidence in direct support of the model; it highlighted the prevalence of wild food remains within neolithic charred assemblages and stated that "on some sites it is possible that hazelnuts were at least as important a source of food as grain" (Moffett et al. 1989, p.252). In citing this evidence Thomas considerably downgrades the role of cereals, stating that "carbonised plant remains suggest only a sporadic use of domesticated cereals alongside a range of wild resources" (Thomas 1996, p.4), and "no more than a partial... dependence upon cultivated cereals... with considerable gathering of wild plants" (Thomas 1999, p.29). As Robinson later noted about the reception to the paper in general, "much less interest was displayed in the conclusion that the use of cereals was part of the neolithic economy over much if not all of England and Wales" (Robinson 2000, p.86). To complicate matters further, the way in which we interpret the charred evidence for cereals and wild foods has also recently been challenged (Legge 1989; Jones 2000; Legge et al. 1998; Rowley-Conwy 2000).

It is here that I wish to interject the contribution I hope this paper will make towards the debate. As Thomas says of subsistence, "The provision of food... is always a process that takes place in the context of a set of social relationships" (Thomas 1999, p.11), however, as Whittle comments of actual studies, "there has been too little discussion of the social context of production" (Whittle 1996, p.6). These quotes highlight the problem with the debate as it stands. While we continue to promote one source of evidence over another we come no closer to understanding subsistence in a social context; a problem that is amplified by simultaneously downplaying the 'conflicting' sources. For by dismissing evidence for cereals, wild foods, hunted animals or pastoralism, we disregard data that might help us understand the social context of neolithic subsistence as a whole.

This paper explores how the activities responsible for the creation of charred assemblages can be used to explain variation within them, and then proceeds to examine how the nature of these activities may reveal something about the social context in which they took place.

22.3 The creation of charred assemblages

22.3.1 The charred record for neolithic Britain

While new neolithic data are emerging all the time, there has been little change in the general patterns recorded by Jones (1980), Hillman (1981a) or Moffett et al. (1989). Hazelnut shells continue to dominate many British Neolithic assemblages and there are also sporadic finds of other wild foods, for example, crab-apple (*Malus sylvestris*) and sloe (*Prunus spinosa*; Robinson 2000). Cereal grains, albeit mostly found in small quantities, are a constant feature of a majority of assemblages, and occasionally larger deposits of charred grain are discovered (Huntley 1996). Free-threshing wheat, emmer and both hulled and naked barley are all commonly represented in the cereal finds.

Neolithic assemblages are distinguished both by the taxa that are present and also by those that are absent. For example, chaff and weed seeds have been recovered from later British and European LBK sites but are conspicuous by their near absence in the British Neolithic.

The contexts in which British Neolithic plant remains occur are varied; some are from palisade or causewayed enclosures (Fairbairn 1997, 1999) that are associated with ceremonial purposes such as feasting, and others are from burial sites such as longbarrows (Straker 1990). Charred cereals and wild plant food remains have also been recovered from house structures (Fairweather and Ralston 1993; Jones 2000). Many of the assemblages, however, come from isolated or small clusters of pits, a majority of which are late neolithic in date, although others from the early and middle Neolithic have been found as a result of more recently undertaken excavations.

22.3.2 Charred grain: absent but not forgotten

While some authors interpret the general dearth of cereal grains from British Neolithic sites as being indicative of the insignificance or near absence of arable farming, others have proposed a variety of reasons to explain their scarcity.

Bradley (1978, p.33) observed (mistakenly) that since emmer does not require parching (commonly quoted as being a source of carbonised cereals), the disparity in the frequency of occurrence of cereals may be due to the presence of emmer in the Neolithic and spelt in the Iron Age. However, the fact that emmer does require parching (Hillman 1981b) and that its grain and glumes were present from later prehistoric sites in northeast England (van der Veen 1992) and LBK sites (Knörzer 1973, 1980, 1988; Bakels 1978; Kreuz 1990) likewise does not account for the lack of grains or chaff in the British Neolithic.

The robust nature of hazelnut shells has been suggested as an explanation for their greater prominence in comparison to cereal grains (Legge 1989; Jones 2000), but this alone cannot be used to account for the near absence of cereal grains on British Neolithic sites in comparison with the large numbers present on other sites. Nor does it explain the predominance of hazelnut shells on neolithic sites but their relative scarcity on later prehistoric sites.

An alternative suggestion, that hazelnut shells are waste while cereal grains are not (Legge 1989; Jones 2000) presents similar problems. Given that many charred grains found on later prehistoric and Roman sites are associated with processing waste (Hillman 1981b; van der Veen 1992; Stevens 2003) it does not satisfactorily explain why these assemblages are generally absent from the British Neolithic.

Cereals are frequently stored in houses on settlement sites where there are excellent opportunities for the charring of grain. Hence, the generally low numbers of settlement sites from the British Neolithic might explain the near absence of grain (Rowley-Conwy 2000; Jones 2000). The problem being, however, that many later British and LBK sites have assemblages with high proportions of glumes, which are characteristic of processing waste rather than stored grain (Stevens 2003; Hillman 1981b, p.140; van der Veen 1992). The presence of waste in the form of hazelnut shells is further evidence that domestic activities associated with food preparation took place on many British Neolithic sites.

The argument that the relative importance of wild and cultivated plant foods cannot be compared using the evidence of charred remains alone (Jones 2000) is not contested. However, the arguments cited above do not address whether the small numbers of cereal grains and chaff from British Neolithic samples, in comparison to the quantities found on other sites, can be taken as a true reflection of their relative lack of importance during this period.

Cereal grains on later British and LBK sites, as stated above, are often associated with glume waste and also weed seeds and it would seem probable, therefore, that many were discarded and charred as processing waste. It is then perhaps more appropriate rather than to ask why neolithic sites produce such small amounts of grain, but why do they have so little evidence for processing waste?

22.3.3 Routines, repetition and probability

Notably many of the explanations given above rely on the physical characteristics of the remains themselves or on the contexts in which they were found. Little attention has been devoted to the fact that the nature of the activities responsible for the creation of charred assemblages on LBK sites and within later prehistoric and Roman periods may have been different in neolithic Britain. Conversely, however, if the archaeobotanical record is a direct reflection of the relative importance of cereals, the implication is that the activities which gave rise to the remains are universally the same across all Europe and throughout these time periods.

The first routine application of flotation on excavations demonstrated the consistency and uniformity of carbonised assemblages, which comprised cereal grains, chaff and seeds of probable weeds, together with wood charcoal (Helbaek 1969; Knörzer 1971). The homogeneity was indicative that similar activities were responsible for the creation of these assemblages in Europe and the Near East; the realisation of which subsequently led to the development of studies of crop processing as a means of interpretation (Hillman 1973, 1981b, 1984; Jones 1984, 1987). Hillman (1981b) pointed out that when traditional methods are used there are relatively few ways in which cereals may be processed and also that there is a logical order to the processing stages, which is unlikely to have varied since the crops were first cultivated.

The processing sequence itself is unlikely to have varied but there can be considerable variation in where, when and how often each stage is conducted. It is this variation that influences whether or not cereal remains become charred and thus are preserved archaeologically.

Time, place and probability

Storage is the single most important factor determining the location, timing and degree of repetitiveness of each processing stage; for example, in temperate climates storage divides the processing of the crop into two groups of activities.

The first group of activities is conducted only once, shortly after harvest and prior to storage. The remaining processing stages are then conducted repeatedly as the crop is taken from storage when clean grain is required. Hillman (1981b, p.155) states that it is common for the waste from these routine stages to be thrown straight onto the fire.

The stages conducted prior to storage may take place in the field and if this is the case the resultant waste is unlikely to be brought to the settlement or to fires where it would be charred (Jones 1985). Alternatively, the waste from these stages, which is produced in bulk (i.e., the heavy chaff: straw and rachis fragments), can be used as fodder and temper in pottery or daub and, as such, it will be transported to the site (Hillman 1981b). The remaining stages will most probably be undertaken at the settlement where crops are stored. In wet areas these stages are more likely to be carried out indoors (Hillman 1981b, p.138) and consequently the waste has a higher probability of being discarded into fires and of subsequently becoming incorporated into archaeological features.

Routine, repetition and probability

The second issue concerning the probability of cereal remains becoming charred and preserved archaeologically is the degree of repetitiveness of the processing activities. The first group of processing activities may occur just once a year in late summer and after the harvest, whereby the resultant waste is generated annually. The second group, however, is conducted routinely, perhaps as often as everyday as clean grain is needed, and so the waste from these stages is generated on a regular basis throughout the year. If charred assemblages relate to the waste from processing after storage they have the potential to reveal in what form the crops were stored.

Differences seen in charred assemblages between later British sites have been explained by variations in storage practice (Stevens 2003). The fact that glume waste dominates many Iron Age and Roman sites (Hillman 1981b, p.154; van der Veen 1992, p.83) suggests that hulled wheats were stored in spikelet form during these periods (Stevens 2003; Hillman 1981b, p.156). A similar trend has been recorded for many LBK sites (Knörzer 1973, 1980, 1988; Bakels 1978; Kreuz 1990).

22.3.4 Probability, neolithic storage practices and the absence of charred grain

It is possible to explain the absence of glumes from neolithic sites if crops were stored as clean grain; in this case after harvest the crops would have been processed in full (i.e., including their dehusking in the field; figure 22.1).

It is as easy to account for the absence of grain from British Neolithic sites. Grains inevitably become incorporated into the waste as crops are processed and greater numbers are likely to be discarded, and thus charred as more processing stages are conducted. This wastage continues almost until the end of the sequence. For the hulled wheats, for example, many grains are lost during pounding the spikelets and separating the glumes. But for clean grain in storage there are few processing stages required before it is ready for preparation into food. Usually the grain will be sieved with a 'wheat-sieve' to remove smaller contaminants and then hand-sorted to remove larger weed seeds (Hillman 1981b, 1984). During these final stages the amount of grain that will become incorporated in the waste and possibly discarded into the hearth will be minimal.

Sedentism, routine repetition and probability

This leads us onto other causal factors that might result in grain being under-represented on British Neolithic sites. As stated above, there is a greater likelihood that evidence of routine processing will be preserved in the archaeological record because it derives from events that are repeated throughout the year. It follows that a site occupied for many years will generate more charred grain.

It is unclear how long charred material survives after having accumulated in surface middens or in settlement soils, however, every time a feature is dug on sites occupied over many years there is a good chance that charred material will become incorporated within the deposits. The production and chances of preservation of charred material must be considerably greater on many later British prehistoric and historic sites, which appear to have been occupied for many generations spanning several centuries (cf. Lambrick 1992).

In contrast, British Neolithic sites are generally not thought to have been occupied for long time periods, indeed it is considered that many were only inhabited seasonally (Whittle 1999, p.64; Thomas 1996). Consequently the number of times that processing activities were conducted and charred material was generated would have been minimal in comparison with the later sites.

So there are several reasons why cereals might be highly under-represented in the British Neolithic. Hopefully this discussion has also revealed something about how differently neolithic peoples and those in later British prehistory and in the LBK treated cereals.

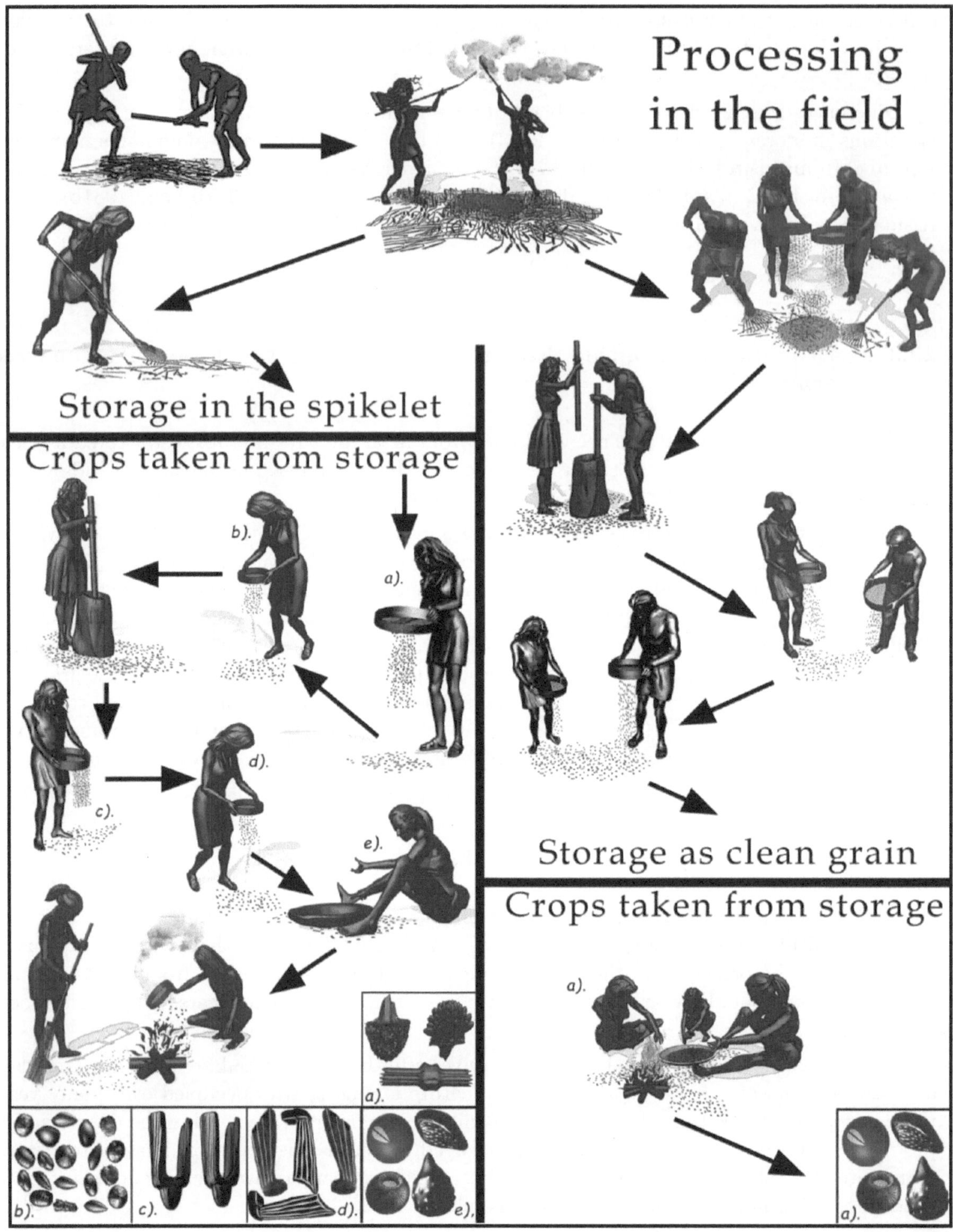

Figure 22.1: The effects of storage on charred assemblages. The left-hand sequence, typical of many British Iron Age sites, shows hulled wheats stored as semi-clean spikelets. The right-hand sequence shows a hypothetical model for the British Neolithic in which hulled wheats are stored fully dehusked, with earlier processing stages conducted in the field. The waste present (i.e., the plant remains that potentially become charred on sites) is removed as crops are taken from storage. This is shown in the boxes at the bottom of each figure showing storage practice.

22.4 The importance of cereals and wild foods

The next issue concerns the under-representation of the cereals in comparison to other subsistence resources and the implications this has with regard to the assessment of their relative importance in the neolithic diet.

22.4.1 The relative importance of cereals

It is not possible to estimate by how much cereals may be under-represented in the British Neolithic, however, we can compare their representation on these sites with sites from other periods and regions.

On the basis of two data sets from northeast and southwest England that span the later prehistoric period (van der Veen 1992; Clapham in Fitzpatrick et al. 1999), it is notable that while samples frequently comprise 10 to 20 grains per litre, approximately half at each site produce only 1 grain or less per litre. The neolithic sites reviewed by Moffett et al. (1989) on average have 0.5 to 1 grain per litre, however, they are comparatively less deficient in grain than might be expected in light of the discussion above. We cannot use the evidence based on the density of carbonised cereal remains, therefore, to state conclusively that peoples of the British Neolithic were processing and consuming considerably less cereals than those on LBK or later British sites.

22.4.2 The relative importance of wild food remains

Fragments of hazelnuts are far from exclusive to British Neolithic assemblages. The difference is that the number of samples containing hazelnut fragments and the number of fragments per litre is considerably and consistently higher on neolithic sites than on later prehistoric and historic British sites (cf. Jones 1980; Hillman 1981b; Moffett et al. 1989; Robinson 2000). The fact that there is a greater representation of hazelnuts does not imply that the wild resources were any more or less important than cultivated plant foods (cf. Jones 1991).

What we can conclude, however, is that wild food remains, and in particular hazelnuts, seem to be of greater importance during the Neolithic than they are in later periods. It should be stressed that the shorter occupation span of neolithic sites will result in the under-representation (and hence under-estimation) of grains and hazelnuts alike. If other wild foods were used that required little to no processing, it would also mean that they are likely to be absent from the archaeobotanical record. The fact that neolithic peoples have been perceived as being mobile may in addition lead to the suggestion that wild foods played an important role in subsistence, although it is possible that their mobility was related more to pastoralist activities (cf. Fleming 1972).

22.4.3 The relative importance of plant foods versus animal foods

A paper by Richards (1996) is another that is cited, in addition to the paper by Moffett et al. (1989), in support of the notion that neolithic peoples 'had no taste for grain'. This paper is based on isotope studies of human bones from Hambledon Hill, and the conclusion reached was that British Neolithic peoples ate predominately meat and that cereals formed little to no part of the diet.

There are several reasons why this paper might be considered as misleading. For example, recent studies of isotopes from modern human hair (O'Connell and Hedges 1999a) showed that even vegetarians with a regular daily intake of dairy products can produce similarly high nitrogen (δ^{15}N) isotope readings to those of the Hambledon Hill individuals. Furthermore, studies of isotope values taken from bone collagen demonstrated that they consistently produced higher nitrogen (^{15}N) values than hair samples taken from the same individuals (O'Connell and Hedges 1999b; O'Connell et al. 2001). Finally, Hambledon Hill was just the first of four neolithic sites studied by Richards; evidence from two of the three other sites suggested that plant protein formed a major part of the diet (Richards 2000, p.132).

Isotope studies undoubtedly have an important role to play in analysing prehistoric diets and perhaps the most significant contribution they have made is to show that there is a great deal of variation in diet both within as well as between neolithic populations (Richards 2000). Richards' initial study hints that animal protein was more important in the diet in the Neolithic compared to the later prehistoric and Roman periods (Richards 1996). Contrary to the conclusions drawn in this study, however, the results do not demonstrate that cereals formed an insignificant or even minor component of the diet of British Neolithic peoples.

22.4.4 The relative importance of various means of subsistence

The aim of this paper has not been to argue that cereals were as important in neolithic Britain as in the LBK or later prehistoric and historic Britain. Nor has it been to argue that the neolithic transition was one that saw the emergence of a sedentary fully agricultural society. Rather the intention has been to show that there is no reason to suspect that cereals were not as important a part of the neolithic diet as other resources. Cereals regularly occur in even the most transient of neolithic features, which at least demonstrates that they were a consistent and significant part of the neolithic diet.

While it is true that British Neolithic fields may have been half the size of those in later prehistory, there is no reason to suppose that people used only half the fields, or that the produce from a single field was shared out amongst twice as many people. This is quite different to the insignificant role that cereals are often given. In Britain, for a period of some three thousand years stretching from the later Bronze Age until the recent present, cereals undoubtedly formed the single most important contribution to the diet. During the Neolithic both the evidence from carbonised sources (Moffett et al. 1989) and isotope studies (Richards 1996) indicate that cereals were not the main dietary staples. This does not automatically imply that they were insignificant as some authors have proposed; that rather, given the time expenditure that cereal cultivation requires they had at least an important role in the neolithic diet.

To conclude, the evidence is best summed up by Whittle 2000 who states that a diverse diet rather than an extensive reliance on a single source of food seems to be the key characteristic of the Neolithic. The data indicate, and to a greater extent than for any other period, that there was a constant reliance on a broad spectrum of resources including cereals, wild-gathered plant foods, meat and dairy products from domestic animals and meat from game.

22.5 Cereal agriculture within the context of neolithic societies

We now turn to the final part of this paper and in this we will briefly examine the implications that the ideas associated with the use of cereals and wild plants have for our understanding of neolithic subsistence within a social context.

22.5.1 The social context of storage within the Neolithic

Earlier in this paper I argued that crops in neolithic Britain were stored as clean grain and that this has several advantages: it diminishes the bulk of the crop by approximately a third and, likewise, decreases the amount of labour needed for processing throughout the year. The storage of relatively clean grain would then make it possible for the crops to be transported more easily (i.e., without the large and heavy equipment needed to process spikelets), or would perhaps enable ease of collection from a centralised store. So the reduced bulk of the crop would have facilitated movement, and the reduction in time spent processing cereals would have allowed for longer periods to be spent on the collection of wild resources.

Querns, which are required to grind the grain to make flour/groats, would have been heavy to transport, and so there is a dilemma about how cereals were consumed on seasonally occupied sites. In the Neolithic querns are rare and the majority have been recovered from causewayed enclosures (Curwen 1937; Pratchett in Wheeler 1943, pp.321–323; Laws 1991), although at some sites such as the longhouse at Yarnton they appear to have been utilised for the preparation of bread (Robinson 2000). There is the possibility that the grain present at many sites would have been consumed after roasting or boiling (cf. Hillman 1981b, 1984), without the need for grinding. This is significant, however, as wear on teeth (the result of attrition caused by grit in milled flour, etc.) has been used to determine the relative importance of cereals in the diet (Hall and Coles 1994, p.90).

The storage of clean grain conflicts with Hillman's suggestion (1981b:138) that in wet climates hulled wheats would have been stored in the spikelet in order to reduce spoilage. Structures suitable for the storage of grain are absent from many neolithic sites (Field et al. 1964; Thomas 1999, fig. 4.1, p.183), which casts doubt on this theory. However, the dearth of appropriate structures may be indicative of the fact that only small amounts of grain were stored at any one time. In this light it must be assumed that cereals may have not been stored or used throughout the entire year.

22.5.2 Communal labour and the scheduling of activities

A second factor that may influence the stage at which crops are stored is the time-consuming nature of harvesting and processing. Hillman (1981b) concluded that climate determined the different patterns of storage for traditional farming communities in Turkey. For example, crops were stored in a less-processed form in wet regions where there were few long dry periods suitable for many of the processing stages that were conducted outdoors.

Certain of the processing stages, (e.g., dehusking in particular) are not only time-consuming but are also demanding in terms of labour investment. In historic times summer was a period when much of the available labour was motivated for the harvesting, threshing and storage of cereals. In the Neolithic in Britain the harvesting and processing of even small quantities of grain in late summer would have been especially demanding on labour. The nature of this demand would of course have been dependant on the scale of cultivation and whether agriculture was organised at the level of the family (e.g., small or extended) or the entire community.

On what scale agriculture was conducted within neolithic Britain is open to debate and resolving this issue is well beyond the scope of the paper. Instead I will present a few of the present thoughts on the subject.

Whether considered to be mobile or sedentary, the prevalent view of agriculture in the British Neolithic is of small fields or plots that were cultivated (possibly using a hoe) by small independent families (e.g., Childe 1940; Hawkes and Hawkes 1958; Megaw and Simpson 1979, p.86; Thomas 1999). Much of this supposition seems to be based on existing preconceptions that are often associated with the paucity of cereal remains. Because the role of cereals has been challenged, however, it follows that the nature of fields should also be questioned.

The notion of small-scale plots associated with shifting cultivation has been convincingly dismissed (Rowley-Conwy 1981b). The ard marks under South-Street long barrow is the only evidence that indicates the cultivation of larger fields (Ashbee et al. 1979). The fact that ard marks have been found to occur under barrows has led to the suggestion that they may be associated with ritual rather than domestic activities (Rowley-Conwy 1981a). In Denmark, however, Kristiansen (1990) has noted in cases where there is preservation of larger landscapes that ard marks stretch far beyond the edge of barrows, thus in conflict with the theory that they were solely ritual in function.

The possible connection of ards with ritual ploughing has led some authors to reject the idea that they were associated with cereal agriculture (Thomas 1996, p.24; Pryor 1998, p.148). However, what ritual significance or ideological association arding could possibly have had in preparation for the burial of the dead within a culture in which supposedly they are not used for cereal cultivation is not explained by either Thomas or Pryor.

Legge (1989) argues convincingly the presence of ard marks and cereal grains must indicate that ards were used in prehistoric Britain presumably for cultivation. Evidence of their presence is important because it contributes to a more realistic impression of neolithic agriculture as one that is far removed from the shifting cultivation of small swidden-like plots. Also it is probable, if ards are equated with ploughing in medieval England, that ards and the specially-trained animals that pulled them would have been shared within the wider community rather than owned by individual farmers. Given the strong emphasis in the literature on the communal aspect of neolithic British society (e.g., Whittle 1997), it is perhaps surprising that the possibility some agricultural activities were undertaken at a larger, more communal scale has not been considered.

22.5.3 The social context of subsistence within neolithic societies

The discussion has provided the opportunity to look at some of the issues surrounding the transition to agriculture. Many aspects of neolithic subsistence that have been described echo traits common to groups of hunter-gatherers. The most obvious of these is the continued use of wild foods, but in addition the ways in which cereals are treated seem to have parallels within hunter-gather traditions. While storage is not unknown in hunter-gatherer societies, it seldom plays such an important role as it does in more fully developed agricultural societies. As hunter-gatherers often processed collected foods for immediate consumption, so the early British Neolithic cultivators seem largely to have processed grain after the crops were gathered in and consumed most before the following year's harvest.

In the absence of large-scale storage, the failure of crops may have been a problem that could have been offset by having a more diverse resource base (Moffett et al. 1989). It could be considered that British Neolithic societies were not structured (i.e., in terms of exchange networks and/or centralised storage) to accommodate risk strategies to cope with crop failures (Halstead and O'Shea 1982). Instead it appears that neolithic peoples relied on risk strategies such as diversity and mobility, which are more commonly associated with hunter-gatherers.

Finally we may speculate on the nature of the agricultural transition itself. For non-literate societies knowledge must be passed by word of mouth or through participation.The implication for British Neolithic societies is that the knowledge concerned with the processing of cereals was acquired during a short period in late summer. As previously argued, peoples of the LBK stored grain as spikelets, performing the remaining operations through the year. The transition to the complete processing of crops may have occurred with the participation of hunter-gatherers in the harvest, perhaps in exchange for some of the crops.

It is hoped that this paper has provided some ideas about the directions which the study of neolithic subsistence might take in the future. It is interesting to note that at least in late neolithic France, dehusked emmer has been recovered from storage pits (Bouby 1998; Huntley and Rowley-Conwy 1996). Bakels (1995) observes of the French Bandkeramik site at Cuiry-lès-Chaudardes, Ainse, that it produced 'unusually' no evidence of glumes in contrast to those sites examined in the Netherlands and Germany, which would seem to provide further evidence for the model proposed and possibilities for its exploration in areas beyond Britain.

Acknowledgments

Many of the ideas within this paper were initially developed and tested while working on a Post-Doctoral Research Project funded by McDonald Institute, Cambridge. I would like to thank Dorian Q. Fuller, Michael Allen and Martin Jones for fruitful conversation regarding some of the issues that emerged in this paper.

References

Ashbee, P., I. F. Smith, and J. G. Evans (1979). Excavations of three long barrows near Avebury. *Proceedings of the Prehistoric Society 45*, 207–300.

Bakels, C. (1978). *Four Linnearbandkeramik Settlements and their Environment: A Palaeoecological Study of Sittard, Stein, Elsoo and Hienheim*, Volume XI of *Analecta Praehistorica Leidensia.* Leiden: Leiden University Press.

Bakels, C. (1995). In search of activity areas within Bandkeramik farmyards: the disposal of burnt chaff. In H. Kroll and R. Pasternak (Eds.), *Res Archaeobotanicae: 9th Symposium IWGP*, pp. 1–4. Kiel: Oetker-Voges Verlag.

Barker, G. and D. P. Webley (1978). Causewayed camps and early neolithic economies in central southern England. *Proceedings of the Prehistoric Society 44*, 161–186.

Barrett, J. C. (1994). *Fragments from Antiquity.* Oxford: Blackwell.

Bouby, L. (1998). Two early finds of gold-of-pleasure (*Camelina* sp.) in middle neolithic and chalcolithic sites in western France. *Antiquity 72*, 391–398.

Bradley, R. (1978). *The Prehistoric Settlement of Britain.* London: Routledge and Kegan.

Bradley, R. (1987). Flint technology and the character of neolithic settlement. In A. G. Brown and M. Edmonds (Eds.), *Lithic Analysis and Later British Prehistory*, BAR British Series 162, pp. 181–185. Oxford: British Archaeological Reports.

Childe, V. G. (1940). *Prehistoric Communities of the British Isles.* London: W. & R. Chambers.

Clark, J. G. D. (1952). *Prehistoric Europe: The Economic Basis.* Cambridge: Cambridge University Press.

Coleridge, S. T. (1836). *The Literary Remains.* London: William Pickering.

Curwen, E. E. (1937). Querns. *Antiquity 11*, 133–151.

Dennell, R. W. (1983). *European Economic Prehistory: A New Approach.* London: Academic Press.

Entwhistle, R. and A. Grant (1989). The evidence for cereal cultivation and animal husbandry in the southern British Neolithic and Bronze Age. In A. Milles, D. Williams, and N. Gardner (Eds.), *The Beginnings of Agriculture*, Symposia of the Association for Environmental Archaeology No. 8. BAR International Series 496, pp. 203–215. Oxford: British Archaeological Reports.

Fairbairn, A. S. (1997). Charred plant remains. In A. Whittle (Ed.), *Sacred Mound, Holy Rings: Silbury Hill and the West Kennet Palisade Enclosures: a Later Neolithic Complex in North Wiltshire*, Oxbow Monograph 74, pp. 134–138. Oxford: Oxbow Books.

Fairbairn, A. S. (1999). Charred plant remains. In A. Whittle, J. Pollard, and C. Grigson (Eds.), *The Harmony of Symbols. The Windmill Hill Causewayed Enclosure*, Cardiff Studies in Archaeology, pp. 139–156. Oxford: Oxbow Books.

Fairweather, A. D. and I. B. M. Ralston (1993). The neolithic timber hall at Balbridie, Grampian Region, Scotland: the building, the date, the plant macrofossils. *Antiquity 67*, 313–323.

Field, N. H., C. L. Mathews, and I. F. Smith (1964). New neolithic sites in Dorset and Bedfordshire with a note on the distribution of neolithic storage pits in Britain. *Proceedings of the Prehistoric Society 30*, 352–381.

Fitzpatrick, A. P., C. A. Butterworth, and J. Grove (1999). *Prehistoric and Roman Sites in East Devon: the A30 Honiton to Exeter Improvement DBFO Scheme, 1996-9: Volume 1. Prehistoric Sites.* Wessex Archaeological Report No. 16. Salisbury: Wessex Archaeology.

Fleming, A. (1972). The genesis of pastoralism in European prehistory. *World Archaeology 4*, 179–191.

Hall, D. and J. Coles (1994). *Fenland Survey: An Essay in Landscape and Persistence.* London: English Heritage.

Halstead, P. and J. O'Shea (1982). A friend in need is a friend indeed: social storage and the origins of social ranking. In C. Renfrew and S. Shennan (Eds.), *Ranking, Resource and Exchange*, pp. 92–99. Cambridge: Cambridge University Press.

Hawkes, C. and J. Hawkes (1943). *Prehistoric Britain.* London: Penguin Books.

Hawkes, J. and C. Hawkes (1958). *Prehistoric Britain* (Revised ed.). London: Penguin Books.

Helbaek, H. (1969). Plant collecting, dry farming, and irrigation agriculture in prehistoric Deh Luran. In K. V. Flannery and J. A. Neely (Eds.), *Prehistory and Human Ecology of the Deh Luran Plain*, pp. 383–426. Ann Arbor: University of Michigan.

Hillman, G. C. (1973). Crop husbandry and food production: modern models for the interpretation of plant remains. *Anatolian Studies 23*, 241–244.

Hillman, G. C. (1981a). Crop husbandry: evidence from macroscopic remains. In I. Simmons and M. Tooley (Eds.), *The Environment in British Prehistory*, pp. 183–191. London: Duckworth.

Hillman, G. C. (1981b). Reconstructing crop husbandry practices from charred remains of crops. In R. J. Mercer (Ed.), *Farming Practice in British Prehistory*, pp. 123–162. Edinburgh: Edinburgh University Press.

Hillman, G. C. (1984). Interpretation of archaeological plant remains: the application of ethnographic models from Turkey. In W. van Zeist and W. A. Casparie (Eds.), *Plants and Ancient Man*, Studies in Palaeoethnobotany. Proceedings of the Sixth Symposium of the International Work Group for Palaeoethnobotany, Groningen, 30 May–3 June 1983, pp. 1–42. Rotterdam: A. A. Balkema.

Huntley, J. P. (1996). The plant remains, in P. Abramson "Excavations Along the Caythorpe Gas Pipeline, North Humberside". *Yorkshire Archaeological Journal 68*, 80–81.

Huntley, J. P. and P. Rowley-Conwy (1996). Neolithic agriculture in western France. *Universities of Durham and University of Newcastle upon Tyne Archaeological Reports for 1995 19*, 1–3.

Jones, G. (1984). Interpretation of archaeological plant remains: ethnographic models from Greece. In W. van Zeist and W. A. Casparie (Eds.), *Plants and Ancient Man*, Studies in Palaeoethnobotany. Proceedings of the Sixth Symposium of the International Work Group for Palaeoethnobotany, Groningen, 30 May–3 June 1983, pp. 43–61. Rotterdam: A. A. Balkema.

Jones, G. (1987). A statistical approach to the archaeological identification of crop processing. *Journal of Archaeological Science 14*, 311–323.

Jones, G. (1991). Numerical analysis in archaeobotany. In W. van Zeist, K. Wasylikowa, and K.-E. Behre (Eds.), *Progress in Old World Palaeoethnobotany: A Retrospective View on the Occassion of 20 Years of the International Work Group for Palaeoethnobotany*, pp. 63–80. Rotterdam: A. A. Balkema.

Jones, G. (2000). Evaluating the importance of cultivation and collecting in neolithic Britain. In A. S. Fairbairn (Ed.), *Plants in Neolithic Britain and Beyond*, Neolithic Studies Group Seminar Paper 5, pp. 79–84. Oxford: Oxbow Books.

Jones, M. (1980). Carbonised cereals from grooved ware contexts. *Proceedings of the Prehistoric Society 46*, 61–63.

Jones, M. (1985). Archaeobotany beyond subsistence reconstruction. In G. Barker and C. Gamble (Eds.), *Beyond Domestication in Prehistoric Europe: Investigations in Subsistence Archaeology and Social Complexity*, pp. 107–128. London: Academic Press.

Knörzer, K. H. (1971). Urgeschichtliche Unkräuter im Rheinland ein Beitrag zur Entstehungsgeschichte der Segetalgesellschaften. *Vegetatio 23*, 89–110.

Knörzer, K.-H. (1973). Pflanzliche Großreste. In J.-P. Farruggia, R. Kuper, J. Lüning, and P. Stehli (Eds.), *Der Bandkeramische Siedlungsplatz Langweiler 2: Gemeinde Aldenhoven, Kreis Düren*, pp. 139–152. Bonn: Rheinland-Verlag.

Knörzer, K.-H. (1980). Pflanzliche Großreste des Siedlungsplatzes Wanlo (Mönchengladbach), Naturwissenschafliche Beiträge zur Archäologie. *Archaeo-Physika 7*, 7–20.

Knörzer, K.-H. (1988). Untersuchungen der Früchte und Samen. In U. Boelicke, D. von Brandt, J. Lüning, P. Stehli, and A. Zimmermann (Eds.), *Der Bandkeramische Siedlungsplatz Langweiler 8: Gemeinde Aldenhoven, Kreis Düren*, pp. 813–852. Köln: Rheinland-Verlag.

Kreuz, A. (1990). *Die ersten Bauern Mitteleuropas: Eine archäobotanische Untersuchung zu Umwelt und Landwirtschaft der Ältesten Bandkeramik.* Analecta Praehistorica Leidensia 23. Leiden: Leiden University Press.

Kristiansen, K. (1990). Ard marks under barrows: a response to Peter Rowley-Conwy. *Antiquity 64*, 322–327.

Lambrick, G. H. (1992). The development of late prehistoric and Roman farming on the Thames gravels. In M. Fulford and E. Nichols (Eds.), *Developing Landscapes of Lowland Britain: The Archaeology of the British Gravels, A Review*, Volume 14 of *Occasional Papers from the Society of Antiquaries of London*, pp. 23–38. London: Society of Antiquaries.

Laws, K. (1991). The foreign stone. In N. M. Sharples (Ed.), *Maiden Castle: Excavations and Field Survey 1985–6*, English Heritage Archaeological Report 19, pp. 229–233. London: English Heritage.

Legge, A. J. (1989). Milking the evidence: a reply to Entwhistle and Grant. In A. Milles, D. Williams, and N. Gardner (Eds.), *The Beginnings of Agriculture*, Symposia of the Association for Environmental Archaeology No. 8/BAR International Series 496, pp. 217–242. Oxford: British Archaeological Reports.

Legge, A. J., S. Payne, and P. Rowley-Conwy (1998). The study of food remains in prehistoric Britain. In J. Bayley (Ed.), *Science in Archaeology: An Agenda for the Future*, pp. 89–94. London: English Heritage.

Megaw, J. V. S. and D. D. A. Simpson (1979). *Introduction to British Prehistory.* Leicester: Leicester University Press.

Mercer, R. J. (1981). Introduction. In R. Mercer (Ed.), *Farming Practice in British Prehistory*, pp. ix–xxvi. Edinburgh: Edinburgh University Press.

Moffett, L., M. A. Robinson, and V. Straker (1989). Cereals, fruit and nuts: charred plant remains from neolithic sites in England and Wales and the neolithic economy. In A. Milles, D. Williams, and N. Gardner (Eds.), *The Beginnings of Agriculture*, Symposia of the Association for Environmental Archaeology No. 8. BAR International Series 496, pp. 243–261. Oxford: British Archaeological Reports.

Nilsson, S. (1868). *The Primitive Inhabitants of Scandinavia.* Longmans Green. Trans. J. Lubbock.

O'Connell, T. C. and R. E. M. Hedges (1999a). Investigations into the effect of diet on modern human hair isotopic values. *Journal of Physical Anthropology 108*, 409–425.

O'Connell, T. C. and R. E. M. Hedges (1999b). Isotopic comparison of hair and bone archaeological analyses. *Journal of Archaeological Science 26*, 661–665.

O'Connell, T. C., R. E. M. Hedges, M. A. Healey, and A. H. R. W. Simpson (2001). Isotopic comparison of hair, bone and nail: modern analyses. *Journal of Archaeological Science 28*, 1247–1255.

Pryor, F. (1998). *Farmers in Prehistoric Britain.* Stroud: Tempus.

Pryor, F. M. M. (1988). Earlier neolithic organised landscapes and ceremonial in lowland Britain. In J. C. Barrett and I. A. Kinnes (Eds.), *The Archaeology of Context in the Neolithic and Bronze Age: Recent Trends*, pp. 63–72. Sheffield: J. R. Collis.

Richards, M. P. (1996). 'First farmers' with no taste for grain. *British Archaeology 12*, 6–7.

Richards, M. P. (2000). Human consumption of plant foods in the British Neolithic: direct evidence from bone stable isotopes. In A. S. Fairbairn (Ed.), *Plants in Neolithic Britain and Beyond*, Neolithic Studies Group Seminar Paper 5, pp. 123–136. Oxford: Oxbow Books.

Richmond, A. (1999). *Preferred Economies: The Nature of the Subsistence Base throughout Mainland Britain during Prehistory.* BAR British Series 290. Oxford: Archaeopress.

Robinson, M. A. (2000). Further considerations of neolithic charred cereals, fruits and nuts. In A. S. Fairbairn (Ed.), *Plants in Neolithic Britain and Beyond*, Neolithic Studies Group Seminar Paper 5, pp. 85–90. Oxford: Oxbow Books.

Rowley-Conwy, P. (1981a). The interpretation of ard marks. *Antiquity 61*, 263–267.

Rowley-Conwy, P. (1981b). Slash and burn in the temperate European neolithic. In R. J. Mercer (Ed.), *Farming Practice in British Prehistory*, pp. 85–96. Edinburgh: Edinburgh University Press.

Rowley-Conwy, P. (2000). Through a taphonomic glass darkly: the importance of cereal cultivation in prehistoric Britain. In J. Huntley and S. Stallisbrass (Eds.), *Taphonomy and Interpretation*, Symposia of the Association for Environmental Archaeology No. 14, pp. 43–53. Oxford: Oxbow Books.

Smith, I. F. (1974). The Neolithic. In C. Renfrew (Ed.), *British Prehistory: A New Outline*, pp. 100–136. Duckworth.

Stevens, C. J. (2003). An investigation of agricultural consumption and production models for prehistoric and Roman Britain. *Environmental Archaeology 8*, 61–76.

Straker, V. (1990). Carbonised plant macrofossils. In A. Saville (Ed.), *Hazleton North, Gloucestershire, 1979-82: The Excavation of a Neolithic Long Cairn of the Cotswold-Severn Group*, English Heritage Archaeological Report 15, pp. 214–218. London: English Heritage.

Thomas, J. (1991). *Rethinking the Neolithic.* Cambridge: Cambridge University Press.

Thomas, J. (1996). Neolithic houses in mainland Britain and Ireland: a sceptical view. In T. Darvill and J. Thomas (Eds.), *Neolithic Houses in Northwest Europe and Beyond*, Oxbow Monograph 57/Neolithic Studies Group Seminar Paper 1, pp. 1–12. Oxford: Oxbow Books.

Thomas, J. (1999). *Understanding the Neolithic.* London: Routledge.

Thorpe, I. J. (1996). *The Origins of Agriculture in Europe.* London: Routledge.

Troels-Smith, J. (1960). Ivy, mistletoe and elm: climate indicators, fodder plants: a contribution to the interpretation of the pollen zone border VII-VIII. *Danmarks Geologiske Undersøgelser 4*(4), 1–32.

Turner, J. F. (1893). The significance of the frontier in American history. *Report of the American Historical Association 1893*, 199–227.

van der Veen, M. (1992). *Crop Husbandry Regimes: An Archaeobotanical Study of Farming in Northern England 1000 BC–AD 500.* Sheffield Archaeological Monographs 3. Sheffield: J. R. Collis.

Wheeler, M. (1943). *Maiden Castle, Dorset.* London: Society of Antiquaries of London.

Whittle, A. W. R. (1996). *Europe in the Neolithic: The Creation of New Worlds.* Cambridge: Cambridge University Press.

Whittle, A. W. R. (1997). *Sacred Mounds and Holy Rings: Silbury Hill and West Kennet.* Number 74 in Oxbow Monographs. Oxford: Oxbow Books.

Whittle, A. W. R. (1999). The Neolithic period: 4000–2500/2200 BC. In J. Hunter and I. Ralston (Eds.), *The Archaeology of Britain*, pp. 58–76. London: Routledge.

Whittle, A. W. R. (2000). Bringing plants into the taskscape. In A. S. Fairbairn (Ed.), *Plants in Neolithic Britain and Beyond*, Neolithic Studies Group Seminar Paper 5, pp. 1–7. Oxford: Oxbow Books.

Chapter 23

On the importance of cereal cultivation in the British Neolithic

Glynis Jones and Peter Rowley-Conwy
University of Sheffield and University of Durham, UK

23.1 Introduction

This paper has two purposes. Firstly, we will present a summary of current evidence for cultivated and wild plants in the Neolithic of Britain. Since the pioneering synthesis by Moffett et al. (1989), comprehensive reviews have been undertaken at a regional level (e.g., Huntley and Stallibrass 1995; Campbell and Straker 2003; Murphy 1997), and we will bring these together with databases of archaeobotanical reports (e.g., English Heritage 2004; Tomlinson and Hall 1996) and other recent work. Secondly, we will discuss the importance of cereal cultivation. Some have argued that the neolithic economy remained essentially mesolithic, and that cereals were 'special' foods consumed only rarely and in 'ritual' contexts. Others have argued that cereals were more widely consumed, and formed the basis of the domestic economy.

This latter debate is part of a wider discussion of the British Neolithic, and of how the archaeological record of this and other periods should be approached and interpreted. The issue has been approached from very different directions by different authors. Our approach is strictly archaeobotanical; specifically (a) the factors that give rise to the remains we find, and (b) the contexts in which they are found. We hope that our discussion will form a useful contribution to the ongoing debate, and that our table of archaeobotanical sites and data will form the basis for periodic updating by subsequent researchers.

23.2 British Neolithic plant remains

Table 23.1 presents a list of sites with charred archaeobotanical items—waterlogged remains and impressions in pottery have not been included. While we have made every effort to track down as many sites as possible, we cannot pretend that this list is exhaustive. Some published and archive reports are certain to have been overlooked, so in addition to the table being updated with future finds we anticipate that material currently available but not known to us will also be added. In most cases, the original archaeobotanical reports have been consulted and, where this is not the case, our secondary source is also cited in table 23.1. We have not systematically included spot finds or reports dated to before 1950. Table 23.2 presents a less compete list for Ireland.

Site	Vol	Barley	Gl-wh	Br-wh	Wh	Total	Glume	Rachis	Weeds	Hazel	Apple	Other fruit	GTotal	References
Early Neolithic														
Balbridie	.	some n	many	few	.	c.20000	few	.	few	some	few	flax	many	1
Bishopstone	.	.	6	.	9	25	few	.	14	.	.	.	39	2
Boghead	1000	305n	38	.	.	411	.	.	119	14	.	.	544	3
Bolam Lake	50	.	1	.	1	4	.	.	2	.	.	1 haw	6	4
Briar Hill	6	1	.	.	.	3	.	.	20	5	.	1 sloe	28	2
Bromfield	7	3	.	.	1	4	.	.	3	some	2	.	9+h	2
Capel Eithen	.	8	.	.	.	8	.	.	12	some	.	.	20+h	5
Claish	.	10h	14	2	7	80	.	.	.	>517	7	.	>604	6
Coneybury	61	.	10	.	.	50	1	.	.	.	.	.	51	7
Coupland	c.125	2	11	.	1	18	307	.	.	many	.	.	243+h	8
Dorney	967	2h	16	3	5	99	.	.	2	53	.	3 sloe/haw	154	9
Etton	.	.	.	.	.	.	.	.	.	some	.	.	some	10
Flagstones	218	10	.	.	10	24	.	.	14	.	.	.	38	11
Gwernvale	.	.	.	.	.	few	.	.	.	many	.	.	many	2
Haddenham	.	393	24	.	46	632	.	1	35	many	1	1 Rubus	669	12
Hambledon Hill (HH)	.	12	178	.	109	358	2	.	1	many	.	.	361+h	13
HH Stepleton	.	456nh	716	.	243	1774	.	.	1	many	.	1 grape	1775+h	13
HH Stepleton conc.	.	.	c.108000	.	.	c.108000	c.54300	.	c.200	some	.	.	162500+h	13
Hazleton	550	3	51	33	.	283	7	.	50	>1000	.	.	>1340	14
Hembury	.	.	many	.	.	many	.	.	.	.	.	.	many	14
Isbister	.	56n	1	.	.	63	.	.	239	.	.	.	302	15
Knap of Howar	.	9	.	.	.	32	.	.	.	1	.	.	33	16
Lismore Fields 1	172	1	2215	63	1296	3658	2350	.	10	many	20	124 flax	6038+h	17
Lismore Fields 2	55.8	.	5	.	8	13	43	.	6	many	27	.	89+h	17
Lismore Fields rest	355	.	559	16	236	858	39	.	5	many	.	.	902+h	17
Lofts Farm	292	.	.	.	.	2	.	.	1	some	.	1 sloe	3+h	2
Maiden Castle	1242	44	many	.	101	312	9	.	62	85	.	.	468	18
Moel y Gerddi	.	.	.	.	.	.	.	.	26	2	.	1 Rubus	28	19
Nympsfield	.	.	.	.	.	.	.	.	2	.	.	.	2	20
Robin Hood's Ball	102	.	.	.	.	1	.	.	.	26	.	.	27	2
Rowden	66	2	8	.	.	41	.	.	.	.	.	.	41+h	21
Spong Hill, UK Spong Hill	15	some	some	.	.	4	few?	.	.	2	.	.	6	2
Springfield Lyons	32	.	.	.	3	5	.	.	3	.	.	1 sloe	8	22, cited in 23
The Stumble 28C	.	.	50	4	.	394	151	.	91	many	3	2 sloe	639+h	24
Weir Bank Stud Farm	.	.	.	.	.	few	.	.	.	.	.	.	few	25
Whitesheet Hill	169	.	1	.	1	few	1	.	3	many	.	1 sloe	few	26
Wilsford Down	10	.	.	.	.	.	.	.	.	2	.	.	2	2
Windmill Hill outer	530	8h	3	.	7	46	.	.	87	1	.	.	134	27
Windmill Hill middle	330	3	13	.	11	76	2	.	45	.	.	.	123	27
Windmill Hill inner	140	12h	2	.	3	45	.	.	97	.	.	.	142	27
Windmill Hill outside	22.5	30nh	3	.	24	121	129	.	.	195	some	sloe,haw	445	28

Column key: Vol=volume of deposit sampled in litres; Barley=barley grains; Gl-wh=emmer, einkorn and possible spelt grains; Br-wh=free-threshing wheat grains; Glume=glume wheat glume bases; Rachis=free-threshing rachis fragments; Weeds=seeds of potential weeds; Hazel=fragments of hazelnutshell; Apple=apple/pear; GTotal=grain+glumes+rachis+weeds+hazel+apple/pear (n=including naked barley; h=including hulled barley; h=hazelnutshell).

Reference key: 1. Fairweather and Ralston (1993), 2. Moffett et al. (1989), 3. Maclean and Rowley-Conwy (1984), 4. Huntley (2002), 5. Williams (1999), 6. Miller and Ramsay (2002), 7. Carruthers (1990), 8. Huntley (nd), 9. Robinson (2000a), 10. Nye and Scaife (1998), 11. Straker (1997), 12. Jones (nda), 13. Jones and Legge (nd), 14. Straker (1990a), 14. Helbaek (1952), 15. Lynch (1983), 16. Dickson (1983), 17. Jones (ndb), 18. Palmer and Jones (1991), 19. Williams (1988), 20. Arthur and Paradine (1979), 21. Carruthers (1991), 22. Murphy (1990b), 23. Murphy (1997), 24. Murphy (nd), 25. Clapham (1995), 26. Hinton (2004), 27. Fairbairn (1999), 28. Fairbairn (2000)

Table 23.1: Archaeobotanical finds from neolithic or possible neolithic sites in Britain

Site	*Vol*	*Barley*	*Gl-wh*	*Br-wh*	*Wh*	*Total*	*Glume*	*Rachis*	*Weeds*	*Hazel*	*Apple*	*Other fruit*	*Total*	*References*
Early/Late Neolithic														
Aston-on-Trent	.	2?	1159	.	.	1159	26?	.	?	84	.	.	1269	29, 30
Beehive	.	.	.	.	.	.	.	.	164	120	.	.	284	31
Brampton	.	2	.	.	2	7	1	.	2	1	.	1 sloe	11	32, cited in 23
Castle Hill	.	.	.	.	.	42	2	.	18	9	.	2 Pspin, 2 Rubus, 2 Samb	few	33
Chigborough Farm	.	4	2	.	3	19	2	1+1br	7	3	.	.	33	34, cited in 23
Cornie Reach	.	1	.	.	.	1	.	.	.	few	.	.	few	35
Cowie Road	.	3	1	2	.	6+	.	.	some	.	.	.	>6	36
Crickley Hill	.	.	.	.	.	.	.	.	.	some	.	.	some	37
Deeping St Nicholas	.	1	.	.	.	4	.	.	9	13	.	4 haw	26	38, cited in 23
Easton Down	190	.	.	.	.	1	.	.	16	1	.	.	25	39
Field Farm	495	.	1	.	.	9	.	.	60	22	.	.	91	40
Garden Farm	.	.	.	.	.	.	.	.	.	many	.	.	many	41, cited in 35
Gatehampton Farm	.	.	.	.	.	.	.	.	.	1	.	.	1	42
Green Park	.	.	.	.	.	.	.	.	.	.	.	.	.	43
Greyhound Yard	.	.	.	.	3	5	.	.	6	.	.	1 Samb	11	44
Holloway Lane	.	.	.	.	.	.	.	.	.	many	.	.	many	41, cited in 35
Hazard Hill	.	.	.	.	.	.	.	.	.	some	.	.	some	45, cited in ABCD
Kinbeachie	.	212nh	13	.	10	452	.	.	8	many	.	1 Prunus	460+h	46
Le Pinnacle	.	.	.	.	.	.	.	.	.	some	.	.	some	47
Leven-Brandesburton	.	.	.	.	.	.	.	.	.	some	.	.	some	48
Long Range	.	1h	.	.	.	1	.	.	.	1	.	.	2	33
Maxey	3320	some	some	some	.	14	.	.	3	5	.	.	22	49
Old Down Farm	.	.	.	.	.	.	.	.	.	2	.	.	2	37
Sewerby Cottage Farm	.	8h	3	12	5	51	.	.	1	26	.	.	78	50
Stacey Bushes	.	.	.	.	.	.	.	.	.	1	.	.	1	51, cited in ABCD
The Stumble 28A	.	7n	158	.	163	801	369	.	279	many	4	14 sloe,12 Rubus	1453+h	24
The Stumble 28B	.	.	29	.	78	326	98	.	259	many	1	11 sloe	684+h	24
The Stumble 28J	.	.	.	.	5	14	2	.	91	many	.	5 sloe	106+h	24
Yarnton	7000	12	23	5	8	201	2	.	10	2716	7	3 sloe, 1 haw	2936	9
Watchfield	.	some	.	2	.	5	.	.	.	.	.	.	5	52
Wath Quarry	.	.	.	.	.	.	.	.	.	few	.	.	few	53
West Heslerton	.	.	.	.	1	1	.	.	11	some	.	.	some	54

Reference key, cont.: 29. Alvey (1964), 30. Reaney (1968), 31. Higgins (2003), 32. Murphy (1992a), 33. Clapham (1999), 34. Murphy (1988b), 35. Giorgi (1996), 36. Holden (1997), 37. Green (1981), 38. Murphy (1994), 39. Fairbairn (1993), 40. Carruthers (1992), 41. Giorgi (1994), 42. Letts (1995), 43. Campbell (2004), 44. Jones and Straker (1993), 45. Taylor (1963), 46. Hastie and Holden (2001), 47. Carruthers (2001), 48. Hall et al. (1994), 49. Green (1985), 50. Huntley (2001), 51. Keepax (1985), 52. Huntley (1993), 53. Jaques et al. (2001), 54. Allen (1986).

Table 23.1, cont.

Site	Vol	Barley	Gl-wh	Br-wh	Wh	Total	Glume	Rachis	Weeds	Hazel	Apple	Other fruit	Total	References
Late Neolithic														
Barrow Hills	260	.	.	.	.	10	.	.	14	307	324	.	655	55
Barton Court Farm	115	3	2	3	9	21	.	.	11	84	23	.	139	56
Beckton Farm	.	few n	.	.	.	few	.	.	.	few	.	.	few	57
Blewbury	25	2	.	6	3	19	.	.	16	76	.	.	111	2
Broom	.	.	.	.	.	.	.	.	4	2277	1234	5 sloe	3511	58
Capel Eithen	.	314nh	2	2	4	323	.	.	15	many	.	1 Prunus, 1 Rubus	338+h	5
Caythorpe	c.30	2	>1000	.	.	>1000	1	.	some	.	40	.	>1041+h	59
Cherviot Quarry	.	.	.	.	.	3	.	.	.	few	.	.	few	60
Coneybury	26	304nh	.	.	.	331	.	.	5	.	.	1haw	336	7
Dog Farm	.	1h	.	1	1	6	.	.	.	1839	1	1 Rubus	1846	9
Down Farm	.	11n	.	.	1	25	.	.	5	2	1	.	33	61
Flagstones	167	23	.	1	25	82	.	1br	52	.	.	.	134	62
Lairg	.	many	many	.	.	many	some	.	.	.	.	.	many	63
King Barrow Ridge	125	.	1	.	.	1	.	.	1	160	.	1sloe	162	7
M.G. Abingdon	28	.	.	.	.	1	.	.	1	66	.	.	68	2
Marton-le-Moor	>1000	>1000	264	.	8	>1272	23?	.	.	many	674	.	>1972+h	59
Mount Farm	.	8	1	1	1	11	.	.	11	425	.	.	447	56
Nosterfield	30	2	.	.	.	2	.	.	.	9	2	.	13	64
Pits Plantation	.	1h	.	.	.	few	.	.	1	some	1	.	few	65
Potlock	.	4n	.	.	5	19	.	.	253	2	.	1 sloe, 2 haw, 46 Rubus	274	66
Prospect Park	.	.	.	.	.	few	.	.	1	10	.	4 haw	few	66
Rectory Farm	.	1n	2	2	1	9	.	.	.	1	.	.	10	23
Rosinish	.	165hn	5	.	.	170	.	.	.	.	.	.	170	68
Scord of Brouster 1	>120	c.62h	.	.	.	62	.	414	193	.	.	.	669	69
Scord of Brouster 2	165	c.549h	.	.	.	549	.	413	326	.	.	.	1288	69
Scord of Brouster 3	.	c.2813h	.	.	.	2813	.	467	105	.	.	.	3385	69
Skara Brae	.	311	28	.	.	355	.	.	.	.	.	.	355	70
Springfield	.	.	3	6	3	21	1	.	21	5	.	2 haws,2 Prunus	47	71
Stones of Stenness*	.	254	.	.	.	312	.	.	.	.	.	.	312	72
Thirlings	168	1	.	.	.	2	.	.	22	1587	.	1 Rubus, 1 haw	1611	73
Trelystan	114	.	.	.	.	.	.	.	3	some	1	1 Rubus	.	2
Trowse	.	.	3	.	1	19	2	1br	1	many	2	.	25+h	74, cited in 23
West Kennet encls*	.	25h	2	9	15	166	.	2br	52	1	.	1sloe	221	75
Whitton Hill	123	4	1	.	.	6	.	.	14	46	.	.	66	73, 76
Willow Farm	250	10	9	.	1	42	.	1	4	102	95+	.	244+	77
Windmill Hill outside	68	.	.	.	.	2	.	.	.	437	.	.	439	28
Woodham Walter	.	.	.	80	.	80	.	.	.	.	.	.	80	78

Reference key, cont.: 55. Moffett (1999), 56. Jones (1980), 57. Boardman (1997), 58. Moffett and Ciaraldi (1999) 59. Huntley and Stallibrass (1995), 60. Hall (2000), 61. Jones (1991), 62. Straker (1997), 63. Holden (1998), 64. Huntley (1996a), 65. Huntley (1996b), 66. Monckton and Moffett (nd), 67. Hinton (1996), 68. Maclean and Rowley-Conwy (1979), 69. Milles (1986b), 70. Rowley-Conwy (nd), 71. Murphy (2001), 72. MacLean (1978), 73. M. Van der Veen (1985), 74. Murphy (1992b), 75. Fairbairn (1997), 76. Huntley (2000), 77. Monckton (2002), 78. Boyd (1987).

Table 23.1, cont.

Site	Vol	Barley	Gl-wh	Br-wh	Wh	Total	Glume	Rachis	Weeds	Hazel	Apple	Other fruit	Total	References
Late Neolithic/Bronze Age														
Brean Down	.	.	.	.	2	2	.	.	7	1	.	1sloe,2haw	10	79
Chilbolton	.	.	.	.	.	8	.	.	3	7	.	.	18	80
Crescent Copse	.	.	.	.	.	.	.	.	.	some	.	.	some	81
Easton Lane	.	3h	.	.	.	31	1	.	.	262	.	.	294	82
Four Crosses	.	3	.	.	.	5	.	.	3	.	.	.	8	83
Gravelly Guy	170	.	.	.	.	8	.	.	18	156	.	.	182	2
Grimes Graves	2763	.	.	.	.	.	.	.	5	.	.	.	5	84
Jaywick	19	.	.	.	.	1	2	.	.	.	.	.	3	85, cited in 23
Longhham	.	2	.	.	.	4	.	.	3	many	.	.	7+h	86, cited in 23
Maiden Castle	880	20	many	.	15	64	9	1	18	12	.	.	104	18
Ness of Gruting	.	4190	.	.	.	4229	.	17	.	.	.	.	4246	83
Oakham	350	9	.	6	11	56	.	.	5	4	.	.	65	87
Pakenham	.	.	.	.	.	.	.	.	.	many	.	.	many	88, in 23
Poundbury	.	3h	1	.	.	7	.	.	2	17	.	.	26	89
Redgate Hill	.	4n	1	6	3	14	.	.	.	4	.	.	18	90, cited in 23
Slough House Farm	.	.	4	.	4	24	3	.	8	4	.	.	39	34, cited in 23

Reference key, cont.: 79. Straker (1990b), 80. Green (1990), 81. Higgins (2000), 82. Carruthers (1989), 83. Milles (1986a), 84. Legge (1981), 85. Wilkinson and Murphy (1995), 86. Murphy (1990a), 87. Monckton (1989), 88. Murphy and Wiltshire (1989), 89. Monk (1987), 90. Murphy (1993).

Table 23.1, cont.

Site	Barley	Gl-wh	Br-wh	Wh	Tot. grain	Glume	Rachis	Weeds	Hazel	Apple	Other fruit	GTotal	References
Early Neolithic													
Ballyharry	some	some	.	.	some	.	.	.	many	.	Rubus	many	1
Cloghers	some	some	some	.	some	.	.	some	some	.	.	some	2
Corbally	few	some?	.	.	many	.	many	.	many	.	.	.	3
Drummenny	many n	.	some	.	lots	.	.	.	.	.	.	many	4
Enagh	.	.	.	.	.	.	.	4	33	.	.	37	5
Knowth	19n	.	1	.	29	.	.	.	?	.	?	.	6
Tankardstown South	.	581	.	.	2317	133	.	10	18	324	.	2802	7
Pepperhill	.	.	.	.	8	.	.	.	86	1	.	95	7
Towneyhall II	1?	many	.	.	.	.	.	.	some	.	.	.	8
Late Neolithic													
Island Magee	1	.	.	.	4	.	.	4	1	.	.	9	9

Column key: see table 23.1.
Reference key: 1. Moore (2003), 2. Kiely (2003), 3. Purcell (2002), 4. Dunne (2003), 5. McSparron (2003), 6. Groenman-van Watteringe (1984), 7. Monk (1988), 8. Helbaek, in Eogan (1963, p.42), 9. Alldritt (2000)

Table 23.2: Archaeobotanical finds from neolithic sites in Ireland

The most obvious development is that the number of entries in table 23.1 has increased from the 26 listed by Moffett et al. (1989) to well over 100. To a major extent this is due simply to the large volume of new work done in the last 15 years. Two other factors also play a part, however. Firstly, table 23.1 includes sites in Scotland, while Moffett et al. covered only England and Wales. Secondly, while we list sites rather than individual samples, it is in some cases logical to list the samples from a site in two or three groups, rather than just sum the whole site as one entry. This is the case for example for the three ditches at Windmill Hill, each of which contained several samples (Fairbairn 1999); for the different structures at Lismore Fields (Jones ndb) and Scord of Brouster (Milles 1986b); and for the different groups of contexts at the Hambledon Hill-Stepleton enclosure complex (Jones and Legge nd).

Following Moffett et al., the list is divided broadly into early neolithic (EN) and late neolithic (LN) samples. The division between them falls at approximately 3300 cal BC, though needless to say datings are not always so secure that a position before or after 3300 cal BC can necessarily be guaranteed. In some cases it is unclear whether the assemblage falls before or after this date, and the date in such instances is given as E/LN. There may also be problems towards the end of the Neolithic. In some cases it was not possible to separate the late neolithic and early bronze age features, and sites dated as 'Beaker' are an added complication. All such cases are listed as LN/EBA.

In listing many new sites, table 23.1 emphasises that interpretations of the British Neolithic may run into problems if they use data only a few years out of date. In terms of sheer quantity (both of sites and of materials) we estimate that the British Neolithic material is now more substantial than the summed total from our nearest European neighbours: Norway, Denmark, Sweden, Belgium, the Netherlands, northern France, and Ireland. Table 23.3 presents a list of the sites with more than 100 cereal grains, chaff fragments and weed seeds from Scandinavia. Britain is by no means any longer the 'poor relation'—a fact that is of importance to what follows, and which should inform future discussions.

23.3 Importance of cereals—the debate

Views of the importance of cereals in the British Neolithic economy have varied. In the middle years of the twentieth century the Neolithic was generally regarded as the result of the arrival of immigrants bringing with them the entire 'neolithic package' of domestic animals, cultivated plants, sedentary residential practices, and technologies such as ceramics and ground stone (e.g., Childe 1957; Clark 1952)—even though direct evidence of cereals was minimal (Jessen and Helbaek 1944; Helbaek 1952). In the 1960s the notion of an autochthonous origin for the Neolithic became a serious possibility (Clark 1966). This was encouraged by the recalibration of radiocarbon dates, which showed that many northwest European traits supposed to derive from southeastern antecedents were in fact older than those antecedents (Renfrew 1972). The adoption of exotic neolithic items by mesolithic indigenes was commonly argued, for example by Higgs and Jarman (1969).

In more recent years the debate has taken on a new cast. The indigenism of Higgs and Jarman, representatives of the British wing of the 'New Archaeology', has crossed the boundary between that form of archaeology and 'Post-Processualism'—one of the very few theoretical perspectives to make this transition. The gradualism inherent in the Higgs/Jarman view has been developed in an interesting way. The adoption of agriculture has become such a long, slow process that the Neolithic is often portrayed as largely based on wild resources; full sedentary agriculture did not become the mainstay of the economy until sometime in the Bronze Age, during the 2nd millennium BC. For Edmonds, for example, the contribution of cereal agriculture in the Neolithic varied from modest to minimal:

Site	Date	Litres	No. grain	No. chaff	No. weeds	No. hazel	No. apple	Total	Reference
Norway									
Hjelle	LN	?	113	4	2	0	0	119	1
Voll	LN	?	639	0	78	84	0	801	1
Denmark									
Bundsø1929	MN A	?	13	0	107	0	0	120	2
Lønt	MN A	70	1191	0	246	many	0	1437+h	3
Østerskov Å hse L	MN A	42	30	47	37	4	1	119	4
Østerskov Å pit AK	MN A	22	134	1237	443	64	0	1878	4
Spodsbjerg BAD	MN A	1.2	182	2439	423	many	1	3045+h	5
Spodsbjerg profile II	MN A	0.6	16	102	139	5	0	262	5
Vasagård	MN A	3-5	149	0	5	0	0	154	6
Klastrup	MN B	?	198	3	10	0	0	211	7
Mortens Sande 2	MN B	58	151	0	3	0	0	154	8
Strandet Hovedgaard S	MN B	c. 200	313	4	279	1	0	597	9
Strandet Hovedgaard N	MN B		200	1	249	1	0	450	9
Glattrup IV hse I	LN	?	881	121	22	0	0	968	10
Glattrup IV hse III	LN	?	223	3	27	0	0	253	10
Glattrup IV hse IV	LN	?	321	119	45	0	0	428	10
Glattrup IV hse V	LN	?	81	2422	54	2	0	1348	10
Glattrup IV hse VII	LN	?	35	24	67	0	0	113	10
Hemmed Kirke pits	LN	?	4058	52	44	3	0	4157	11
Hemmed Plantage hse I	LN	?	340	301	116	135	0	892	11
Hemmed Plantage hse III	LN	?	1000	373	230	46	0	1647	11
Lodbjerg	LN	?	251	3	10	0	0	264	7
Sweden									
Alvastra	MN A	?	8614	562	633	110	86	9849	12
Lockarp 7B pit 8	MN A	?	76	72	1	0	0	149	13
Piledal pit	MN A	?	989	0	24	0	0	989	14
Stävie 242	MN A	?	300	4	3	0	0	305	15
Fosie IV hse 92	LN	420	755	0	18	1	0	774	16
Fosie IV hse 95	LN	260	77	0	35	0	0	112	16
Fosie IV pit 3695	LN	20	80 ml	0	0	0	0	80 ml	16
Petersborg 6 hse 3	LN	?	315	0	1	0	0	316	17
Piledal 141	LN	?	113	0	11	0	0	124	18

Column key: see table 23.1.
Reference key: 1. Soltvedt (2000), 2. Jessen (1939), 3. Robinson (1996), 4. Robinson and Westphal (2001), 5. Robinson (1998), 6. Robinson and Harild (1994), 7. Robinson (2003), 8. Robinson and Kempfner (1987), 9. Robinson and Boldsen (2000), 10. Henriksen (2001), 11. Henriksen (2000), 12. Göransson (1995), 13. Gustafsson (1999), 14. Hjelmqvist (1985), 15. Hjelmqvist (1982), 16. Gustafsson (1995), 17. Gustafsson (2002), 18. Hjelmqvist (1992)

Table 23.3: Archaeobotanical finds from neolithic sites in Scandinavian countries

> For many, stock husbandry and limited cultivation made a significant contribution to the rhythm of daily and seasonal life. For others, the pattern of the year owed more to the character and availability of other species and, in many regions, hunting and collecting remained important. (Edmonds 1999, p.16)

In his most recent book, Whittle also sees cereal cultivation as sporadic and usually of minor importance:

> In many situations, cereal cultivation may have been small-scale, even episodic, since relatively small amounts survive and because other plant remains can also be found. However, this supposition need not exclude the possibility of times when labour was invested on a greater scale. (Whittle 2003, p.157)

Pollard and Reynolds (2002, p.42) concur that "the scale of cereal cultivation and consumption may have been quite low", and similar sentiments have been expressed by for example Bradley (1998, p.52), Fairbairn (1999, 2000), King (2003, p.32), Richmond (1999, pp.32–34), Robinson (2000b), Thomas (1999, p.29), Waddington (2000, p.41), and others.

Some of these authors have gone even further, and removed what little cereal agriculture there was entirely from the domestic economy, placing it in the context of 'ritual' consumption:

> ...much of the evidence of early neolithic crop utilisation comes from sites which are of a special character.... Crops were present, but the discussion has hinted that they were utilised by the select few, sometimes for a specialised purpose. (Richmond 1999, p.34)

Thomas underlines the non-domestic nature of cereals:

> Domestic resources, both plant and animal, had an importance in Neolithic Britain which was primarily symbolic. They were deployed in ritual, exchange and feasting.... Neither played a major part in feeding people from day to day, and these people were, from an economic point of view, still formally mesolithic. (Thomas 1993, p.388)

Not all have concurred with this view. Some have argued that domestic plants may have played a more important role in the economy (e.g., Barclay 1997; Cooney 2000, 2003). Some arguments along these lines have been based on the plants themselves (Legge 1989; Monk 2000), others on the isotopes in the bones of the people who consumed them (Richards and Hedges 1999; Schulting 2000; Schulting and Richards 2002a, b). We are aligned with this latter point of view (Jones 2000; Rowley-Conwy 2000, 2003, 2004), and we take the opportunity offered by the construction of table 23.1 to revisit the issue. Clearly, no one disputes the basic fact that some cultivated plants are found from the earliest phase of the Neolithic; the question is how we interpret them and establish their role and importance. The next two sections therefore consider two key aspects: first, how taphonomy may influence our understanding of the frequency of cereals; and second, how context may influence our understanding of the 'special' role of cereals.

23.4 Taphonomy and the 'rarity' of cereals

Extracting information from the archaeological record is an arduous and difficult business. It must be forcibly stressed that *the archaeological record was made unintentionally—not for our gratification as archaeologists.* Our first responsibility is therefore to seek to understand the making of the record, ever bearing in mind the factors that distort our easy understanding of it. In short, we must try to understand the taphonomy of the on-site macrobotanical record. Consider the following contrast: (1) the off-site pollen record results from *natural* processes, and once these are understood the vegetational record may be read straightforwardly; (2) the on-site macrobotanical record results from *cultural* processes, and these are (a) more difficult to understand, but (b) are the very object of our study, because they result from the human behaviour that it is our goal to understand. Excavators, identifiers and analysts all know this; so do many synthesists and theorists, although others sometimes appear not fully to grasp it.

In this section we examine two ways in which taphonomic factors distort our straightforward pollen-type reading of the macrobotanical record: the relative importance of hazelnuts and cereals, and the significance of low-density cereal samples.

23.4.1 The importance of hazelnuts

Hazelnuts are very common at British Neolithic sites, being probably the most frequently represented of the 'other plants' mentioned by Whittle (see the quote above). Nutshell fragments are often not counted but are just mentioned as being rare or numerous, and in such instances we record 'few', 'some' or 'many' in table 23.1.

How should the relative frequency of hazelnuts and cereals be assessed? The easiest way is simply to sum the nutshell fragments and compare the total to that of cereal grains. This method gives the result that on many sites in table 23.1, nuts were more important than cereals. Taken at face value, this supports the argument of those quoted above who suggest that cereals were only of modest importance. However, this is a classic case in which the taphonomy of the various types of plant remains must be understood before we can interpret the archaeological record. These items are not in the archaeological record due to natural processes, and their frequencies cannot be read off as if the archaeobotanical record was a pollen diagram.

The first consideration is that only items that have been charred will survive to get into the archaeological record and ultimately into table 23.1. This is a very tight taphonomic filter, and we must consider the exigencies of charring very carefully in order to avoid drawing incorrect conclusions (Jones 2000). Hazel nut shell and cereal grains are profoundly different in one way: nut shell is a waste product, while cereal grains are food. All other things being equal, nut shell will therefore be discarded, while cereals will be processed and eaten. This will immediately have a major effect: people will make a considerable effort to *avoid* charring cereal grains, which will thus only be preserved as the result of an accident. They will however make no such effort for nut shell, and may indeed discard it straight into the nearest fire, where it stands a fair chance of being charred and entering the archaeological record. People may even make a deliberate effort to burn nut shell: when dried, it is a useful kindling material for lighting fires. This would of course make it even more likely to become charred. When we compare nut shell with cereal grains we are thus not comparing like with like, and this difference has a major effect on charring frequency. Two further points are worth making. First, cereal grains were nevertheless charred in substantial numbers despite people's best efforts. The surviving charred grains are the "tip of the iceberg" testifying to the importance of this foodstuff in the Neolithic—while virtually no hazel *nuts* (as opposed to fragments of shell) survive from this period. Second, another way to quantify the importance of the two foodstuffs is to see how often they occur in burnt stores—the most substantial of the accidents causing charring. Some burnt cereal stores have been excavated (see below), while no store of intact but burnt hazelnuts has been found (Jones 2000). This suggests that cereals were of greater importance than nuts.

Survival and recovery may be another factor causing hazelnut shell to be over-represented (Jones 2000). Nut shell is hard and robust; charring experiments do not appear to have been carried out, but it is likely that hazelnut shell survives charring better and under a wider variety of circumstances (e.g., temperature) than do cereal grains. The fragments in which charred nut shell survives are also often larger than cereal grains, and are thus more visible during trowelling. This we believe has an effect on sampling for botanical remains. Hard-pressed excavators with limited budgets of both time and money are most likely to take a botanical sample when something relevant is visible in the trench—and this is more likely to be nut shell.

Cereals and nut shell thus enter the archaeological record through very different cultural processes. They cannot be directly compared. For the reasons outlined above we conclude that cereals were probably more important than hazelnuts in neolithic economies.

23.4.2 Low-density cereal samples

The low-density of cereal grains in British Neolithic samples is sometimes advanced as an argument for the low importance of cereals in the economy as a whole, in contrast to the record from neighbouring countries, where large and dense grain samples are relatively more common (see references above, e.g., Moffett et al. 1989, pp.254–255), and to the more abundant cereal finds in the British Iron Age (Entwhistle and Grant 1989, p.204). Many British samples do indeed contain rather few grains, and this sort of sample was virtually the only one in the table published by Moffett et al. (1989). But table 23.1 reveals that while many more of the same type have been added subsequently, so have several higher density finds—e.g., Balbridie, Lismore Fields, the Stepleton concentration—so the situation is by no means so clear as it was 15 years ago.

One may legitimately ask why in neighbouring countries low-density samples are rare, and high-density samples more common. There are two possible answers to this. The first is that flotation is more common in Britain than it is in some of the neighbouring countries. Low-density samples are a common product of flotation, because samples may be taken when plant remains are not visible to the excavators—i.e. when they are at low densities. Such samples are therefore common in Britain, while in some neighbouring countries the high-density samples, clearly evident during normal excavation, are relatively more common and are therefore more prominent in archaeological interpretations. Paradoxically, our greater efforts to find neolithic cereals in Britain may have led to a diminished perception of their importance here. The later use of flotation in countries like Denmark has revealed that low-density samples are common there too (Robinson 2003; Rowley-Conwy 2000).

The second answer is that high-density samples from burnt cereal stores may really be more common on the continent—not because cereals were more important there, but because of a difference in storage. Rectangular timber houses, such as those found on the continent, provide ideal cereal storage in the roof space; timber houses also contain domestic fires, which from time to time would cause catastrophic conflagrations resulting in charred stores entering the archaeological record. Round houses, being much lighter structures, offer less roof storage potential, so storage was probably outside the domestic residence, possibly in pits. At all events, while round houses burnt down as frequently as their rectangular counterparts, they did not create high-density cereal samples when they did (Rowley-Conwy 2000). Early neolithic houses in Britain were rectangular, while later ones were round (Darvill 1996); and indeed where early neolithic rectangular buildings have been found, they sometimes do contain burnt grain, e.g., Balbridie (Fairweather and Ralston 1993) and Lismore Fields (Garton 1991; Jones ndb).

The rarity of grain is therefore less marked than it was just a few years ago, and indeed the efforts of British archaeobotanists have meant that our neolithic cereal record is better than that of our immediate neighbours. When flotation is employed, low density samples are the norm. By the same token, the quantity of cereal remains from British Neolithic sites is comparable to that from iron age sites, as a comparison with a recent study of British cereal finds shows (T. Mills, pers. comm.). While the latter study does not claim to provide a comprehensive record of all iron age cereals, it attempts to include all published sites with quantified plant remains totalling more than 1000 cereal items plus all sites with more than 100 items if they include 30 or more samples. When the cereal finds from table 23.1 are summed (with or without samples with <100 cereal items and samples of possible bronze age date), they amount to considerably more than those from the iron age study (with or without samples of possible Roman date) (table 23.4). This is partly due to the huge find of emmer wheat from the Stepleton enclosure at Hambledon Hill but, when this is excluded from the calculation, there are still twice as many neolithic cereal

Phase and context	Grains	Chaff items*
Neolithic		
Total including or excluding samples with <100 items and samples of possible BA date	150,000	60,000
Excluding the Stepleton concentration	40,000	5,000
Excluding the Stepleton concentration and Balbridie	20,000	5,000
Excluding the Stepleton concentration, Balbridie and Lismore Fields	15,000	2,500
Iron Age**		
Excluding samples of possible Roman date	20,000	15,000
Including samples of possible Roman date	25,000	20,000

*glume bases of glume wheats plus rachis internodes of free threshing cereals; **sites with 10+ samples and >1000 cereal items plus sites with 30+ samples and >100 cereal items.

Table 23.4: Approximate numbers of cereal items from neolithic sites and from iron age sites in a recent study (T. Mills, pers. comm.)

grains as there are in the iron age study, though the number of chaff fragments is lower in the Neolithic. It is not until the timber houses at Balbridie and Lismore Fields are excluded that the number of grains falls significantly below that found in the iron age study. These figures are unstable and can easily be reversed by a few very large grain samples (the cereal find from Hembury for example has not been included in the calculations because the numbers of grains are unknown). They are presented merely to indicate that, contrary to popular belief, it is no longer the case that cereals are rare in neolithic contexts compared with the Iron Age.

As an illustration of the contextual variation that might be expected for archaeobotanical samples resulting from an agricultural economy, samples from the late iron age settlement of Thorpe Thewles are plotted in figure 23.1 alongside samples from a burnt granary at South Shields Roman fort (van der Veen 1992). At the former, chaff and particularly weed-rich samples predominate. These have been interpreted as the waste products of a cereal-dominated economy but no high-density samples suggesting storage, such as those at South Shields, were found. The economies that produced the two types of sample were both cereal based; what differs are the cultural processes that led to different aspects of this economy entering the archaeological record.

23.5 Contexts: 'special' or domestic?

We now turn our attention to contexts where cereals are found in quantity. Some authors have questioned the domestic status of large cereal finds, partly on the grounds of their very abundance and partly on the status of the buildings in which they have been found (see, for example, Thomas 1996a, b, 1999, 2003; Richmond 1999). The timber hall at Balbridie (Scotland), in particular, has been highlighted both for the substantial size of the building (24-m long—Fairweather and Ralston 1993) and for the plentiful cereal remains contained therein which, it is argued, suggest storage rather than processing (Thomas 1996b, p.9). The grain storage function of the Balbridie

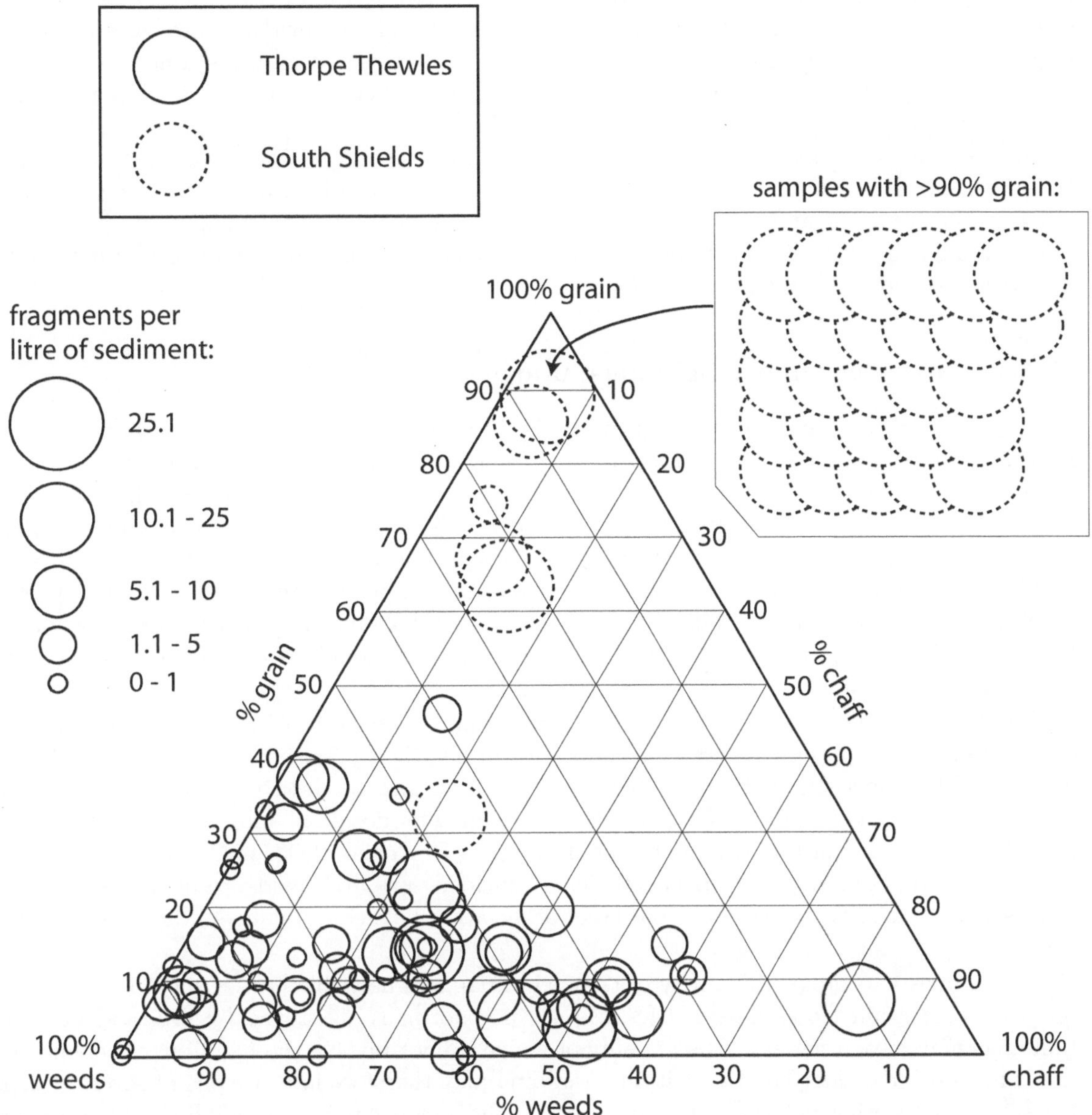

Figure 23.1: Triangular graph plotting the proportions of cereal grains, chaff, and weeds from Thorpe Thewles iron age settlement and South Shields Roman fort. Redrawn and combined from van der Veen (1992, figs. 8.3 and 8.4).

Hall has then been extrapolated to other neolithic timber buildings (Thomas 1996a, 1999, 2003; Richmond 1999).

There are, however, several problems with the dismissal of these buildings as possibly representing "specialised storage, consumption or redistributive locations" and their contents as "a very special kind of food" (Thomas 1999, p.25). This suggestion entails two basic requirements: (1) that buildings of this type were universally used for storage and (2) that this storage was in some way 'unusual'. Neither of these is supported by the existing evidence. To take the second requirement first, cereals are a seasonal resource and so storage of grain is a necessity for any community dependent primarily on cereals for subsistence; evidence of storage is therefore not, in itself, evidence of 'special status'. No evidence has been presented to indicate that the type of grain storage at Balbridie or other timber structures was in any way unusual. Nevertheless, the prevalence of grain storage, and the absence of cereal processing, in the Neolithic have been cited as the principal archaeobotanical evidence for cereals being an unusual or special food (e.g., Richmond 1999). In this section we question this assertion through a consideration of the architectural and archaeobotancal evidence.

23.5.1 Neolithic timber buildings: unusual contexts?

The 'archetypal' timber hall at Balbridie is, in fact, not particularly representative of other neolithic timber buildings. Buildings are much more common than is often assumed, even in the early Neolithic (see the review in Rowley-Conwy 2003), and most are smaller than Balbridie. Amongst the neolithic structures described as large timber 'houses' by Thomas (1999, p.25), Building 1 at Lismore Fields (Buxton, Derbyshire) is thought actually to be two adjacent buildings each approximately one-third the length of the Balbridie timber hall (c. 7.5 m—Garton 1991; Thomas 1996b), thereby falling into the size category of "much smaller rectangular houses" found in Ireland (e.g., those at Ballygalley and Tankardstown South) "for which a rather stronger case can be made as dwellings" (Thomas 2003, p.71).

Furthermore, the majority of neolithic timber buildings have provided little or no evidence of cereal storage, or indeed of cereals at all. For example, the timber building at Yarnton (upper Thames Valley, measuring 20 m × 10 m) is similar in size to the Balbridie timber hall but produced no charred cereals "despite near 100% sampling for flotation" indicating that "activities resulting in charring of grain were not taking place" (Hey et al. 2003, p.82). Similarly, a third timber building (Building 2) at Lismore Fields provided very little evidence of charred remains (Jones ndb). There is therefore no consistent association between timber buildings and charred grain stores.

In a domestic context, this rarity of charred remains in neolithic houses should not surprise us given that grain stores intended for use are not normally destroyed by fire, and domestic buildings may have a variety of uses, many of which do not involve the handling of cereals near fire. The absence of charred grain within a building may therefore indicate one of several things: (1) storage and other activities involving cereals did not occur in the building, (2) grain *was* stored in the building but there was no catastrophic fire which preserved the grain, or (3) cereal processing *did* occur within the building but the resulting by-products were either not charred or the charred remains were cleared away and dumped elsewhere. It is worth remembering that the Balbridie cereals did not come into existence in a cleaned state, but must have come through the processing sequence; it is merely that the waste products from that processing have not survived. As argued above, in a domestic context preservation by charring is nearly always an accident: even when cereal residues are deliberately disposed of on fires, the purpose is to destroy not to preserve them, and so charred processing residues as well as charred grain stores are found only when circumstances conspire to bring about their preservation. It is therefore extremely difficult to draw conclusions from the *absence* of charred remains. We should therefore concentrate on interpreting the evidence provided when cereals *are* preserved by charring and so the following discussion focuses on sites which have produced more than 100 cereal items.

23.5.2 Cereal deposits: special food?

Timber buildings

Large neolithic cereal finds from timber buildings cannot be universally interpreted as stored grain, and the distinction between Balbridie and other timber buildings is reflected in the cereal remains themselves. Although, at Balbridie, the cereal remains (predominantly emmer wheat) clearly represent cleaned, stored grain (Fairweather and Ralston 1993), it is acknowledged that the Lismore Fields cereal grain was found "together with chaff fragments" (Thomas 1996b, p.9). In fact, the two adjacent buildings at Lismore Fields (Building 1) provided archaeobotanical samples of rather different compositions: of the larger samples (for which reliable proportions can be calculated), those in the western building were composed primarily of chaff fragments and those in the eastern building primarily of grain (both predominantly emmer wheat) with more mixed deposits at the junction of the two buildings (Jones ndb). The 'density' of charred remains may also be lower than at Balbridie where cereals were found in "varying concentrations" (Fairweather and Ralston 1993). Concentrations also vary at Lismore Fields but the highest concentration was c. 200 grains per litre of deposit sampled, and was usually much lower. So, while the grain (or spikelets) in the eastern building may represent the remnants of storage, there is also evidence for cereal processing in the western building. (Note that, as cereal grain tends to survive better than chaff (Boardman and Jones 1990), the grain could represent spikelets or dehusked grain, but it is unlikely that the chaff derives from stored spikelets from which the grain has subsequently been lost.)

The finds at Lismore Fields are not unusual when compared to similar buildings in Ireland. There is evidence for crop processing at Ballygalley (Co. Antrim; Simpson 1993, 1996; Thomas 1996b, p.9) in pits outside the buildings. Grains only were recovered from the slot trenches of House 1 itself, and these are described as "more fragmentary than samples from elsewhere on the site" and "may result from trampling and brushing of floor waste" (Simpson 1996, p.129). This is not particularly consistent with the use of the building for storage (although it does not rule it out), and trampling or brushing of charred remains is likely to result in the destruction of any chaff originally associated with the damaged grain, chaff being more fragile than grain (Boardman and Jones 1990). Equally, it is not unexpected that grain processing might take place outside and that "only 'clean' grain would have been introduced into the house" (Simpson 1996, p.134). At the Tankardstown timber building (Co. Limerick), the cereal grains in most samples were again fragmented possibly due to "trampling on the floor surfaces within the house" after accidental charring; in a few samples the grains were well-preserved. These remains have been interpreted as follows: "There is little doubt of the fact that these remains relate to neolithic domestic food-producing activity on the site" (Monk 1988, p.186). The representation of grain and chaff at the three neolithic houses at Corbally (Co. Kildare) is apparently the reverse of that at Ballygalley, with chaff most abundant in the internal features. The richest of these assemblages "may therefore represent crop-processing by-products, with cereal waste accumulating in these cut features" leading to the conclusion that "it is likely that at least some crop processing was carried out indoors" (Purcell 2002, p.67). At Kinbeachie, in Scotland, a small concentration of charred barley grain in one corner of a timber building has been interpreted as accidentally charred, possibly during corn-drying, as there is no evidence that the building itself had burnt down (Hastie and Holden 2001).

This variety of cereal (by-)products presents a very different picture to "the lack of chaff remains from such sites" (including Lismore Fields) which, it is claimed, "suggests that no processing took place" (Richmond 1999, p.33). Rather it seems that cereal remains recovered from neolithic timber buildings in both Britain and Ireland represent a mixture of cleaned grain and crop processing by-products consistent with a purely domestic function, and provides no grounds for suggesting a 'special' kind of food. This is contrary to the prevailing view that, at least in Britain, neolithic timber buildings were built for specialised storage, possibly with a ceremonial significance. It could even be argued that, if burnt grain is found in timber buildings because they are 'special' or ceremonial, and particularly if they were intentionally burned as tentatively suggested by Thomas (2003, p.71), they might be expected *always* to contain quantities of charred cereal grain. But if the burnt grain in timber buildings is the result of domestic activity, it would be expected *sometimes* (by accident), but not consistently.

Other neolithic contexts

Quantities of cereal remains have also been found at other types of settlement site, and in a variety of contexts. At Scord of Brouster (Shetland, Scotland), for example, several stages of cereal (barley) processing have been identified inside the three circular or oval stone-built houses, including assemblages interpreted as threshing waste in two of the houses, sieving waste in one of these, and grain drying prior to storage in a third house (Milles 1986b). Cereal remains (mostly emmer) from the intertidal settlement site, The Stumble (Essex), were tentatively interpreted as "a background scatter of material across the site, produced during such domestic activities as spikelet parching and grain roasting", and one sample contains "a proportion of crop cleaning waste" giving "some indication that the earlier stages of crop processing were taking place" (Murphy 1988a).

Plant remains have come from a variety of other neolithic contexts. Non-crop species and chaff have been recovered from lynchets at Scord of Brouster, suggesting the spreading of ash from domestic hearths onto cultivated fields (Milles 1986b). Cereals and/or waste products have come from middens, for example between two of the houses at Skara Brae (Rowley-Conwy nd), and from below the cairn at Hazleton North (Straker 1990a, p.215). At Aston-on-Trent they were recovered from a hearth and some gulleys (Alvey 1964). Concentrations of barley grain have been reported from Ness of Gruting (Milles 1986b) and Rosinish (Maclean and Rowley-Conwy 1979).

Many archaeobotanical samples come from pits, and this raises the interesting question of whether pits are indeed sometimes 'special' contexts. Some are inside ritual monuments, and are discussed in the following section. Others are outside such monuments, and may be quite close by (e.g., at Windmill Hill—Fairbairn 2000, pp.169–171) or at a considerable distance (e.g., Mount Farm—Jones 1980, p.61). How close to a monument does a pit have to be for a 'special' interpretation to be suggested? Others are from settlements (e.g., Thirlings—M. Van der Veen 1985). Still others, probably the majority, are of uncertain context. Pits may originally have been dug for a variety of purposes, though their final contents may not reveal this—a pit dug for a 'special' purpose may nevertheless finally be back-filled with domestic refuse. A pit containing 'special' deposits may also contain domestic items that have got in by accident, perhaps particularly likely in the case of tiny grains, weed seeds or chaff fragments. Such items cannot be found even by highly skilled excavators who are specifically looking for them, but have to be recovered by flotation. Neolithic diggers and fillers of pits are unlikely either to have noticed the infiltration of low-density plant remains, or to have cared if they did notice them. Perhaps this explanation applies to pits where other finds suggest a 'special' context while the plant remains do not. For example, at Down Farm, pits near the Dorset Cursus are identified as *non-domestic* on the basis of their animal bones (Legge 1991, pp.68, 72), stone tools (Brown 1991, pp.111-114) and pottery (Cleal 1991, pp.140–141); the plant remains however "appeared to be domestic refuse" (Jones 1980, p.61). Pits are thus problematic. Those at Caythorpe and Marton-le Moor (Yorkshire) have produced "considerable evidence for cereal use" as well as wild resources and are interpreted as representing storage and domestic rubbish respectively (Huntley and Stallibrass 1995), while at Capel Eithen clean barley grain was found (Williams 1999), but the majority of cereal finds are at low-density.

The widespread low-density samples from neolithic sites are likely to result from crop processing (the grain surviving more commonly than the fragile chaff), and may again be taken as evidence of the central importance of cereals. As with hazelnut shell, we are seeing the deliberately burnt waste product—but in this case the waste product derives from cereal agriculture. We interpret such low-density samples as further evidence for the economic importance of cereal agriculture.

Thus, at a broad level, the variety of neolithic contexts from which plant remains are routinely recovered, and the range of agricultural activities represented by the archaeobotanical remains, conforms rather well to what would be expected in domestic circumstances. In any other archaeological context, be it the British Iron Age or the Central European Neolithic, this variety of contexts would not (and has not) raised any questions of 'special' meanings.

Causewayed enclosures and other 'ritual' sites

A stronger case could perhaps be made for a 'special' status of plant remains found at causewayed enclosures, though here too 'everyday' foods may have been prepared and consumed. At Hambledon Hill, a large quantity of emmer wheat spikelets was tipped into a pit outside the Stepleton enclosure ditch (Jones and Legge nd). Clearly this deposit represents cereals that were not eaten but whether they were deliberately burned, or were discarded after accidental charring during bulk processing, is more difficult to establish (see below). Cereal grain (both emmer wheat and barley) was widely distributed (in much smaller quantities) in the interior pits of both the main and Stepleton enclosures together with quantities of hazelnut shell of which the largest concentrations were again found near or just outside the enclosure ditches (Jones and Legge nd). It is therefore difficult to see why the cereals should be classified as special while the hazelnuts should not be. At Yarnton, for example, it was suggested that the plant food remains (mostly hazelnut, sloes and apples) found in pits "had been specially selected and may not be representative of an everyday diet" (Hey et al. 2003, p.86).

At the Windmill Hill causewayed enclosure, charred cereals and wild plants were again found throughout the fills of the enclosure ditches and, due to their association with wood charcoal and ash, were thought likely to derive from burnt debris collected from fireplaces (Fairbairn 1999, p.149). The association of charred plant remains with animal bone has been interpreted as evidence that they result from food processing prior to consumption, and the lack of chaff remains to indicate that "large-scale processing of cereal grain from a standing crop" was an unlikely source. This contrasts with the large deposit of unprocessed spikelets at Hambledon Hill and the cereal remains (wheat and barley) from pits outside the Windmill Hill enclosure where both grain and chaff were preserved. At Windmill Hill these cereals were thought to be deliberately burnt as a form of ceremonial consumption, in association with wild plants which it is suggested could have been either deliberately burnt or derived from burnt waste (Fairbairn 2000, p.174). It is, however, extremely difficult, for assemblages of these kinds, to distinguish cereal products deliberately burnt as part of a ceremony from those accidentally charred during processing (e.g., parching or drying).

At the Maiden Castle causewayed enclosure, a mixture of grain, fruits and nuts was encountered (Palmer and Jones 1991) and at Haddenham one large concentration of grain (mostly barley with some emmer and weed seeds) was recovered from a pit cut into the enclosure ditch, with a few scattered cereal grains and wild plants in other contexts (Jones nda). Concentrations of cereal grain were also found at the Hembury causewayed enclosure (glume wheat with some chaff—Helbaek 1952; Moffett et al. 1989), in the ditch of the Coneybury henge (clean barley grain—Carruthers 1990) where it was concluded that the deposit may have been of ritual significance or waste from a domestic accident, at the Boghead mound (barley and emmer with some weeds—Maclean and Rowley-Conwy 1984).

Whether the plant remains at causewayed enclosures and other 'ritual' sites result from ceremonial burning and deposition, or from the disposal of everyday burnt waste and parching accidents, thus remains an open question. There is certainly no evidence to single out the cultivated cereals as the 'special' food and the wild plant remains as 'everyday'; it could equally be the other way round. Perhaps more likely, both were everyday foods which were sometimes used in ceremonies.

23.6 Conclusions

We have argued that cereal cultivation was of central importance during the British Neolithic, and that there is no reason to suggest that cereal consumption was confined to 'special' occasions or circumstances. Indeed the domestic status of similar assemblages from iron age contexts has never been questioned (see above). We believe that the tendency to under-emphasise the importance of neolithic cereals stems from the prevailing view of the British Neolithic as a mobile society, dependent largely on wild resources or pastoralism, and that it is for this reason the botanical evidence has been interpreted differently from that of the Iron Age. In other words the archaeobotanical evidence has been accommodated to fit current interpretations of the Neolithic rather than providing empirical evidence to support them.

This has come about, we believe, for two main reasons. First, there has been a general tendency to interpret archaeobotanical evidence at 'face value' rather than engaging with the minutiae of the archaeobotanical record. Second, some interpretations have involved insufficiently detailed consideration of site formation processes, taphonomy and archaeological context. Archaeobotanical specialists, however, conducting sample-by-sample analyses, provide evidence for a wide range of domestic activities, which are not necessarily apparent at the gross 'site' level. If this contribution has demonstrated the value of such analytical treatment of the archaeobotanical record, it has served its purpose.

Acknowledgements

We are deeply indebted to Gill Campbell, Jacqui Huntley, Peter Murphy and Vanessa Straker for their regional reviews of plant remains, and to all of them as well as Allan Hall, Lisa Moffett, Angela Monckton and Amy Bogaard for access to their records. We are also grateful to Tim Mills for access to his database of iron age cereal assemblages, and to Stefan Gustafsson (Umeå), Fredrik Hallgren (Uppsala), Peter Steen Henriksen (Copenhagen), Torben Sarauw (Aalborg) and Morten Steineke (Malmö) for information concerning Scandinavian sites. We are extremely grateful to Allan Hall for comments on an earlier draft of this paper.

References

Alldritt, D. M. (2000). Charcoal and plant macrofossils, in P. Duffy and H. F. James "Flint scatters and burnt mounds: results from an archaeological watching brief on Island Magee, County Antrim in 1995 and 1996, Scotland to Northern Ireland gas pipeline". *Ulster Journal of Archaeology 59*, 25.

Allen, M. (1986). Plant remains, in D. Powlesland, "Excavations at Heslerton, North Yorkshire 1978–82". *Archaeological Journal 143*, fiche M2/51–2.

Alvey, R. C. (1964). Carbonized cereals from Aston-on-Trent, Derbyshire. Unpublished archive report for Derby Museum.

Arthur, J. R. and P. Paradine (1979). Seeds, in A. Saville, "Further excavations at Nympsfield chambered tomb Gloucestershire". *Proceedings of the Prehistoric Society 45*, 84.

Barclay, G. J. (1997). The Neolithic. In K. J. Edwards and I. B. M. Ralston (Eds.), *Scotland After the Ice Age: Environment, Archaeology and History, 8000 BC-AD 1000*, pp. 127–149. Chichester: Wiley.

Boardman, S. (1997). Plant remains, in T. Pollard, "Excavation of a neolithic settlement and ritual complex at Beckton Farm, Lockerbie, Dumfries and Galloway". *Proceedings of the Society of Antiquarians of Scotland 127*, 107–109.

Boardman, S. and G. Jones (1990). Experiments on the effects of charring on cereal plant components. *Journal of Archaeological Science 17*, 1–11.

Boyd, P. (1987). Carbonised seeds, in D. G. Buckley and J. D. Hedges, "Excavation of a crop-mark enclosure complex at Woodham Walter, Essex, 1976". *East Anglian Archaeology 33*, 41.

Bradley, R. (1998). *The Significance of Monuments.* London: Routledge.

Brown, A. (1991). Structured deposition and technological change among the flaked stone artefacts from Cranborne Chase. In J. Barrett, R. Bradley, and M. Hall (Eds.), *Papers on the Prehistoric Archaeology of Cranbourne Chase*, Oxbow Monograph 11, pp. 101–133. Oxford: Oxbow Books.

Campbell, G. (2004). Charred plant remains. In *Green Park (Reading Business Park) Phase 2 Excavations 1999—Neolithic and Bronze Age Sites. Thames Valley Landscapes*, Monograph 19. Oxford: Oxford Archaeology/Oxford University School of Archaeology.

Campbell, G. and V. Straker (2003). Prehistoric crop husbandry and plant use in southern England: development and regionality. In K. A. Robson Brown (Ed.), *Archaeological Sciences 1999. Proceedings of the Archaeological Sciences Conference, University of Bristol, 1999*, BAR S1111, pp. 14–30. Oxford: Archaeopress.

Carruthers, D. (1989). The carbonised plant remains. In P. J. Fasham, D. E. Farwell, and R. J. B. Whinney (Eds.), *The Archaeological Site at Easton Lane, Winchester*, Hampshire Field Club Monograph 6, pp. 131–134. Winchester: Hampshire Field Club Society.

Carruthers, D. (1990). Carbonised plant remains. In J. Richards (Ed.), *The Stonehenge Environs Project*, pp. 250–252. London: Historic Buildings and Monuments Commission for England.

Carruthers, D. (1991). The carbonised plant remains from Rowden. In P. J. Woodward (Ed.), *The South Dorset Ridgeway Survey and Excavation 1977-1984*, Dorset Natural History and Archaeological Society Monograph 8, pp. 106–111. Dorchester: Dorset Natural History and Archaeological Society.

Carruthers, D. (1992). Plant remains. In C. A. Butterworth and S. J. Lobb (Eds.), *Excavations in the Burghfield Area, Berkshire. Developments in the Bronze Age and Saxon Landscapes*, Wessex Archaeology Report 1, pp. 63–65. Salisbury: Wessex Archaeology.

Carruthers, D. (2001). Charred plant remains from the neolithic horizon, in M. Patton, "Le Pinacle, Jersey: a reassessment of the neolithic, chalcolithic and bronze-age horizons". *Archaeological Journal 158*, 26–27.

Childe, V. G. (1957). *The Dawn of European Civilization* (6th ed.). St. Albans: Paladin.

Clapham, A. J. (1995). Plant remains. In I. Barnes, W. Boismier, R. M. J. Cleal, A. P. Fitxpatrick, and M. R. Roberts (Eds.), *Early Settlement in Berkshire: Mesolithic-Roman occupation Sites in the Thames and Kennet Valleys*, Wessex Archaeology Report 6, pp. 35–45. Salisbury: Wessex Archaeology.

Clapham, A. J. (1999). Charred plant remains. In A. Fitzpatrick, C. A. Butterworth, and J. Grove (Eds.), *Prehistoric Sites in East Devon: The A30 Honiton to Exter Improvements DBFO Scheme, 1996-9*, Wessex Archaeology Report 16, pp. 51–59, 152–155. Salisbury: Wessex Archaeology.

Clark, J. G. D. (1952). *Prehistoric Europe: The Economic Basis.* Cambridge: Cambridge University Press.

Clark, J. G. D. (1966). The invasion hypothesis in British archaeology. *Antiquity 40*, 172–189.

Cleal, R. M. J. (1991). Cranbourne Chase—the earlier prehistoric pottery. In J. Barrett, R. Bradley, and M. Hall (Eds.), *Papers on the Prehistoric Archaeology of Cranbourne Chase*, Oxbow Monograph 11, pp. 134–200. Oxford: Oxbow Books.

Cooney, G. (2000). *Landscapes in Neolithic Ireland.* London: Routledge.

Cooney, G. (2003). Rooted or routed? Landscapes of neolithic settlement in Ireland. In I. Armit, E. Murphy, E. Nelis, and D. Simpson (Eds.), *Neolithic Settlement in Ireland and Western Britain*, pp. 47–55. Oxford: Oxbow Books.

Darvill, T. (1996). Neolithic buildings in England, Wales and the Isle of Man. In T. Darvill and J. Thomas (Eds.), *Neolithic Houses in Northwest Europe and Beyond*, Oxbow Monograph 57/Neolithic Studies Group Seminar Paper 1, pp. 77–111. Oxford: Oxbow Books.

Dickson, C. A. (1983). Macroscopic plant remains from Knap of Howar, Orkney, in A. Ritchie, "Excavations of a neolithic farmstead at Knap of Howar, Papa Westray, Orkney". *Proceedings of the Society of Antiquaries of Scotland 113*, 114–115.

Dunne, C. (2003). Neolithic structure at Drummenny Lower, Co. Donegal: an environmental perspective. In I. Armit, E. Murohy, E. Nelis, and D. Simpson (Eds.), *Neolithic Settlement in Ireland and Western Britain*, pp. 164–171. Oxford: Oxbow Books.

Edmonds, M. (1999). *Ancestral Geographies of the Neolithic. Landscapes, Monuments and Memory.* London: Routledge.

English Heritage (2004). Environmental archaeology bibliography. `http://ads.ahds.ac.uk/catalogue/specColl/eab_eh_2004/` Accessed 2005-03-16.

Entwhistle, R. and A. Grant (1989). The evidence for cereal cultivation and animal husbandry in the southern British Neolithic and Bronze Age. In A. Milles, D. Williams, and N. Gardner (Eds.), *The Beginnings of Agriculture*, Symposia of the Association for Environmental Archaeology No. 8. BAR International Series 496, pp. 203–215. Oxford: British Archaeological Reports.

Eogan, G. (1963). A neolithic habitation-site and megalithic tomb in Townleyhall townland, Co. Louth. *Journal of the Royal Society of Antiquaries of Ireland 93*, 37–81.

Fairbairn, A. S. (1993). Charred plant remains, in A. Whittle, A. J. Rouse and J. G. Evans, "A neolithic downland monument in its environment: excavations at the Easton Down Long Barrow, Bishops Cannings, North Wiltshire". *Proceedings of the Prehistoric Society 59*, 221.

Fairbairn, A. S. (1997). Charred plant remains. In A. Whittle (Ed.), *Sacred Mound, Holy Rings: Silbury Hill and the West Kennet Palisade Enclosures: a Later Neolithic Complex in North Wiltshire*, Oxbow Monograph 74, pp. 134–138. Oxford: Oxbow Books.

Fairbairn, A. S. (1999). Charred plant remains. In A. Whittle, J. Pollard, and C. Grigson (Eds.), *The Harmony of Symbols. The Windmill Hill Causewayed Enclosure*, Cardiff Studies in Archaeology, pp. 139–156. Oxford: Oxbow Books.

Fairbairn, A. S. (2000). Charred seeds, fruits and tubers, in A. Whittle, J. J. Davies, I. Dennis, A. S. Fairbairn and M. A. Hamilton, "Neolithic activity and occupation outside Windmill Hill causewayed enclosure, Wiltshire: survey and excavation 1992-93". *Wiltshire Archaeological and Natural History Magazine 93*, 131–180.

Fairweather, A. D. and I. B. M. Ralston (1993). The neolithic timber hall at Balbridie, Grampian Region, Scotland: the building, the date, the plant macrofossils. *Antiquity 67*, 313–323.

Garton, D. (1991). Neolithic settlements in the Peak District: perspective and prospects. In R. Hodges and K. Smith (Eds.), *Recent Developments in the Archaeology of the Peak District*, Sheffield Archaeological Monographs 2, pp. 3–21. Sheffield: Department of Archaeology aand Prehistory.

Giorgi, J. (1994). The West London Gravels Assesment Report: the environmental samples. Unpublished MoLAS Archive Report.

Giorgi, J. (1996). The prehistoric plant remains, in D. Lakin, "Excavations at Corney Reach, Chiswick W4, 1989-1995". *Transactions of the London and Middlesex Archaeological Society 47*, 70–71.

Göransson, H. (1995). *Alvastra Pile Dwelling. Palaeoethnobotanical Studies.* Theses and Papers in Archaeology New Series 6. Stockholm: Lund University Press.

Green, F. J. (1981). Seeds, in S. M. Davies, "Excavations at Old Down farm, Andover. Part II: Prehistoric and Roman". *Proceedings of the Hampshire Field Club Archaeological Society 37*, 96.

Green, F. J. (1985). Evidence for domestic cereal use at Maxey. In F. Pryor, C. French, D. Crowther, D. Gurney, G. Simpson, and M. Taylor (Eds.), *The Fenland Project No. 1: Archaeology and Environment in the Lower Welland Valley*, East Anglian Archaeology 27, pp. 224–232. Cambridge: Cambridgeshire Archaeological Committee.

Green, F. J. (1990). Charred plant remains, in A. D. Russell, "Two Beaker burials from Chilbolton, Hampshire". *Proceedings of the Prehistoric Society 56*, 168–169.

Groenman-van Watteringe, W. (1984). Appendix two: pollen and seed analyses. In G. Eogan (Ed.), *Excavations at Knowth 1: Smaller Passage Tombs, Neolithic Occupation and Beaker Activity*, pp. 325–329. Dublin: Royal Irish Academy.

Gustafsson, S. (1995). *Fosie IV. Jordbrukets Förändring och Utveckling från Senneolitikum till yngre Järnålde.* Stadsantikvariska avdelingen Rapport 5. Malmö: Malmö Museer.

Gustafsson, S. (1999). *Rapport över Arkeobotaniska Analyser från Lockarp 7B, Öresundsförbindelsen.* Umeå: Arkeologiska Institutionen, Miljöarkeologiska Laboratoriet, Umeå University.

Gustafsson, S. (2002). Odling och växter kring Petersborg. In S. Siech and Å. Berggren (Eds.), *Öresundsförbindelsen. Petersborg 6*, Rapport över Arkeologisk Slutundersökning 15, pp. 78–81. Malmö: Malmö Kulturmiljö.

Hall, A. (2000). *Assessment of Plant Remains from Excavations at Cheviot Quarry, Milfield, Northumberland (Site Code 02-02-00).* Reports from the Environmental Archaeology Unit 2000/78. York: Environmental Archaeology Unit, University of York.

Hall, A., H. Kenward, M. Hill, F. Large, D. Jaques, K. Dobney, M. Issitt, and S. Lancaster (1994). *Technical Report: Biological Remains from Excavations on the Leven-Brandesburton By-pass, N. Humberside.* Reports from the Environmental Archaeology Unit 94/15. York: Environmental Archaeology Unit, University of York.

Hastie, M. and T. G. Holden (2001). Carbonised plant remains, in G. J. Barclay, S. P. Carter, M. M. Dalland, M. T. Hastie, G. Holden, A. MacSween and C. R. Wickham-Jones, "A possible neolithic settlement at Kinbeachie, Black Isle, Highland". *Proceedings of the Society of Antiquities of Scotland 131*, 65–69.

Helbaek, H. (1952). Early crops in southern England. *Proceedings of the Prehistoric Society 18*, 194–233.

Henriksen, P. S. (2000). *Agerbrug i Senneolitikium og Bronzealderen på Djursland.* NNU Rapport 7. Copenhagen: Nationalmuseet Denmark.

Henriksen, P. S. (2001). *Arkæobotanisk undersøgelse af materiale fra fire bopladser fra Dolktid og Ældre Bronzealder ved Skive.* NNU Rapport 11. Copenhagen: Nationalmuseet Denmark.

Hey, G., J. Mulville, and M. Robinson (2003). Diet and culture in southern Britain: the evidence from Yarnton. In M. Parker Pearson (Ed.), *Food Culture and Identity in the Neolithic and Early Bronze Age*, BAR International Series 1117, pp. 79–88. Oxford: Archaeopress.

Higgins, P. (2000). Plant, insect and mollusc remains, in M. Heaton and R. M. J. Cleal, "Beaker pits at Crescent Copse, near Shrewton, Wiltshire, and the effects of arboreal fungi on archaeological remains". *Wiltshire Archaeological and Natural History Magazine 93*, 78.

Higgins, P. (2003). Plant, mollusc and insect remains, in M. Heaton, "Neolithic pits at the Beehive". *Wiltshire Archaeological and Natural History Magazine 96*, 59–60.

Higgs, E. S. and M. R. Jarman (1969). The origins of agriculture: a reconsidertation. *Antiquity 43*, 31–41.

Hinton, P. (1996). Plant remains, in P. Andrews (ed.) "Prospect Park, Harmondsworth, London Borough of Hillingdon: settlement and early burial from the Neolithic to the Early Saxon periods". In P. Andrews and A. Crockett (Eds.), *Three Excavations along the Thames and its Tributary, 1994*, Wessex Archaeology Report 10, pp. 43–47. Salisbury: Wessex Archaeology.

Hinton, P. (2004). Plant remains, in M. Rawlings, M. J. Allen and F. Healy, "Investigation of the Whitesheet Down environs 1989-90: Neolithic causewayed enclosure and iron age settlement". *Wiltshire Archaeological and Natural History Magazine 97*, 177–179.

Hjelmqvist, H. (1982). Economic plants from a middle neolithic site in Scania. *Meddelanden från Lunds Universitets Historiska Museum 4*, 108–113.

Hjelmqvist, H. (1985). Economic plants from two stone age settlements in southernmost Scania. *Acta Archaeologica 54*, 57–63. (for 1983).

Hjelmqvist, H. (1992). Some economic plants from the prehistoric and mediaeval periods in southern Scania. In L. Larsson, J. Callmer, and B. Stjernquist (Eds.), *The Archaeology of the Cultural Landscape*, Acta Archaeologica Lundensia 4(19), pp. 359–367. Lund: University of Lund.

Holden, T. (1997). Macroplant assessment, in J. S. Rideout, "Excavation of neolithic enclosures at Cowie Road, Bannockburn, Stirling, 1984–5". *Proceedings of the Society of Antiquities of Scotland 127*, 53–54.

Holden, T. (1998). Charred plant remains. In R. MacCullagh and R. Tipping (Eds.), *The Lairg Project. The Evolution of an Archaeological Landscape in N Scotland 1988-1996*, pp. 165–172. Edinburgh: Scottish Trust for Archaeological Research.

Huntley, J. P. (1993). Plant remains, in C. Scull, "Excavation and survey at Watchfield, Oxfordshire, 1983-92". *Archaeological Journal 149*, 145.

Huntley, J. P. (1996a). Nosterfield, nr. Ripon: NON95. The charred plant remains. Durham Environmental Archaeology Report 13/96.

Huntley, J. P. (1996b). Pits Plantation, Rudston: PPR965. An assessment of the environmental samples. Durham Environmental Archaeology Report 3/96.

Huntley, J. P. (2000). Evaluation report of: The Milfield Basin Archaeology Project, Northumberland: plant macrofossil, waterlogged wood and charcoal assessment, by J. Cotton. Archaeological Services University of Durham Report 678.

Huntley, J. P. (2001). Sewerby cottage farm, martongate, bridlington, east yorkshire (scf): Assessment of charred plant remains from neolithic deposits.

Huntley, J. P. (2002). Palaeoenvironmental samples, in C. Waddington and J. Davies, "An early neolithic settlement and late bronze age burial cairn near Bolam Lake, Northumberland: fieldwalking, excavation and reconstruction. *Archaeologia Aeliana 5/30*, 42–43.

Huntley, J. P. (n.d.). The plant remains. In C. Waddington and D. Passmore (Eds.), *Landscape Archaeology: A case study of the Milfield Basin, Northumberland.* English Heritage. In Press.

Huntley, J. P. and S. Stallibrass (1995). *Plant and Vertebrate Remains from Archaeological Sites in Northern England: Data Reviews and Future Directions.* Research Report 4. Durham: Architectural and Archaeological Society of Durham and Northumberland.

Jaques, D., J. Carrott, A. Hall, and S. Rowland (2001). *Assessment of biological remains from excavations at Wath Quarry, Wath, North Yorkshire (Site Code 07-08-00).* Reports from the Environmental Archaeology Unit, York, 2001/34. York: Environmental Archaeology Unit, University of York.

Jessen, K. (1939). III. Kornfund, in T. Mathiassen, "Bundsø. En Yngre Sternalders Boplads paa Als Aarbøger for Nordiske Oldkyndighed og Historie". *Aarbøger for Nordiske Olkyndighed of Historie 1939*, 65–84.

Jessen, K. and H. Helbaek (1944). *Cereals in Great Britain and Ireland in Prehistoric and Early Historic Times.* Biologiske Skrifter III, no. 2. Copenhagen: Det Kongelige Danske Videnskabernes Selskab.

Jones, G. (2000). Evaluating the importance of cultivation and collecting in neolithic Britain. In A. S. Fairbairn (Ed.), *Plants in Neolithic Britain and Beyond*, Neolithic Studies Group Seminar Paper 5, pp. 79–84. Oxford: Oxbow Books.

Jones, G. (n.d.a). Charred plant remains from the neolithic causewayed enclosure at Haddenham, Cambridgeshire. In *The Emergence of a Fen Edge Landscape.* English Heritage and the McDonald Institute for Archaeological Research. In press.

Jones, G. (n.d.b). Evidence for the importance of cereals in the neolithic: charred plant remains from the neolithic settlement at Lismore Fields, Buxton. Forthcoming.

Jones, G. and A. J. Legge (n.d.). Evaluating the role of cereal cultivation in the neolithic: charred plant remains from hambledon hill, dorset. In R. Mercer and F. Healy (Eds.), *Hambledon Hill, Dorset. Excavation and Survey of a Neolithic Monument Complex and its Surrounding Landscape*, English Heritage Monograph. English Heritage. In press.

Jones, J. and S. Straker (1993). Macroscopic plant remains. In P. J. Woodward, S. M. Davies, and A. H. Graham (Eds.), *Excavations at Greyhound Yard, Dorchester 1981-4*, Monograph Series 12, pp. 349–350. Dorset Natural History Archaeological Society.

Jones, M. (1980). Carbonised cereals from grooved ware contexts. *Proceedings of the Prehistoric Society 46*, 61–63.

Jones, M. (1991). Sampling in paleoethnobotany. In W. van Zeist, K. Wasylikowa, and K.-E. Behre (Eds.), *Progress in Old World Palaeoethnobotany: A Retrospective View on the Occassion of 20 Years of the International Work Group for Palaeoethnobotany*, pp. 53–62. Rotterdam: A. A. Balkema.

Keepax, C. (1985). The charcoals, in Green and Sofranoff, "A neolithic settlement at Stacey Bushes, Milton Keynes". *Rec. Buckinghamshire 27*, 19–20.

Kiely, J. (2003). A neolithic house at Cloghers, Co. Kerry. In I. Armit, E. Murohy, E. Nelis, and D. Simpson (Eds.), *Neolithic Settlement in Ireland and Western Britain*, pp. 182–187. Oxford: Oxbow Books.

King, M. (2003). *Unparalleled Behaviour: Britain and Ireland during the 'Mesolithic' and 'Neolithic'.* British Archaeological Reports International Series 355. Oxford: Archaeopress.

Legge, A. J. (1981). The agricultural economy. In R. J. Mercer (Ed.), *Grimes Graves, Norfolk. Excavations 1971-2*, Volume 1 of *Research Reports Series 11*, pp. 79–103. London: HMSO.

Legge, A. J. (1989). Milking the evidence: a reply to Entwhistle and Grant. In A. Milles, D. Williams, and N. Gardner (Eds.), *The Beginnings of Agriculture*, Symposia of the Association for Environmental Archaeology No. 8/BAR International Series 496, pp. 217–242. Oxford: British Archaeological Reports.

Legge, A. J. (1991). The animal remains from six sites at Down Farm, Woodcutts. In J. Barrett, R. Bradley, and M. Hall (Eds.), *Papers on the Prehistoric Archaeology of Cranbourne Chase*, Oxbow Monograph 11, pp. 54–100. Oxford: Oxbow Books.

Letts, J. (1995). Charred plant remains. In T. G. Allen (Ed.), *Lithics and Landscape: Archaeological Discoveries on the Thames Water Pipeline at Gatehampton Farm, Goring, Oxfordshire 1985-92*, Thames Valley Landscapes Monograph 7, pp. 107–109. Oxford: Oxford University Committee for Archaeology.

Lynch, A. (1983). The seed remains. In J. W. Hedges (Ed.), *Isbister. A Chambered Tomb in Orkney*, BAR British Series 115, pp. 171–175. Oxford: British Archaeological Reports.

M. Van der Veen (1985). Evidence for crop plants from north-east England: an interim overview with discussion of new results. In N. R. J. Fieller, D. D. Gilbertson, and N. G. A. Ralph (Eds.), *Palaeobiological Investigations*, BAR International Series 266, pp. 197–220. Oxford: British Archaeological Reports.

Maclean, A. C. and P. Rowley-Conwy (1979). Cereal identification, in I. A. G. Shepherd and A. N. Tuckwell, "Traces of Beaker-period cultivation at Rosinish". *Proceedings of the Society of Antiquities of Scotland 108*, 131.

Maclean, A. C. and P. Rowley-Conwy (1984). The carbonized material from Boghead, Fochabers. *Proceedings of the Society of Antiquities of Scotland 114*, 69–71.

MacLean, C. (1978). Cereals from Pits A–C, in J. N. G. Ritchie "The stones of Stenness, Orkney". *Proceedings of the Society of Antiquities of Scotland 107*, 43–44.

McSparron, C. (2003). The excavation of a neolithic house in Enagh townland, Co. Derry. In I. Armit, E. Murohy, E. Nelis, and D. Simpson (Eds.), *Neolithic Settlement in Ireland and Western Britain*, pp. 172–175. Oxford: Oxbow Books.

Miller, J. and S. Ramsay (2002). Plant macrofossils, in G. J. Barclay, K. Brophy and G. MacGregor "Claish, Stirling: an early neolithic structure and its context". *Proceedings of the Society of Antiquities of Scotland 132*, 90–96.

Milles, A. (1986a). Charred plant remains from a Beaker pit at Site 2, Four Crosses, in W. Britnell, "Four ring-ditches at Four Crosses, Llandysilio, Powys, 1981–1985". *Proceedings of the Prehistoric Society 52*, microfiche 10.

Milles, A. (1986b). Charred remains of barley and other plants from Scord of Brouster. In A. Whittle (Ed.), *Scord of Brouster. An Early Agricultural Settlement on Shetland*, Oxford University Committee for Archaeology Monograph 9, pp. 119–122. Oxford: Oxford University Committee for Archaeology.

Moffett, L. (1999). The prehistoric use of plant resources. In A. Barclay and C. Halpin (Eds.), *Excavations at Barrow Hills, Radley, Oxfordshire. Volume I. The Neolithic and Bronze Age Monument Complex*, Thames Valley Landscapes Monograph 11, pp. 243–247. Oxford: Oxford Archaeological Unit.

Moffett, L. and M. Ciaraldi (1999). Charred plant remains, in S. C. Palmer, "Archaeological excavations in the Arrow Valley, Warwickshire". *Birmingham and Warwickshire Archaeological Society Transactions 103*, 32–34.

Moffett, L., M. A. Robinson, and V. Straker (1989). Cereals, fruit and nuts: charred plant remains from neolithic sites in England and Wales and the neolithic economy. In A. Milles, D. Williams, and N. Gardner (Eds.), *The Beginnings of Agriculture*, Symposia of the Association for Environmental Archaeology No. 8. BAR International Series 496, pp. 243–261. Oxford: British Archaeological Reports.

Monckton, A. (1989). Plant remains, in P. Clay, "Neolithic /early bronze age pit circles and their environs at Oakham, Rutland". *Proceedings of the Prehistoric Society 64*, 323–324.

Monckton, A. (2002). Charred plant remains from bronze age features and a burnt mound at Willow Farm, Castle Donington, Leicestershire (xA14.97). Unpublished Report for University of Leicester Archaeological Services.

Monckton, A. and L. Moffett (n.d.). Charred plant remains from Potlock, Derbyshire. Unpublished manuscript.

Monk, M. (1987). Archaeobotanical studies at Poundbury. In C. S. Green (Ed.), *Excavations at Poundbury, Dorset 1966-82. Vol. 1: The Settlement*, pp. 132–137. Dorchester: Dorset Natural History Archaeology Society Monograph 7.

Monk, M. (1988). Archaeobotanical study of samples from pipeline sites. In M. Gowan (Ed.), *Three Irish Gas Pipelines: New Archaeological Evidence in Munster*, pp. 185–191. Dublin: Wordwell.

Monk, M. (2000). Seeds and soils of discontent: an environmental archaeological contribution to the nature of the early neolithic. In A. Desmond, G. Johnson, M. McCarthy, J. Sheehan, and E. Shee Twohig (Eds.), *New Agendas in Irish Prehistory. Papers in Commemoration of Liz Anderson*, pp. 67–87. Co. Wicklow, Ireland: Wordwell.

Moore, D. G. (2003). Neolithic houses in Ballyharry townland, Islandmagee, Co. Antrim. In I. Armit, E. Murohy, E. Nelis, and D. Simpson (Eds.), *Neolithic Settlement in Ireland and Western Britain*, pp. 156–163. Oxford: Oxbow Books.

Murphy, P. (1988a). Carbonised neolithic plant remains from The Stumble, an intertidal site in the Blackwater estuary, Essex, England. *Circaea 6*, 21–38.

Murphy, P. (1988b). Charred plant remains: wood, in S. Wallis and M. Waughman, "Archaeology and the landscape in the lower Blackwater valley". *East Anglian Archaeology 82*, 196–205.

Murphy, P. (1990a). *Longham, Norfolk: Carbonized Plant Remains from Beaker, Bronze Age and Iron Age Contexts.* Ancient Monuments Laboratory Report 61/91. London: English Heritage.

Murphy, P. (1990b). *Springfield Lyons, Chelmsford, Essex: Carbonised Plant Remains from a Bronze Age Cremation Cemetery.* Ancient Monuments Laboratory Report 122/90. London: English Heritage.

Murphy, P. (1992a). *Brampton, Cambridgeshire: Caronised Plant Remains from a Neolithic Mortuary Enclosure and Associated Features.* Ancient Monuments Laboratory Report 61/92. London: English Heritage.

Murphy, P. (1992b). *Norwich Southern By-Pass: Plant Remains from Beaker, Bronze Age, Iron Age, Romano-British and Late Saxon Contexts; River Valley Sediments.* Ancient Monuments Laboratory Report 20/92. London: English Heritage.

Murphy, P. (1993). Mollusca and plant macrofossils, in F. R. Healy, M. J. Cleal and I. Kinnes, "Excavations on Redgate Hill, Hunstanton, 1970 and 1971". *East Anglian Archaeology 59*, 65–69.

Murphy, P. (1994). *Fenland Project Report 1: Little Duke Farm, Deeping St Nicholas, Lincs.: Plant Macrofossils and Molluscs from a Bronze Age Barrow and Pre-Barrow Contexts.* Ancient Monuments Laboratory Report 32/93. London: English Heritage.

Murphy, P. (1997). Environment and economy. In J. Glazenbrook (Ed.), *Research and Archaeology: A Framework for the Eastern Counties. 1: Resource Assessment*, East Anglian Archaeology Occassional Paper 3, pp. 3, 7–8, 10, 12–22, 30–31, 42–43, 54–55, 63–64. Chelmsford: Essex County Council.

Murphy, P. (2001). Plant remains from neolithic and bronze age contexts, in D. G. Buckley, J. D. Hedges and N. Brown "Excavations of a Neolithic cursus, Springfield, Essex, 1979-85". *Proceedings of the Prehistoric Society 67*, 147–149.

Murphy, P. (n.d.). Charred plant remains and palaeoeconomy. In *Archaeology of the Essex Coast, Volume 2: Investigations at the Intertidal Site of The Stumble, Blackwater Estuary*, Volume East Anglian Archaeology. Chelmsford: Essex County Council.

Murphy, P. and P. E. J. Wiltshire (1989). *Pakenham, Suffolk (PKM 027): Environmental and Economic Studies.* Ancient Monuments Laboratory Report 99/89. London: English Heritage.

Nye, S. and R. Scaife (1998). Plant macrofossil remains. In F. Pryor (Ed.), *Etton: Excavations at a Neolithic Causewayed Enclosure near Maxey, Cambridgeshire 1982–7*, pp. 289–300. London: English Heritage.

Palmer, C. and M. Jones (1991). Plant resources. In N. M. Sharples (Ed.), *Maiden Castle: Excavations and Field Survey 1985–1986*, English Heritage Archaeological Report 19, pp. 129–139. London: Historic Buildings and Monuments Commission for England.

Pollard, J. and A. Reynolds (2002). *Avebury: The Biography of a Landscape.* Stroud: Tempus.

Purcell, A. (2002). Excavation of three neolithic houses at Corbally, Kiluclen, Co. Kildare. *Journal of Irish Archaeology 11*, 31–74.

Reaney, D. (1968). Beaker burials in south Derbyshire. *Derbyshire Archaeological Journal 88*, 68–81.

Renfrew, C. (1972). *Before Civilization: The Radiocarbon Revolution and Prehistoric Europe.* London: Penguin Books.

Richards, M. P. and R. E. M. Hedges (1999). A neolithic revolution? New evidence of diet in the British Neolithic. *Antiquity 73*, 891–897.

Richmond, A. (1999). *Preferred Economies: The Nature of the Subsistence Base throughout Mainland Britain during Prehistory.* BAR British Series 290. Oxford: Archaeopress.

Robinson, D. E. (1996). *Analyse af Forkullet Korn og Kornaftryk fra det Neolitiske Anlæg (MNA I/II) ved Lønt, Starup Sogn, Sønderjylland.* NNU Rapport 21. Copenhagen: Nationalmuseet Denmark.

Robinson, D. E. (1998). Plantemakrofossiler fra Spodsbjergbopladsen. In *Spodsbjerg—en Yngre Stenalders Boplads på Langeland*, pp. 175–189. Rudkøbing: Langelands Museum.

Robinson, D. E. (2000a). Det slesvigske agerbrug i yngre stenalder og bronzealder. Arkæobotanikkens udsagn. In P. Ethelberg, E. Jørgensen, D. Meier, and D. Robinson (Eds.), *Det Sønderjyske Landbrugs Historie. Sten- og Bronzealder*, pp. 281–298. Haderslev: Haderslev Museum and Historisk Samfund for Sønderjylland.

Robinson, D. E. (2003). Neolithic and bronze age agriculture in southern Scandinavia: recent archaeobotanical evidence from Denmark. *Environmental Archaeology 8*, 145–166.

Robinson, D. E. and I. Boldsen (2000). *Arkæobotaniske Undersøgelser af Materiale fra Enkeltgravshustomter ved Strandet Hovedgaard.* NNU Rapport 5. Copenhagen: Nationalmuseet Denmark.

Robinson, D. E. and J. A. Harild (1994). *Arkæobotaniske Analyser af Materiale fra Vasagård, Bornholm.* NNU Rapport 7. Copenhagen: Nationalmuseet Denmark.

Robinson, D. E. and D. Kempfner (1987). Carbonised grain from Mortens Sande 2. *Journal of Danish Archaeology 6*, 125–129.

Robinson, D. E. and J. Westphal (2001). *Arkæobotaniske Undersøgelser af Materiale fra en Brnadtomt og Tilhørende Grube fra MNIA ved Østerskov Å, Humble, Langeland.* NNU Rapport 30. Copenhagen: Nationalmuseet Denmark.

Robinson, M. A. (2000b). Further considerations of neolithic charred cereals, fruits and nuts. In A. S. Fairbairn (Ed.), *Plants in Neolithic Britain and Beyond*, Neolithic Studies Group Seminar Paper 5, pp. 85–90. Oxford: Oxbow Books.

Rowley-Conwy, P. (2000). Through a taphonomic glass darkly: the importance of cereal cultivation in prehistoric Britain. In J. Huntley and S. Stallisbrass (Eds.), *Taphonomy and Interpretation*, Symposia of the Association for Environmental Archaeology No. 14, pp. 43–53. Oxford: Oxbow Books.

Rowley-Conwy, P. (2003). No fixed abode? Nomadism in the Northwest European neolithic. In G. Burenhult and S. Westergaard (Eds.), *Formal Disposal of the Dead in Atlantic Europe during the Mesolithic–Neolithic Interface 6000–3000 BC*, BAR International Series 1201, pp. 115–144. Oxford: Archaeopress.

Rowley-Conwy, P. (2004). How the west was lost. A reconsideration of agricultural origins in Britain, Ireland and southern Scandinavia. *Current Anthropology 45*, S83–S113.

Rowley-Conwy, P. (n.d.). The charred plant remains. In D. V. Clarke and A. N. Shepherd (Eds.), *Skara Brae: A Full Compendium of the Site.* In preparation.

Schulting, R. (2000). New AMS dates from the Lambourn long barrow and the question of the earliest neolithic in southern England: repacking the neolithic package. *Oxford Journal of Archaeology 19*, 25–35.

Schulting, R. and M. P. Richards (2002a). Finding the coastal mesolithic in southwest Britain: AMS dates and stable isotope results on human remains from Caldey Island, South Wales. *Antiquity 76*, 1011–1025.

Schulting, R. and M. P. Richards (2002b). The wet, the wild and the domesticated: the Mesolithic-Neolithic transition on the west coast of Scotland. *European Journal of Archaeology 5*, 147–189.

Simpson, D. (1993). Ballygally. *Current Archaeology 134*, 60–62.

Simpson, D. (1996). The Ballygalley houses, Co. Antrim, Ireland. In T. Darvill and J. Thomas (Eds.), *Neolithic Houses in Northwest Europe and Beyond*, Oxbow Monograph 57/Neolithic Studies Group Seminar Paper 1, pp. 123–132. Oxford: Oxbow Books.

Soltvedt, E.-C. (2000). Carbonised cereal from three late neolithic and two early bronze age sites in western Norway. *Environmental Archaeology 5*, 49–62.

Straker, V. (1990a). Carbonised plant macrofossils. In A. Saville (Ed.), *Hazleton North, Gloucestershire, 1979-82: The Excavation of a Neolithic Long Cairn of the Cotswold-Severn Group*, English Heritage Archaeological Report 15, pp. 214–218. London: English Heritage.

Straker, V. (1990b). Charred plant macrofossils. In M. Bell (Ed.), *Brean Down: Excavations 1983–1987*, English Heritage Archaeological Report 15, pp. 211–219. London: English Heritage.

Straker, V. (1997). Charred plant remains. In R. J. C. Smith, F. Healy, M. J. Allen, A. L. Morris, I. Barnes, and P. J. Woodward (Eds.), *Excavations Along the Route of the Dorchester By-Pass, Dorset, 1986-1988*, Wessex Archaeology Report 11, pp. 184–190. Salisbury: Trust for Wessex Archaeology.

Taylor, G. (1963). Charcoal, in C. H. Houlder, "A neolithic settlement on Hazard Hill, Totnes". *Proceedings of the Devon Archaeological Exploration Society 21*, 30.

Thomas, J. (1993). Discourse, totalization and 'The Neolithic'. In C. Tilley (Ed.), *Interpretative Archaeology*, pp. 357–394. Oxford: Berg.

Thomas, J. (1996a). The cultural context of the first use of domesticates in continental Central and Northwest Europe. In D. R. Harris (Ed.), *The Origins and Spread of Agriculture and Pastoralism in Eurasia*, pp. 310–322. London: UCL Press.

Thomas, J. (1996b). Neolithic houses in mainland Britain and Ireland: a sceptical view. In T. Darvill and J. Thomas (Eds.), *Neolithic Houses in Northwest Europe and Beyond*, Oxbow Monograph 57/Neolithic Studies Group Seminar Paper 1, pp. 1–12. Oxford: Oxbow Books.

Thomas, J. (1999). *Understanding the Neolithic.* London: Routledge.

Thomas, J. (2003). Thoughts on the 'repacked' neolithic revolution. *Antiquity 77*, 67–74.

Tomlinson, P. R. and A. R. Hall (1996). A review of the archaeological evidence for food plants from the British Isles: an example of the use of the Archaeobotanical Computer Database (ABCD). *Internet Archaeology 1.* `http://intarch.ac.uk/journal/issue1/tomlinson/` Accessed 2005-03-25.

van der Veen, M. (1992). *Crop Husbandry Regimes: An Archaeobotanical Study of Farming in Northern England 1000 BC–AD 500.* Sheffield Archaeological Monographs 3. Sheffield: J. R. Collis.

Waddington, C. (2000). The neolithic that never happened? In J. Harding and R. Johnston (Eds.), *Northern Pasts. Interpretations of the Later Prehistory of Northern England and Southern Scotland*, BAR British Series 302, pp. 33–44. Oxford: Archaeopress.

Whittle, A. (2003). *The Archaeology of People: Dimensions of Neolithic Life.* London: Routledge.

Wilkinson, T. J. and P. Murphy (1995). *The Archaeology of the Essex Coast, Volume 1: The Hullbridge Survey.* East Anglian Archaeology 71. Chelmsford: Essex County Council.

Williams, D. (1988). Plant macrofossils, in R. S. Kelly, "Two late prehistoric stone circular enclosures near Harlech, Gwynedd". *Proceedings of the Prehistoric Society 54*, 142–144.

Williams, D. (1999). Plant macrofossils. In S. I. White and G. Smith (Eds.), *A Funerary and Ceremonial Centre at Capel Eithin, Gaerwen, Anglesey: Excavations of Neolithic, Bronze Age, Roman and Early Medieval Features in 1980 and 1981*, Transactions of the Anglesey Antiquarian Society 1999, pp. 109–112. Anglesey: Anglesey Antiquarian Society.

Index of plant names

Index of common plant names (for crop/economic taxa only)

Index of site names

General index